LEAD IN THE ENVIRONMENT

Prepared for
The National Science Foundation

William R. Boggess, Editor
Professor Emeritus of Forestry
University of Illinois at Urbana-Champaign

Bobby G. Wixson, Assistant Editor
Professor of Environmental Health
University of Missouri at Rolla

CHP

CASTLE HOUSE PUBLICATIONS LTD.

Any opinions, findings and conclusions or recommendations expressed in this publication are those of the Author(s) and do not necessarily reflect the views of the National Science Foundation.

ISBN 0 7194 0024 4

CASTLE HOUSE PUBLICATIONS LTD. 1979

Foreword

For the past 5 years the National Science Foundation, through its Research Applied to National Needs Program, has sponsored research at Colorado State University, the University of Illinois at Urbana-Champaign, and the University of Missouri at Rolla, concerned with the occurence, transport, distribution, and possible environmental effects of lead. Although funded under separate institutional grants, project scientists maintained close and active liaison on problems of mutual interest and concern. This major effort, both interdisciplinary and interinstitutional in character, has climaxed in the publication of this volume, *Lead in the Environment.* Although results of project research form the central core of the book, it is well documented with references from both classic and current literature.

As a group, the three projects have developed analytical techniques and methodologies for determining lead and following it from source to ultimate sink through both living and nonliving ecosystem components. Individually, each project had its unique focus. Colorado State University concentrated on characterizing the atmospheric component of lead emissions from automotive sources and determining its movement in the environment. Also considered were: the possible effects on living organisms, the role of lead in inadvertent weather modification, the effect of city geometry on dispersion, and the technical aspects of various control strategies.

The University of Illinois and the University of Missouri studies were executed in an ecosystem context. At the University of Illinois, lead inputs, accumulation, transport, and outputs were determined in an 86-sq. mi. watershed where intensive agriculture (corn and soybeans) is the dominant land use. The watershed also includes an urban component in the twin cities of Champaign and Urbana. The primary source of lead is automotive emissions from the combustion of leaded gasoline. The accumulation of lead along busy highways and city streets and its possible movement through both terrestrial and aquatic food chains have received particular attention. Lead-bearing dust that might be tracked from streets into homes and thus form a source of contamination for small children has received major emphasis.

The primary lead source in the University of Missouri study is the mining, milling, and smelting operations in one of the newest and largest lead-producing industries of the world. The study is unique in that research began soon after industrial operations were initiated. Thus it was possible to observe and document the initial buildup of lead in both terrestrial and aquatic environments. Comparisons were also possible between developing conditions in the New Lead Belt with those from a nearby area where lead operations have existed for almost 100 years. In contrast to the University of Illinois location, operations in the New Lead Belt are contained in an area that is almost wholly forested and where watershed and recreational values are very high. The University of Illinois and the University of Missouri studies provide extreme contrasts in land use, lead sources, and the length of time that contamination has taken place.

The book is divided into six logical parts. Each part will stand alone as a unit, although all are closely related in the total discussion. Part I provides a general background for research related to the environmental effects of lead. A discussion of the physical and chemical characteristics of lead in chapter 1 presents information essential to analytical techniques, source identification, and the behavior of lead under a variety of environmental conditions. Chapter 2 is concerned with the collection, handling, preparation, and analysis of samples from a variety of sources. Special attention is given to those aspects of monitoring and sample handling that might affect analytical results. The advantages and disadvantages of different analytical methods are discussed, and recommendations are made regarding the suitability of a particular method for samples from different origins.

Part II, concerned with the transport and distribution of lead in an ecosystem context, contains a major part of the research results, especially from the University of Illinois and the University of Missouri projects. Each ecosystem component is discussed separately from an overall standpoint and, in some cases, as it relates to some specific aspect of the two widely different ecosystems.

The effect of lead on living organisms is discussed in part III. Short sections are also provided on two important effects of lead: human health and inadvertent weather modification. Health implications, based on lead levels found in the various ecosystem components, are discussed by Paul Hammond of the University of Cincinnati College of Medicine.

The control of lead emissions from industrial and automotive exhaust sources is examined in part IV. Many of the industrial aspects are based on direct experiences of the University of Missouri research team with industry in the New Lead Belt of the state. An excellent analysis of problems related to automotive emissions is based on research from Colorado State University. This subject is approached both from the technical aspects of direct emission control and the indirect effects of city layout and geometry on the dispersion and accumulation of lead.

The economic aspects of controlling emissions from the combustion of leaded gasoline are analyzed in part V. Although positive action has been taken to phase out the use of lead in gasoline eventually, the situation remains confused by litigation, arguments over the use of catalytic converters, and the tradeoffs between environmental and economic considerations.

A general summary is provided in part VI, along with recommendations related to the possible environmental effects of lead. These recommendations are based on interpretations of research results and the best judgments of those who have been most closely concerned with the projects during the past 5 years. Even so, judgments made today may be obsolete tomorrow as industrial technologies change, and latent or otherwise unnoticed environmental effects appear.

Preparation of this volume has not been an easy task. Authors of individual chapters were often in separate institutions, and with full teaching and research programs it was difficult for them to execute the concentrated effort necessary to meet deadlines. Delays were experienced because of serious illness of key personnel. Other staff members changed their places of employment during the period of preparation and revision of manuscripts. In spite of these difficulties, all those concerned with *Lead in the Environment* have been most cooperative, and the editor is grateful for their patience and understanding.

Special thanks must go to the principal investigators: Dr. Harry Edwards at Colorado State University, Dr. Gary L. Rolfe at the University of Illinois at Urbana-Champaign, and Dr. Bobby G. Wixson at the University of Missouri at Rolla. It should be pointed out that Dr. Wixson served as editor during the period when first drafts of manuscripts were being prepared. He has continued to lend valuable assistance and support as an assistant editor. Dr. Ronald S. Goor and Dr. Marvin E. Stephenson, Program Managers for NSF, have been generous with their support and encouragement.

Lead in the Environment represents a landmark example of interinstitutional cooperation. Science, its supporters, and clientele would benefit greatly if other research efforts were so pooled and coordinated.

William R. Boggess, *Editor*
Austin, Tex.

Contents

PART I

Characteristics, Monitoring, and Analysis

CHAPTER 1

PHYSICAL AND CHEMICAL CHARACTERISTICS OF ENVIRONMENTAL LEAD

Myron L. Corrin
Department of Atmospheric Science
Colorado State University

*David F. S. Natusch**
Department of Chemistry
Colorado State University

INTRODUCTION

This chapter deals with the physical and chemical nature of lead-containing species in the environment. It is specifically concerned with the activities of the research groups at Colorado State University, the University of Illinois at Urbana-Champaign, and the University of Missouri at Rolla that bear upon this subject. Reference will be made to the work of other investigators only if this work provided the impetus for further study by the three university groups, supplied information required in the formulation of research plans, yielded data that could be used to check the validity of experimental results and their interpretation, or discussed techniques that are of general value.

The term "physical characterization" bears upon those physical properties of lead-containing species that are significant in terms of transport, physical interactions with other substances, deposition in various environmental "compartments", physiological activities, removal mechanisms, etc. As such, the discussion is concerned with such general physical properties as particle size and distribution, phase relations, solubility, surface characteristics, washout efficiency, morphology, crystal structure, and the experimental methods for the determination of physical properties.

The term "chemical characterization" bears mainly upon the chemical composition of lead-containing materials in the environment. The chemical form of lead will determine solubility in water and biological fluids, the extent to which lead is fixed in soils, and the types of chemical reactions occurring in the atmospheric, aquatic, and soil environment. It is, therefore, of primary importance to determine the physical nature and chemical forms of lead (or other trace elements) in the environment if its behavior is to be understood and its environmental impact evaluated and predicted.

The characterization of environmental lead will be considered in terms of lead-containing material (1) emitted into the atmosphere, water, or soil; (2) existing in the atmosphere; (3) existing in water, either in solution or suspension; (4) occurring in soils; and (5) present in biological organisms. Consideration will be given to specific effects of environmental lead in terms of such processes as atmospheric washout.

LEAD EMISSIONS INTO THE ENVIRONMENT

Lead-containing material is introduced into the environment from a variety of sources. The principal input, however, is from the combustion of lead-containing fuels used in internal combustion engines, the combustion of lead-containing coal or fuel oil, and activities in lead mining and refining. The extent and geographical distribution of such emissions are considered elsewhere. A major source of lead input into the environment is the exhaust from automobiles. Most of the research at Colorado State University and a large portion of that at the University of Illinois are devoted to problems associated with this lead source. Work has also been reported from the University of Illinois bearing upon environmental lead introduced from the combustion of coal. Investigators at the University of Missouri are primarily concerned with lead originating from mining, smelting, and refining. The physical and chemical characterization of such emissions, before physical or chemical changes that occur in transport or deposition, will be considered in the following discussion.

*Formerly Department of Chemistry, University of Illinois at Urbana-Champaign.

Automotive Emissions

This section deals with the characterization of lead-containing gaseous and particular matter emitted from internal combustion engines fueled with leaded gasoline and with a study of the physical and chemical nature, under controlled laboratory conditions, of simulated automotive emissions.

Physical Characterization: Purdue et al. [1] have reported that a minimum of 90 percent of the lead emissions in the exhausts of internal combustion engines using leaded fuel are particulate, and the remaining 10 percent may be organic lead vapors (presumably uncombusted lead alkyls). The lead antiknock formulation may include, in addition to organic halogen compounds acting as scavengers, any or all of $Pb(CH_3)_4$, $Pb(CH_3)_3(C_2H_5)$, $Pb(CH_3)_2(C_2H_5)_2$, $Pb(CH_3)(C_2H_5)_3$, and $Pb(C_2H_5)_4$. Menne and Corrin [2] found that in ambient air adjacent to a heavily traveled highway, the only lead alkyl detectable was $Pb(CH_3)_4$; concentrations ranged from 1 to 20 percent of the total lead concentration. The implications of these results will be discussed later. Lead alkyl concentrations in air up to 2 $\mu g/m^3$ have been reported by Laveskog. [3]

The particle size distributions of lead-containing particulate matter in fresh automotive exhaust have been investigated by Habibi et al., [4] Habibi, [5] Hirschler et al., [6] Hirschler and Gilbert, [7] Mueller et al., [8] and Ter Haar et al. [9] The particle size distribution varies with engine operating conditions. The chemical composition of the lead-containing particles also varies with particle size. In a general sense, however, from 20 to 30 percent of the total lead is found in particles greater than 5 μm; from 50 to 70 percent in the 1- to 5-μm range, and 5 percent or less in submicrometer particles. More than 90 percent of the emitted particles possess a diameter less than 2 μm. The sizes quoted are aerodynamic sizes. It should be pointed out that the size refers to the particle itself, which contains species other than lead compounds. It is customary to express the particle size distribution of lead-containing species in terms of the fraction of total lead contained in particles of a given size range. It is thus impossible to calculate the number of particles in a given size range from the distribution function so expressed. The settling velocity of a 5-μm particle (aerodynamic size) in still air is 0.1 cm/s; given an emission height of perhaps 3 m, the lifetime of such a particle in the still atmosphere would be 50 min.

A Digression on Particle Size: Particle sizes determined by visual observation or light and electron microscopy are geometrical sizes deduced by various techniques from a two-dimensional display. They are generally reported in terms of the radius (or diameter) of an "equivalent" sphere. Particle sizes as measured by impactors, air centrifuges (such as the Goetz spectrometer), or techniques involving the transport behavior of particles in an applied field are aerodynamic sizes and are generally expressed in terms of a sphere of unit density that possesses the same aerodynamic behavior as the particle in question. The two size descriptions are not identical and cannot be interchanged without additional information on such parameters as shape and density. Most of the work reported on lead-containing particles involves aerodynamic sizing. This is an appropriate description, because the important factors of atmospheric transport, fallout, washout, and deposition in the lungs are determined by aerodynamic properties (and hence aerodynamic size) rather than particle geometry as expressed in geometrical size.

Exhaust Simulation: Zimdahl [10] at Colorado State University has constructed a generator for various lead salts used in plant studies with simulated lead-containing aerosol. A similar generator has been used by Grant and Vardiman [11] to investigate possible weather modifications produced by lead aerosols. However, the major study on automotive exhaust simulation has been conducted at the University of Illinois.

In order to simulate the reactions and transport of automotive exhaust particles with pure and known compounds in the laboratory, Boyer and Laitenen [12] at the University of Illinois have developed methods for the production and characterization of artificial lead-containing aerosols. Such model particulates should have a morphology and particle size distribution as similar as possible to the exhaust particles found in the environment.

Lead bromide and lead bromochloride aerosols were produced from controlled temperature melts of the pure compounds in the apparatus illustrated in figure 1-1. Through adjustment of the voltage applied to the heating coil, the melt temperature was maintained at 400° C for lead bromide (melting point of 373° C) and 450° C for lead bromochloride (melting point of 430° C). At these temperatures the va-

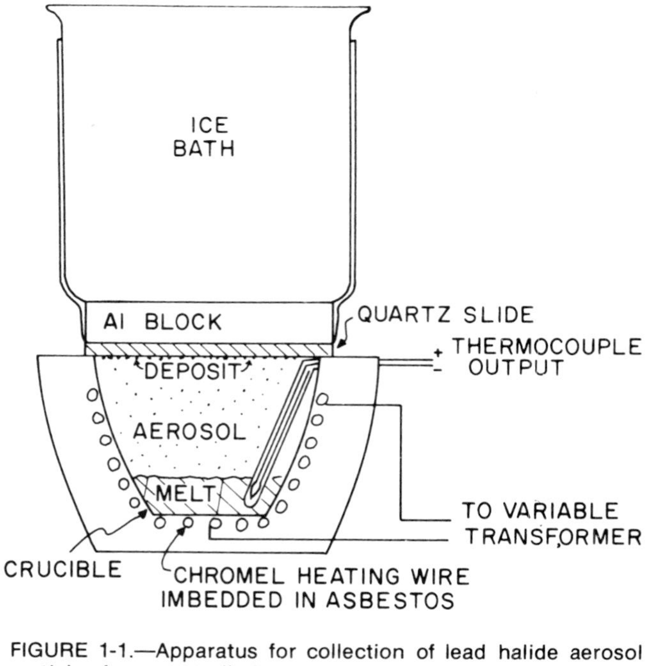

FIGURE 1-1.—Apparatus for collection of lead halide aerosol particles from controlled temperature melts.

por pressure over the two lead halides is sufficient to produce upon condensation a visible white "smoke" in the cooler region above the melt. The particles forming this smoke were collected on quartz slides for examination by electron microscopy and X-ray diffraction. The reaction chamber depicted in figure 1-2 was connected in series with the generating crucible to permit a study of the gas phase reactions of these lead-containing aerosols.

Varying the temperature of the generating crucible and reaction chamber produced changes not only in particle size distribution and morphology (as determined with the electron microscope), but also in the chloride-to-bromide ratios of the $PbBr_xCl_y$ aerosols. This behavior is shown in table 1-1. It has been demonstrated by Calingaert et al. [13] that lead bromide and lead chloride form a complete series of solid solutions with a preferential formation of the isomorphic 50-mole-percent compound lead bromochloride. The interplanar spacings of pure $PbBr_2$, $PbCl_2$, and stoichiometric PbBrCl are plotted in figures 1-3 and 1-4; a definite change of slope occurs at the 50-mole-percent composition of PbBrCl. The additional points plotted represent interplanar spacings of aerosols generated at different temperatures. These points indicate that nonstoichiometric Pb-Br-Cl ratios can be obtained in essentially stable aerosols from PbBrCl heated either above or below its melting point. The evolution of gaseous bromine from PbBrCl melts has been reported by Zimdahl. [14]

In low-speed stop-and-go city driving the exhaust system is relatively cool; under such conditions PbBrCl could be deposited in the exhaust system. In high-speed freeway driving this deposited PbBrCl is subjected to higher temperatures, and nonstoichiometric bromide-rich compounds could be selectively volatilized from the exhaust system. At least one recent study by Moran and Manary [15] has reported unidentified bromide- and chloride-rich species in exhaust lead-containing particulates. The results depicted in figures 1-3 and 1-4 suggest, according to Boyer and Laitenen, that previous reports dealing with the existence of stoichiometric PbBrCl could, with minor errors in interplanar spacing measurements, be interpreted in terms of nonstoichiometric compounds.

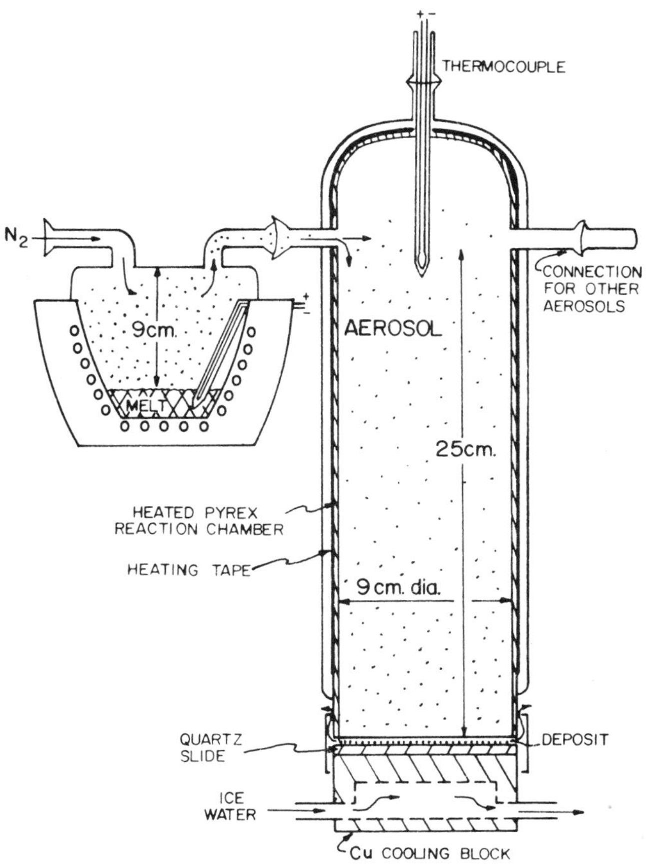

FIGURE 1-2.—Apparatus for the reaction of lead halide aerosols with other aerosols.

Figures 1-5 through 1-9 show progressive changes in morphology and geometric particle size of the aerosols with increasing Br/Cl ratios and decreasing generation temperatures. Thus at 510° C, well above the melting point of PbBrCl, the aerosols are well-defined spheres with a mean particle diameter of about 3 μm. As the generation temperature is lowered to a few degrees above the melting point, the particles become progressively more elongated, with a cross-sectional diameter of about 1.5 μm. Needlelike crystals with a cross-sectional diameter of 0.5 μm or less result when the generation temperature is well below the melting point.

Table 1-1.—Composition of lead halide aerosols produced at several temperatures

Aerosol No.	Generation temperatures (°C)		Aerosol composition (Mole ratio $PbBr_xCl_y$)		
	Crucible	Chamber	x	y	$x + y$
1	370	318	1.78	0.35	2.13
2	366	370	1.64	0.48	2.12
3	412	403	1.58	0.47	2.05
4	450	450	1.23	0.83	2.06
5	505	365	1.28	0.75	2.03
6	510	370	1.25	0.74	1.99
7	500	450	1.24	0.80	2.04
8	505	450	1.21	0.85	2.06

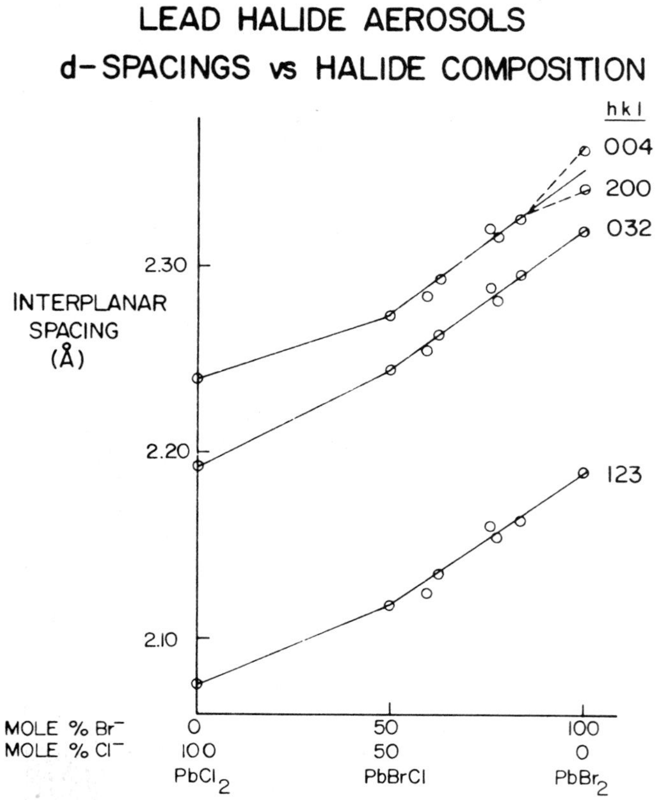

FIGURE 1-3.

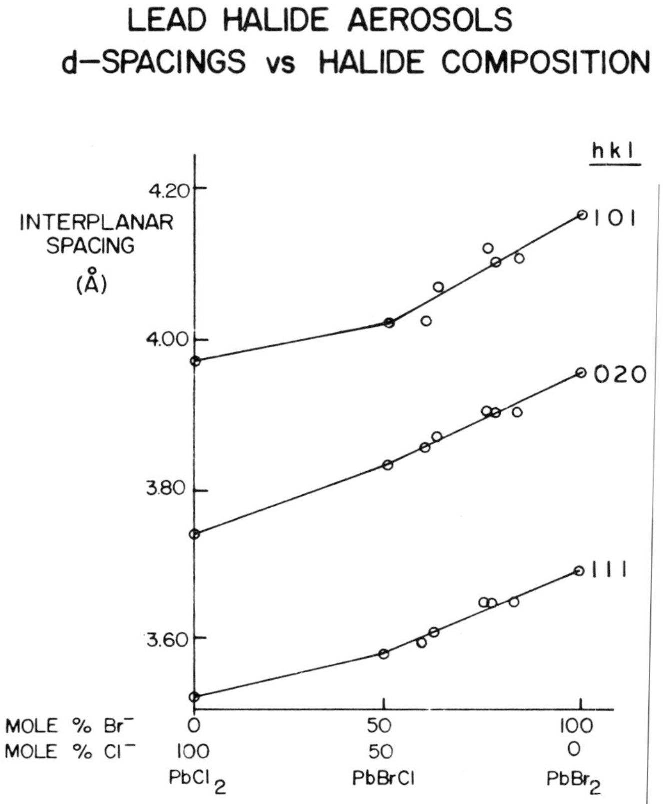

FIGURE 1-4.

FIGURE 1-5.—Scanning electron micrograph of $PbBr_{1.25}Cl_{.74}$ aerosol particulates. Crucible temperature is 510° C; chamber temperature is 370° C.

FIGURE 1-6.—Scanning electron micrograph of aerosol particulates. $PbBr_{1.28}C1_{.75}$. Crucible temperature is 505 °C; chamber temperature is 365 °C.

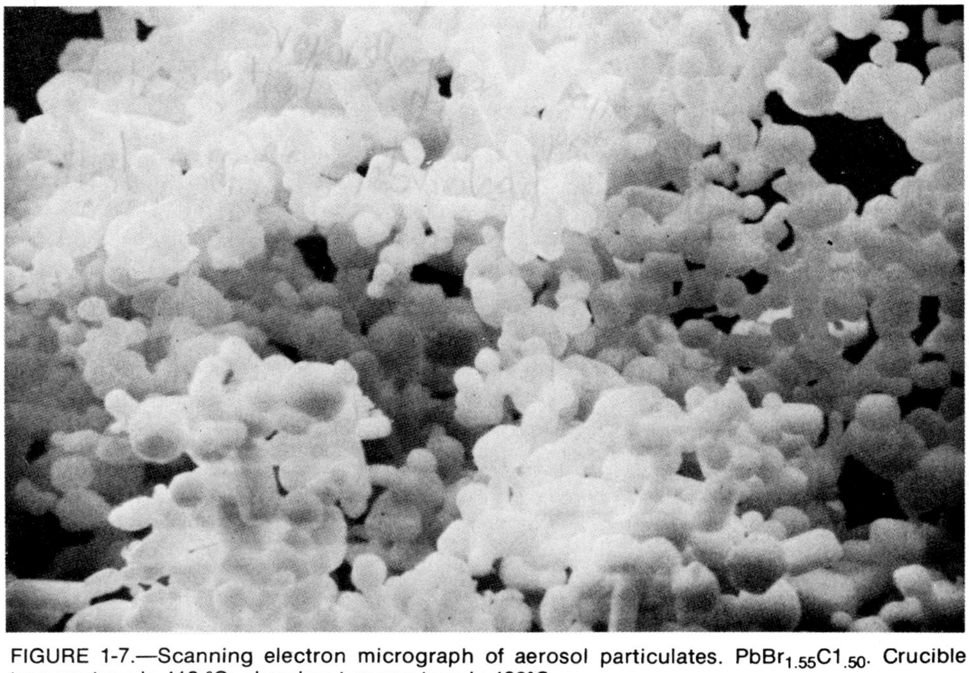

FIGURE 1-7.—Scanning electron micrograph of aerosol particulates. $PbBr_{1.55}C1_{.50}$. Crucible temperature is 412 °C; chamber temperature is 403°C.

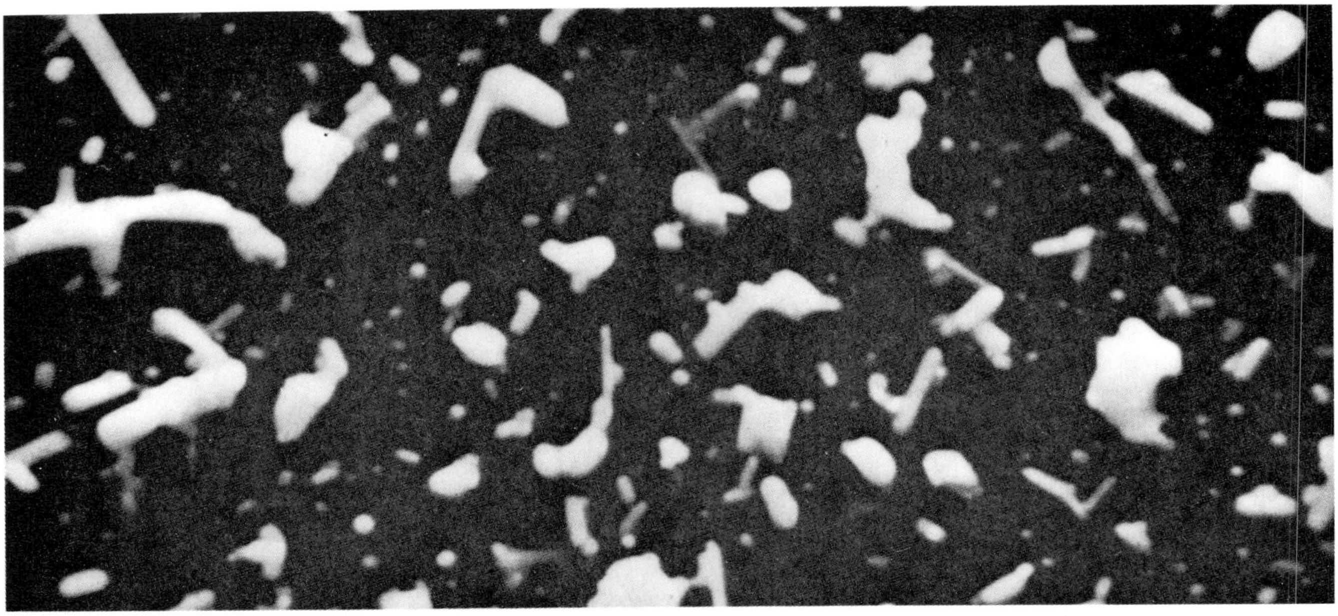

FIGURE 1-8.—Scanning electron micrograph of aerosol particulates. $PbBr_{1.64}C1_{.48}$. Crucible temperature is 366 °C; chamber temperature is 370 °C.

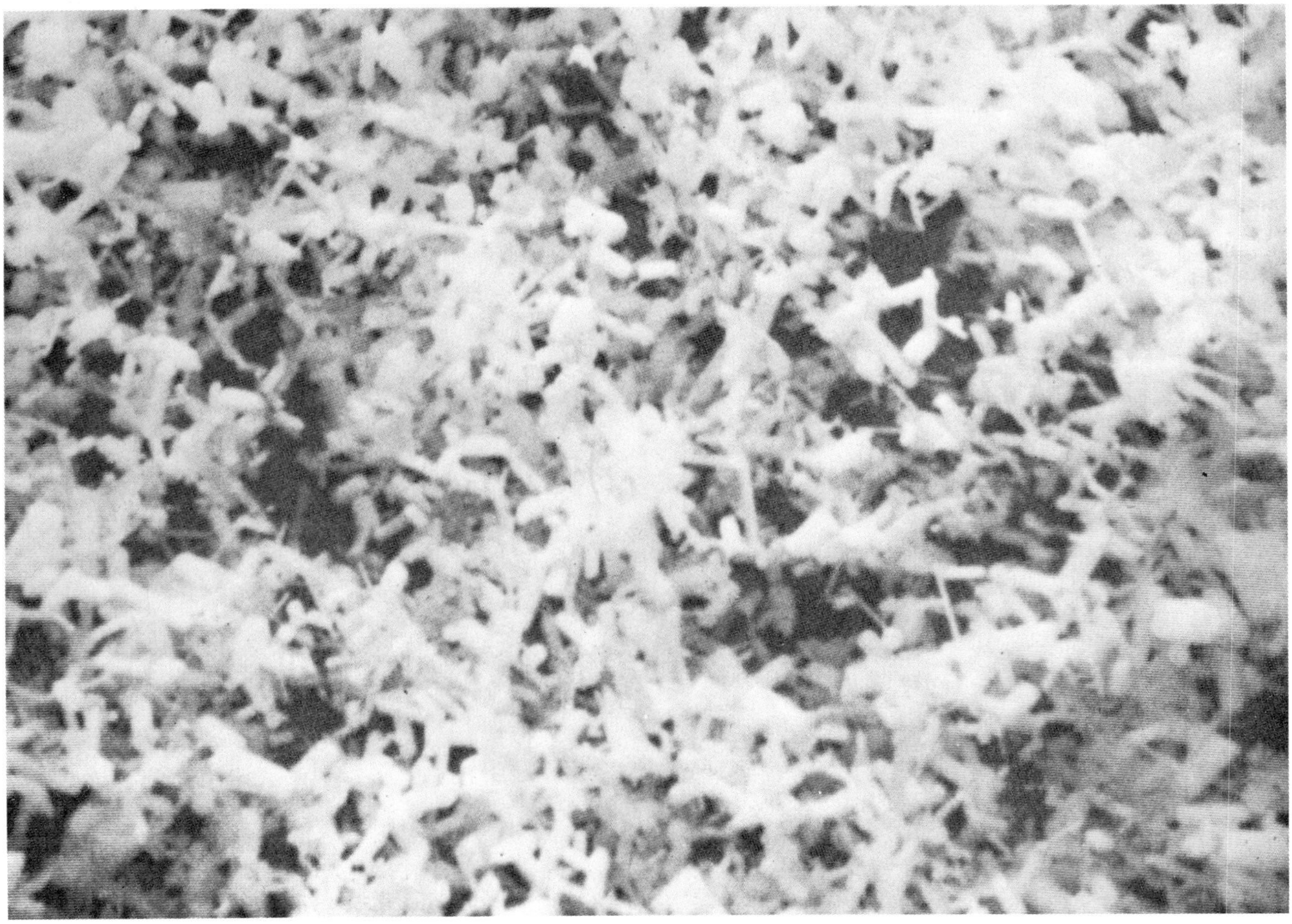

FIGURE 1-9.—Scanning electron micrograph of $PbBr_{1.78}Cl_{.35}$ aerosol particulates. Crucible temperature is 370 °C; chamber temperature is 318 °C.

TABLE 1-2.—Exposure of lead halide aerosols to high humidity, high carbon dioxide atmospheres, and filtered automobile exhaust

Exposure conditions	Mole ratio $PbBr_xCl_y$				Halide remaining (percent)
	Before		After		
	x	*y*	*x*	*y*	
3 percent H_2O vapor, 7 percent CO_2, 25°C, 24 h	([1])	([1])	1.33	0.68	–
12 percent H_2O vapor, 7 percent CO_2, 50°C, 90 h	([1])	([1])	1.36	0.65	–
Filtered unleaded auto exhaust, 105°C, 1 h	1.24	0.85	1.24	0.79	99.5
Filtered leaded auto exhaust, ultraviolet (UV) light, 105°C, 2 h	1.21	0.85	1.20	0.84	99.0

[1] Not analyzed.

Although these results establish the character of the lead halide aerosols that have been artificially generated in the manner previously described, they do not necessarily imply that lead-containing exhaust particles behave similarily as a function of temperature. They do, however, indicate the feasibility of both compositional and morphological changes occurring in the exhaust system.

Further studies were conducted at the University of Illinois to determine the behavior of these model lead-containing aerosols on exposure to high humidity, to high carbon dioxide concentrations, and to automotive exhaust gases from which the particulate matter had been partially removed. Automobile exhaust gas was obtained by inserting a probe into the exhaust line of a 1966 six-cylinder Ford engine; the probe was placed about 1 ft. downstrem from the exhaust manifold. This gas was then passed through heated copper tubing into a temperature-controlled enclosure containing a 3-ft. temperature equilization line, a stainless steel filter assembly fitted with glass fiber filters, and a reaction chamber (fig. 1-2) containing the lead halide aerosol. The gas flow rate was 1.7 l/min; the chamber temperature was maintained at 105° C to prevent condensation of water.

The results of these experiments are given in tables 1-2 and 1-3. It is clear that "laboratory pure" lead halide aerosols are quite stable toward hydroly-

FIGURE 1-10.—Scanning Electron Micrograph of Exhaust Particulates from Leaded Gasoline.

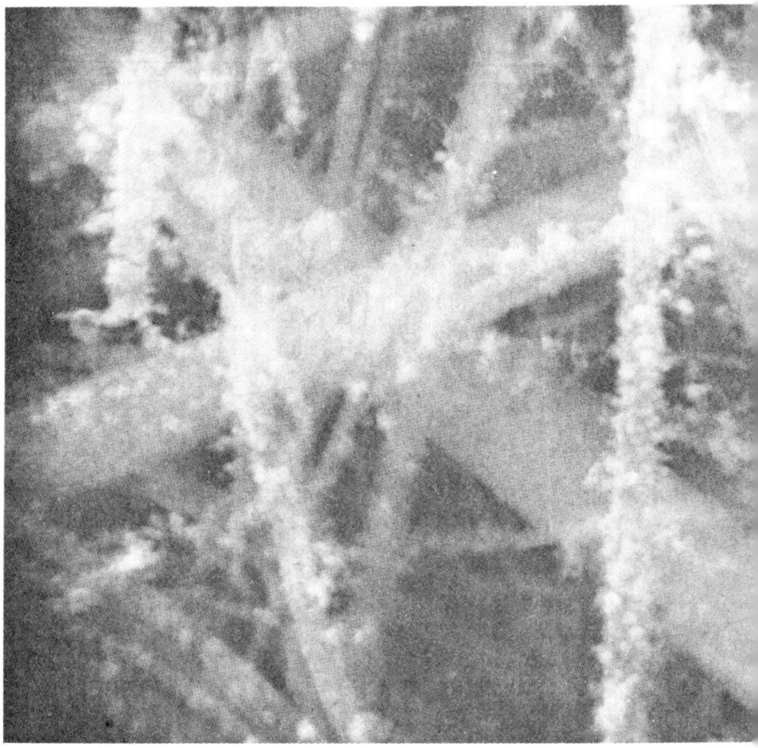

FIGURE 1-11.—Scanning Electron Micrograph of Exhaust Particulates from Unleaded Gasoline.

TABLE 1-3.–Decomposition of lead halides subjected to various conditions

Sample and conditions	Halide left after decomposition (μeq)	Halide lost during decomposition (μeq)	Total halide recovered (μeq)	Theoretical Pb and halide available (μeq)	Pb found by atomic absorption (μeq)
"Pure" $PbBr_2$	–	–	69.2	65.8	68.4
$PbBr_2$, 6.2 mW/cm², dry CO_2-free air, 32° C, 25 h	([1])	7.94 (30.1 percent)	([1])	([1])	26.4
$PbBr_2$, 9 percent CO_2, moist air, 25° C, 24 h	18.0	<0.05 (<0.3 percent)	18.0	18.4	18.0
"Pure" PbBrCl	–	–	292	297	294
PbBrCl, 9 percent CO_2, moist air, 50° C, 72 h	282	1.1 (0.4 percent)	283	288	288
PbBrCl, 6.2 mW/cm², dry N_2, 32° C, 51.5 h	176	9.6 (9.2 as Br_2; 0.4 as Cl_2; 5.1 percent of total)	186	194	190
PbBrCl, 6.2 mW/cm², dry CO_2-free air, 32° C, 50 h	116	5.8 (5.4 as Br_2; 0.4 as Cl_2; 4.8 percent of total)	122	124	121

[1] Not anlayzed.

tic exchange with water or to reaction with carbon dioxide under these experimental conditions. Furthermore, although Robbins and Snitz [16] have reported that up to 75 percent of the bromine is lost from fresh automobile exhaust particles in the first 20 min. after emission, no evidence of such losses was obtained for these model lead halide aerosols in the simulated exhaust gas exposure.

Scanning electron microscope studies of the particulates collected on the glass fiber filters with the experimental setup previously described show quite dramatically the difference between particles collected under dynamic conditions from the exhaust of an automobile burning leaded and nonleaded gasolines. (figs. 1-10 and 1-11). Figure 1-10 shows that instead of the filter collecting well-defined particles, the lead halides actually "grow" on the filter strands. Figure 1-11 for nonleaded gasoline shows that the particles collected have a mean diameter considerably smaller than the 0.3 μm value normally quoted for automotive emissions.

Chemical Characterization: The lead halides ($PbBr_2$, $PbCl_2$, and PbBrCl) and the alpha and beta forms of the double salt lead bromochloride ammonium chloride ($2PbBrCl{\cdot}NH_4Cl$), account for 80 to 100 percent of the lead-containing compounds found in the exhaust of automobiles burning gasoline containing a lead antiknock additive, according to Habibi et al., [4] Hirschler et al., [6] and Bayard and Ter Haar. [17] The exact composition depends upon the fuel and driving conditions. Habibi [5] has noted that large particles (greater than 200 μm) in fresh exhaust contain PbBrCl, 2PbO · PbBrCl, and small amounts of $PbSO_4$ and $Pb_3(PO_4)_2$. These particles would fall out very rapidly. (The settling velocity is about 70 cm/s). In the 2- to 10-μm range the principal constituent is PbBrCl. The submicrometer particles contain PbBrCl and $2PbBrCl{\cdot}NH_4Cl$; the latter is not a primary product but is formed as the exhaust is diluted and mixed with outside air.

Natusch and associates [18] at the University of Illinois have examined the dependence of elemental composition in automobile exhaust particles on depth into the individual particles. Relatively large particles (10 μm) were obtained from a cyclone collector attached to an automobile operated under test cycle conditions by Universal Oil Products Co. Two surface analysis techniques were employed. In the first, the particles were etched on one side with a stream of positively charged argon ions; both the etched and unetched sides were analyzed with an electron microprobe. The second technique utilized Auger electron microprobe spectroscopy with simultaneous ion etching.

Both techniques showed that lead, bromine, and chlorine occur at much higher levels on the particle surfaces than in their interior (fig. 1-12), indicating that surface deposition (perhaps condensation) takes place in the exhaust system. The particles studied all had small, submicrometer particles fused to their surfaces (fig. 1-13), and there is marginal evidence from these microprobe studies indicating that these attached particles may be highly enriched in lead, bromine, and chlorine.

These results suggest that lead bromochloride vapor may either condense directly from the vapor phase or be deposited as small particles onto large, partially molten particles that are rich in Al, Fe, Ca, Zn, P, and S (fig. 1-13). It is not clear whether particles of this composition are derived from the fuel itself, from engine or exhaust materials, or from the intake air. Essentially similar matrix compositions were found for particles obtained from unleaded fuel.

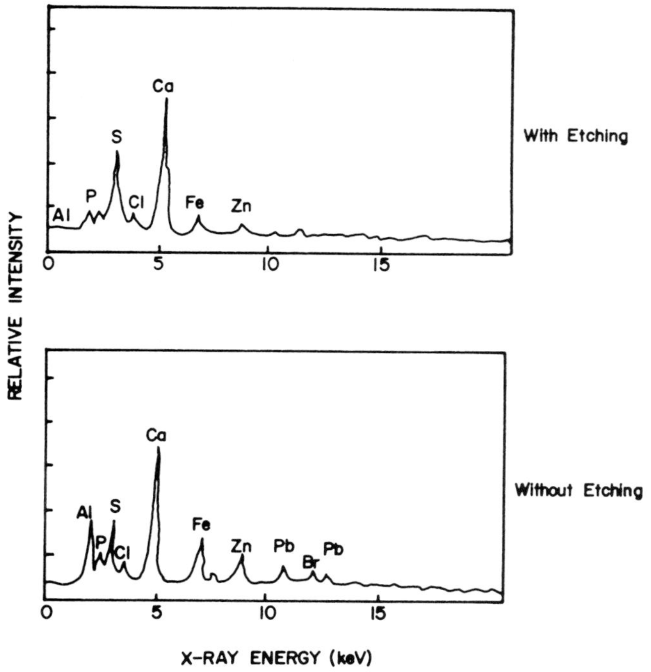

FIGURE 1-12.—X-ray spectra of (*a*) unetched and (*b*) etched portions of a single automobile exhaust particulate.

Lead Emissions from Coal Combustion

In addition to the automobile, a major source of airborne lead is the emission from coal burned in industrial furnaces and in electrical powerplants. Indeed, in countries where there is little consumption of leaded automotive fuels, the combustion of coal probably constitutes the major source of airborne lead.

Natusch, Evans, Wallace, and Davison [19] at the University of Illinois have studied the trace element content of fly ash derived from eight coal-fired powerplants in the United States and three from overseas. Fly ash samples were collected from precipitator systems and differentiated, both physically by sieving and aerodynamically with a Roller particle size analyzer, into a range of particle size fractions. Airborne fly ash samples were collected inside the plant stack and size differentiated aerodynamically *in situ* with the use of an Andersen stack sampler. These two types of samples represent size-fractionated material both retained in the plant and emitted into the air.

Elemental analyses were performed on the solid samples by spark source mass spectrometry, DC arc emission spectrometry, and wavelength dispersive X-ray fluorescence spectrometry. After bomb diges-

FIGURE 1-13.—Scanning electron micrograph of a large automobile exhaust particle showing small (probably PbBrCl) particles fused to its surface.

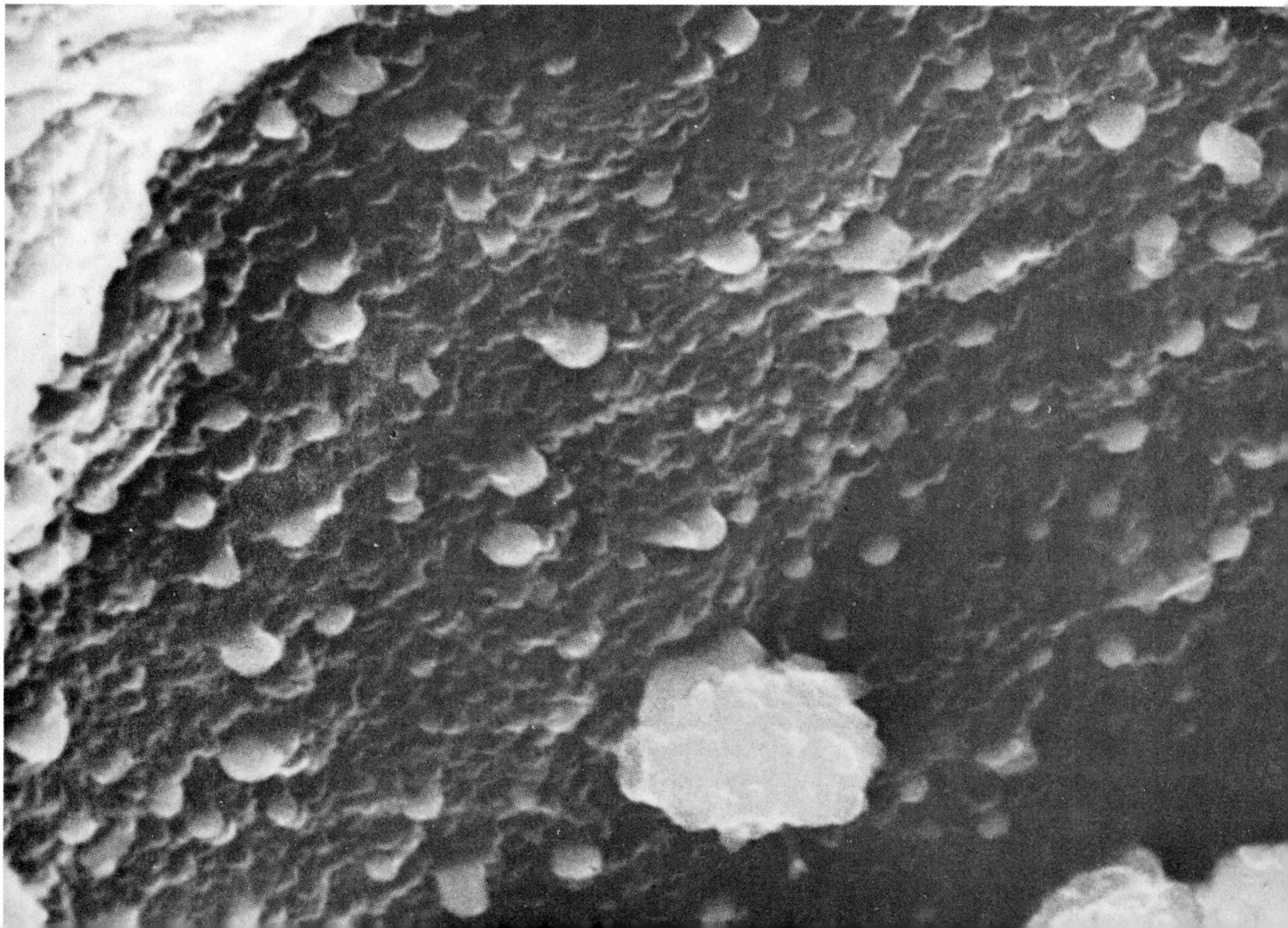

tion of the particles in hydrofluoric acid and aqua regia mixture, the resulting solutions were analyzed by atomic absorption spectrometry and differential pulsed anodic voltammetry; see Natusch and Wallace. [20] Results for the lead content of fly ash derived from one representative plant are presented in table 1-4 together with analytical dala for seven other elements. The concentration depends markedly on particle size.

In addition to the elements listed, V, Cr, Fe, Mn, Si, Mg, C, Be, and Al showed concentration dependence on particle size in some, but not all, of the samples studied; but the elements Bi, Sn, Cu, Co, Ti, Ca, and K show no convincing trends.

These results clearly show that many of the most toxic elements, including lead, are most concentrated in the smallest (respirable) fly ash particles. The results further imply that

(1) although existing particle collection devices may effectively remove the bulk of the fly ash, they are relatively inefficient for removing the most toxic elements.

(2) estimates of lead emissions based on analyses of undifferentiated samples of fly ash retained in the plant may grossly underestimate actual emissions.

(3) the highest concentrations of lead are emitted in particles that deposit in the lungs.

The mechanisms by which lead and other toxic metals become concentrated in small airborne particles are not entirely clear. One attractive explanation is that the element (or one of its compounds) is volatilized in the high-temperature (1300° to 1600°C) combustion zone and then recondenses onto the large (per unit mass) available surface areas provided by the small particles.

This surface deposition hypothesis is supported by analyses by ion microprobe mass spectrometry of external and internal surfaces of individual fly ash particles. With this technique the particle surface is sputtered away by a stream of negatively charged, molecular oxygen ions. The sputtered ions, which are characteristic of the surface, are then separated and determined by mass spectrometry. By observation of the concentration of any element as a function of sputtering time, a profile of concentrations versus depth into the particle surface can be obtained. A typical example of a depth profile of lead in fly ash is depicted in figure 1-14. Similar surface deposition behavior has been observed by Linton et al. [21] for Tl, Sb, As, Ni, Cr, Zn, S, V, Mn, C, Be, Na, and K.

Mathematical models based upon this volatilization-adsorption concept predict, according to Natusch, [22] that the average element concentration, $\overline{C_x}$, should depend on the particle diameter, *d*, according to an equation in the form.

$$\overline{C_x} = \overline{C_0} + \overline{C_A} d^{-1}$$

in which $\overline{C_0}$ and $\overline{C_A}$ are, respectively, related to the average concentration of element X intrinsic to the particle and the average surface concentration of the absorbed element. This equation is based upon the somewhat questionable assumption that different size particles have the same bulk density and, thus, the same matrix composition. However, within the sampling and analytical errors reported, the model does qualitatively predict the observed trends, as is illustrated in figure 1-15 for several elements.

TABLE 1-4.–Elements showing pronounced concentration trends

Particle diam, μm	Pb	Ti	Sb	Cd	Se	As	Ni	Cr	Zn	S, wt%	Mass fraction %
	μg/g										
					A. Fly Ash Retained in Plant						
					Sieved fractions						
>74	140	7	1.5	<10	<12	180	100	100	500	. . .	66.30
44–74	160	9	7	<10	<20	500	140	90	411	1.3	22.89
					Aerodynamically sized fractions						
>40	90	5	8	<10	<15	120	300	70	730	<0.01	2.50
30–40	300	5	9	<10	<15	160	130	140	570	0.01	3.54
20–30	430	9	8	<10	<15	200	160	150	480	. . .	3.25
15–20	520	12	19	<10	<30	300	200	170	720	. . .	0.80
10–15	430	15	12	<10	<30	400	210	170	770	4.4	0.31
5–10	820	20	25	<10	<50	800	230	160	1100	7.8	0.33
<5	980	45	31	<10	<50	370	260	130	1400	. . .	0.08
					Analytical method						
	a	*a*	*a*	*a*	*a*	*a*	*b*		*a*	*a*	
					B. Airborne Fly Ash						
>11.3	1100	29	17	13	13	680	460	740	8100	8.3	
7.3–11.3	1200	40	27	15	11	800	400	290	9000	. . .	
4.7–07.3	1500	62	34	18	16	1000	440	460	6600	7.9	
3.3–04.7	1550	67	34	22	16	900	540	470	3800	. . .	
2.1–03.3	1500	65	37	26	19	1200	900	1500	15000	25.0	
1.1–02.1	1600	76	53	35	59	1700	1600	3300	13000	. . .	
0.65–1.1	. . .	. . .	. . .	. . .	. . .	. . .	. . .	. . .	. . .	48.8	
					Analytical method						
	d	*a*	*a*	*d*	*d*	*d*	*d*	*d*	*a*	*c*	

[a] dc arc emission spectrometry. [b] Atomic absorption spectrometry. [c] X-ray fluorescence spectrometry. [d] Spark source mass spectrometry.

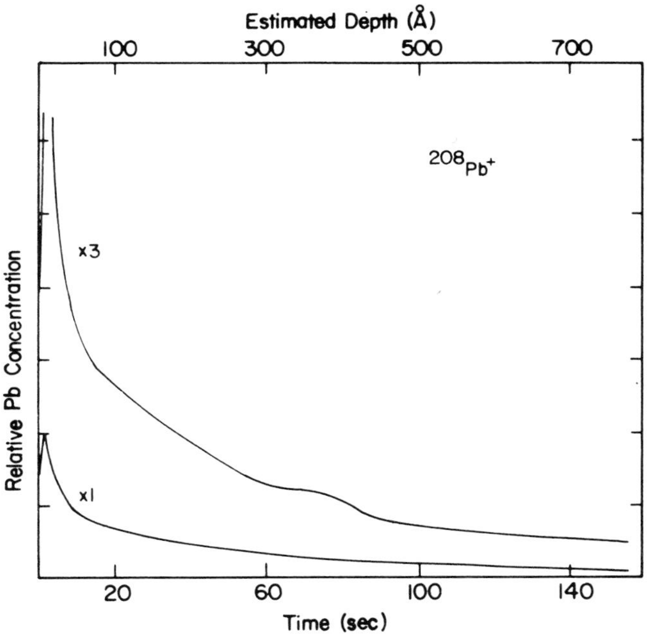

FIGURE 1-14.—Dependence of lead concentration on depth into a single particle of coal fly ash.

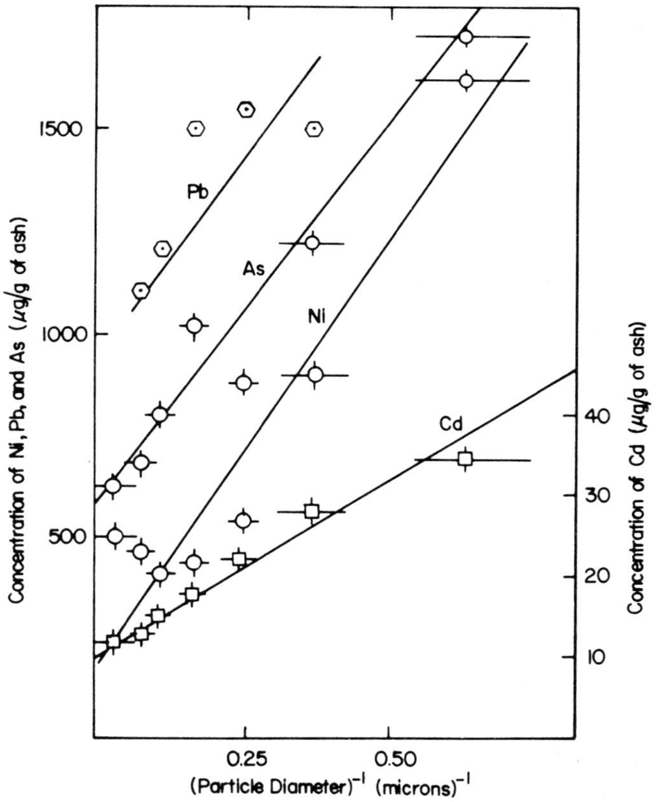

FIGURE 1-15.—Metal concentrations versus reciprocal of particle diameter.

If the proposed mechanism of preferential lead partitioning among particle size fractions is basically correct, one would expect all high-temperature combustion-source particles to exhibit similar concentration trends. Sources include metallurgical smelters and blast furnaces, cement plants, municipal incinerators, and automobiles. As mentioned earlier, lead does indeed show a strong tendency to be most concentrated on the surface of automobile exhaust particulates, and preliminary results show that the same is true for municipal incinerator ash.

The significance of these results is twofold: first, they establish the importance of determining concentrations of lead (and other trace elements) as a function of particle size; and second, they indicate that the highest concentrations are present on particle surfaces that most readily come in contact with biological extracting fluids and tissues.

Emissions From Smelting and Refining

The physical characterization of emissions from smelting and refining is largely confined to particle size distribution studies. Thus Vandegrift [23] reported that emissions from lead blast furnaces ranged in size from 0.03 to 0.3 μm. The distribution of lead was not measured as a function of particle size. According to Duprey, [24] particulate emissions from secondary lead-smelting operations contained 95 percent of particles smaller than 5 μm in diameter.

Purushothaman [25] at the University of Missouri has measured the particle size distribution of lead-containing particulate emissions from a baghouse filter on the stack of a primary lead smelter. An eight-stage Andersen impactor was employed with isokinetic sampling at 32 traverse points in the main stack. Results are given in table 1-5. About 54 percent of the total particulate mass was accounted for by particles with an ECD (corrected effective cutoff diameter at 50 percent impaction efficiency) of less than 2.4 μm; 60 percent of the particulate lead was associated with particles smaller than 2.4 μm. The cumulative size distribution given in figure 1-16 indicates a mass median equivalent diameter of 1.5 μm

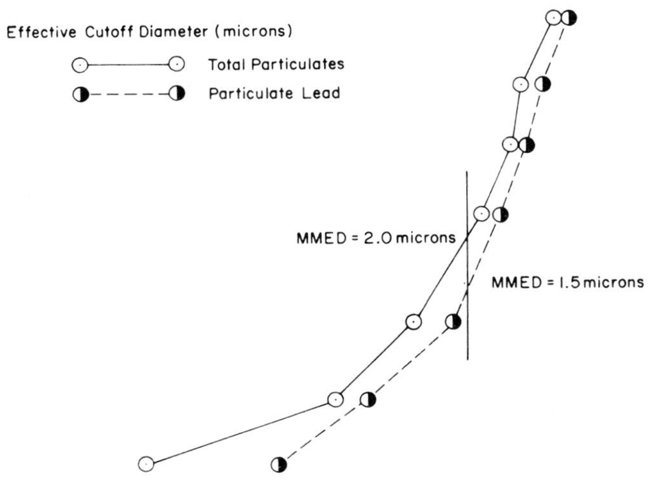

FIGURE 1-16.—Cumulative Size Distribution of Smelter Stack Emissions.

TABLE 1-5.–Particle sizing of stack emissions from a primary lead smelter

State	ECD[1] (μm)	Amount collected (mg)		Percent of total mass	
		Particulates	Lead	Particulates	Lead
0	12.7	1.9	0.540	12.2	10.8
1	8.0	1.9	0.505	12.2	10.1
2	5.3	1.4	0.365	9.0	7.3
3	3.75	0.5	0.265	3.2	5.3
4	2.35	1.4	0.335	9.0	6.6
5	1.2	3.4	0.720	21.8	14.4
6	0.74	2.8	1.235	17.9	24.6
7	0.49	2.2	0.665	14.1	13.3
Backup filter	<0.49	0.1	0.380	0.6	7.6

[1] Corrected effective cutoff diameter at 50 percent impaction efficiency.

for lead-containing aerosols, with a geometric standard deviation of 5 μm. Because 86 percent of the lead-containing aerosols are less than 10 μm in diameter, their fallout rate should be small; and long-range transport can occur.

Studies by Tibbs [26] indicate that the chemical species in baghouse effluent from smelting and refining activities include PbS, $PbSO_4$, and elemental lead. Bolter [27] at the University of Missouri has confirmed these results on the basis of X-ray diffraction techniques. The chemical species found in baghouse effluent and stack flue dust are listed in table 1-6.

ATMOSPHERIC LEAD

Physical Characterization

The particle size distribution of urban lead-containing aerosols has been investigated by a number of workers, including Gladney et al., [28], Colvos et al., [29] Lee et al., [30, 31] and Robinson and Ludwig. [32] Most of the lead is contained in the submicrometer particles. The mass median diameter of lead-containing particles (referred here not to the mass of the particle but to its lead content) is in the range of 0.15 to 0.45 μm. Robinson and Ludwig [32] reported that 75 percent of the lead was found in particles smaller than 0.7 μm. It was further noted that the lead particle size distributions are essentially independent of the locations measured. These observations support the concept that the collision rate between small particles is low and that, after fallout of the larger particles, a steady state distribution is soon established. (See Bayard and Ter Haar [17] and Robinson and Ludwig. [32]

The deposition and retention of particulate matter in the respiratory system have been extensively investigated. [33] For a mass median aerodynamic diameter of 0.15 μm, approximately 1 percent of the total inhaled particulate matter is deposited in the nasopharyngeal compartment, 44 percent in the tracheobronchial compartment, and 55 percent in the pulmonary compartment. For a mass median diameter of 0.45 μm, 10 percent is deposited in the nasopharyngeal compartment, 40 percent in the tracheobronchial compartment, and 50 percent in the pulmonary compartment. Thus a large fraction of the urban airborne lead may be deposited in the lung alveoli, where the long residence time insures sufficient extraction and absorption into the bloodstream. The efficiency of this process (60 to 80 percent) may be contrasted with the efficiency of lead uptake from the stomach (5 to 15 percent). [34, 35] Even though about 50 percent of the inhaled lead is washed into the stomach from the upper portion of the respiratory tract, only about 10 percent of this reaches the bloodstream (5 percent of the inhaled lead), but approximately 35 percent of the total lead inhaled [36] enters the blood.

A second consequence of the small aerodynamic size of lead-containing particulates is their long residence time in the atmosphere. It is now generally accepted that lead-containing particles are globally distributed, as demonstrated by Patterson [37] with respect to deposition on and in the Greenland ice cap. His results are shown in figure 1-17 and illustrate the significant correlation between increased lead deposition and the use of lead alkyl additives in gasoline.

TABLE 1-6.–Chemical composition of lead-containing dusts from smelting operations

Sample description	Composition (percent)			
	PbS	$PbSO_4$	PbO $PbSO_4$	Elemental Pb
AMAX B.F.[1] baghouse (cellar)[2]	70	10	–	20
AMAX B.F. baghouse (cellar)[3]	80	5	–	15
AMAX main stack flue bottom	∿95	5	–	–
ASARCO sinter baghouse (cellar)	–	∿100	–	–
ASARCO B.F. baghouse (cellar)	–	30	70	–

[1] Blast furnace.
[2] Sample taken on 2/6/73 afternoon.
[3] Sample taken on 6/15/73 morning.

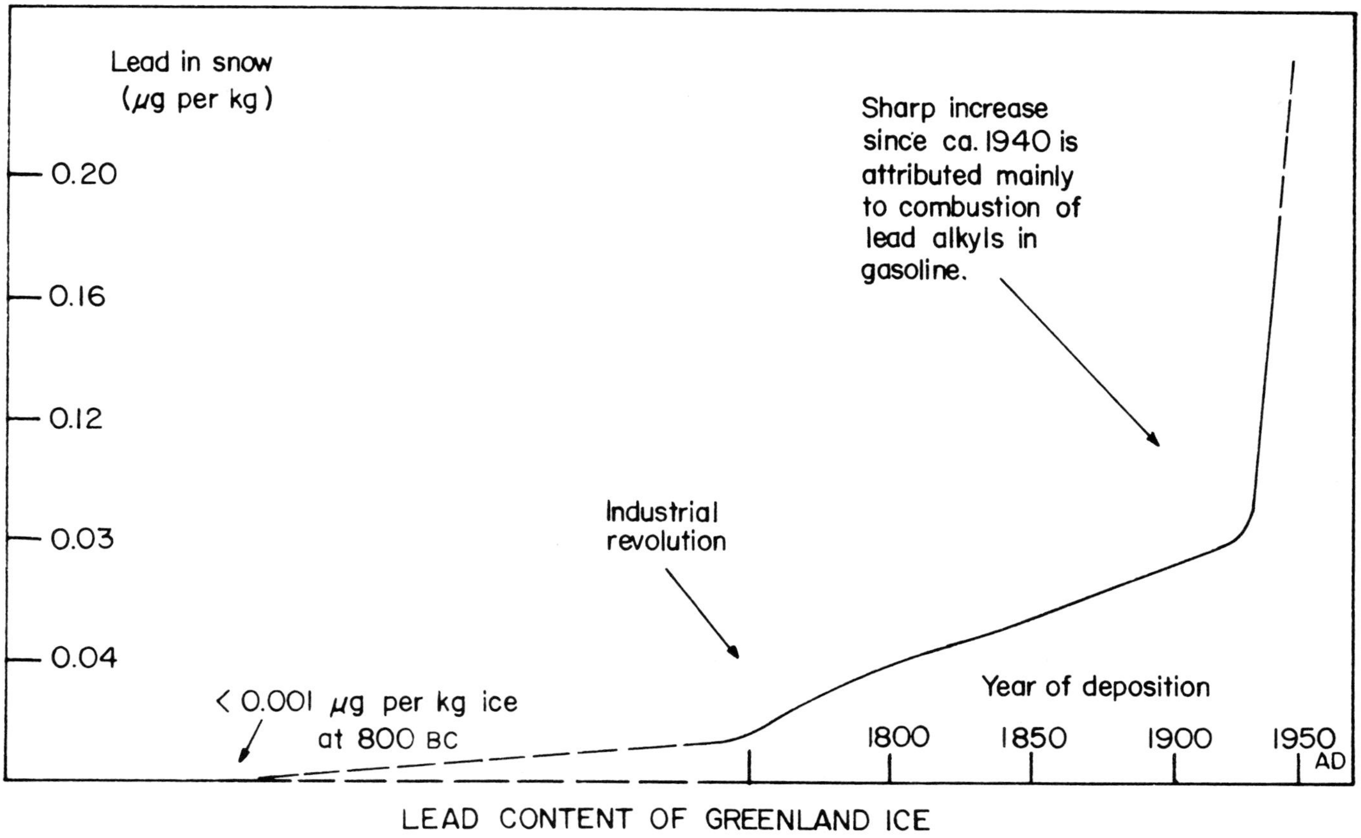

FIGURE 1-17.—Lead content of Greenland ice.

Interest has recently been focused on airborne and deposited dust collected in urban households. Preliminary studies by Krueger [38] and by Natusch et al. [39] have indicated surprisingly high levels of lead (600-6,000 ppm) in vacuum cleaner dusts collected in Boston, Chicago, New York, and Urbana. It seems likely the lead is introduced into the house mainly from roadside deposits that adhere to the feet or shoes rather than by airborne particles. For example, households with a pet dog consistently show an approximately 50 percent higher lead levels. However, the possibility is of considerable significance in households that lead may gain access to the body by inhalation of airborne dust.

Corrin and Menne at Colorado State University have investigated the particle size distribution of lead-containing particulates as a function of distance from a highway. Some preliminary experiments raise an important question discussed below. In a typical experiment, two eight-stage Andersen impactors fitted with 5 μm Gelman membrane backup filters were employed. One impactor was placed just off the highway and the other 60 m. downwind. The flow rate was 28 l/min. and the total sampling time was 42 min. The Teflon collector plates (employed to minimize Pb background corrections) and the backup filter were analyzed for lead content. The results are plotted in figure 1-18. Corrections have been applied for plate collection efficiency. Tabulated data are presented in table 1-7.

O'Donnell et al. [40] have observed that cascade impactors cannot be used to obtain total airborne concentrations; large particles are lost in the top jet sieve and do not reach the first collector plate. There is a small similar loss at the second collector sieve, but no loss at the following stages. A greater quantity of lead was collected in the backup filter off road than on road (table 1-7); this filter collects particles presumably less than 0.1 μm. In the experiment previously cited, carbon monoxide concentrations were also determined (by infrared absorption) at roadside and 60 m. downwind; thus dilution factors could be calculated for an inert contaminant originating from automobile exhaust (inert, at least, for the time frame of the measurements). The CO concentrations were 6.3 and 4.0 mg/km, and the corre-

TABLE 1-7.–Andersen impactor results

Impactor stage	Cutoff limits (μm)	Lead collected (μg) On road	Off road
0	8-14	0.268	0.323
1	5-8	0.368	0.023
2	4-5	0.178	0.473
3	2.5-4.0	0.063	
4	1.2-2.5	0.033	
5	0.8-1.2	0.368	0.063
6	0.4-0.8		0.043
7	0.1-0.4	0.088	0.418
Filter		0.140	0.290

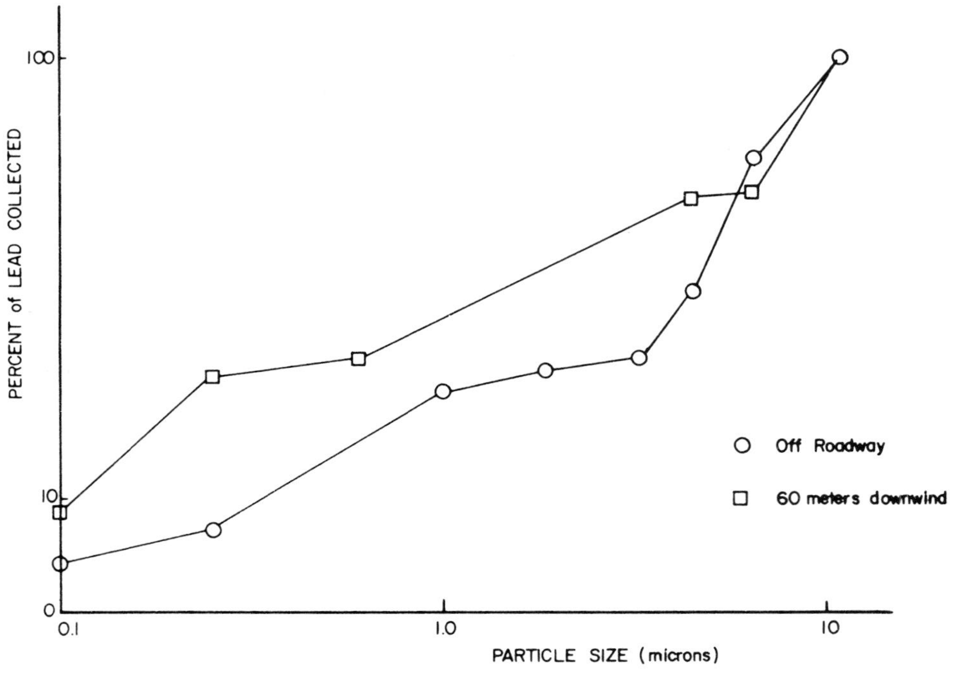

FIGURE 1-18.—Size distributions of particulate lead in relation to distance from a road.

sponding dilution factor for the 60-m ransport was 1.6. If no additional lead-containing particles less than 0.1 μm were formed in transit from the highway and if no particles in this size range were lost in transit, the quantity of lead picked up by the backup filter 60 m. downwind of the highway should be $0.140/1.6 = 0.09\ \mu g$. The measured value was 0.29.

These results and additional confirmatory evidence may be interpreted in terms of an increase in the absolute number of small lead-containing particles during the transport process. This increase may be due to (1) entrainment of small particles from the ground, (2) breakup of larger particles, (3) synthesis of small lead-containing particles by a gas-to-particle conversion process, or (4) the growth by agglomeration of particles originally too small to be collected at the roadside filter to sizes sufficiently large to be collected by the downwind filter.

It is tempting to ascribe the observations either to a gas-to-particle conversion process or to the growth of lead-containing particles. This latter explanation would be in accord with observations by Skogerboe [41] on the collection efficiency of various filter media with respect to lead-containing particles. The gas-to-particle conversion process requires the existence of lead in gaseous form in appreciable amounts at the roadway. Studies of lead alkyl concentrations in ambient air close to a source by Purdue et al. [1] indicate concentrations up to 10 percent of total lead in gaseous form; a number of higher measured values were rejected because of lack of reproducibility. Laveskog [3] found lead alkyl concentrations up to 18 percent of total lead in underground parking garages. The problem of lead alkyl concentrations close to heavily traveled roadways has not been solved. It is a problem of considerable importance, however, not only in the interpretation of lead transport but also in terms of health effects, because the toxicity of lead alkyls is considerably greater in terms of mass than that of lead-containing particulates. The existence of appreciable concentrations of gaseous lead species would explain the results of Olson and Skogerboe [42] in that a greater fraction of total lead is collected in the backup graphite crucible to a filter with increasing distance of the collector from the source. The question then becomes: Is the material passing through the filter gaseous or in particles too small to be collected?

In view of this difficulty in interpreting the results of filter experiments and because of the possible detrimental health effects that would arise from high concentrations of gaseous lead in ambient air close to a source, a program was undertaken at Colorado State University to develop a rapid and accurate technique for the determination of lead alkyls in air. The approach involved the use of chilled hexane scrubbers (-78° C) for collection, concentration by evaporation under reduced pressure, and analysis by gas chromatography with an electron capture detector. Suitable columns were developed, a backflush system was designed to eliminate effects on the detector of halide-containing species, and the measurement system was calibrated with lead alkyl standards supplied by the Ethyl Corp.

This measurement technique avoids the difficulties inherent in distinguishing between gaseous and particulate lead on the basis of retention or lack of retention by a filter. Some results reported by Menne and Corrin [2] indicate that within 30 m. of a heavily traveled highway the only lead alkyl detectable (detection limit approximately 0.01 $\mu g/m^3$) was lead tetramethyl. With one exception, the concentration of this species was 1 to 2 percent of the total ambient lead concentration; the one exception gave a value of 20 percent. The absence of other lead alkyls and the fact that lead tetramethyl possesses the highest vapor pressure in the lead alkyl series suggest that the presence of lead tetramethyl is due to evaporation from leaded fuels rather than emission as the result of use in the automobile engine.

Adsorption of Lead Tetraethyl Vapor by Simulated Atmospheric Dust Particles: It has been suggested that if appreciable quantities of lead alkyls are emitted into the atmosphere, a sink may exist for these species in their absorption on the surfaces of atmospheric particulate matter. Edwards and Rosenvold [43] and Edwards et al. [44] have studied the uptake of lead tetramethyl from the vapor phase on four finely divided solids that, in toto, simulate the composition of atmospheric dust. These authors found appreciable uptake of the lead alkyl vapor both in the absence and presence of water vapor. For equal surface areas of the solids, the order of increasing uptake was (1) silica, (2) alumina, (3) carbon black, and (4) graphitized carbon black. The presence of sulfur dioxide in trace amounts increases the quantity of vapor sorbed. It was further noted that only a portion of the sorbed lead could be recovered from the aged samples by mild heat or hexane extraction. It was concluded that sorption of organic gaseous lead compounds by atmospheric dust components may represent a significant scavenging mechanism and also that atmospheric particulates may serve as substrates for conversion of the sorbed lead species into solid inorganic lead compounds.

These conclusions are subject to two questions:

(1) Is the effect observed significant at concentrations of lead alkyls likely to be encountered in the real atmosphere (massive doses of $Pb(C_2H_5)_4$ were employed in the experiments)?

(2) Do the various solids properly simulate atmospheric dust?

Solubility of Lead Bromochloride in Water: Much of the work at Colorado State University and considerable work at the University of Illinois has involved simulation of lead-containing atmospheric particulate matter by PbBrC1, the principal lead-containing constituent of fresh automobile exhaust. This compound is prepared in the laboratory by heating an equimolar mixture of lead bromide and lead chloride to 250°C for several hours. [45] The identity of this product with PbBrC1 has been demonstrated by comparison of its X-ray diffraction pattern with that obtained by Julien and Ogilvie. [46]

The solubility of PbBrC1 in pure water is an important factor in terms of atmospheric washout and rainout processes, solution transport through soil, and physiological behavior following lung deposition. Consequently, Corrin and Springborn [47] have measured the solubility of PbBrCl in water at temperatures of 19°, 33°, and 43°C. Samples containing an excess of solid were equilibrated at the desired temperature, and the lead content of the resulting saturated solutions was determined by atomic absorption spectrometry. The results are given in table 1-8. The logarithm of the solubility is plotted in figure 1-19 against the reciprocal of the absolute temperature together with data for $PbBr_2$ and $PbCl_2$. A question arises concerning the nature of the solid phase in equilibrium with the solution: Is the solid phase PbBrCl or a mixture of $PbBr_2$ and $PbCl_2$? Two facts argue for the first alternative: The solubility of PbBrCl is less than that of either $PbBr_2$ or $PbCl_2$, and the plot in figure 1-19 is linear. If this conclusion is valid, some thermodynamic arguments may be pursued.

TABLE 1-8.–Solubility of PbBrCl in water

Temperature (° C)	Solubility (g/l)
19	6.62
33	8.10
43	10.3

Some Thermodynamic Properties of PbBrCl: Consider the following set of reactions:

(1) $PbCl_2(s) \rightleftharpoons Pb^{2+}(aq) + 2Cl^-(aq)$
$$\Delta G_1^o = -RT \ln K_1$$

(2) $PbBr_2(s) \rightleftharpoons Pb^{2+}(aq) + 2Br^-(aq)$
$$\Delta G_2^o = -RT \ln K_2$$

(3) $PbBrCl(s) \rightleftharpoons Pb^{2+}(aq) + Br^-(aq) + Cl^-(aq)$
$$\Delta G_3^o = -RT \ln K_3$$

in which ΔG^o is the change in Gibbs Free Energy for the standard state reaction and K_i is the solubility product. If ideal solution behavior is assumed, the

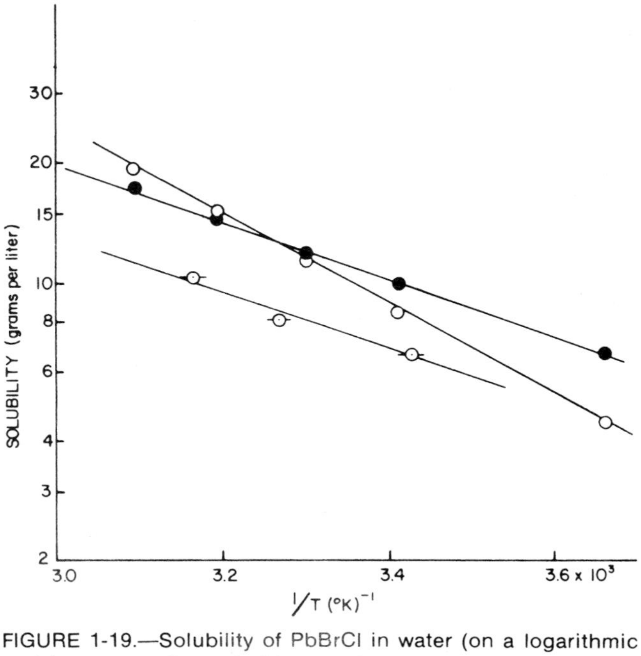

FIGURE 1-19.—Solubility of PbBrCl in water (on a logarithmic scale) versus reciprocal of temperature.

TABLE 1-9.–Stability calculations for PbBrCl equilibrium

Temperature (° C)	Compound	Solubility (g/l)	Solubility (moles/l)	K	ΔG^o (kcal)	K_{eq}
40	$PbCl_2$	14.5	5.21×10^{-2}	5.67×10^{-4}	1.7	0.065
	$PbBr_2$	15.3	4.17×10^{-2}	2.90×10^{-4}		
	PbBrCl	9.55	2.96×10^{-2}	1.04×10^{-4}		
20	$PbCl_2$	9.9	3.56×10^{-2}	1.80×10^{-4}	1.2	.127
	$PbBr_2$	8.5	2.31×10^{-2}	4.79×10^{-5}		
	PbBrCl	6.64	2.06×10^{-2}	3.49×10^{-5}		
0	$PbCl_2$	6.73	2.42×10^{-3}	5.67×10^{-5}	0.8	.228
	$PbBr_2$	4.55	1.24×10^{-3}	7.63×10^{-6}		
	PbBrCl	4.38	1.36×10^{-3}	1.00×10^{-5}		

solubility products may be written in terms of concentrations rather than activities. These three questions may be combined to give

$$(4)\ 2PbBrCl(s) \rightleftharpoons PbCl_2(s) + PbBr_2(s)$$
$$\Delta G^o = RT \ln (K_1 K_2 / K_3^2)$$

This process provides a measure of the stability of PbBrCl in solid form toward disproportionation into $PbBr_2$ and $PbCl_2$. Calculations have been made at three temperatures for the equilibrium constant and standard free energy for the reaction given in equation (4). These data are given in table 1-9. The solubility values for PbBrCl were calculated from a least-square linear plot of the information in table 1-8. An error of 10 percent in the solubility of PbBrCl at 0° C would still yield a positive value for ΔG^o. The stable solid species in the reaction given by equation (4) is thus PbBrCl.

It is also possible to calculate the standard free energy of the formation of PbBrCl at 298 K. Calculations similar to those employed in table 1-9 yield a value of +1.27 kcal for the reaction between PbBrCl(s) to form $PbBr_2$ and $PbCl_2$. Then

$$(5)\ Pb(s) + Cl_2(g) \rightleftharpoons PbCl_2(s) \quad \Delta G_f^o = -68.51 \text{ kcal}$$

$$(6)\ Pb(s) + Br_2(l) \rightleftharpoons PbBr_2(s) \quad \Delta G_f^o = -66.21 \text{ kcal}$$

$$(7)\ 2PbBrCl(s) \rightleftharpoons PbCl_2(s) + PbBr_2(s)$$
$$\Delta G^o = 1.27 \text{ kcal}$$

These equations may be combined to give

$$(8)\ 2Pb(s) + Cl_2(g) + Br_2(l) \rightleftharpoons 2PbBrCl(s)$$
$$\Delta G^o = -136 \text{ kcal}$$

The standard free energy of formation of PbBrCl(s) at 298 K is thus −68 kcal. Note that $PbCl_2$ is more stable at this temperature than PbBrCl, but PbBrCl is more stable than a 1:1 molar mixture of solid $PbBr_2$ and $PbCl_2$.

Further Studies on the Solubility of PbBrCl: When present as a solid species in airborne particles, lead must transfer to the solution phase before it can exert a toxic effect on living organisms. This process may occur in essentially pure water (as in a raindrop); in unpolluted waters with a wide range of solutes, ionic strengths, and ligand content; in highly polluted waters; or in body fluids such as saliva, lung mucuous, and stomach fluids. Clearly, it is of considerable practical interest to determine both the capacity and rate of solution of a range of environmentally significant solutions systems. Some preliminary experiments in this area have been performed by Laitenen and Fry [48] at the University of Illinois. These workers examined the effect of an organic molecule, in this case the indicator dye "wool violet," on the solubility of PbBrCl crystals in pure water. The dye was quantitatively adsorbed on the surface of the PbBrCl crystals, but it in no way altered the rate or extent of solubility when these crystals were reintroduced into distilled water.

Washout: Washout is defined as that process by which both gaseous and particulate material are removed from the atmosphere by incorporation into precipitation below a cloud base. This is in contrast to rainout, which occurs within the cloud as precipitation is formed. Dahl and Corrin [49] at Colorado State University have investigated the washout of lead containing aerosols by a simulation technique fully described by Dahl. [50]

The significant parameter in washout is "washout efficiency," which is a function both of the size of the collecting raindrop and the particle collected. Washout efficiency is defined as the ratio of particles removed to the number of particles swept out by the falling raindrop and, hence, is a product of collision number and collision efficiency toward capture. In the simulation experiment because analysis is by lead content in the collected precipitation, washout efficiency (WE) can be expressed as

$$WE = (4C_w r)/3C_a L$$

where C_w is the concentration (g/cm^3) in the collected precipitation; C_a the original concentration of lead in air; r the radius of the falling drop; and L the fall length. All lead analyses in this work were corrected for lead blanks in the water and analytical reagents.

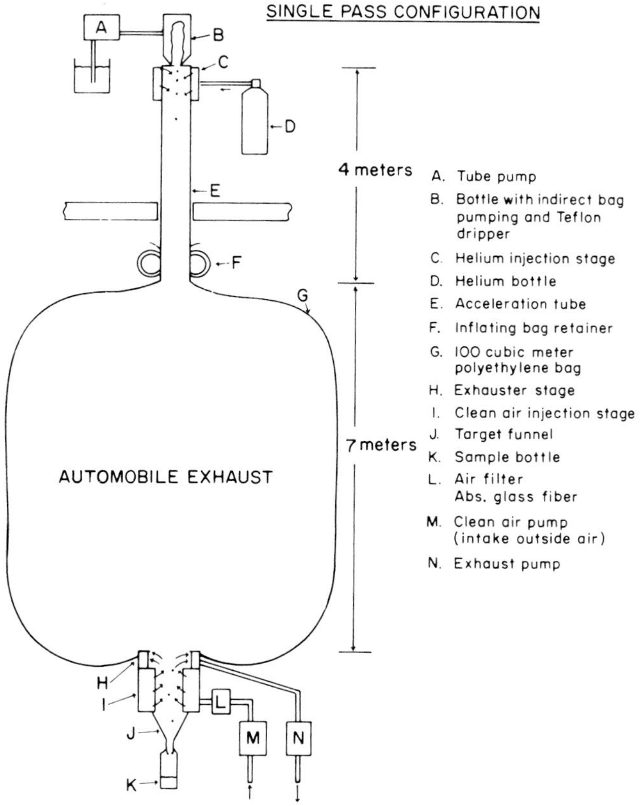

FIGURE 1-20.—Experimental setup for washout studies.

The experimental setup for the washout studies is shown in figure 1-20. Automobile exhaust was collected in large plastic bags, and the total lead concentration and particle size distribution (cascade impactors) were determined as a function of storage time. Results are presented in figures 1-21 and 1-22. Only a single water drop size, 4 mm. was examined. Under the conditions of these experiments, a limiting value of the washout efficiency for aged exhaust

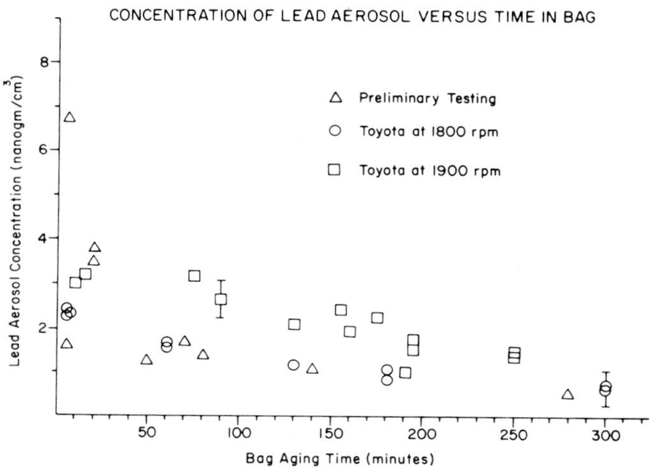

FIGURE 1-21—Concentration of lead aerosol versus time in bag.

particles was measured as 0.005 (fig. 1-23). This efficiency varies with the size of drop and of the particulate; the washout efficiency reported is for aged exhaust.

An illustrative calculation may be in order. The area of metropolitan Denver is approximately 450 mi.^{2} Assuming a mean lead concentration below the cloud base of 1 μg/m^3, the total amount of atmospheric lead removed through the washout process by 1 in. of precipitation may be calculated. The lead content of the precipitation may be determined from

$$C_W = \frac{3\, C_a L \, WE}{4r}$$

Assuming a cloud base 1,000 m. above ground level and r = 0.4 cm, C_W = 9 × 10^{-10} g/cm^3. The total amount of water in an inch of precipitation over a 450-mi^2 area is 3 × 10^{13} cm^3, and thus the total lead washed out in the precipitation = 2.7 × 10^4 g. The yearly precipitation in Denver is approximately 15

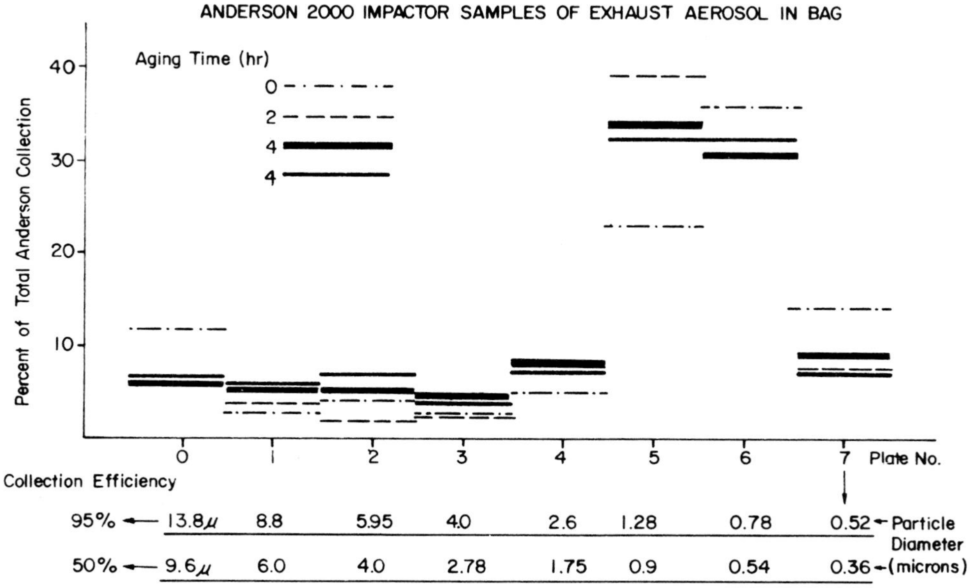

FIGURE 1-22.—Andersen 2000 impactor samples of exhaust aerosol in bag.

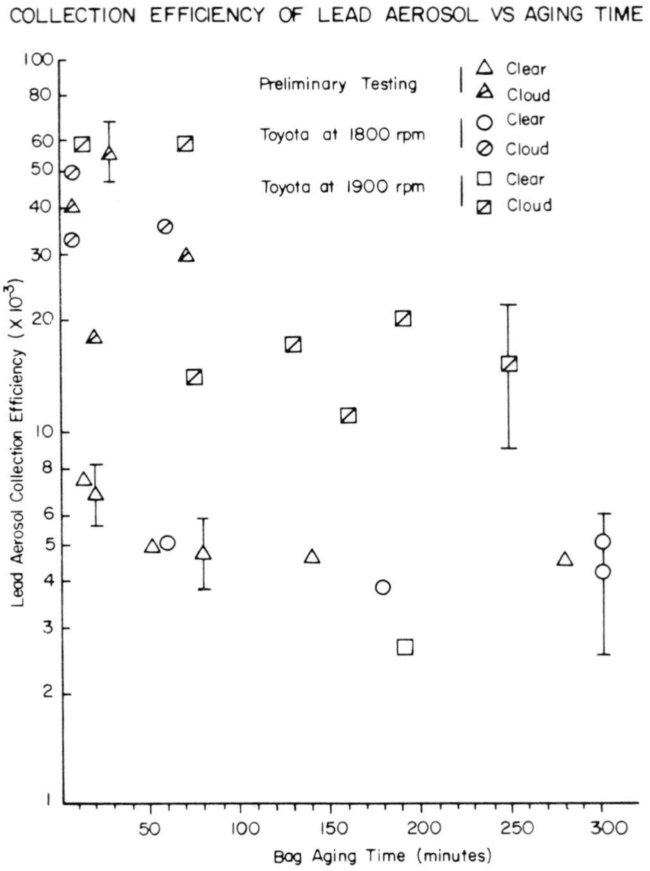

FIGURE 1-23.—Collection efficiency of lead aerosol versus aging time.

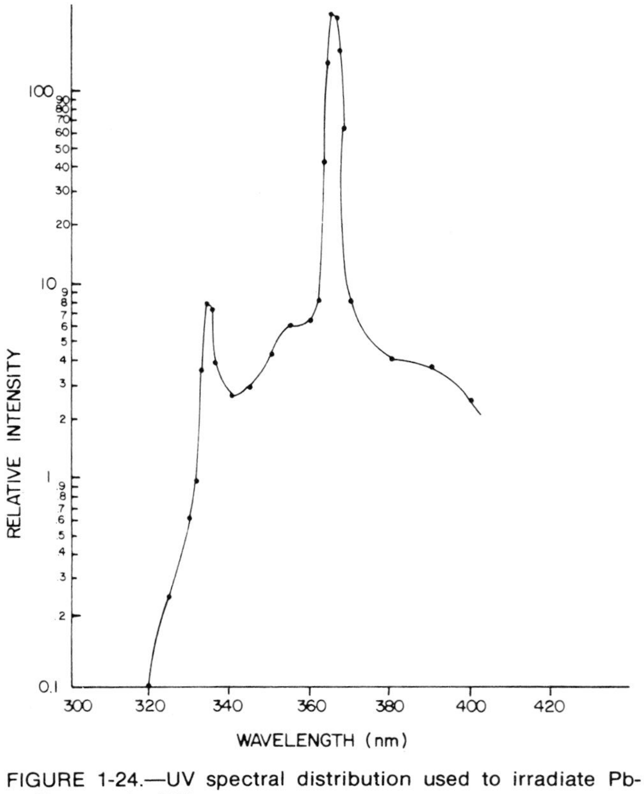

FIGURE 1-24.—UV spectral distribution used to irradiate PbBrCl aerosols. [53]

in., and the amount of lead removed by washout per year = 4 × 10^5 g. The number of automobile miles per year in Denver has been estimated as 3 × 10^{10}. With an average mileage of 12 mi/gal. (2.5 × 10^9 gal./year) and a lead emission of 1 g/gal., the total lead input into the atmosphere is 2.5 × 10^9 g/year. Thus the quantity of lead removed by the washout process is on the order of 0.02 percent.

Lazrus et al. [51] have observed that the average concentration of lead in rainfall collected at widely ranging sites in the United States is 3.4 × 10^{-2} μg/cm^3; for the Chicago area this value was 1.4 × 10^3 μg/cm^3. When contrasted with the value of 9 × 10^{-10} μg/cm^3 calculated above for the washout process, these values strongly suggest that the washout process is not a significant mechanism for the removal of atmospheric lead (and presumably other trace metals). A much more important removal mechanism is the in-cloud rainout process.

Chemical Characterization

The primary lead emissions are converted through chemical reactions occurring in the atmosphere into a large variety of lead species. The work of Habibi [5], utilizing electron probe and X-ray diffraction techniques, has indicated the existence in the atmosphere of the following lead species: $PbSO_4$, PbBrCl, $PbCO_3$, $PbBr_2$, $PbCl_2$, PbO_x, Pb(OH)Cl, Pb(OH)Br, $Pb_3(PO_4)_2$, $2PbBrCl \cdot NH_4Cl$, $3PbO \cdot PbSO_4$, $2PbO \cdot PbCl_2$, and $2PbO \cdot PbBrCl$. The major atmospheric species were identified as $2PbBrCl \cdot NH_4Cl$, $PbSO_4$, and $PbCO_3$. Little is known of the details of the atmospheric chemistry of lead. Yet it is the chemical composition, not of the primary lead-containing species as emitted, but rather of the species existing in the atmosphere, that determines such parameters as solubility, uptake, mobility in soil, and plant uptake.

It has been shown by Pierrard [52] that lead halide aerosols in automobile exhaust can lose halogen by photochemical decomposition when the aerosol is exposed to UV radiation. This same reaction has been investigated by Laitenen and Boyer at the University of Illinois. [53] These workers studied the effects of UV irradiation on lead halide aerosols generated from molten salts. The UV spectrum used is shown in figure 1-24. Figures 1-25, 1-26, and 1-27 illustrate, respectively, the rate of loss of halogen from $PbBr_2$ and a series of $PbBr_xCl_y$ aerosols, the dependence of halogen loss on the Br/Pb content of the aerosol, and the dependence of halogen loss on the partial pressure of oxygen.

The mechanisms for the photodecomposition of lead halide have been discussed by Kaldor and Somorjai, [54] as well as by Pierrard. They involve the absorption of a photon to release an electron and a hole which diffuse to trapping sites within the crystal lattice. Neutralization of the trapped electrons by lead ions forms photolytic lead, while neutralization of holes by bromide and chloride ions produces free bromine and chlorine atoms, which diffuse to the surface and escape.

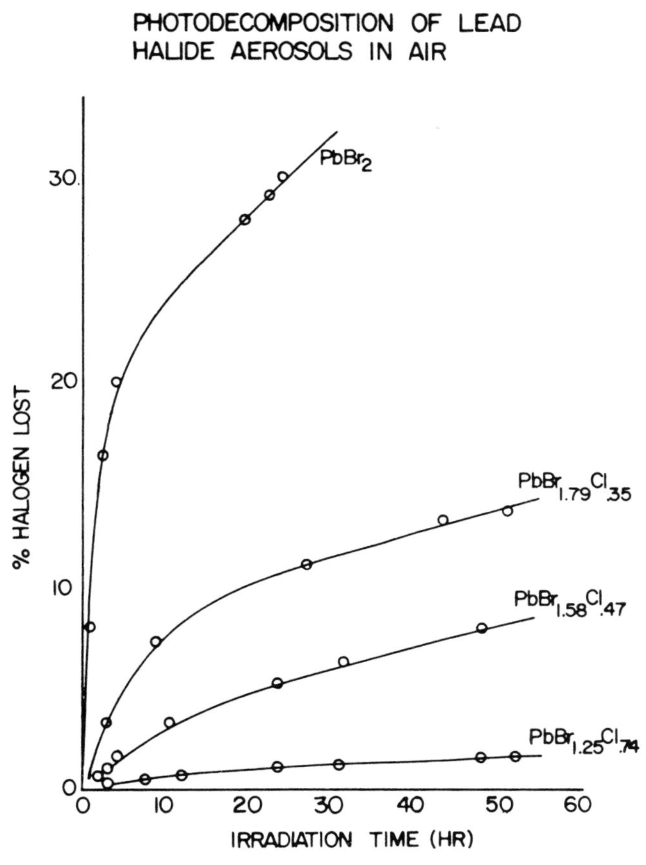

FIGURE 1-25.—Photodecomposition of lead halide aerosols in air.

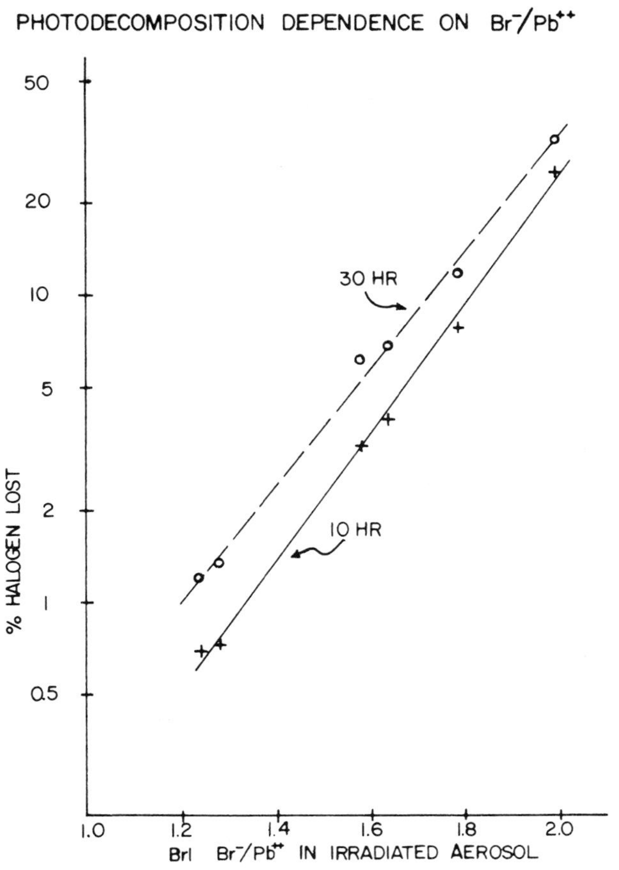

FIGURE 1-26.—Photodecomposition dependence on Br^-/Pb^{2+}

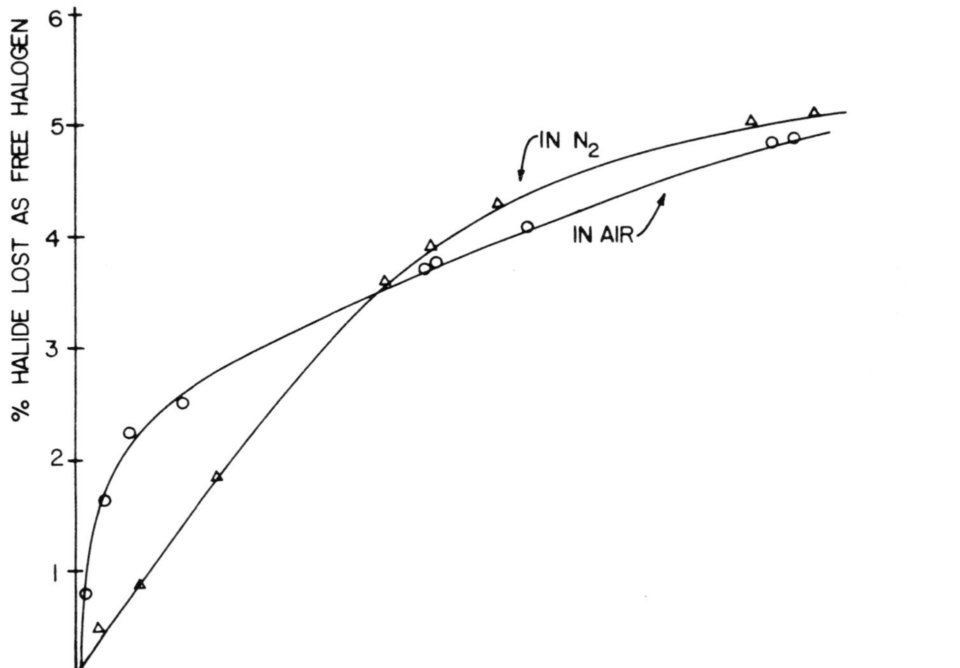

FIGURE 1-27.—Photodecomposition of PbBrCl.

Figure 1-26 shows that the degree of photodecomposition (defined here as the percent of available halide ion lost as free halogen) increases exponentially with the Br^-/Pb^{2+} ratio. One possible explanation is an exponential increase in electron-hole production due to a linear increase in the absorptivity of the lead halide for photons with increasing bromide ion concentration. However, without further information, no definite reason can be given for the dependence of the degree of photodecomposition on the ratio of bromide ion to lead ion.

ENVIRONMENTAL LEAD IN WATER

Physical Characterization

The parameters of physical characterization would include solubility of lead compounds, the physical properties of solutions of lead compounds in various liquids, and the physical nature (such as particle size and surface area) of suspended particulates. The work of the environmental lead groups at Colorado State University and the University of Illinois has not borne upon these studies. The absorption of lead from aqueous solution, however, has been examined at the University of Illinois.

The clay fractions (defined here in terms of a size less than 1 μm rather than in terms of composition) of soils and sediments are known to contain appreciable amounts of the hydrous oxides of iron and manganese. [55] These hydrous oxides are amorphous in nature and have a very high sorption capacity, according to Stumm and Morgan [56], for certain heavy metal ions. They thus play a significant role in determining the concentrations of such heavy metal ions in their environment. Although there are considerable data on the cation sorption by hydrous oxides, there is no simple theory that can satisfactorily explain the sorption behavior of different cations, nor are there any published reports on lead sorption. (See, however, the work of Zimdahl in this regard in the section on soils.) The following work by Laitenen and Gadde [57] provides the only definite information on the uptake of lead by hydrous oxides.

Studies were made of the effect of lead ion concentration, pH, and age of the oxide on the percent sorption of lead by synthetic hydrous ferric oxide (HFO). The results are shown in figures 1-28 and 1-29 and are listed in table 1-10. HFO was prepared by adjusting the pH of $0.1M$ ferric nitrate solution to 6.0 with $0.1M$ NaOH. After the precipitate had been allowed to settle for 2 hours, it was filtered and washed repeatedly with distilled water to remove Na^+ and $NO^-{}_3$. The HFO was then aged overnight in distilled water with pH=6. The amorphous nature of the HFO thus prepared was confirmed by its lack of X-ray powder diffraction lines. Lead uptake by the HFO was determined by measuring the loss of lead from solution with both pulsed and dc polarography. All experiments involved 0.625 μ moles of iron as HFO in a 100-ml solution.

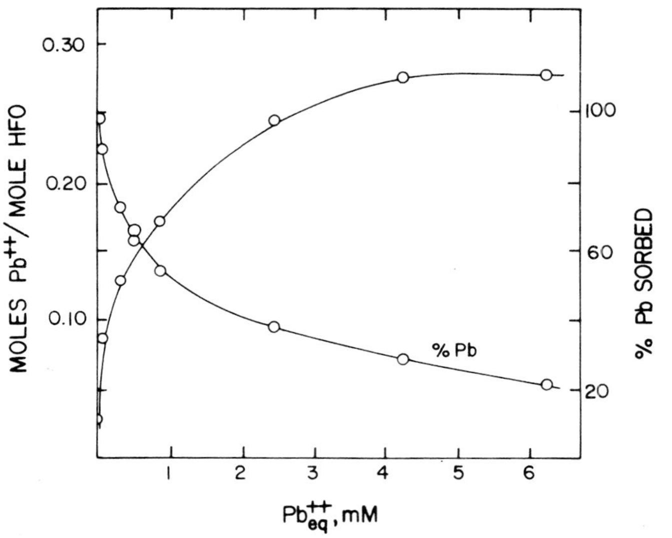

FIGURE 1-28.—Lead sorption as function of lead ion concentration.

As much as 0.28 moles of lead per mole of iron in HFO was found to be sorbed at PH=6 (figure 1-30). The sorption involved release of H^+ from the HFO. This process was found to be reversible with respect to hydrogen ion. The ratio of moles of H^+ released per mole of Pb^{2+} sorbed was found to decrease from 1.59 at a pH of 6 to 1.18 at a pH of 5. Calculations of the concentrations of hydroxy complexes of lead in solution at different pH values show that those complexes are known not to be active in sorption below a pH of 6. Consequently, it is assumed that free or hydrated Pb^{2+} ions are sorbed. Furthermore, because the sorption of lead ions (or a hydroxy complex above pH=6) occurs to a considerable extent values well below a pH of 8.5, given by Parks [58] as the zero point of charge, this adsorption appears to be specific rather than the non-specific or counter-ion type of adsorption discussed by Fuerstenau. [59]

The non-integral ratio of hydrogen ion released per lead ion sorbed is not unique to lead and HFO; it is also found for other cation-adsorbent systems. For example, ratios ranging from 1.0 to 1.7 were found for Mn^{2+} sorption on MnO_2 by Stumm and

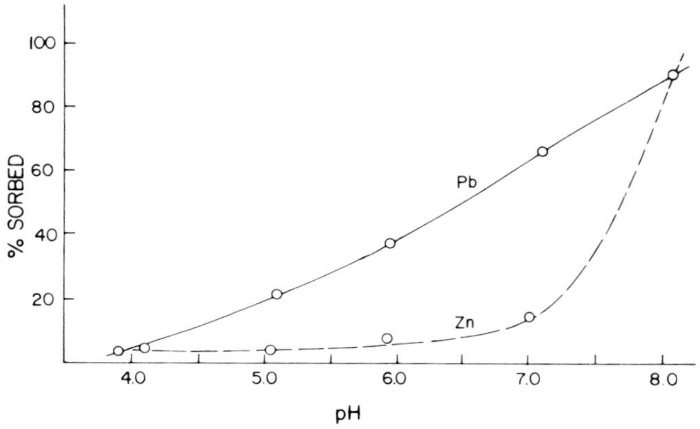

FIGURE 1-29.—Percent lead sorption as a function of pH.

TABLE 1-10.—Sorption and desorption behavior of lead on fresh and aged hydrous ferric oxide[1]

Pb^{2+} initial mM	Percent total lead sorbed at pH 6			Percent lead on HFO recovered at pH 4		
	HFO-1	HFO-2	HFO-3	HFO-1	HFO-2	HFO-3
0.5	99.4	98.1	95	63	72	78
1.0	99.4	95.8	89	74	76	81
1.5	96.6	87.6	66	81	81	86
2.0	84.9	64.4	58	83	84	88
2.5	76.4	52.7	52	85	87	91

HFO-1: Lead was present during precipitation of HFO.
HFO-2: Lead was added immediately after precipitation of HFO.
HFO-3: Lead was added after aging the HFO for 24 h at pH 6.

[1] Hydrous ferric oxide (HFO) — 0.625 μmoles (as Fe); solution volume — 100 ml; equilibration time — 3 h.

Morgan [56], and Kozawa [60] found ratios as high as 4 for adsorption of Cu^{2+} and Zn^{2+} for ammoniacal solution onto SiO_2. In general, cation sorption on hydrous oxides is interpreted in terms of surface complexation, the extent of which determines the ratio of hydrogen ion released to cation uptake.

The high sorption capacity for lead by HFO and the reversible nature of lead sorption with respect to pH have important environmental consequences. Thus pH changes in the range 4 to 9 can occur in soils and surface waters with the addition of fertilizers and chemical wastes or, as shown by Urone and Schroeder, [61] by rainfall rendered acid by the atmospheric conversion of sulfur dioxide to sulfuric acid. Sudden, especially acidic changes in pH may, therefore, release large amounts of soluble lead into aquatic environments.

The foregoing study clearly demonstrates the lead scavenging capacity of HFO and the potential danger of lead release in waters subject to changes in pH. In contrast to clays (here used in the compositional sense), which take up heavy metals by ion exchange, the HFO system is insensitive to the release of lead by exposure to high concentrations of neutral salts.

Relatively large amounts of heavy metals, including lead, have been found to be associated with algae and other aquatic plants in the tailing ponds, meanders, and receiving streams of the New Lead Belt of southeast Missouri. The amount of heavy metals associated with aquatic vegetation was found to be inversely proportional to the distance downstream from the mine and mill. Highest values for lead associated with algae were found in close proximity to the mine effluent discharge, where small particles of ore were brought to the surface along with excess water. The presence of galena (PbS) in close association with profuse algal growth at the discharge point was shown conclusively by microscopic examination of polished thin sections. It is very probable that much of the lead found in the aquatic ecosystem is derived from the mine in small

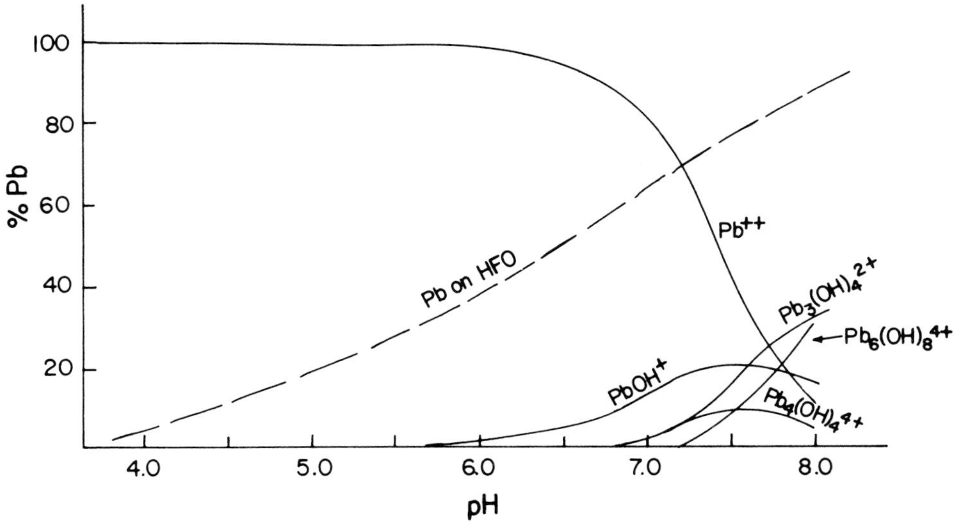

FIGURE 1-30.—Comparison of the percentage of total lead absorbed on hydrous ferric oxide (HFO) with the calculated percentages of several lead species.

galena particles that escape the milling process and even from the tailing ponds. It is also possible that ion exchange may occur at the surface of cells [62-64] to efficiently bind the lead to the algae and other microorganisms. Indeed, thorough washing in the stream water failed to remove more than approximately 10 percent of the bound lead. However, no definitive studies have been performed to determine the exact nature of the actual anions associated with lead in these downstream aquatic plants.

Chemical Characterization

Although it is now generally accepted that soluble lead and other complexing heavy metals are present in natural and polluted waters as complexes with organic ligands, the exact nature of the ligands actually complexed with lead is unknown. The existence and behavior of metal (chiefly lead, copper, and cadmium) organic "complexes" have been the subject of several studies by Matson, [65] Allen et al., [66] Bender et al., [67] and Singer. [68]

Even though the nature of lead-organic complexes is unknown, it is of considerable significance to establish the overall "binding" capacity of different aqueous systems for lead. This problem has been addressed by A.M. Hartley at the University of Illinois. The basic technique utilized has been anodic stripping voltammetry (ASV) applied directly to natural water systems.

Results to date show that (1) natural water samples show an apparent increase in lead content upon acidification, and this increase is reversible on subsequent neutralization; (2) freezing and thawing of the preserved samples (up to six cycles) do not affect either lead content or the ratio of lead concentration in acidified and natural samples; (3) when standard additions of lead are made, the ratio of lead in acidified and natural samples is constant; and (4) this ratio is independent of electrolysis time.

The inference to be drawn from these observations is that natural waters contain complexing species capable of binding lead reversibly such that conditions are close to equilibrium. Griffin [69] has shown that natural waters that exhibit this binding capacity can be quantitatively titrated with lead by the use of ASV as the indicator.

Examination of the data obtained from some 1,500 stream water samples has shown no pronounced pattern in this binding capacity. It appears that the day-to-day variation is greater than that ascribed to lake water samples investigated by Allen et al. [66] There is some marginal indication that the magnitude of the effect, as measured by the ratio of "natural" to "acidified" lead values, increases as the stream is sampled in the urban portion of the ecosystem under study as compared to the more rural areas.

ENVIRONMENTAL LEAD IN SOILS

Physical Characterization

Zimdahl and Arvik at Colorado State University have studied the mobility of lead in various soils. Their experiments involved the analyses of soils taken at various depths as a function of distance from a highway (Interstate 25), as well as simulated laboratory studies. The lead content of soils was indistinguishable from background at distances greater than 300 ft and at depths greater than 18 in. Leaching experiments in the laboratory with soil packs and lead nitrate solutions indicated a removal of less than 5 percent of the retained lead when the pack was washed with a quantity of water equivalent to 7 years' rainfall in the Fort Collins area. The lead is tightly held, at least in the kinetic sense. Seeley et. al., [70] have shown that the natural soil adjacent to Interstate 25 has considerable capacity for additional lead uptake. The sorption studies of Zimdahl are described elsewhere.

Chemical Characterization

Preconcentration techniques for the lead species present in soil have been investigated by Skogerboe and associates [71] at Colorado State University. They have shown, in two soils from urban areas containing 2,300 and 960 mg/kg of lead (far above the natural background of 20 to 25 mg/kg), that 20 to 25 percent by weight of each soil could be separated with a small magnet. The magnetic fraction contained 65 to 75 percent by weight of the total lead. A density gradient technique was employed on the magnetic fraction to concentrate the lead further and to produce a series of soil fractions, each representing a given density range. The fraction greater than 3.32 in density was again separated into a magnetic and nonmagnetic fraction; from 75 to 79 percent of the lead was found in this fraction. This concentration procedure permitted the use of X-ray diffraction techniques to identify specific lead compounds. The major lead-containing species was $PbSO_4$, and secondary species included $PbO{\cdot}PbSO_4$ and PbO_2. These three compounds accounted for 40 to 50 percent of the total lead content of the soil.

Studies conducted at the University of Missouri have indicated that soil and decaying vegetable matter located about 1/4 mi from a lead smelter contained the same lead species as those found in the smelter baghouse filter. Lead sulfide was observed in the 15- to 20-μm size range, and lead sulfate and elemental lead in the 45- to 125-μm range.

The mobility of lead in soil containing an organic-rich top layer of decaying leaf litter is influenced by the presence of humic acids. The chemistry of humic acids and their reactions with metal ions have been discussed by Schnitzer and Khan. [72] Lead is absorbed by humic acids and thus becomes concentrated in humus-rich material. [73] A recent study [74] of the solubility of lead minerals in humic acids reports a sharp increase of solubility in 0.1 percent humic acids extracted from soil, compared with the solubility in carbon dioxide saturated water. Although, most of the metal humates displayed their "classical insolubility" in water, they are readily mobilized in a solution of 0.1 percent humic acids. These data, as recent work [75] in the structure and

TABLE 1-11.–Concentration of 17 elements in size-separated dust

Element	Fraction (µg/g, unless noted as percent) 2	3	4	5	6	7
Pb	817	2,178	3,139	3,522	2,447	2,308
Cd	0.54	1.45	6.09	3.52	2.85	3.46
As	3.7	3.9	8.1	8.5	6.1	9.2
Eu	0.05	0.06	0.2	0.2	0.1	0.3
Ga	2.2	2.3	2.4	3.1	2.3	3.7
K	0.31%	0.46%	0.47%	0.42%	0.47%	0.58%
La	10.3	8.9	14.3	14.2	19.1	40.7
Mn	391	612	1,004	1,156	785	998
Na	0.25%	0.32%	0.36%	0.28%	0.27%	0.29%
Sb	0.5	1.4	2.5	34.5	3.3	2.7
Sm	1.8	1.7	2.6	3.2	3.3	7.3
Ba	180	349	490	304	421	606
Ca	4.34%	5.74%	5.21%	5.99%	4.61%	4.02%
Dy	1.1	2.1	3.4	2.7	0.5	3.1
Sr	248	390	411	283	394	356
U	2.6	1.9	2.5	1.2	1.2	4.7
Br	2.2	38.3	67.5	52.7	17.1	15.6

reactivity of fulvic acid (a humic material of relatively low molecular weight), indicate the possibility of aquatic mobilization and partial removal of otherwise insoluble lead compounds from polluted areas.

The physical and chemical distributions of some 17 heavy metals in roadside dusts have been studied by Natusch, Lamb, and Loh at the University of Illinois (table 1-11). Dust collected at a moderately busy (20,000 cars per day) street intersection in Urbana was separated sequentially by physical size, density, and magnetism. The size fraction greater than 500 µm in physical diameter was discarded. The concentrations and mass fractions of lead found in each sample are presented in tables 1-12 and 1-13.

The data in table 1-11 show clearly that lead is preferentially associated with magnetic particles of high density and that it predominates in particles with diameter of 45 to 100 µm. However, most of the mass of lead (table 1-13) is associated with nonmagnetic particles of density 2.2 to 2.9 g/cm^3 and greater than 2.9 g/cm^3.

Electron microscopic analysis of a set of 58 magnetic particles of size 75 to 100 µm and density greater than 2.9 g/cm^3 (figure 1-31) showed that 11 particles contained observable lead concentrations. In five of these particles both Br and Cl were observed, but none contained Cr (derived from lead chromate road-marking paint). It seems probable, therefore, that high concentrations of lead were present in the sub-samples studied originated primarily from automobile exhausts, and that the conversion of PbBrCl to other nonhalide species takes place after deposition on the roadway. This is in accord with the findings of Skogerboe et al [71] in roadside soils.

By means of gas evolution analysis, preliminary identification of the actual lead compounds in roadside dust indicates that most of the lead is present as $PbSO_4$ and also as either the single or double halide salts.

LEAD IN BIOLOGICAL SYSTEMS

Passow et al. [75] have postulated that the leakage of K^+ from red blood cells (RBC) treated with lead either *in vivo* or *in vitro* is a result of lead interaction with the cell membrane. Since RBC is a target organ for lead, according to Chisolm [76] and since its membrane is less complex than that of other tissues, this organ may well serve as a model for the interaction of lead with other cell systems. Furthermore, the lipid extracts of these cells can be obtained and made into model membranes; i.e., vesicular dispersions of the lipid in water and bilayer lipid membranes. [77] Physical-chemical measurements can be made with these model systems; measurements include permeability, electric conductivity, and nuclear magnetic resonance (NMR).

Lead affects on heme synthesis are discussed elsewhere in this volume. But quite apart from specific interest in the interaction between lead and the RBC membrane, there is a more general interest in mem-

TABLE 1-12.–Mass balance of lead in separated dust

Size (µm)	Fraction (percent), Density (g/cm^3): <1.6	1.6-2.2	2.2-2.9 Nonmagnetic	2.2-2.9 Magnetic	>2.9 Nonmagnetic	>2.9 Magnetic
250-500	2.31	0.74	6.64	0.24	5.80	7.14
100-250	2.23	1.00	7.56	0.21	17.89	6.73
75-100	0.69	1.17	8.33	0.97	1.80	11.34
45-75	0.74	0.39	2.71	0.26	6.25	3.49
20-45	0.12	0.28	1.77	1.37	0.01	0.13

TABLE 1-13.–Concentration of lead in separated dust (μg/g)

Size (μm)	Concentration (μg/g)					
	Density (g/cm³)					
	<1.6	1.6-2.2	2.2-2.9		>2.9	
			Nonmagnetic	Magnetic	Nonmagnetic	Magnetic
250-500	2,829	2,610	321	1,696	1,398	9,155
100-250	4,331	2,963	513	2,177	4,214	13,023
75-100	3,857	3,429	1,276	3,422	3,656	22,131
45-75	4,272	5,060	1,018	2,360	4,494	19,030
20-45	3,839	3,557	1,327	1,649	8,957	13,314

brane interactions. One can postulate that the degree of toxicity of a lead-containing species might be partially controlled by its rate of transport across a membrane. Consequently, work at the University of Illinois has been directed toward elucidation of the ways in which metal ions, lead in particular, interact with both model and real membranes *in vitro.*

Natusch, Li, and Heisler at the University of Illinois have employed 220-MHz proton and 40-MHz phosphorus NMR and dual-wavelength UV/visible spectroscopy to study interactions of metal ions with model phospholipid membranes sonically dispersed in bilayer vesicular form in solution. The dual-wavelength spectroscopic studies were undertaken by Heisler, Li, and Natusch [78] with a dye, murexide (ammonium purpurate), as a spectroscopic probe. Murexide, according to Mela, [79] forms a weak complex with a variety of metal ions, including lead. Complex formation produces a shift in the electronic absorption spectrum of the dye, permitting computation of the amount of lead complexed to murexide, complexed to the liquid membrane, and free in solution.

Results of this study (given in detail elsewhere in this volume) suggest that lead and other metal ions do bind to lipid membranes, thereby reducing the mobility of individual lipids in the membrane. The binding cannot be of the normal ligand coordination type in the case of membrane as a whole, but rather binds to individual lipids. The correlation with cationic charge and, to a lesser extent with cationic charge density, suggests that the binding is electrostatic in nature and, therefore, nonspecific, as in the case of lead binding to hydrous metal oxides. The actual transport of lead compounds across model membranes was shown to depend markedly on the nature of the anion or ligand associated with lead. Nonetheless, in general, relatively little is known about the physical and chemical forms of lead present in animal or plant tissues.

REFERENCES

1. L. J. Purdue, R. E. Enrione, R. J. Thompson, and B. A. Bonfield. Anal. Chem. *45*: 527 (1973).
2. M. Menne and M. L. Corrin. Paper presented at the 2nd Annual NSF-RANN Trace Contaminants Conference, Asilomar, CA (August 1974).
3. A. Laveskog. Organolead Compounds in Auto Exhaust and Street Air. Institute Anal. Chem., University of Stockholm. TPM-BIL-64. (1972).
4. K. Habibi, E. S. Jacobs, W. G. Kunz, Jr., and D. L. Pastell. Paper presented to the Air Pollution Control Association, West Coast Section. (October 1970).
5. K. Habibi. Environ. Sci. Technol. *4*: 239 (1970).
6. D. A. Hirschler, L. F. Gilbert, F. W. Lamb, and L. M. Niebylski. Ind. Eng. Chem. *49*:1131 (1957).
7. D. A. Hirschler and L. F. Gilbert. Arch. Environ. Health. *8*:297 (1962).
8. P. K. Mueller, H. L. Helwig, A. E. Alcocer, W. K. Gong, and E. E. Jones. American Society Testing Materials. Special Technical Publication No. *352:* 60 (1962)
9. G. L. Ter Haar, D. L. Lenane, J. N. Hu, and M. Brandt. J. Air Pollut. Contr. Ass. *22*: 39 (1972).
10. R. Zimdahl. Interim report to NSF. Colorado State University. 112 (1972)
11. L. O. Grant and S. Vardiman. Personal communication.
12. K. W. Boyer and H. A. Laitenen. Environ. Sci. Technol. *8*:1093 (1974).
13. G. Calingaert, F. W. Lamb, and F. Meter. J. Amer. Chem. Soc. *71*:3709 (1949).
14. R. Zimdahl. Personal communication.
15. J. B. Moran and O. J. Manary, Interim Report to NAPCA. NTIS Report PB-196783 (1970).
16. J. A. Robbins and F. L. Snitz. Environ. Sci. Technol. *6*:164 (1972).
17. M. Bayard and G. L. Ter Haar. Nature. *232*:553 (1971).
18. D. F. S. Natusch, J. R. Wallace, and C. A. Evans. Science. *183*:202 (1974).
19. R. L. Davison, D. F.S. Natusch, J. R. Wallace, and C. A. Evans. Environ. Sci. Technol. *8*:1107 (1974).
20. D. F. S. Natusch and J. R. Wallace. Science. *186*:695 (1974).
21. R. Linton, A. Loh, D. F. S. Natusch, C. A. Evans, and P. Williams. Science *191*:852 (1976).
22. D. F. S. Natusch. Personal communication.
23. A. E. Vandegrift L.J. Shannon, E.W. Lawless, P.G. Garman and E.E. Sailee. Particulate Pollutant System Study, Vol. III. Handbook of Emission Properties NTIS PB 203522 (1971).
24. Duprey R.L. Compilation of Air Pollutant Emission Factors. EPA Publication No. AP-42 (1972).
25. C. Purushothaman, Personal communication.
26. N. H. Tibbs. Personal communication.
27. E. Bolter Personal communication.
28. E. S. Gladney, W. H. Zoller, A. G. Jones, and G. E. Gordon. Environ. Sci. Technol. *8*:551 (1974).
29. G. Colvos, C. S. Wilson, and J. L. Moyers. Paper presented to the Athens-Halifax Symposium on Recent Advances in the Analytical Chemistry of Pollutants, Athens, Ga (May 14, 1973).
30. R. E. Lee, R. K. Patterson, and J. Wagman. Environ. Sci. Technol. *2*:288 (1968).
31. R. E. Lee, S. S. Goranson, R. E. Enrione, and G. B. Morgan. Environ. Sci. Technol. *6*:1025 (1972).

32. E. Robinson and F. L. Ludwig. J. Air Pollut. Contr. Ass. *17*:664 (1968).
33. P. E. Morrow. Amer. Ind. Hyg. Ass. J. *25*:213 (1964).
34. C. C. Patterson. Arch. Environ. Health. *11*:344 (1965).
35. L. Dautrebard.Microaerosols. Academic Press, New York, N.Y. (1962).
36. D. F. S. Natusch. Personal communication.
37. C. C. Patterson. Scientist and Citizen. *10*:66 (1968).
38. H. W. Krueger. Lead content of Dirt from Urban Households. Report of Project 72-4 to EPA (1972).
39. D.F.S. Natusch, R. E. Lamb, and A. Loh. Personal communication.
40. H. O'Donnell, T. L. Montgomery, and M. Corn. Atmos. Environ. *4*:1 (1970).
41. R. K. Skogerboe. Personal communication.
42. K. W. Olson and R. K. Skogerboe. Environ. Sci. Technol. *9*:227 (1975).
43. H. W. Edwards and R. J. Rosenvold. Paper presented at the 2nd NSF-RANN Trace Contaminants Conference, Asilomar, CA (August 1974).
44. H. W. Edwards, R. J. Rosenvold, and H. G. Wheat. Paper presented at the 9th Annual Conference on Trace Substances in Environmental Health, Columbia, MO (June 20, 1975).
45. F. W. Lamb and L. M. Niebylski. Anal. Chem. *23*:1388 (1951).
46. H. P. Julien and R. E. Ogilvie. J. Amer. Chem. Soc. *82*:293 (1960).
47. M. L. Corrin and R. Springborn. Personal communication.
48. H. Laitenen. Personal communication.
49. D. Dahl and M. L. Corrin. Paper presented at the Precipitation Scavenging Symposium, Urbana, IL (August 1974).
50. D. Dahl. M. S. thesis, Colorado State University. (1973).
51. A. L. Lazrus, E. Lorange, and J. P. Lodge, Jr. Environ. Sci. Technol. *4*:55 (1970).
52. J. M. Pierrard. Environ. Sci. Technol. *3*:48 (1969).
53. H. Laitenen. Personal communication.
54. A. Kaldor and G. A. Somorjai. J. Phys. Chem. *70*:3538 (1966).
55. T. L. Desphande, D. J. Greenland, and J. P. Quirk. J. Soil Sci. *19*:108 (1968).
56. W. Stumm and J. J. Morgan. Aquatic Chemistry. Wiley-Interscience, New York, N. Y. p. 452 (1970).
57. H. Laitenen. Personal communication.
58. G. A. Parks. Chem. Rev. *65*:177 (1965).
59. D. W. Fuerstenau, Pure Appl. Chem. *24*:135 (1970).
60. A. Kozawa. J. Inorg. Nucl. Chem. *21*:315 (1961).
61. P. Urone and W. H. Schroeder. Environ. Sci. Technol. *3*:436 (1969).
62. N. L. Gale, M. G. Hardie. J. C. Jennett, and A. Aleti. Proceedings of 6th Annual Conference on Trace Substances in Environmental Health, Columbia, MO p. 95 (1972).
63. T. G. Tornabene and H. W. Edwards. Science. *176*:1334 (1972).
64. Y. Tanaka and S. C. Skoryna. Proceedings of the 7th International Seaweed Symposium, Science Council of Japan. p. 131 (1971).
65. W. R. Matson, Ph.D. thesis, Massachusetts Institute of Technology (1968).
66. H. E. Allen, W. R. Matson, and K. H. Mancy. J. Water Pollut. Contr. Fed. *42*:572 (1970).
67. M. E. Bender, W. R. Matson, and R. Jordan. Environ. Sci. Technol. *4*:520 (1970).
68. P. C. Singer. Trace Metals and Metal-Organic Interactions in Natural Waters. Ann Arbor Science Publishers, Inc., Ann Arbor, MI (1973).
69. R. Griffin. Ph.D. thesis, University of Michigan. (1970).
70. J. L. Seeley, D. Dick, J. H. Arvik, R. L. Zimdahl, and R. K. Skogerboe. Appl. Spectros. *26*:456 (1972).
71. R. K. Skogerboe. Personal communication.
72. M. Schnitzer and S. U. Khan. Humic Substances in the Environment. Marcel Dekker, Inc. New York, N.Y. (1972).
73. A. Scalay. Geochim. Cosmochim. Acta. *28*: 1605 (1964).
74. W. E. Baker. Geochim. Cosmochim. Acta. *37*:269 (1973).
75. H. Passow, A., Rothstein, and T. W. Clarkson. Pharm. Rev. *13*:105 (1961).
76. J.J. Chisolm. Scient. Amer. *224*:15 (1971).
77. D. Chapman, V.B. Kamat, J. De Gier, and S. A. Penkett. J. Mol. Biol. *31*:101 (1960).
78. S. Heisler, K. P. Li, and D. F. S. Natusch. Personal communication.
79. L. Mela. Arch. Biochem. Biophys. *123*:286 (1968).

CHAPTER 2

MONITORING FOR LEAD IN THE ENVIRONMENT

R. K. Skogerboe
Department of Chemistry
Colorado State University

A.M. Hartley and *R.S. Vogel*
Institute for Environmental Studies
University of Illinois

S.R. Koirtyohann
Environmental Trace Substances Center
University of Missouri

INTRODUCTION

Sampling, sample preparation, analytical measurement, and data interpretation are distinct but inseparable aspects of environmental measurements; hence monitoring rather than analysis is used in the title of this chapter which deals with the current status of methods for monitoring lead in the environment. Monitoring problems include inadequate sampling or selection of improper samples, fallacious interpretations of measurements, and improper or incorrect uses of analytical methods. One need only examine the results of interlaboratory or collaborative test programs to realize that serious problems exist in the use of analytical methods.

Because such problems exist, a critical evaluation of methods of monitoring for lead has been prepared with emphasis on those methods and practices that are most widely used. It appears that cost and simplicity have often served as prominent selection criteria for these widely used methods. Other methods requiring expensive equipment or highly specialized operator training are considered primarily to acquaint readers with their capabilties. If such capabilities are needed, adequate references are provided as a basis for a literature search.

In 1933, the well-known analytical chemist, G.E.F. Lundell, published a perceptive and informative paper entitled "The Chemical Analysis of Things as They Are." [1] Dr. Lundell pointed out that there is an increasing tendency to consider chemical analysis as dealing with one or two variables instead of the dozen or more that are often involved. Moreover, he indicates, "there is no dearth of methods that are entirely satisfactory for the determination of elements when they occur alone. The rub comes in because elements never occur alone, for nature and man frown on celibacy." These comments aptly apply to environmental monitoring and analysis, and environmental scientists should heed Dr. Lundell's evaluation of the problems they face.

Sampling, sample storage and preparation, sample analysis, and interpretation of the analytical results are integral phases of an environmental monitoring procedure. The validity (accuracy) of an analytical measurement may be negated by errors (inaccuracies) which originate in one or more of the other phases of procedure. [2] The dozen or more variables mentioned by Lundell [1] that may affect a measurement are distributed over these phases and must be carefully considered. In the following discussion, sampling and sample storage and preparation are jointly considered for convenience. Analysis, validation of analyses, and interpretation of results are considered separately, since several approches may be used in the solution of any one problem.

SAMPLING FOR ANALYSIS AND MONITORING OF ENVIRONMENTAL SYSTEMS

While many publications stress that random sampling is to be preferred, it should be emphasized that what might be called "organized" random sampling is most appropriate and practical for environmental studies. It is impossible to formulate a valid sampling plan outside the context of a particular problem. Such plans must always be based on the goals or objectives involved. The majority of environmental contamination studies appear to deal with the sources, sinks, and routes of movement of the contaminant through the segments or compartments of an ecosystem. A reasonably practical sampling plan in such instances may often be based on a conceptual model. A carefully designed conceptual model will generally prove invaluable in deciding systematically what organizational levels of an ecosystem should be sampled to determine mobility of the contaminant between these levels. Once the compartments of the environment to be sampled have been defined, the spatial and/or temporal sampling frequencies required for each must be defined with particular reference to the purpose of the study. Sampling programs of this type are among the more difficult to plan, perform, and assess. [3] Cholak and Story [4] have emphasized that the monitoring of biological systems is unusually involved because of their inherently greater complexity and larger compositional variations. A preliminary systematic survey of the system(s) in question may be a necessity for establishing a sound experimental (sampling) design. The use of such survey results to formulate the final design may bring other benefits. The preliminary data may often be used to stipulate the spatial and temporal sampling frequencies required and, consequently, to formulate the best experimental design techniques for the objective evaluation of the results. The data may also be used to estimate the total number of samples required for the purpose at hand, and economies may be realized therefrom. The results may also indicate the feasibility of utilizing the evolutionary operations concept [5] to enhance the efficiency of the study. The environmental researcher would do well to familiarize himself with the basic concepts of sampling theory and practice. Texts such as those by Cochran [6] and Yates [7] may be recommended as good sources of information for this purpose.

Sampling and sample preservation are areas of considerable weakness in many environmental studies. Continued investigations directed at the development of improved methods and/or techniques are urgently needed. It is not sensible to devote large programs to the development of improved analytical measurements, if those measurements are to be expended on poorly selected or inadequately prepared samples. Current practices in sampling, storage, and preparation for each type of environmental sample are summarized in the following sections. The references given are not comprehensive but are cited as primary examples.

Atmospheric Samples

Comprehensive reports of recent developments in the sampling and analysis of atmospheric pollutants are given in reviews by Altshuller [8] and by Mueller et al. [9] A three-volume monograph edited by Stern [10] covers the cause, effect, transport, measurement, and control of air pollution; summarizes capabilities of various devices and techniques; and lists current problem areas. These are largely re-emphasized by Saltsman et al. [11] in a discussion of the standardization of procedures that utilize existing measurement technology.

Most atmospheric sampling for particulates is based on drawing the air through filter materials such as fiber glass, asbestos, cellulose paper, or porous plastics. [12] The fact that these filter materials exhibit significant residual blanks for a number of trace elements is of interest. [13-15] If long-term, high-volume sampling is used to collect a sufficient sample, the blank contribution is negligible. Similarly, if the trace element content of the atmosphere is sufficiently high, shorter sampling periods may be used without a significant blank contribution to the analysis results. However, in relatively clean atmospheres, the use of such collection materials may necessitate inordinately long sampling periods, i.e., large samples.

If statistically significant collection data are to be obtained with the more commonly used filter media, samples must be large enough until the amount of material collected for analysis exceeds the amounts of those elements present in the filters as residual impurities. A typical example of this problem is shown in table 2-1, which gives data taken for a 10-h air sample collected on a 0.8-μm Nuclepore filter. The results indicate that the sample filter analyses were not significantly different from the blanks; therefore no useful data were obtained from the experiment. Gandrud and Lazrus [13] have reported the presence of sodium, potassium, magnesium and calcium in IPC-1478 filter papers at residual levels in the fractional milligram range. Luke et al. [14] and Kometani et al. [15] state that glass fiber filters contain such prohibitively high and variable amounts of residual impurities that their use for air sampling should be avoided except for highly contaminated areas or long-term sampling.

Short sampling periods are required if atmospheric transport models are to be developed that account for the effects of short-term meteorological and climatological changes. [16] At present, the modeling procedures used to assess the total effect of numerous sources and meteorological conditions on air quality in a region are quite inadequate to provide a solid basis for developing programs to solve pollution problems. Thus there is a need for an improved

TABLE 2-1.—Analysis of 0.8-micrometer Nuclepore air collections

Element	Filter blank[1] (μg/filter)	Analyses: Run 1[2] (μg/filter)	Run 2[2] (μg/filter)
Al	7.8 ± 1.5	8.4	8.8
Ca	55 ± 4	54	50
Mg	3.8 ± 1.5	9.9	3.4
Pb	0.4 ± 0.2	0.7	0.6

[1] Average concentrations and standard deviations calculated from analyses of three filters from same production lot.

[2] Collection period of 10 h at l/min.

atmospheric particulate collection system based upon short-term sampling, which generally dictates the need for a high-purity filter material.

Practically, a sampling approach cannot be reliably formulated without considering the analytical requirements. Because contamination and loss of trace elements can occur, the sampling approach and medium should be amenable to analysis with a minimum of handling or sample pretreatment. The analytical methods should permit the accurate determination of as many trace elements as possible with a high degree of specificity at low absolute levels. With this type of capability, the determination of elemental mass ratios in aerosol samples, as demonstrated by Winchester [17] and Hoffman et al., [18] can provide valuable evidence indicative of the origins of pollutants and of atmospheric processes. Several authors [19-22] have reported that the concentration ratios of elements in two samples lend themselves better than the individual concentration levels to the establishment of their common origin. A statistical technique discussed by Anders [23] involves matching the elemental concentration ratios of a sample with the equivalent ratios of other samples to show that the resulting correlations may indicate existing generic relationships. Dams et al. [24] directed attention to an analogous method for identifying in discrete sources the origin of trace elements that were contributed simultaneously to a given set of atmospheric particulate samples. Although single constituent analysis techniques are valuable, there is a definite need for analytical methods capable of determining multiple elements in any one sample. If all aforementioned criteria are satisfied, the sampling-analysis method then approaches the ideal in terms of ultimate utility.

Particulate lead is commonly defined as that retained by a filter with a nominal pore size of 0.45 μm. [25] Lead passed by the filter is defined as nonparticulate, molecular, or organic lead. [25] The ASTM tentative method [25] for sampling and determining atmospheric lead includes separate procedures for particulate and organic lead. Air is drawn through the 0.45 μm filter and then through a sampling tube containing crystalline iodine. The membrane filter is digested in nitric, sulfuric, and perchloric acids to degrade the organic matter and solubilize the lead. Lead is determined by a colorimetric dithizone procedure or by atomic absorption spectrophotometry. (See below.) To determine organic lead, the iodine crystals are dissolved in acidic potassium iodide solution, the excess iodine is reduced with sodium sulfite, and the lead is measured as above.

The ASTM definition of particulate lead appears to be very subjective, as no studies have been reported unequivocally indicating that the filters used are absolutely efficient for particulate collection. Rather, studies have shown the contrary. [26-28] In addition, Snyder and Henderson [29] have shown that crystalline iodine traps are not completely efficient in collecting lead passed by filters into the traps. While gaseous tetraethyl lead was trapped effectively, the collection efficiency for other organic lead compounds varied from 20 to 80 percent of the total, depending on the sampling rate. Low efficiencies were also observed for aqueous iodine solutions. As a result, Snyder [30] has substituted 30 to 50 mesh activated charcoal for the collection of organic lead. Recently, Purdue et al. [31] have used scrubbers containing ICl solutions behind filters to trap "organic" lead, and their data suggest that this serves as an efficient collection device.

As indicated, the high-volume sampler used for sampling and determining suspended particulate matter in the atmosphere can be applied to the collection of particulate lead species. Their use and operational factors involved for sampling particulate matter in ambient air have been studied. [32-35] In operation, air is drawn into a covered housing and through a horizontally oriented filter by means of a high-flow-rate blower operated at an average flow rate of 1.5 m^3/min. [33] Reports indicate that glass-fiber filters have a collection efficiency of at least 99 percent for particles as small as 0.3 μm. [35] A sampling period of 24h is usually recommended for determining average particulate matter concentrations. The sampling period may be much less if particulate levels are unusually high.

The accuracy with which the samples reflect the true average concentration over the sampling period depends upon the degree of constancy of air flow maintained through the filter and the collection efficiency of the filter. Both the air-flow rate and the collection efficiency are affected by the concentration and the nature of the suspended particulates. Under some conditions, the error in the measured concentration may be $\pm$50 percent or more of the true concentration. [36] Means for obtaining constancy of flow rate within $\pm$10 percent over week-long sampling periods were presented by Harrison et al. [36] and Kneip et al. [37]

A primary use of the high-volume sampler has been to determine mass concentrations of particulates in ambient air. A relatively large (8 in. by 10 in.) filter is needed to accumulate sufficient particulate matter over the test period for weighing with conventional facilities. If mass determination of total particulates is not required, the membrane-type filter system provides greater freedom in experimental design, and less analytical problems are associated with the high and variable blanks characteristic of the glass or cellulose filters often used in high-volume air samplers.

The reports by Spurny et al. [26, 27] emphasize the collection efficiency problem associated with the membrane filters. Their measurements show that collection efficiency depends on the pore size of the filters, the impingement velocity of the particles on the filter face (the face velocity), the sizes of aerosol particles, and the clogging rate. For any one filter pore size, the efficiency is good for larger particles (above approximately 0.5 μm) and those below approximately 0.002 μm. For the intermediate sizes, however, efficiency drops dramatically, depending on the face velocity, to levels as low as 5 to 10 percent of the total present in a particular size range. Cohen [28] presented comparative measurements that reflect these effects but did not refer to the studies of Spurny et al. [27] For the average suggested flow rate of 1.5 m^3/min. with an 8 in. by 10 in. filter, [33] the collection efficiency of a 0.45-μm membrane filter for particles in the 0.05- to 0.5-μm range can be expected to be in the 5 to 80 percent collection efficiency range, depending on the actual particle sizes and the densities. Data from the National Air Surveillance Network [38] indicate that the mass median diameter of lead particulates is typically 0.5 μm. Thus from 30 to 70 percent of the total mass of lead particulates are present in the size ranges where the filter efficiencies are lowest for the face velocities most often used. Consequently, it cannot be argued that the amounts of lead passed by the filters are negligible in comparison with the total mass collected. Clearly, the practitioners of air sampling must become aware of the filter efficiency problem and, at the very least, must adjust collection conditions (face velocities in particular) to maximize the collection capabilities of the filters.

The use of spectroscopically pure graphite cups for particulate filters has been described by Woodriff and Lech. [39] Porous graphite, by its nature, has been described as ideal for filtration of micrometer and submicrometer particles, [40] although adsorption of molecularly dispersed lead is possible. Seeley and Skogerboe [41] have shown that such filters are subject to minimal blank problems and may be used for short-term as well as long-term sample collections. Their results indicate that graphite is a more efficient collector of particulate lead than paper, glass fiber, or membrane filters. Through the use of an emission spectrographic analysis method, several atmospheric trace elements may be simultaneously determined from any one sample. [41] Thus the graphite sampler system combined with the spectrographic analysis method offers the promise of eliminating some of the sampling problems and expanding atmospheric monitoring capabilities.

Storage of filters prior to analysis usually involves closure in a clean, airtight container. Freezing for long-term storage has been used, but data indicating the necessity for this are not available. In the case of lead, there is a time-dependent conversion of the lead halides to other compounds that must depend on storage conditions. [42] While such conversions should have little effect on the determination of total lead, they must be considered if identification of specific lead compounds is required.

Most sample preparation procedures for trace metal analysis involve the conversion of the filter and the sample materials to soluble forms. The tentative ASTM method [25] recommends digestion in a hot ternary acid (HNO_3, H_2SO_4, and $HClO_4$) mixture with heating until SO_3 fumes appear. While this digestion method is rigorous enough to dissolve most refractory compounds, the failure to eliminate all sulfate ions by continued heating may be a serious source of error for the lead determination. The vast majority of atmospheric particulate samples contain much more calcium than lead. While lead may not be precipitated directly as the insoluble sulfate, it may very well be coprecipitated with calcium sulfate. Such coprecipitation methods have historically been used in the determination of trace amounts of lead, and the formation of colloidal nuclei of calcium sulfate has been documented even at relatively low calcium concentrations. [43] In essence, this digestive approach should be used with caution. It is certain that rigorous wet digestion procedures are required to completely solubilize the particulate species contained in many atmospheric collections. In comparing direct X-ray fluorescence analyses of filter collections from incinerators with the acid digests from the same filters, Birks et al. [44] noted that the latter gave consistently lower results. Failure to obtain complete solubilization was indicated as the cause of the discrepancies. [44]

Studies at Colorado State University [45] have indicated that rigorous digestion in 80 percent HNO_3/20 percent $HClO_4$, with evaporation to perchlorate fumes and subsequent takeup in 3M HNO_3, serves as an effective means for the solubilization of lead contained on air filters. Others [46] have recommended that the filter sample be dry ashed in a muffle furnace prior to takeup in acid; later, however, they reported [47] that direct ashing at 500° C resulted in losses of lead ranging from 12 to 47 percent of the amount present. Thus while dry ashing may serve as a convenient means for providing a completely acid-soluble trace metal matrix, the precautions discussed later in the dry oxidation section should be judiciously observed.

Particle Size Sampling

The accuracy of particulate matter size evaluation depends to a large degree on the representative character of the sample being evaluated. Particles less than approximately 5 to 10 μm in diameter usually have a relatively low settling velocity. Consequently, samples collected by gravity or settlement are generally not representative of the atmospheric composition at any one sampling point. Moreover, fallout samples cannot be related to a particular volume of air for concentration estimates. For these reasons, fallout measurements should be considered in relative terms only, and it must be recognized that such samples represent only those particles that, because of their size and/or density, have high settling velocities.

Particle count or size analyses can be based on filtration collection techniques. The use of filters for particle collection and the attendant advantages and limitations have been discussed. [10] Electrical and thermal precipitation techniques have also been used for characterizing particulate dispersions, but they appear to be primarily applicable to specialized cases. The three-volume series edited by Stern [10] summarizes the capabilities and limitations associated with these techniques.

The most widely used methods for sizing or counting particulates are based on the principle of inertial impingement. Among these, the cascade impactor is widely used. The instrument consists of a series of impingement stages of decreasing jet width or diameter. Thus the air velocity increases with decreasing jet diameter such that progressively smaller particles are collected on each successive stage. The efficiency of these units is best for the larger particle size ranges, although unusually large particles may not be collected. [10] Small particles may have insufficient momentum to impact on the collecting plates; depending on operating conditions, those of approximately 0.1-μm aerodynamic diameter or less will pass all stages. Those particles that pass the impactor may be (at least partially) collected on a filter in series with it. The collection efficiency will, of course, depend on the filter chosen. The mass median diameter for lead particulates is typically around 0.5 μm; [38] a significant percentage of the lead particulates will consequently not be collected by the impactor. If total particulate lead is to be determined, backing the impactor with a filter is essential. The previous questions raised about filter collection efficiencies should be borne in mind.

Most impactor measurements of particulate dispersions of lead are based on the determination of the mass collected on each stage. Removal of lead from the impactor plates may require acid digestion. The blank contributions from the plates will depend on their identities. The glass microscope slides or stainless steel plates often used may contribute several micrograms of lead, depending on the digestion method. Such blank contributions may be eliminated by covering each impactor plate with a thin film of Mylar, but precautions must be taken to maintain a smooth, wrinkle-free surface so that the airflow pattern is not changed. The use of plates fabricated from Teflon permits the use of rigorous digestion methods without the attendant blank problems. [45] Comments made previously concerning the storage of filter samples and the preparation procedures used prior to analysis also apply to impactor plates.

In summary, any atmospheric sampling plan must consider the analytical data requirements. Inefficiencies in collection, sample contamination, and loss of trace elements can (and do) occur. The sampling approach and the medium used should permit analysis with a minimum of sample handling and/or pretreatment. Multielement analytical methods offering a high degree of specificity over broad concentration ranges are to be preferred for a wide variety of atmospheric monitoring purposes. If these idealized analytical criteria are satisfied and the sampling problems mentioned above are recognized, elucidation of atmospheric phenomena relating to lead can proceed on a concrete basis.

Collection, Preservation, and Preparation of Water Samples

Natural waters are complex systems of aqueous and solid phases (suspended matter) in which heavy metals may be distributed in ionic form, as complexes in solution, as adsorbed species on suspended matter, or as species included in suspended matter. Factors to be considered in the selection of sampling sites for streams and lakes, discussions of sampling equipment, storage recommendations, and analytical methods are included in the literature. (See references [48-54] for examples.)

Water sampling devices range from relatively simple glass or plastic containers for "grab" sampling to the more sophisticated Van Dorn or Nansen bottles. [49, 50, 52, 53] Since the latter are often constructed of brass, their use in sampling waters for lead, copper, or zinc determinations should be avoided. Automated sampling devices that can be programmed to collect at regular intervals or at variable rates, thus obtaining a composite sample that represents an integrated average over the sampling period have been developed. It should be recognized that converging streams of water often undergo slow mixing and that less dynamic water systems may tend to stratify due to lack of mixing. Natural waters may thus be heterogeneous with respect to their trace element distributions, and sampling plans should be devised to take this into account. Again, systematic sampling on a preliminary survey basis may be required to optimize the sampling plan.

The sampling of natural waters must consider the distribution of trace elements between the aqueous and solid phases. Although valid arguments can be made against it, common practice is to define suspended matter as that material retained on a 0.45-μm-pore-size membrane filter following vacuum filtration. [48-50] Consequently, all species that pass the filter are classed as dissolved materials.

The average suspended and dissolved concentrations of several elements for numerous U.S. streams and lakes are summarized in table 2-2. [51] There is no apparent trend with respect to the distributions of these elements between the two phases. In contrast, data for lead distribution between the two phases in an Illinois stream sampled at various locations and flow rates (table 2-3) show prominent concentration ratio changes and emphasize the heterogeneous nature of trace element distributions and their dependence on the aquatic conditions. [55] Experiments

TABLE 2-2.—Average concentrations of suspended and dissolved trace metals in U.S. surface waters [51]

Element	Average concentration (μg/l)	
	Suspended solids	Dissolved solids
Al	3,900	74
Ba	38	43
Bi	0.34	0.19
Cu	26	15
Cr	30	9.7
Fe	3,000	52
Mn	105	58
Ni	29	19
Pb	120	23
Sr	58	217
Zn	62	64

TABLE 2-3.–Comparison of lead in suspended solids and the aqueous phase of stream water samples [55]

Sample No.	Lead (μg/l) Suspended solids[1]	Aqueous phase[2]
818	4.0	3.6
819	14	2.7
820	22	1.7
822	410	2.5
824	10	4.0
827	360	35
829	12	0.6
830	19	1.1
834	9.6	0.3
836	5.6	0.3
839	5.6	1.7
842	4.0	1.1
844	7.2	0.3

[1] Retained on type HA Millipore filter of 5.0-μm pore size.
[2] Filtrate through the same filter.

have also shown that the filtration rate can be increased by a factor of 10 by using a 5-μm filter rather than a 0.45-μm filter. [55] Over 90 percent of the lead in suspended matter retained by the smaller filter could also be retained by the larger one. [55]

The formation of precipitates has been reported in previously filtered samples that have been stored. It has been suggested that nucleation of precipitates can be induced by passage through membrane filters. [56] Possibly pH and Eh changes induce precipitation as the dissolved gases change in storage.

Based on the expectation that water purification systems should remove particulates, the official methods of The U.S. Environmental Protection Agency (EPA) and The American Public Health Association (APHA) recommend that the sample be filtered in the field and analyses be carried out only on the dissolved species. [49,50] This rationale is difficult to defend because their are obvious variations in the trace metal concentration with regard to particle size distribution, and water purification procedures are widely divergent. Certainly, some particulates, through a variety of natural processes, may be passed upwards through food chains or be converted to a form more likely to have environmental significance. Thus it is easier to rationalize that separate determination of both suspended and dissolved species are more likely to provide a basis for the evaluation of heavy metal effects in aquatic systems.

Some methods recommend acidification of water samples to a pH of approximately 3.5 at the time of sampling. [48,50] This obviously minimizes losses by precipitation or by adsorption on container walls. Filtration should be performed prior to the acid addition if the distribution of heavy metals between the solid and aqueous phases is to be determined. The loss of trace amounts of metal ions through adsorption on container walls is not a trivial problem. [57] Robertson [58] and Eichholz et al. [59] have shown that such processes account for major changes in the ionic concentrations. In the case of lead, Struempler [60] has indicated that immediate acidification to a pH of 2 is required to prevent adsorptive losses.

An uncertainty in handling methodology for trace metal analysis involves the preservation of the water sample in its original state until analyses can be performed. The distribution of trace metals between the aqueous phase and suspended matter is reported to be sensitive to sample storage conditions. However, information is meager on the relative changes in lead distribution between solid and liquid phases in natural water samples, the effect of containers, and time and temperature of storage. Some researchers advocate field filtration immediately on sampling. [49,50] Others recommend freezing water samples with liquid nitrogen immediately on collection as a means of maintaining the original composition of the sample. [57] The degree of success in reconstituting frozen samples has also been questioned, [56] but investigations at the University of Illinois have not detected a difference between original and reconstituted samples with respect to elemental distribution of lead between the aqueous phase and suspended matter. [55] Since lead transport studies require information on both the elemental distribution of lead and the characterization of lead species, preservation of the original state of the water sample prior to analysis is essential to the validity of the analytical results. This problem has not yet been resolved.

As a general recommendation, samples of natural waters should be placed in acid-leached and thoroughly rinsed polypropylene bottles and filtered promptly through 0.45-μm membrane filters. Both filters and filtrate should be reserved for separate analyses but the latter should be acidified. [50] Filtered samples awaiting analysis should be stored at a temperature of 4° C or less.

The sampling of precipitation presents a unique problem. Since sampling schedules cannot be controlled, a statistical sampling scheme must be employed. For economic and practical reasons, unattended sampling stations are exposed to ambient atmospheric conditions during intervals between storms. For such stations, a rainfall sample is usually a mixture of rainfall and atmospheric fallout over the time interval that the collector was exposed.

The principal problem lies in the variable degree of contamination of the precipitation sample with heavy metals from atmospheric fallout and also from insects and other extraneous material that accumulate on the collecting surfaces and become part of the sample. Unattended polyethylene collecting bags are particularly vulnerable to such contamination. Sensing devices that are activated by precipitation to remove a shield from the collecting surfaces provide some reduction in contamination.

Instrumentation has been developed for the specific collection of dustfall and rainfall. A prototype system described by Semonin [61] consists essentially of a pair of polyethylene collection containers and a motor-driven sliding shield controlled by a rainfall sensor. During dry periods, the shield covers the rainfall collector, and the dustfall collector is exposed. At the onset of precipitation, a sensing device activated by the electrical conductivity of water droplets across a grid causes the shield to be transferred from the rain collection container to the dustfall container. At the cessation of precipitation, the shield returns to the original position. Through the use of this apparatus, relatively uncontaminated rainfall samples can be obtained from an unattended station.

To obtain valid analytical results the sampling, sample filtration, and sample storage methods mentioned above should be applied to precipitation. [62,63]

Investigators interested in ecosystem studies should become familiar with current practices and principal sources of information. The Environmental Instrumentation Group of the Lawrence-Berkeley Laboratory at the University of California has prepared, under sponsorship of the RANN Program of the National Science Foundation, a compendium containing a comprehensive survey of instrumentation and methods for monitoring water quality. [64] The survey includes descriptions of the operating characteristics of available instruments, critical comparisons of instrumental methods, and discussions of promising methodologies and new instrumentation. Of at least equal value are the works of the principal agencies and institutions engaged in monitoring natural waters. [49, 50, 52, 53]

Sampling Aquatic Communities

The sampling of fish, benthos, periphyton, stream drift, and rooted plants that populate aquatic environments is usually for chemical analysis of the various organisms or to estimate the biomass of the community components. The sampling plans and collection methods differ according to the objectives. Mackenthun [65] gives a comprehensive discussion of plans and methods for sampling aquatic communities. Sampling equipment includes the Peterson dredge, core samplers, and "stove pipe" samplers for collecting bottom life; drift nets for collecting suspended biota; seines and electrofishing devices for collecting fish and other free-swimming organisms; and artificial substrates for collecting periphyton. [49, 65, 66]

Since periphyton populations are the primary producers in aquatic environments, considerable attention has been given to methods for the collection and assessment of periphyton population densities. Most workers use one of several artificial substrates including fiberboard, wood, concrete, plexiglass, polyethylene, and glass. [65, 66] A review prepared by Sladeckova [66] lists 448 references to methods of investigating periphytic communities. The reviewer states that no single method or device can be universally used; in some cases ecological factors might make methods for the evaluation of periphyton on natural substrates preferable. However, the use of artificial substrates is considered essential for determining the rates of periphyton formation.

To sample stream systems at the University of Illinois, benthic populations are collected by means of hand-operated-grab or piston-coring devices. The choice depends on the character of the stream bottom. Stream drift is collected by straining a measured volume of water through a 30-mesh net. Aquatic plants are sampled by collecting the whole plants from a measured area of the stream. Fish are collected by using two seines concurrently in a measured sampling area. Sampling of periphyton populations is accomplished by introducing artificial substrates on which periphyton accumulate. A convenient procedure is to anchor sealed 64-oz polyethylene bottles filled with water at selected positions along the stream bed. After 12 to 14 days immersion, the samplers are recovered from the stream bed, and the accumulated periphytic growth is scraped from the exterior surfaces with a suitable squeegee. [67] Sample treatment and storage follow the procedures outlined for each particular material. [65]

Collection and Preparation of Soil Samples

Methods for soil sampling vary according to analytical objectives, topography, and land use. Land use compartments in ecosystem studies may include cultivated cropland, unused areas of varied structure and topography, forest and woodlots, interfacial areas (roadside, fencerows, shorelines), and residential and urban areas. Literature references dealing with methods of soil sampling and analysis are almost exclusively concerned with food production. [68-71] The classical methods so well developed over the past century for this purpose are not necessarily the best for research dealing with heavy metals transport throughout the environment.

The tendency of lead to be relatively immobile in soil leads to an extreme vertical-concentration gradient when the lead is from external sources, e.g., from atmospheric fallout. In cases where the soil surface is covered with forest litter, lead contamination may not reach the soil surface at all, but rather is retained by the organic horizons. [72]

The following general practices are suggested as guidelines in field sampling soils for lead transport studies. [6, 7, 68]

(1) Use judgmental rather than purely random sampling. Select a ground area, preferably without heavy vegetation cover, that is typical of the general area under study.
(2) Avoid areas with large stones at or near the surface.
(3) Carefully control the depth of sampling. Since most of the externally introduced lead will be within a few centimeters of the surface, considerable care should be taken to preserve the vertical integrity of the sample until the portion selected for analysis is taken.

Soils in cultivated areas are best sampled by well-documented statistical sampling procedures. [68, 69] Sampling by numerical depth or by soil horizons depend on whether the analytical objectives involve mineralogical or metals transport studies. [68, 69]

Samples of soil are generally air-dried, sieved through a 2-mm aperture screen to remove gravel, plant litter, and other extraneous material, and agglomerates are broken up. A representative sample of approximately 100 g is taken by quartering or with a riffle-type splitter, and the selected portion is stored in a sealed container. If samples are to be stored over extended periods of time, air-drying at or near ambient temperatures is recommended to arrest the growth of organisms. Soils tend to stratify on storage, so remixing should be carried out and analytical samples should be selected by quartering or riffling procedures.

In selecting the analytical or the sample preparation procedures used for soil analysis, the question must be addressed whether "total" or "available" lead is to be determined. Methods that provide data on total lead may be less appropriate to typical environmental studies than those that estimate only the available lead. It is reasonable to expect, for example, that mineralogically bound lead will not be available to plants on a relatively short time scale.

The matter of defining available versus nonavailable lead constitutes a problem of major concern with regard to soil-plant transfer processes. Soil scientists do not agree on a general means for defining available fractions of trace elements in soils; a plethora of different defining techniques have been used. Most involve extraction of the trace element with a dilute mineral acid or buffer medium and designation of the extracted portion as availble to plants. Clearly, the diversity of extraction systems is indicative of the uncertainty in the definition. The definition should be based on the degree of correlation between the available trace element (defined in any one of several ways) and the amount or rate of uptake by plants, i.e., the best correlation should define the best extraction medium for determining availability. Such studies should consider possible differences related to plant species, soil types, and even weather. Successful studies of this nature have been carried out for copper and zinc. Studies with the same purpose have not been reported for lead, but they are presently being cooperatively carried out at Illinois, Missouri, and Colorado State. Until these studies are complete, the matter must be regarded as unresolved.

Preparation of soil samples for determining total lead may be based on wet digestion in oxidizing acid media, dry ashing, or fusion procedures as discussed below.

Plant Tissue Collection and Treatment

Little mention appears in the literature of methods for the field sampling of plant life for environmental studies. The standard methods and procedures usually are focused on statistical sampling for the determination of nutrient elements in food crops.

Sampling of plant materials in environmental monitoring is generally carried out by random selection of the indigenous species representative of a given area of interest. Where the entire plant is not collected, emphasis is usually placed on the portion of the plant consumed by herbivores or harvested for market. Developing sampling plans requires close coordination between plant and animal sampling groups, especially where food chains are involved.

Prior to analysis, a decision must be made whether the plant material should be washed to remove surface contamination from fallout and soil particles. The following general guidelines may be used: If the plants are sampled in a study of total lead contamination, or if they serve as animal food sources, washing should be avoided. If the effect of lead on plant processes is being studied, or if the plant is a source of human food, the plant samples should be washed. In either case, the decision must be made at the time of sampling, as washing is not effective after the plant materials have dried. Nor can fresh plant samples be stored for any length of time in a tightly closed container before washing because molds and enzymatic action may affect the distribution of lead on and in the plant tissues. Freshly picked leaves stored in sealed polyethylene bags at room temperature generally mold in a few days. Storage time may be increased to approximately 2 weeks by refrigeration.

Methods reported in the literature for removing surface contamination vary considerably, ranging from mechanical wiping with a camels-hair brush to leaching in mineral acids or EDTA. [73] Surface contamination can generally be removed with minimum leaching of constituents from leaf tissue by using dilute solutions of selected synthetic detergents, followed by rinsing in deionized water. A recent paper [74] provides the best data available on the problem of removing trace metal contaminants from leaf surfaces and should be consulted by those concerned with the problem.

After collection, plant samples should be dried as rapidly as possible to minimize chemical and biological changes. Samples that are to be stored for extended periods of time or to be ground should be oven dried; at least 4 h at 70° C is required to arrest enzymatic reactions and render the plant tissue amenable to the grinding process. Storage in sealed containers is always advisable.

If replicate reproducible portions of plant materials are required, grinding in a Wiley mill with a 20-mesh screen provides a suitable means for reducing particle size without generating gross amounts of fine material that tend to segregate on standing. Stored samples of ground plant tissue should be remixed by tumbling or other suitable means before taking a portion for analysis. Skogerboe [45] has found a dependency between the size of leaf particles and lead concentration in ground orchard leaf samples. (table 2-4) Particles less than about 75 μm in size are primarily leaf tissue, while the larger sized fractions contain progressively more leaf veins and petioles. The lead distribution varies between leaf veins and petiole tissues. Thus the comminution and sampling processes should be designed to compensate for these factors.

TABLE 2-4.–Segregation of lead in ground tree leaves [45]

Particle size (μm)	Lead (μg/g)							
	> 420	250-420	177-250	125-177	74-125	< 74	Weighted average	Grab Sample
Sample								
A	–	440	470	420	370	750	550	550
B	–	820	920	1,200	1,500	2,000	1,300	1,600
C	2.5	22	6.2	16	20	44	18	23

The digestion of plant leaf samples for analysis may be based on either wet or dry oxidation procedures. The general precautions to be taken are discussed below.

Collection, Storage, and Preparation of Animal Tissue Samples

Objectives and methods for animal tissue sampling in the three-University investigations have been in two principal categories: (1) random collection of small rodents (mice, shrews) at specific locations for the subsequent determination of the total lead burden in the animal and (2) laboratory studies of transport and effects of lead ingested by small rodents under controlled environmental conditions. Small rodents are usually caught by snap traps located at selected sites, then collected at regular intervals and placed in tightly capped jars. The whole body samples may then be freeze-dried to constant weight and stored at room temperature.

Laboratory studies of the effect of lead ingestion on animal processes and on transport of the ingested lead require analytical information on specific tissues or organs. The following general procedure is recommended for the sampling of specific organic tissue of small mammals to minimize contamination from adjacent tissue or ambient sources. Prior to dissection, the whole animal is thoroughly washed with distilled water. The entire organ or tissue section is removed by dissection. Certain organs, especially kidneys, contain relatively large amounts of blood, which should be removed from the organ to minimize the effect of blood constituents on the analytical results. Upon removal from the animal, a tissue sample is held with stainless steel forceps and rinsed with distilled water to remove bits of extraneous tissue and blood. The sample is rinsed in approximately 0.1M HNO_3, followed by thorough rinsing with deionized water. The samples are then individually sealed in polyethylene bags and stored frozen; alternatively, they may be freeze-dried and stored at room temperature.

The analysis of bone tissue for heavy metals usually involves small mammals. Soft tissue should be removed from the excised bone with a previously cleaned scalpel, and the sample should be well rinsed with deionized water. If the bone tissue is to be differentiated from marrow and blood, the bone sample should be longitudinally fractured, and the marrow and blood should be removed. The bone sample should then be dried at 105° C and stored in a sealed container.

Blood samples are conventionally obtained by venipuncture in volumes of 5 to 10 ml by means of clean stainless steel needle-plastic syringes, or by 5-ml "Vacutainer" tubes containing a suitable amount (approximtely 10 IU/5 ml) of heparin to prevent clotting. [75, 76]

The advent of microvolume trace analysis techniques such as the Delves Cup, carbon rod, tantalum strip, and carbon furnace (see below) permit microanalyses for trace constituents such as lead in blood-sample volumes as small as 10 μl. This capability makes possible the determination of lead at basal levels in humans on "finger prick," rather than venal puncture, samples and also permits time sequential analyses of blood lead in mice and other small experimental animals with minimal disturbances of life processes. In a procedure reported by Mitchell et al., [77] blood samples were collected in hemocrit tubes, and 50-μl aliquots of a mixture of blood (50 μl) and deionized water (200 μl) were dispensed into Delves Cups for analysis by atomic absorption (AA) spectroscopy. The extremely high sensitivity of such methods and the small sample sizes require extreme care to control contamination. The sampling site on a human subject or small animal must be scrupulously cleaned prior to drawing the blood sample. Aldous [78] has successfully controlled contamination during sampling by spray coating the sampling site with a thin film of collodion from an aerosol dispenser and immediately making the puncture through the film.

Blood samples to be stored in a nonhemolized state should be either freeze-dried or frozen. The digestion of blood and other animal tissues for analysis may proceed by either wet or dry oxidation methods with the procedures discussed in a later section.

The levels of lead, cadmium, and arsenic in human hair have been found to accurately reflect environmental exposure to those elements. [79-82] Hair analysis thus offers promise as an epidemiological tool. The relative concentrations of lead, as well as arsenic and cadmium referred to base levels, have been useful in the diagnosis of clinical poisoning. [80-82] Scalp hair has several of the characteristics of an ideal tissue for epidemiological study in that it is easily collected without discomfort and is normally discarded.

Lead may be accumulated on hair by external deposition, e.g., hair preparations and ambient conditions, or endogenously from ingested or inhaled metal. Differentiation of external and internal metal components in hair can be made by stringent wash procedures. [79, 80, 83] Hammer et al. [79] employed sequential washings in detergent, distilled water, alcohol, and ethylene diaminetetraacetate (EDTA) to remove exogenously bonded metals. However, chemical binding of exogenously and endogenously deposited metals in hair is not fully understood, and the washing procedure before analysis is suspected of leaching trace metal components weakly bound at either site. [79, 80, 83] A concentration gradient is normally observed along the length of a hair, which reflects a time-exposure relationship. [80] This factor should be considered in selecting the sample and may be useful in establishing the time at which major exposures (ingestion) occurred.

For the determination of heavy metals in a hair sample, the prewashed hair specimen is dried, weighed, wet-ashed with nitric-perchloric acids, and analyzed by atomic absorption spectroscopy or other appropriate techniques.

Insects and Other Arthropods

Devices and methods used for collecting insects and other arthropods vary according to the type of interest. The suction sampler [84, 85] provides a convenient method for collecting foliar insect populations. The sweep net is particularly useful for survey sampling over extended areas. [86] Pitfall traps are used in sampling ground populations. Black-light traps may be used to attract night-flying insects. [87] If time-related sampling schedules are required, interval separation devices are recommended. [88] Estimates of insect biomass per unit area may be obtained as described by Horsfall. [88]

Wet Versus Dry Oxidation Preparative Methods

The literature on methodology for sample decomposition is immense, and the procedures cited use both wet and dry oxidation methods extensively. However, dry ashing methods are usually implicated when analytical problems in lead recovery or losses are reported. Their use is considered by many to be less reliable than wet oxidation. Yet, conclusive data are often lacking and authors disagree on experiments applied to a given material.

Comparing wet and dry ashing methods is difficult in the absence of data specific to real-life samples. Hence it is inappropriate to state categorically that one method is superior to another. Some generalizations can be made, however, on procedural differences. Wet oxidation has the advantage of requiring a minimum of apparatus and is less prone than dry ashing to volatilization and retention losses. A disadvantage of wet ashing is that the relatively large amounts of reagents added, manipulations required, and contact with glassware bring a higher risk of contamination than dry ashing. Wet ashing requires more manipulations than dry ashing and does not conveniently accomodate itself to large samples, i.e., 20 to 50 g.

Dry ashing exhibits advantages and disadvantages directly opposed to wet ashing. Dry ashing requires few or no reagents, manipulation is minimized, bulky samples present less of a problem, and little operator attention is required. A primary disadvantage is the risk of volatilization, convection, and retention losses, unless conditions are carefully controlled and care is continuously exercised to maintain the proper time-temperature programming of the furnaces. Volatilization losses of lead are likely to be particularly severe in the presence of high concentrations of halides. However, with suitable controls, either dry ashing or wet ashing can be successfully used to determine lead in plant and animal tissue or other environmental samples. In most cases, the choice can be made on the basis of operational factors.

One of the most useful compilations of information on methods and techniques for the oxidation of organic matter is provided by Gorsuch [89] in his book "The Destruction of Organic Matter." The following abstracted comments are given as a summary of current practices.

Wet Oxidation: Nitric acid is the universally used primary oxidant for the destruction of organic matter. The azeotrope boils at 120° C, a factor that assists in its removal after oxidation, but, correspondingly, limits its effectiveness in completing the oxidation process. The most effective medium for wet oxidation of organic material is a mixture of nitric and perchloric acids. However, more than ordinary care is required in the use of perchloric acid mixtures, especially with unfamiliar materials. The well-known precaution of having sufficient nitric acid present until easily oxidizable material has reacted (indicated by cessation of brown fumes [90]), is especially important in the wet digestion of animal tissue with a high lipid content. For some materials, such as whole blood, hydrogen peroxide can be used as an alternative to perchloric acid to complete the oxidation.

Completeness of oxidation, in this sense, implies breaking down organic matter into acid-soluble material to release the lead for analytical measurement. Evans and Morrison [91] have observed, in the spark-source mass spectrometric investigation of trace elements in biological materials, that wet oxidation with nitric-perchloric acid mixtures did not eliminate all of the organic fragment ions in the spectrum, but left significant amounts of small, acid-soluble organic molecules. These are probably of little concern in most analyses, but the possibility of chelating fragments should not be ignored in procedures where separations or electrochemical methods are used.

In trace metals analysis, several factors concerning the use of wet oxidation methods should be emphasized. For most determinations, it is essential that reagent blanks be run on a systematic basis.

Experience has shown that the lead blank for analytical reagent grade acids may be as high as several mg/kg. For analyses at trace levels of 1 mg/kg or less in sample materials, redistillation of the acid from quartz or exhaustive electrolysis may be required to reduce the blank to acceptable levels. Specially purified reagents are available, but the degree of control of high-purity acids available commercially at premium cost has often been disappointing. Since significant blanks may originate from glassware, all vessels must be scrupulously cleaned with hot nitric or hydrochloric acids immediately before use. Control of contamination and reagent purification are further discussed in a later section.

Some environmental materials (e.g., air filter samples from incinerators) are refractory in nature and require unusually rigorous wet digestion to solubilize the lead species present. While this problem can usually be solved by repeated digestions in nitric-perchloric acid media, the increased reagent requirements increase the blank problem. The tentative standard method from ASTM [25] recommends the use of a ternary acid mixture for wet ashing. The mixture contains sulfuric acid, which may cause interferences if lead is to be determined by atomic absorption. (See below.)

Examination of the analytical literature on lead shows that wet oxidation methods are widely used. However, the methods are subject to a variety of problems that must be carefully considered when in use. The most serious of these have been pointed out. The publications by Gorsuch [89] and Tölg [92] should be required reading for anyone who is determining lead in environmental materials and wants to avoid serious errors.

Dry Oxidation: The most common dry oxidation methods fall into two categories: conventional oxidation in air at temperatures approximating 500° C and so-called low-temperature ashing (LTA) employing excited oxygen.

Dry oxidation methods all involve the following sequential processes:

(1) Evaporation of physically and chemically bound water.
(2) Evaporation of volatile materials and progressive thermal destruction.
(3) Continued oxidation of the nonvolatile residue until all organic matter is converted to inorganic.

Gorsuch [89] and Tölg [92] present evidence that some elements, including lead, may be lost under certain conditions by furnace dry ashing techniques. While the addition of fluxing materials such as sulfuric acid and other "ashing aids" reduces the losses in many cases, the risk of raising the blank level is increased. Gleit and Holland [93] have shown that lead contained in a blood sample was completely lost when ashing was carried out at 400° C for 24 h. In dry ashing air filters for 4 h at 400° C, Skogerboe [45] has shown that from 11 to 75 percent of the lead was lost, depending on the collection and storage history of the filters. Filters ashed shortly after field collections show much greater losses than those stored for several days to weeks. It is certain that the tendency to lose lead on dry ashing will be highly dependent on the nature of the sample, as well as the ashing conditions. Samples high in halide content may lose lead through the formation and/or direct vaporization of the volatile lead halides at temperatures of 150° C or less. [93] Thus "fresh" air-filter samples with appreciable quantities of lead halides present may show more loss than samples that have been aged, with the concomitant conversion of the lead halides to less volatile compounds such as the oxides, carbonates, and sulfates. [94] An alternate method for preventing the loss of volatile halides involves programming the temperature rise of the furnace over a period of approximately 4 h to a maximum of 450° to 500° C [55] This prevents flash burning and allows conversion of the halides to the less volatile oxides; both factors prevent vaporization losses. *Any direct ashing without a gradual temperature rise should be avoided because of the potential losses.*

Another factor to be considered in dry ashing is the possibility of retention losses [89-92] Some materials ashed in borosilicate or silica ware may react with the ashing vessel to produce a complex silicate that might not be decomposed by subsequent digestion with acid. The problem is intensified by the tendency of such containers to become etched on repeated use, thereby increasing the surface area.

Another form of dry ashing, the LTA method, [93] utilizes reactive oxygen species to oxidize a sample of organic material at temperatures of typically 150° C or less. A gas plasma of radio-frequency-excited oxygen species is formed about the sample contained in a chamber at a pressure of approximately 1 torr. Because of the relatively low temperatures involved, this technique also permits the determination of metals such as selenium and arsenic, ordinarily lost during conventional high-temperature ashing methods. It should be emphasized that under certain conditions the LTA method also can result in trace element losses. For example, the presence of chloride in biological materials can result in the formation of volatile metal halides (including lead) with losses through vaporization even at the low temperatures involved. [93]

A significant disadvantage of the LTA technique for routine use is the complexity of equipment and the slow oxidation rate typically obtained. A tissue sample of several grams may require 34 h for complete oxidation. Other materials of a more refractory nature may not be satisfactorily ashed by the LTA at all. However, current developments in equipment design and methods may significantly alter this situation.

It is not the purpose of this report to stipulate which method of oxidation should be used. Rather, it is to emphasize that both methods are subject to problems of contamination or loss of lead. The extent of losses and contamination will obviously depend on the methods used, the nature of the samples, and the degree of caution and expertise exercised by the analyst. Sample preparation methods should always be carefully evaluated in order to determine their utility for the problem at hand. Where possible, standards should simulate the nature of the samples reasonably well and should be carried through the full sample preparation and analysis procedures, as one means of compensating for contamination losses. In any suite of samples analyzed, it is also advisable to routinely run aliquots of one or more samples that have been spiked with known amounts of lead to verify that the recovery is complete. While such recovery experiments do not always provide absolute proof that the problems are minimal, they do serve well as indicators of the general validity of the method(s) used. Repetitive check analyses of standard materials provide a better test.

REAGENT PURIFICATION, CONTAMINATION CONTROL, AND CHEMICAL SEPARATIONS

Reagent Purification

The lower limits at which trace concentrations of elements can be measured in a sample are ultimately determined by the precision in establishing the value of the analyte element in the "blank." The reliability of analytical results depends greatly upon minimizing the level and the variability of the blank. Although the instrumental threshold of detection for trace elements has been lowered to picogram levels by analytical techniques and instrumentation that have come into common usage, the practical limits of detection in the analysis of real life samples are also established by the level and variability of contaminants. Variability of the blank level is principally the combined effect of statistical variations in measurement and random contamination from various sources in the laboratory environment.

Impurities in reagents such as water, mineral acids and organic solvents used in large relative amounts (10 to 100 fold) with respect to sample weight are responsible for the systematic contaminant level of the sample and blank. The desirability of controlling the blank contribution to less than 10 percent of the analyte element concentration for naturally occurring lead levels in biological materials usually requires reagent purity significantly greater than the American Chemical Society specifications. For example, consider a typical wet-oxidation preparation of a 1-g sample of animal tissue with a lead concentration of 1 mg/kg. A lead contribution of less than 10 percent by reagents that would be equivalent to 0.1 μg of Pb in the approximately 30 ml of mineral acids used in digestion and final solution — or an average lead concentration of 3 μg/kg in the reagents. This is approximately two orders of magnitude below the ACS reagent specification for lead. Specifically purified grades of mineral acids, marketed "for Hg determinations" and containing less than 3 μg/kg lead, are suitable for use without additional purification. Often it is sufficient or remove only the interfering impurity. The lead concentration in ordinary reagent grade hydrochloric or nitric acid can be reduced from 500 μg/kg to 10 μg/kg or less by distillation of the acid-water azeotrope from well-leached borosilicate glass apparatus. [55] Proper precautions must be taken in apparatus assembly and operation to minimize spray carryover into the Teflon or polypropylene receiving vessel.

Water of approximately 1 μg/l lead or less can be conveniently prepared by the passage of tap water through tandem columns of commercial, mixed bed ion-exchange resins, or by double distillation from acid-leached quartz or borosilicate glass. Quartz is generally preferable for distillation of acids and water because of the surfaces where no further exchange takes place. However, for long-term usage, Pyrex glass (which reaches a steady state after about 100 h of continued use) is quite satisfactory for redistilling acids used in determinations of lead at trace to ultratrace levels.

Maintaining the "as prepared" purity of mineral acids and water depends on the storage containers used and conditions of storage. Recommended container materials, in decreasing order of preference, are Teflon, polypropylene, linear polyethylene, quartz, and borosilicate glass. Regardless of the choice, proper preconditioning by acid leaching and repeated rinsing with deionized or distilled water is mandatory. Where the need exists, ultrapure mineral acid reagents and water can be prepared by subboiling distillation in which infrared heaters vaporize the surface of a liquid without boiling and attendant production of spray. [95-96] Hydrochloric, nitric, and perchloric acids produced in experimental quantities by the National Bureau of Standards [NBS] with subboiling techniques have lead concentrations estimated at less than 0.02 μg/l.

Special purification procedures for other reagents based on a combination of distillation, coprecipitation, liquid-liquid extraction, and absorption are referenced in numerous publications and reviews. [92, 97-100]

As a general practice, Tölg [92] advises that the only way to be sure of the purity of a reagent is by regular checking. It is best, therefore, to use only a few easily purified reagents and monitor their purity regularly. Laboratory cleanliness is essential to maintain reagent purity, and such practices cannot be too strongly emphasized.

Contamination Control

Any procedure for the determination of lead may involve contamination from the time of sampling through the final analytical measurement. Considerable care must be taken to minimize inadvertent lead additions from sampling equipment, grinders, containers, glassware, and reagents, as well as from the ambient atmosphere at the sampling site and in the laboratory. Mechanical grinding equipment such as impact and rotary-type mills and high-speed blenders may contaminate the sample to some degee with the materials of construction. When grinding with a Wiley mill is required, Koirtyohann [72] advises a preliminary inspection of the screen for evidence of soft-solder used in fabrication; if necessary, it should be replaced with a stainless steel screen attached with epoxy cement.

Contamination from containers can be minimized by the use of relatively inert materials such as Teflon, linear polyethylene, or polypropylene. Widemouth screwcap containers can be cleaned more conveniently than the narrow-mouth types. Before use, containers and glassware should be washed in water containing a detergent, leached in HCl or HNO_3, and then thoroughly rinsed with successive portions of deionized water. An inverted beaker placed over the neck of the reagent bottles provides a convenient dust cover to reduce contamination.

Glassware may sometimes require more reactive cleansing methods. Other agents such as hot 1:1 nitric acid:water, nitric acid:sulfuric acid mixtures, and chromic-sulfuric acid are effective. A disadvantage of chromic-sulfuric acid mixtures is that chromium is readily absorbed on glassware and silica ware and is exceedingly difficult to remove. Clean containers should be stored with the closure attached. Glassware should be kept in a cabinet protected from laboratory atmospheres. Containers, utensils, and glassware used to determine trace concentrations of lead should be segregated from general-purpose equipment.

Borosilicate glassware can be used throughout the procedures for HCl, HNO_3 and $HClO_4$ digestion, oven drying at 105° C, and ashing up to 500° C. However, scrupulous attention must be given to reactions that might occur between samples and glassware, especially in dry-ashing operations. Etched glassware should be discarded, and the sample preparation that caused the etching should be suspect and rechecked by a different procedure.

All reagents contaminate the sample to some degree. When contamination is measurable by the analytical method, it becomes significant, and the researchers must adhere to careful parallel treatment of blank and sample.

Ambient air is a "reagent" used in large quantities in drying and ashing operations. Hence, it can be a source of significant contamination. The composition of air in the laboratory usually approximates that of the surrounding atmosphere. In mechanical-convection drying ovens and in muffle furnaces that use forced auxiliary air, it is advisable to filter input air through absorption tubes or plugs of glasswool backed with 0.45-μm membrane filters.

Lead fallout from ambient air onto benchtops in laboratories can cause high blanks and random errors in microsamples techniques. The results of a survey made in a typical laboratory devoted to trace metals analysis indicated lead fallout of 15 to 90 μg $Pb/m^2/24$ h on laboratory benchtops. [55] In the same survey, Delves Cup planchets exposed for 15 min. to ambient atmospheres frequently acquired amounts of lead in excess of basal blood levels. Subsequent test samples prepared under protective covers showed no detectable contamination by the same methods. [55]

The use of microtechniques for trace element analysis demands control of ambient air contaminants approaching clean-room conditions, if the full capabilities of the methods are to be realized. A substitute for overall air control is local control of specific work areas. Commercially available, portable benchtop work stations are commercially available that employ particulate air filters to substantially reduce the levels of airborne contaminants. Analysis of particulates in ambient air before and after passage through a typical system shows a thousandfold decrease in the amount of Pb, and a tenfold decrease in the Cd level. [101]

Chemical Separations

Separations can often be used to increase the sensitivity of an analysis by concentrating the element(s) of interest and simultaneously reducing interferences by eliminating the sample matrix. The cost of these advantages is always a longer procedure. Blank problems, brought about by the additional reagents and manipulations, are frequently a severe limitation. Trace components can be separated by solvent extraction, coprecipitation, ion exchange, distillation, or electrodeposition. Precipitations cannot generally be used to remove a matrix element and retain the trace component in solution. Bulky precipitates carry traces down with them by coprecipitation or absorption.

The separation method chosen depends on the analyte, the matrix, and the final measurement method. For lead in environmental samples, solvent extraction prior to an atomic absorption determination is particularly attractive because measurements can be made directly on the organic phase. In addition to the chemical enrichment, a twofold increase in response can be obtained with a properly chosen organic solvent, due to improved nebulizer efficiency. Many solvents are satisfactory, but some that are popular for certain extractions e.g., chloroform, do not burn well and must be avoided for the flame analysis methods. For atomic absorption (AA) measurements, the extraction must be complete, but it need not be specific. Indeed, it is often advantageous to simultaneously concentrate several trace elements, while rejecting the major sample components.

Extraction of metal chelates, formed with ammonium pyrrolidine-carbodithioate (APDC)[1] into methylisobutyl ketone (MIBK), prior to AA measurements was first described by Allan [102] and has since become very popular. Lead and several other metals are efficiently extracted over a rather wide pH range. [103, 104] All reagents required are inexpensive, and they are readily purified. The major components in most biological and environmental samples are left behind, and improvements by factors of 10 to 50 in the ability to measure low lead levels are

[1]This reagent is popularly but incorrectly known as ammonium pyrrolidinedithiocarbamate (APDC).

attainable. The extraction procedure is simple, with pH adjustment probably the most time-consuming step. Preparation, extraction, and measurement of 50 sample solutions per day are a reasonable load for a single technician without automated equipment.

The APDC-MIBK extraction is a good preconcentration step, but its advantages over other similar extractants are probably not as great as its overwhelming popularity would indicate. Other satisfactory chelating agents for lead include dithizone, [105] sodium diethyldithiocarbamate, [106] diethylammonium diethyldithiocarbamate, [107] and tributylphosphate. [108] Recently, extraction of lead with liquid ion exchanges [109, 110] or as the tetraiodoplumbate [111, 112] has been used for preconcentration.

Solvent extraction can also be used prior to other measurement methods, and it is an inseparable part of the dithizone colorimetric method for lead. Additional steps may be needed with some methods to get the concentrate in a suitable form. Anodic stripping methods probably should not be used following solvent extraction because traces of chelating agents and solvents are likely to interfere. Also, blank problems would be serious at the extremely low levels at which anodic stripping is generally used.

Ion exchange separations involving lead are well-known, but they have never become popular for the types of samples encountered in environmental work. Ion exchange membranes [113] and ion-exchange resin loaded filters [114] place the sample in a physical form particularly attractive for X-ray fluorescence methods. Trace components are concentrated by repeated filtration through, or prolonged contact with, the exchanger disc, which is then placed in the X-ray beam without further treatment. Since multielement analyses are possible, the method will probably see increased use in the future.

Coprecipitation of a trace component with an appropriate carrier present in relatively large amount can give useful separations. Lead has been precipitated as the sulfate with strontium as a carrier, [115] as the hydroxide with bismuth carrier, [116] as phosphate with thorium carrier, [117] and as sulfide with a copper carrier. [118] Organic precipitants can also be used as in the method developed by Mitchell. [119] Trace metal 8-hydroxyquinolates are precipitated along with the aluminum complex as a carrier. Ignition of the precipitate yields the trace metals in an alumina matrix suitable dor arc emission spectroscopy. The procedure concentrates about 16 metals and therefore allows the multielement capability of the spectrograph to be exploited. Such considerations need not restrict the use of coprecipitation methods because subsequent treatment of a precipitate is limited only by compatibility of the carrier with the measurement method used.

Electrodeposition as PbO_2 on the anode can be used to separate and determine lead. [118, 120] Electrochemical determination methods are discussed elsewhere in this chapter. However, work is known to have been done in which electrodeposition has been used in environmental studies where the primary goal was separation of the lead from the sample matrix.

ANALYTICAL METHODS FOR THE DETERMINATION OF LEAD

Spectrophotometry.

Spectrophotometric methods for the determination of small amounts of lead are dominated by a single reagent and a few basic approaches to using it. Dithizone (diphenylthiocarbazone) methods have been used for many years. [121] Until recently they were the only ones listed in manuals of official methods published by organizations such as the Association of Official Analytical Chemists, [122] the American Public Health Association, [49] or the Society for Analytical Chemistry. [123] Dithizone extraction was also recommended as the primary method in a recent comprehensive report on lead. [124] The method has the advantage of a long, well-known history and, in the hands of a skilled analyst, it remains the one against which others may be judged. Unfortunately, it also suffers from some significant limitations.

Dithizone is an intensely colored compound that reacts with about 17 metals to form chelates that can be extracted into chloroform or carbon tetrachloride. [125] In spite of the large number of metals that can react, considerable specificity can be gained by careful control of pH and the use of masking agents.

The wavelength of maximum absorption is different for the various chelates and for the reagent itself. Excess reagent normally absorbs to some extent at the measurement wavelength. However, errors from this source can be controlled by selecting conditions to minimize the amount of excess reagent present and to assure that it is constant; errors can also be controlled by measuring the color at two wavelengths to assess the amount of excess reagent and applying a correction.

Lead dithizonate is usually extracted from slightly basic solution (pH 8.5 to 9.0) in the presence of citrate, cyanide, and a reducing agent such as hydroxylamine. The citrate prevents precipitation of alkaline earth and transition metal hydroxides and phosphates; cyanide is a masking agent to prevent extraction of metals such as copper and zinc; and the reducing agent prevents oxidation of the dithizone. Under these conditions only Sn^{2+}, Bi^{2+}, and Tl^{+} are extracted with the lead. Interference from absorption by excess reagent is minimized for lead because the absorption maximum of the chelate at 510 nm occurs near a minimum in the reagent absorption curve. Also, unreacted dithizone tends to be soluble in basic aqueous solution, thereby reducing the amount of excess reagent extracted.

Lead dithizonate is readily decomposed and the metal transferred to the aqueous phase when the organic solvent is shaken with dilute acid. This property has led to the development of double extraction methods that are most trustworthy in complex samples. Lead is first extracted from the sample solution with a large excess of reagent, then back extracted into a definite volume of dilute nitric acid. The bulk of the sample matrix remains in the original aqueous solution. The pH of the nitric acid solution is then adjusted to 8.5 to 9.0, and lead is extracted with a known volume of standard dithizone solution for the actual absorption measurement.

Detailed procedures have been published [121, 125, 126] but a brief outline of the necessary steps follows:

(1) Add ammonium citrate, hydroxylamine, and potassium cyanide solutions to the sample aliquot and adjust pH to 8.5 to 9.0.
(2) Extract with aliquots of dithizone solution in chloroform or carbon tetrachloride until all lead is extracted, as indicated by the green color of the excess reagent in the organic phase. Combine all extracts. Discard the aqueous phase.
(3) Shake the combined organic phase with a measured volume of dilute HNO_3. Discard the organic phase.
(4) Make the solution basic by adding a measured volume of a buffer solution (plus masking and reducing agents) and extract with a measured volume of dithizone solution.
(5) Measure the absorbance at 510 nm.
(6) Prepare standards by adding known amounts of lead to the same volume of HNO_3 as used in step 4 and performing steps 5 and 6. The zero standard should be set to zero absorbance on the spectrophotometer to compensate for absorption by the excess dithizone.

For relatively simple samples, the double extraction is not necessary. The first extraction is made with a standard volume of dithizone solution measured at 510 nm. Complex samples and those about which the analyst has little information require the double extraction.

Advantages of the dithizone method include:

(1) Its acceptance as "official" by organizations such as the AOAC and APHA. This acceptance indicates established reliability if published instructions are carefully followed.
(2) Only simple and inexpensive apparatus and reagents are required.
(3) The determination is well-behaved spectrophotometrically, and gives linear working curves if the amount of excess reagent is controlled.
(4) The method is sensitive to a few micrograms of lead.
(5) Large samples can be used readily.
(6) Few matrix problems are encountered in the double extraction procedure because of the preliminary separation in the first extraction.

Most of the disadvantages of the dithizone method are associated with the fact that many steps are required, and numerous reagents must be added. Each step is a potential source of error, and each reagent is a source of contamination. Reagents must be carefully purified and glassware scrupulously clean. Fortunately, all reagents can be cleaned by a preliminary dithizone extraction, and dithizone itself may be freed of trace contaminants by taking advantage of its solubility in basic aqueous solution. [125] The entire procedure must be in the hands of a skilled and patient worker if reliable results are to be obtained. In unskilled hands the method is likely to yield disastrous results. The requirement for a skilled worker and the rather lengthy procedure result in a high labor cost per sample.

Other disadvantages include interference from Bi and Tl unless special precautions are observed (a still longer procedure). [125] Bi and Tl are not likely to be present in samples of biological materials, and the interference can often be ignored. Large samples are required (10 ml of blood, for example), and a dangerous chemical (CN^-) must be used as a masking agent.

Only two other colorimetric reagents have been suggested for the determination of small amounts of lead: diethyldithiocarbamate [127] and tetramethyldiaminodiphenylmethane. [128] Neither method seems to offer significant advantages over dithizone, and their use on practical environmental samples has not been reported.

Atomic Absorption Spectroscopy

Analytical AA began in 1955 with independent papers by Walsh [129] and Alkemade and Milatz. [130, 131] The growth in popularity of the method since the early 1960's has been almost explosive, and it now ranks among the most extensively used techniques for trace analysis. The method resembles ordinary solution spectrophotometry in that light of the proper wavelength is passed through an absorption "cell," and an attenuation of the beam, which is logarithmically related to the concentration of the absorbing species, is measured. Two important differences must be recognized. First the "cell" for AA must provide conditions to decompose the sample into gaseous metal atoms. Once formed, the atoms can absorb characteristic "resonance" radiation, but the wavelength region over which atoms of a given element absorb is only a few thousandths of a nanometer. Small bench-top monochromators do not approach this resolution. The most useful approach to atom formation is to aspirate sample solutions into a flame; for many elements, atoms are formed quite efficiently. [132-135] A special light source is used to overcome the resolution problem. A hollow cathode lamp containing the analyte element produces resonance radiation of precisely the correct wavelength to be absorbed. The monochromator then needs only to separate the resonance line of interest from other lines in the source, a job done quite adequately by small bench-top units.

The flame is normally formed at a long, narrow slot in the burner head, with the long axis parallel to the optical axis. The oxidant gas (air or N_2O) is used to operate a pneumatic nebulizer that aspirates about

3 ml/min, about 10 percent of which is reduced to sufficiently small droplets to pass through the mixing chamber and be delivered to the flame. The fuel gas, usually acetylene, is mixed with the oxidant in the burner chamber, and the combustion mixture is passed to the flame under laminar flow conditions.

Commercial instruments are available from several manufacturers. [136] Some are simple, inexpensive units, while others offer greater capability at higher cost and increased complexity. Costs range from about $4,000 to $16,000. Some features of more expensive units include:

(1) Double-beam optical systems that allow electronic compensation for source instability.
(2) Dual channels for measurements on two elements simultaneously.
(3) Signal integration for increased precision.
(4) Automation of such things as flame ignition and instrument zero, which makes operation more convenient.
(5) Background compensation systems to correct for any nonatomic absorption.
(6) Automatic sample changing and data processing.

Atomic absorption spectroscopy can be used to determine about 60 elements at the level of a few μg/g or less in solution. More complete treatments of the theory and practice of the method are available. [137-139]

Flame AA Analysis for Lead: Lead has resonance lines at two wavelengths, 217.0 and 283.3 nm. Both are sometimes recommended for the determination, the line chosen depending to some extent on the individual instrument. The line at 217.0 nm gives greater absorption for a given lead concentration. On many instruments the signal is noisier, however, and the 283.3 line gives more satisfactory overall performance. Background absorption problems (see below) are generally more severe at 217.0 nm.

Lead atoms are readily formed in an air-acetylene flame, and if the concentration in the sample solution is a few tenths of a μg/g or higher, it is quite easily determined by simply aspirating the sample solution into the flame. The sample solution must be reasonably dilute (<1 to 2 percent total solids), and it should be free of suspended solids that might clog the aspirator. Small amounts of soluble organic matter can be tolerated in the sample solution because it is rapidly destroyed by the flame.

AA determinations by direct aspiration are quite rapid. A single reading requires only about 30 sec., standardization is simple and straightforward, and no elaborate calculations are required. Without automation, an experienced operator can handle readings and calculations for several hundred solutions in an 8-h day.

Precision and accuracy are typically 1 to 3 percent but maximum precision and the great speed can seldom be attained simultaneously. Frequently standardization may be necessary to correct for minor changes in flame conditions or nebulizer operation. Precision can be improved quite significantly, with relatively little effort, by taking multiple readings on samples and standards. To demonstrate the improvement, five solutions were prepared; each was split into two bottles that were then given random numbers and submitted to the University of Missouri laboratory. Lead was determined with a single instrumental measurement and also by taking four readings on the sample alternated with four readings on a standard of similar lead content. The average standard reading was used to establish the working curve in the latter case, and the average sample value was read from it. With single readings, the average relative deviation between duplicates was 2.2 percent and the maximum deviation was 4.4 percent. When the average of four sample and standard readings was used, the values for average and maximum relative deviations were 0.7 percent and 1.0 percent, respectively. It is important to note that multiple standard readings are an integral part of this improvement. There is little value in knowing a sample reading more precisely than one knows the calibration data.

The detection limit for the AA determination of lead by direct aspiration is in the range of 0.02 to 0.1 μg/g, depending on the instrument. The lower limit for practical quantitative analytical work is 5 to 10 times the detection limit. For many types of environmental work, this sensitivity is not good enough, and AA by direct aspiration will fail. This is especially evident when the dilution necessary to prepare solid samples for analysis is considered.

Lead determinations by flame AA suffer from relatively few interferences, but serious errors can still occur. The only specific chemical interference that has been reported is a suppression of absorption in the presence of soluble silicates dissolved in nearly neutral solutions. [140] Errors can be avoided by working in 3*N* HNO_3. One of the authors (SRK) was unable to duplicate the silicate suppression, which leads to the conclusion that it may not occur in all flames or with all burner systems. Until more is known, analysts should be cautious in reporting lead values if soluble silicates are likely to be present. Spectral interferences are potentially more serious, especially when working at low lead levels with concentrated solutions of complex samples. Light losses due to scattering by particles in the flame and/or absorption by molecules formed in the flame from the sample matrix can cause errors. [141] Special care is required because in the usual AA procedure the background absorption and the error it causes can easily go undetected. The situation is made more hazardous by repeated statements in the early AA literature that the method was completely free of spectral interferences. [138]

The magnitude of background light losses depends on the type of flame used, the general composition of the sample, the method of sample digestion used, and the total solids content of the analytical solution. The general extent of the problem may be inferred from figure 2-1. A whole blood sample was digested in the $HNO_3/H_2SO_4/HClO_4$ acid mixture as

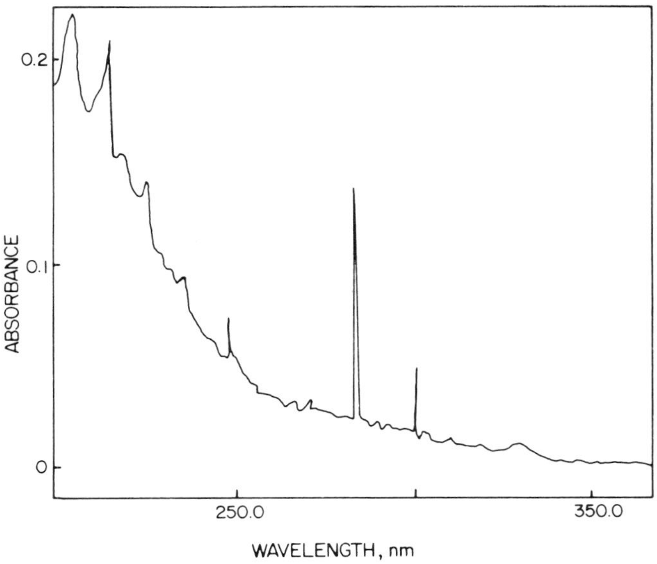

FIGURE 2-1.—Background losses in Flame AA studies of whole blood.

stipulated by the tentative ASTM procedure. [124] Adequate sample size was used to obtain a final solution containing 2 percent dissolved solids. A hydrogen hollow cathode lamp was used to measure the light scattering and molecular absorption characteristic of the sample. The zero absorbance level at each wavelength was set by using distilled water, the background was measured at each wavelength, and the spectrum was plotted from point by point measurements. The sharp band spectra that degraded in the ultraviolet between 200 to 250 nm are attributed to NO absorption from nitric acid in the sample solutions. These bands are superimposed on a broadband background absorption that is probably due primarily to light scattering by particles in the flame, although absorption by other molecular species may contribute to it. Regardless of the exact mechanism, the light loss always tends to be more serious at shorter wavelengths. At 350 nm the loss is nearly negligible, but its magnitude is clearly significant for both samples at the two lead absorption wavelengths. The sharp absorption lines at 285.2 and 303.3 nm are due to atomic absorption by the relatively large amounts of sodium present in both sample solutions. These spectra define what can be serious sources of error in the analysis of lead or other trace elements with absorption wavelengths below approximately 300 nm. Certainly, if lead is determined at the 217.0 nm line, the scattering and molecular absorption strikingly enhance the absorption signal determined and cause a serious positive error in the analysis. Although the magnitude of the background absorbance is much lower at the 283.3 nm line, it should be emphasized that if one is working at low lead levels and using instrument scale expansion, the relative error in concentration can be several hundred percent, and it is a positive error. At higher lead levels, the error is much less serious. Results [142] and procedures [124] of highly questionable value for the determination of lead have been published because the authors were not aware of the background problem.

Other investigators [143] who should be aware of the problem have apparently assumed that calibration by the method of standard additions compensates for the background effects. It can be readily shown, however, that this is not true. The background contribution to the sample and the standard additions measurements are essentially constant, and the curve extrapolation to determine the same concentration can consequently be in error by a large amount. Background corrections are essential for a wide variety of environmental samples if these errors in the AA determination of lead are to be avoided.

It should also be noted that the magnitude of background light loss in a flame is typically enhanced by the use of sulfuric acid in the sample digestion process. Many elements contained in environmental samples (e.g., Ca, Mg, and Pb) form insoluble sulfates that may cause light scattering and sulfuric acid causes very significant molecular absorption. [141] Moreover, even though the concentration of lead in a sample is low, sufficient water must be added to assure dissolution of the sulfates of major sample components (e.g., Ca) in order to prevent coprecipitation of the lead. Coprecipitation of lead on strontium sulfate is efficient enough to be used analytically. [144] Because of these factors, the authors cannot accept the ternary acid digestion method recommended by the ASTM procedure. [124]

Lest the authors be accused of over-emphasizing the background problem, it should be pointed that for many sample solutions with low total solids, background light losses are negligible at the 283.3 nm wavelength, especially if higher lead levels are being measured. However, this potentially serious problem is too often ignored in AA work, and it is emphasized here in the hope of reducing the frequency of future errors. Use of the 217.0 nm line for complex samples is not recommended unless background corrections are made.

Background corrections are made by measuring the light loss at wavelengths in the vicinity of the resonance line where no elemental absorption is expected. The absorption is subtracted from that at the resonance line to yield a corrected value. A nonabsorbing line from the hollow cathode lamp may be used in a separate measurement for the correction. Lead has nonabsorbing lines at 280.2 and 287.3 nm, either of which can be used to correct for background in the vicinity of the 283.3 nm resonance line. No suitable nonabsorbing line is available for use in the vicinity of the 217.0 nm resonance lead line. A more elegant method of correcting for background employs a continuum source such as a hydrogen or deuterium arc lamp and a hollow cathode lamp. [141] Both background and elemental absorption are measured with the hollow cathode source, while only background is measured with the continuum source. The difference in absorption with the two sources is the net elemental signal. The correction is made at wavelengths immediately adjacent to the resonance line and is applicable at most wavelengths of interest in AA. Instruments are commercially available, [136] and others can be modified [145] to obtain the correction automatically with a single measurement.

Flame AA measurements by direct aspiration do not provide the necessary sensitivity for many samples of environmental interest. Extraction of a lead chelate into a suitable organic solvent, as discussed earlier in this chapter, is one way to improve sensitivity and simultaneously reduce interference problems.

Alternate methods of introducing samples into a flame have been suggested to improve sensitivity. Two, which are quite similar in principle, are the tantalum sampling boat [146] and the Delves Cup methods. [147] In each method the sample is pipetted into a tantalum or nickel container, the solvent and the bulk of the organic matter are driven off either near the flame or on a hot plate, and the container is inserted into the flame. The lead is quickly vaporized and produces a high but short-lived atom population in the flame. The effective rate of sample introduction is much higher than by direct aspiration, and improves sensitivity. A few nanograms of lead can be measured in up to 1 ml of sample. The signal appears as a "spike" on the recorder, the height or area of which is a measure of the lead in the sample.

Disadvantages of these methods include lower precision (±5 to 15 percent), [148] more severe matrix effects, and lower sample output compared with direct aspiration. A greater problem is that various laboratories have produced a wide range of results with these methods. One does not find this in the AA literature. However, on the basis of their experience and informal discussions with many technicians, the authors conclude that the methods are reliable in some hands but not others. [148-149] Even within the same laboratory, different operators sometimes produce results that vary greatly in reliability.

Improved results have been obtained at the University of Illinois by modifying the commercial Delves Cup equipment to assure that the positioning of the cup can be exactly reproduced and also by making careful background corrections. Mitchell et al. [77] have described a semiautomated procedure to measure lead in blood. In this procedure a small computer is used to integrate the signal and obtain a peak area, rather than to use the height of the peak. The area depends on fewer variables, and the correlation coefficient between analyses by the cup procedure and an accepted solvent extraction method improved from 0.41 to 0.92 when integration was employed. Peak areas have been shown to also improve results with the sampling boat. [148]

Additional work is apparently needed to identify and control more of the variables in the Delves Cup and sampling boat methods. If carefully applied to samples in which the matrix content is low (water), very predictable (blood), or removed by a prior separation, the methods should provide satisfactory results. In other cases they should be used with caution.

Nonflame AA Analysis for Lead: In all flame methods, the sample vapor is diluted by the large volume of flame gases ($\simeq$100 l/min at 2,400K) and is rapidly carried through the optical path (10^{-4} to 10^{-3} sec.) of the instrument. L'Vov [150] first described the use of an electrically heated carbon furnace for AA in which atoms could be formed in a nearly static atmosphere and remain in the light path for a relatively long time. Since that time, carbon furnace devices have been described by Woodriff and Ramelow, [151] West and coworkers, [152, 153] Massmann, [154] Matousek and Stevens, [155] and Norval and Butler. [156] Furnaces are now commercially available from at least two companies. All provide very high sensitivity ($\simeq 10^{-11}$ g for lead). They are operated in an inert gas (Ar or N_2) atmosphere to prevent oxidation of th carbon, and all provide a transient (spike) signal. Relative standard deviations in the range of 2 to 5 percent are usually claimed. The sample solution (1 to 100 μl) is placed in or on the heating element. Voltage is then applied in increasing steps to evaporate the solvent, ash the sample, and finally to atomize the analyte. Quite high currents (up to 300 A) are used in the atomizing step to heat the device from around 500° C to as high as 2,500° C in a few seconds. The Woodriff furnace is exceptional in that it is maintained at a high temperature and the sample is inserted into the hot zone through a side arm.

A nonflame method that is similar in many respects to the carbon filament of West and coworkers is based on an electrically heated tantalum strip. [157] The reducing power of carbon is lost, but for many elements this does not appear to be a serious problem. For lead determination, the capabilities and limitations of the tantalum strip appear to be similar to the furnaces.

The furnaces have come into widespread use only recently, and it is not now possible to make a full assessment of their utility for lead determinations. A few general statements are possible. The most attractive feature of the furnaces is their sensitivity. Part-per-billion lead determinations should be possible on samples of only a few microliters; the ease and convenience are at least roughly comparable to flame AA work. However, many variables must be kept under control to avoid errors, and these will vary somewhat for different furnace types.

Background absorption caused by particles and/or molecules from the sample matrix is much more serious than with flames. Figure 2-2 shows background absorption due to sodium chloride in one of the furnaces. If the analyte and the matrix differ in volatility, this problem can be reduced by careful control of the heating rates, thereby causing analyte and matrix absorption signals to appear at different times at the output. Lead and most biological matrices are of roughly equal volatility. This makes such separations difficult if not impossible.

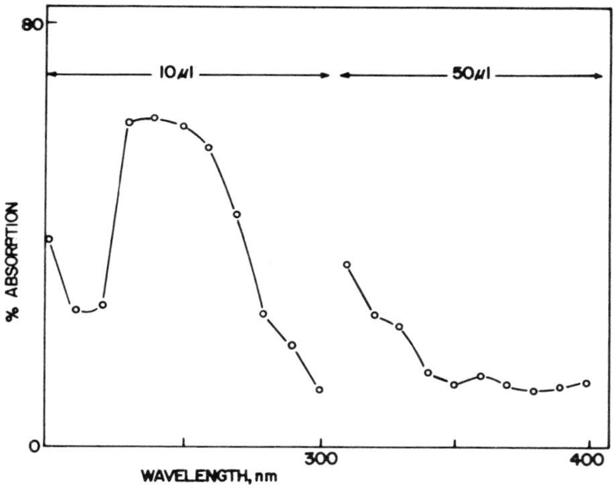

FIGURE 2-2.—Background absorption due to sodium chloride in nonflame AA.

Kahn and Manning [158] have pointed out that nonspecific absorption due to the simultaneous vaporization of the analytical species and sample matrix often causes the measurement of inordinately high analyte concentrations. Hwang et al. [159] found that the most significant problem in using the nonflame tantalum ribbon technique was background absorption during the drying and pyrolysis of the sample. This leads to an incomplete pyrolysis of the blood, and a "white smoke" is produced during atomization. The magnitude of the problem is illustrated for both lead absorption lines in figure 2-3. There is no doubt that background correction is essential and will provide more precise, as well as more accurate results.

Careful optical alignment of the light beams from the two sources is essential for efficient background correction, and the range of background absorption over which the particular instrument will make good corrections must not be exceeded. The analyst must be aware of the amount of background absorption present for each sample type. The efficiency of the background correction for lead can quite easily be tested by making measurements at a nonabsorbing line (280.2 or 287.3 nm) with the corrector on. Any signal that appears is indicative of error in the background correction.

The sample matrix also influences the size of the signal observed for a given amount of lead, sometimes to an extreme degree, as shown in table 2-5. The standard additions method may be used to correct the slope of the response curve if it is linear. At least one report [159] has implied by example that linearity is not essential. Examination of the analytical curves for the determination of lead in blood in this report indicates that the nonlinearity can cause positive errors of 10 to 40 percent through the use of the standard additions method. The fact that a transient signal is being observed dictates a short instrument time constant if linear response is to be expected. Baseline noise levels are higher, of course, with the short time constant. Matrix effects can be avoided by chemical separation of the analyte. Speed and convenience are sacrificed, and blank problems in the nanogram-to-picogram-working range are likely to be overwhelming.

The utility of the graphite furnaces lies in extending the range of atomic absorption to lower concentrations of lead and to very small samples. More work is needed on definition and control of matrix effects, a fact that dictates careful work and cautious interpretation on the part of the present users of these devices.

A different type of furnace based on induction heating has been described by Talmi and Morrison [160] and by Headridge and Smith. [161] Sensitivities are considerably poorer on an absolute basis than for the furnaces described above, but this fact is at least partially compensated for by the ability to handle larger samples. Induction heating has also been

TABLE 2-5.—Effect of matrix on working curve slope for determination of lead with a carbon rod analyzer

Matrix	Curve slope, (absorbance units/ng)
Water	0.40
Blood (1:5 dilution)	0.32
Bone digest	0.10
Synthetic bone ($Ca_3(PO_4)_2$)	0.04

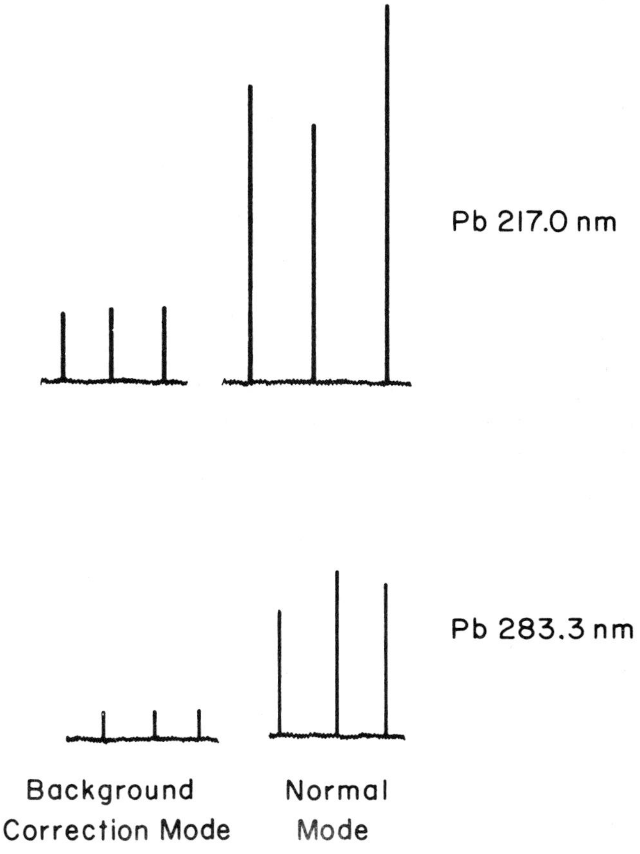

FIGURE 2-3.—Background correction mode and normal mode for two lead absorption lines.

applied for the continuous determination of lead in air by Loftin et al. [162] The air was passed over hot carbon in a radiofrequency (RF) field to decompose the lead compounds. The resulting atoms passed into a heated absorption tube in the light path. Sensitivities to a few tenths of a μ/m^3 were achieved.

In summary, AA spectroscopy is one of the more generally useful methods for the measurement of lead in samples of interest in environmental studies. Instruments are readily available, relatively inexpensive, and simple to operate. Direct flame aspiration gives very rapid analyses that are quite precise and accurate if appropriate precautions are observed. Sensitivity is easily improved by chemical enrichment, use of the sampling boat or cup, or with nonflame atomizers, although there is often a sacrifice in ease and convenience as well as precision and accuracy. The method is highly specific and suffers from few interferences, especially when flames are used. Disadvantages are that the sample must be in solution, and it is usually necessary to destroy organic matter prior to the determination. Background absorption can be troublesome when working at low concentrations and with nonflame atomizers. Perhaps the most severe problem in AA is a direct result of its many advantages. The method has become very popular and is being used by many people with little training in chemistry and none in spectroscopy. When problems do arise in the normally reliable method, they are usually subtle and are likely to go undetected by an analyst who lacks the proper background and training. The problem is intensified by the fact that major contributions to the development of AA have come from instrument manufacturers who, understandably, tend to emphasize capabilities and avoid dwelling on limitations.

Flame Emission and Fluorescence

Atoms formed in a flame, in addition to absorbing light as discussed in the previous section, may become thermally excited and emit characteristic wavelengths that can be measured, or they may emit radiation following excitation by light absorption. These latter two processes form the basis for flame atomic emission and flame atomic fluorescence, respectively. Both are very useful methods of operating with flames for certain elements. Lead is not among them, however. Flame emission works best for elements with resonance lines above about 400 nm; its sensitivity is lower by about a factor of 10 compared with AA for lead. [163] There appear to be no advantages to making measurements in emission that would compensate for the loss in sensitivity. Atomic fluorescence measurements for lead have been reported by several workers, [164-166] and the best detection limits are comparable with AA. The method has not become popular because of a lack of distinct advantages over AA and the fact that simple, inexpensive instruments designed for atomic fluorescence are not available commercially.

Flamelike plasmas that have very high excitation temperatures are being investigated as emission sources by several workers. [167-170] The most promising design for lead appears to be an induction coupled plasma [168] operated in argon at about 30 MHz. Detection limits somewhat superior to those reported for AA have been attained. The added cost and complexity would seem to make the gain of questionable practical value so long as a single element determination is being considered. Plasma emission is very promising as a multielement method, where the capability for lead determinations can add to its general attractiveness. More work is needed to fully assess the potential limitations of this promising source.

Spectrographic Methods

In spectrographic methods the sample is usually vaporized, decomposed, and the constituent elements excited by an electrical arc or a high voltage spark. The light is passed through a spectrometer or moderate dispersion (2 to 5 Å/mm), and the resulting spectrum is recorded photographically. For routine quantitative analysis the emulsion may be replaced with an array of exit slits located at the wavelengths of analyte lines. A photomultiplier tube behind each slit serves as the detector for that element and provides rapid measurements.

Historically, spectrographic methods have contributed greatly to our knowledge concerning trace elements, with the efforts of workers such as Ahrens [171] and Mitchell [172] being particularly noteworthy. Prior to about 1960 the emission spectrograph generally held the most prominent place among available methods in trace element laboratories. Since that time, developments in other areas, especially AA, have changed the relative importance of the spectrograph. In many laboratories, however, the change has been too large, and the unique capabilities of the method are being overlooked in favor of less desirable alternatives.

Many applications of the emission spectrograph to environmental samples attest to the value of the method and its general utility. A partial list includes the analysis of plant materials, [172, 173] tissue samples, [174] soils, [140] and air filters. [175] A complete list would contain hundreds of references. Sacks and Brewer [176] recently published a review containing 139 references devoted to the emission spectrographic analysis of particulate pollutants. Lead determinations occupy a prominent place in that review.

Spectrographic methods are placed at a rather serious disadvantage when they are considered with respect to a single element. The equipment is expensive; the methods are relatively slow when photographic detection is used; very careful standardization is required for quantitative work; and a high level of training and skill is required of the operator. These disadvantages are often offset by the capability to simultaneously measure up to about 60 elements, many at the $\mu g/g$ level, in a few milligrams of sample. For example, Tipton et al. [174] measured 27 elements in tissue ash using two exposures, and

Braverman et al. [177] determined 24 elements in air filters. In this discussion, lead analysis will be considered as only a part of the information one might be seeking from a sample.

For qualitative analysis, sample preparation is usually quite simple. The amount and the physical form in which the sample is introduced are not critical; it is usually sufficient to place a small amount of powder, chips, filings, or solution residue in an electrode cup and burn it in a DC arc. The photographic emulsion is exposed to dispersed light from the arc during sample vaporization, and the presence or absence of lines at wavelengths where the various elements are known to emit energy yields the qualitative information. The intensity of the lines can provide a rough estimate of the amounts of the various elements present even under these poorly controlled conditions. A single exposure can establish qualitatively the sample composition for most metals and some nonmetals much more quickly and easily than any other method. Thus one learns not only if a quantitative analysis for lead is worth running, but also what concentration range to expect and what other elements accompany lead in that particular sample. This information can be quite helpful in selecting the proper approach for the quantitative work. Detection limits for lead for typically a few μg/g in the solid, depending on the matrix.

A logical extension of qualitative spectrographic analysis is its use in survey type sampling programs in which a large number of samples must be examined for trends or anomolous values among the trace elements. The samples are burned under more carefully controlled conditions than for qualitative work, but the principles are very similar. Visual examination of the plates reveals differences larger than a factor of 2 to 3 in the concentration of most of the metals in the samples, again with speed, ease, and convenience seldom attained with other methods. Further refinements such as the addition of an internal standard, more careful control of source conditions, and measurement of line densities on the photoplate can lead to full quantitative determinations.

In the hands of a skilled operator, quantitative spectrographic analysis gives quite reliable results, but it probably should not be attempted by analysts without some specialized training in the area. Complete discussions of the method are available, [178, 179] and only a few basic principles and pitfalls are described here.

The light energy emitted in a given line under a given set of conditions is proportional to the amount of that element in the source. With photoelectric detection, the photocurrent, properly corrected for background emission, is a direct measure of the concentration. With photographic detection, the degree of blackening measured is not a simple function of light intensity causing the line, and calibration of the emulsion response to known relative changes in light intensity is required. This is a rather laborious task, but it need not be done frequently because the calibration is generally rather constant for a given lot number of emulsion, assuming proper storage. Computer programs have been written to aid in the calculations. [180] Photographic detection has the advantage that all wavelengths within the range covered may be examined rather than only a few that were preselected. Problem samples are much more readily handled with the added information that is available.

Line intensities are affected by the exact conditions in the source which include the physical as well as the chemical makeup of the sample. Internal standardization is necessary because of the inherent instability of the sources. An element which is chemically and spectrographically similar to the analytes is added to samples and standards in constant amount, and the ratio of the intensity of the analyte line to that of an internal standard line is used in the analysis.

Powdered samples such as plant ash, soils, or solution residues are often mixed with graphite which contains the internal standard and may also contain a "radiation buffer" to reduce sample matrix effects. The resulting mixture is carefully loaded into a cupped graphite electrode and burned in a DC arc of 8 to 20-A current. The central arc column is most frequently viewed, although some workers prefer to view only in the vicinity of the cathode. [172]

Solutions can be handled by high voltage spark excitation of a residue after evaporation, [181] from the wetted surface of a rotating disk electrode, [182] or from the bottom of a porous cupshaped electrode. [183] Recent work by Gordon [184-186] and by Hambidge [187] points out advantages of arc excitation of solution residues in an inert atmosphere.

The preparation of standards for spectrographic analysis presents some special demands because the light intensity depends on the exact physical and chemical conditions in the source. Matching the physical properties of samples and standards is generally more troublesome than matching the chemical properties. If the samples are dissolved at any stage in the procedure, standards may be started at that point and carried through the remaining steps without physical matching problems. Secondary standards prepared by careful, independent analysis of a few actual samples of the same type that are to be run spectrographically can also serve for calibration purposes. Standards for arc excitation may also be prepared by mixing (as dry powder) the trace elements to be determined with a synthetic matrix prepared to match the major constituent composition of the sample. Dilution and mixing are continued in steps until the desired concentration range is reached. The latter method is quite popular, but it is far from being free of problems. The particle size for the trace components must be quite small if inhomogeneity problems are to be avoided at low concentrations, and matrix contamination can limit the low end of the calibration range. Perhaps the most important point is that true physical matching of samples and standards is impossible. The analytical components of the samples are generally dispersed on the molecular scale, whereas those in the standards are dispersed as discrete particles. Good results are often obtained with the "dry grinding" method of standard preparation. Errors are encountered often enough, however, to warrant more than the usual degree of suspicion until the results are verified.

Advantages of spectrographic analysis are primarily the ease and reliability of qualitative analysis, the ability to easily do multielement surveys, and the inherent multielement capability for quantitative work. If photoelectric detection is used, speed can be another important advantage. Equipment costs vary considerably, depending upon their sophistication, but about $40,000 to $100,000 would be needed for a complete laboratory, and personnel must be adequately trained. The precision of spectrographic measurements can be as good as 1 percent in very favorable cases, but these are seldom reached with environmental samples; for arc excitation, ±10 percent is more common. Accuracy is controlled largely by the quality of the standards and skill of the operator. Spectrographic methods become more attractive as the amount of information needed from the sample increases, and as the number of samples of a given type goes up. The effort invested in standardization is then spread over more samples.

X-Ray Methods

Some of the comments used to introduce emission spectroscopy are equally applicable here. X-ray methods lend themselves to multielement analyses. They are readily automated, and, with important exceptions, they appear most attractive in large installations with heavy sample loads. Rather extensive specialized training is needed for an analyst in this area. Important principles and pitfalls will again be emphasized, and potential users are urged to consult more complete works. [188-191]

Absorption: Lead absorbs X-rays very efficiently, and in certain matrices, gasoline for example, this absorption is readily used for analysis. [190] In more complex samples with low lead concentrations, X-ray fluorescence methods are much more useful.

Fluorescence: Characteristic X-ray emission arises when an ion is formed by ejection of an inner electron from an atom by interaction with a high-energy photon or charged particle. An electron from an outer shell then falls into the vacancy, giving up the energy difference in a single photon. Energy differences between shells are a function of Z, the atomic number, and the energy or wavelength of the photon is therefore characteristic of the element from which it came. For each element , X-ray spectra consist of only a few lines that fall into definite series. The K series arises from ejection of an electron from the innermost shell (principal quantum number = 1) and with moderate dispersion appears as two lines, K_β and K_α, depending upon whether the electron that fills the vacancy comes from the second or third shell. Higher dispersion reveals two components in the K_α line corresponding to either a *2s* or *2p* electron filling the vacancy. A corresponding L series of lower energy and lower intensity arises when the primary energy ejects an electron from the second shell. For a given series, wavelengths may be predicted reasonably well by the formula:

$$Z = K\sqrt{1/\lambda}$$

where Z is the atomic number, K is a constant for a given series, and λ is the wavelength of the X-ray.

X-ray fluorescence methods may be classified according to the energy or wavelength selection method (wavelength dispersive and energy dispersive) and by the source of primary excitation energy (X-ray tube, isotope sources, charged particles). Wavelength dispersive systems use a crystal monochromator for energy selection and require the high intensity of an X-ray tube source to offset the loss of fluorescent intensity in the monochromator. X-ray monochromators are based on the Bragg equation:

$$n\lambda = 2d\sin\theta$$

where n is a small integer, λ the wavelength, d the spacing between adjacent planes in the crystal, and θ the angle between the collimated beam and the crystal. By using a crystal with known d spacing, θ can be selected to give the desired wavelength, with energy or wavelength resolution roughly 10 times better than with energy dispersion systems. Detectors with some inherent energy discrimination, such as proportional or scintillation counters, are commonly used to avoid order overlap problems with crystal monochromators.

Energy dispersive X-ray fluorescent systems are based on solid state lithium-drifted silicon or lithium-drifted germanium detectors. Si(Li) is preferred for work at low energies and Ge(Li) at higher energies. In favorable cases, energy resolution of about 150 to 200 eV can be attained with these detectors when used with a suitable multichannel analyzer. The K lines of adjacent elements are resolved except at low Z, but problems can be encountered due to the inability of these detectors to resolve a K_α line of one element from the $K\beta$ line of the next lower atomic number element. [191] The primary advantages of the solid state detectors are that a range of energies (several elements) can be measured at a single setting, and that the intensity losses inherent in an X-ray monochromator are avoided. Much lower intensity primary sources such as radioisotopic sources and accelerated charged particles then become practical. Poorer resolution, limited counting rates, and the need to maintain the detector at liquid nitrogen temperature at all times are the primary disadvantages.

X-ray spectroscopy with isotopic sources is discussed in detail by Kneip and Laurer. [192] Isotopic sources may be selected that have X-rays of sufficient energy to excite K lines of heavy elements. Background problems are reduced by the monoenergetic nature of the source, and portable X-ray instruments become practical. The choice of energies is limited to some extent by the availability of isotopes with suitable half-lives, and safety considerations limit the source intensity. Background intensities can be reduced by careful control of the geometry of the source, sample, and detector. Nondispersive instruments are available for X-ray fluorescence measurements. The use of such instruments for the analysis of environmental samples is very limited at present.

High-energy protons from a Van de Graaff accelerator or cyclotron produce X-ray spectra with very low background, if the sample and its support are sufficiently thin. Solution residues on carbon foil or plastic film supports fulfill this need. Detection limits better than for X-ray excitation by about an order of magnitude are reported by some authors, [193] and measurements in the sub-part-per-billion range have been reported for water samples. [194, 195] Cooper [196] reports, however, that proton excitation does not give detection limits superior to photon excitation. The fact that access to an accelerator is required severely limits the development and common use of charged particle excitation. The special thin substrates required are likely to be inconvenient for large-scale routine use.

As applied specifically to lead, X-ray fluorescence methods of the past have often lacked the sensitivity required in most environmental studies because of the lower inherent sensitivity of the *L* spectrum. The energy required to excite the *K* lines of elements above $Z \simeq 50$ exceeds that normally available in photons from X-ray tubes. The methods may be quite attractive when higher concentrations are expected, such as for contaminated air filters. [197] Background intensities generally limit the sensitivity attainable, and the improvements brought about by careful optimization of geometry with isotopic sources [44] make the methods look attractive for lower concentrations.

Advantages of X-ray fluorescence include inherent multielement capability, broad elemental coverage, similar sensitivity for many elements, and the ease with which the methods may be automated. The fact that the sample need not be in solution is often an important advantage. In special cases additional features are evident. The portability of some instruments that use isotope excitation is an example.

X-ray methods generally require rather sophisticated and expensive instrumentation, although special purpose instruments for as little as $5,000 may be available. [198] Expense and inconvenience are incurred with solid state detectors that must be maintained at liquid nitrogen temperatures. Matrix effects on X-ray fluorescent intensities are generally quite complex although more predictable than for emission spectroscopy. Even with thin samples such as an air filter, particle size has a significant effect. [191] The existence of such effects and methods of correcting for them have been discussed. [192]

Mass Spectrometric Determination of Lead

Three general approaches have been used for the mass spectrometric determination of trace elements. These include: volatilization of the trace elements as metal chelates or coordination compounds into an "organic" mass spectrometer, direct and indirect analyses with spark source (or inorganic) mass spectrometers, and isotope dilution analyses with stable isotopes. While these techniques generally offer high levels of sensitivity, the equipment required is more complex than that associated with the more conventional analytical techniques. Consequently, mass spectrometric methods are primarily applicable to specialized problems or determinations not readily solved by other analytical techniques, such as those requiring unusually high analytical sensitivity. Because such problems are encountered in the determination of environmental lead, the three techniques are described briefly below.

Metal Chelate Mass Spectrometry: The ideal metal chelate for mass spectrometric analysis must be relatively nonvolatile at ambient temperatures and at pressures ranging down to 10^{-7} torr. At the same time, it must be readily vaporized without appreciable decomposition at temperatures that can be achieved in the source compartment of an organic mass spectrometer, i.e., below approximately 500° C. Additionally, the metal chelate should be easily prepared by simple techniques by using reagents that are sufficiently pure to avoid blank problems. Given these properties, the mass spectra of the metal chelates may be recorded and used for both qualitative and quantitative determinations on several elements. The reports by Isenhour et al. [199-202] provide convenient access to the literature on this subject and include particularly useful approaches for improving measurement precision.

The only report that deals specifically with the determination of lead at trace to ultratrace levels is that by Wood and Skogerboe. [203] Lead was extracted from acid solutions as the diethyl dithiocarbamate into methyl isobutyl ketone. Aliquots of the extract were vaporized into an electron bombardment spectrometer from a direct insertion probe to obtain the mass spectra. Detection limits were at the nanogram level, but reagent blank and instrument contamination problems were reported. [203] While the technique clearly shows analytical promise, more work needs to be done to refine its utilization for trace analysis.

Solids Mass Spectroscopy: In terms of overall analytical sensitivity, few techniques can compete with spark source mass spectrometry. Nearly every element can be determined simultaneously at the ng/g concentration level on samples limited in size. While the fact that only a small amount of sample is consumed is a distinct advantage in many instances, it is accompanied by an inherent problem. That is, the samples must be homogeneous if precise (and accurate) results are to be obtained. [204] Papers by Morrison et al. [205-207] may be used to generalize the applicability of solid mass spectrometry to environmental analyses. In essence, the technique is capable of providing comprehensive survey data for a large number of elements in any single sample. For complex matrices, e.g., biological tissues and soils, organic matter must be removed by ashing to avoid spectral interferences; even then, interferences may originate from oxide spectra. Through the use of careful sample preparation methods and instrumental improvements, concentrations can be determined with precision and accuracy in the 10 to 30 percent range. [205-207] The time required for a sample analysis depends on the number of elements to be determined, as well as on the sample type. Determination of a single element typically requires at least 1h/sample. Thus the use of the technique for the large-scale analysis programs common to environmental research is likely to be justifiable only when the concentrations involved are extremely low or when multielement analyses are required.

Isotope Dilution Mass Spectrometry: Because isotope ratios can be measured with high precision and accuracy by mass spectrometry, isotope dilution methods have played an important role in solving environmental analysis problems. Such methods have been used extensively with spectrometers employing thermal sources for solids and electron impact sources for gases [208] and metal chelates. [201] Their use with spark source mass spectrometers dates to 1965. [209] For each element to be determined, a known amount of spike (tracer), with an isotopic composition different from that of the element itself, is mixed with the sample. The changed isotopic ratios are then measured, and the concentration of the element in the original sample is calculated. [208-212] The method is applicable to lead. In addition to the improved sensitivity and accuracy associated with the technique, relative freedom from interferences may also be realized. For these reasons, the technique should enjoy broad usage in environmental research in spite of a rather high cost per analysis.

ELECTROCHEMICAL METHODS

Prior to the advent of AA, electrochemical methods were prime competitors with the dithizone method for determining lead at low concentrations. Lead is perhaps an ideal element for many electrochemical determinations as its aqueous complexes are moderately stable and labile. Notable exceptions are the complexes formed with hydroxyacids such as citrate and tartrate.

Most commonly, the electrochemistry of lead is that of lead (II) and metallic lead. The plumbous ion behaves reversibly in acidic solutions and has a reduction potential near -0.4 V versus the standard calomel electrode (SCE). [213-215] The fact that this potential lies in the middle of those for the common electrochemical elements, coupled with the tendency of lead to show only moderate reactions with common complexing agents, limits the ability to selectively isolate lead by electrochemical means. Consequently, the less selective methods such as controlled potential electrolysis and direct potentiometry have proven useful for the determination of lead only in limited, specialized cases. One classical method for the determination of lead has been the electrodeposition of PbO_2. The present discussion considers only those techniques judged most consistent with the general requirements associated with environmental analyses.

Polarography

Conventional polarography is generally suitable for the analysis of lead at concentrations down to 2 μg/g. The extensive literature on the method extends back 50 years; a solid foundation of theoretical knowledge about the fundamental processes exists, and commercial instrumentation is readily available. The measurement step requires 10 to 20 min. In neutral or alkaline solutions, lead interacts with oxygen. This results in violation of the current additivity requirement; therefore analytical solutions must be deaerated. Surface active maximum suppressors are occasionally required to prevent erratic behavior at the dropping mercury electrode (DME). If so, the concentrations of the suppressors must be controlled to obtain precise and accurate results. Unfortunately, the organic constituents in many environmental samples act as surface active suppressors. [216] Experience at the University of Illinois [55] has shown that the presence of such materials in natural waters often causes intolerable errors, particularly at low concentration levels. These substances can usually be removed by passage through activated charcoal, but may result in an increased lead blank from the charcoal. There is always the risk that some lead-containing species will be absorbed.

Classical polarography is based on measuring the net current flow through the DME throughout the drop life. The current is composed of a capacitative (non-Faradaic) current proportional to the rate of change electrode area over time of the ($t^{-1/3}$), plus a diffusion-controlled Faradaic current which increases with time according to the rate of diffusion to a growing sphere ($t^{1/6}$). The capacitative current serves as a fundamental limit on the sensitivity of classical polarography. The pulse polarographic method largely overcomes this limitation. [217]

Pulse Polarography

The pulse techniques circumvent the above mentioned difficulties by time-resolving the Faradaic and non-Faradaic currents. The potential is applied only during a short selected interval in the drop life of a DME or at a solid (stationary) electrode. Further, the current is measured within a narrow time increment of this interval. The location and duration of the measurement time increment during the pulse interval are a compromise between the time delay necessary to allow the capacitative current to fall to a negligible value and the appropriate time for measuring the more slowly decaying Faradaic component. Since the drop area is essentially constant during the pulse, the analytical current changes at $t^{-1/2}$. The measurement may be made routinely with commercially available equipment. [218] Detection limits as low as $10^{-8}M$ can be obtained but this capability depends on many factors related to both the samples and the instrumentation.

Pulse polarography operated in the most common incremental potential step mode has the advantage of enhanced resolution. Damaskin [217] defines resolution as a measure of the ability of a method to determine small concentrations of a substance in the presence of one or more substances that are more readily reducible and therefore yield currents in the potential region of interest. In DC polarography, this condition requires suppression or compensation of the large currents arising from the interfering substances and amplification of the net remaining (analytical) signal. As Lingane points out, [214] such methods of compensation have definite limits. The pulse method can tolerate larger concentrations of interferences because the initial potential for each pulse can be selected so that the interfering substances can be constantly reduced. In essence, pulse methods often permit pre-electrolysis elimination of interfering substances. Damaskin calculates that the resolution is more than 20 times that of classical polarography. [217]

By appropriate choice of timing and pulse circuits, the method is capable of producing high signal-to-noise ratios that permit extraordinary amplification. Polarograms with peak currents of a few nanoamperes can be routinely obtained with commercial equipment. [218] Clearly, this method will find extensive use in environmental applications.

Anodic Stripping Voltammetry

Anodic stripping voltammetry (ASV) is a technique in which a suitable analyte solution, normally aqueous and containing either adventitious or deliberately added electrolyte, is electrolyzed under fixed conditions of solution stirring rate, electrode area, and time at a controlled DC voltage. At the end of the pre-electrolysis (plating) period, the metals reducible at the selected potential and soluble in mercury are sequentially stripped from the electrode by application of a linearly time-dependent sweep potential superimposed on the electrolysis potential. The linear time function of the stripping potential, coupled with the finite amount of material available to be reoxidized, results in data that appear in the form of peaks in the order of ease of reoxidation. [219]

ASV is quite sensitive. Current practice with commercially available equipment allows determinations at the 1-ng/g level with routine 5 to 10 percent relative precision. Extension to lower, e.g., 0.1-ng/g levels, is readily attainable with the same instrumentation, requiring only that blanks and contamination be reduced to acceptable values. These levels are well within present average values for such metals in natural water samples. [220]

Sensitivity in ASV primarily results from the large preconcentration obtained during the plating step. With a mercury-coated graphite electrode that has an effective surface area of 5 cm^2, a 5,000-Å film thickness, and a typical solution volume of 10 ml, an exhaustive electrolysis of a 10-ng/g solution will produce an amalgam concentration in the mercury film of some 30 μg/g, for a net preconcentration of approximately 3,000. The current measured during the stripping process is from the preconcentrated analytical species. For this reason, the instrumentation necessary to control and record these processes is relatively simple. The development of reliable methods for preparing the mercury-coated graphite electrodes has permitted an increase in sensitivity over that obtained with the hanging mercury drop electrode (HMDE). [221] A typical HMDE has surface area of 1 to 2 mm^2, while thin film graphite electrodes may have areas in excess of 5 cm^2. [222] In theory, this difference would increase the anodic stripping current by a factor of about 250; about half of this expected increase is realized in practice.

Two prevalent modes of ASV operation are currently in common use. Matson et al. [222] have utilized exhaustive electrolysis to deposit more than 95 percent of the total analyte available in the mercury film. While this typically requires plating times of 30 min. or more, it can reduce the effects of small experimental and analytical solution variations on the pre-electrolysis. With four separate pre-electrolysis cells, it is possible to analyze 8 samples/h. [55] Frequent calibration checks of the systems are necessary, however, so that a resulting output of 20 to 30 samples per man-day is most typical. [55] The alternate mode of operation is to shorten the pre-electrolysis time under closely controlled experimental conditions. A 2-min. pre-electrolysis mode can be used for determinations at lead levels as low as 0.1 ng/g. [55] When duplicate standard addition calibration measurements are made on every sixth sample, more than 60 samples per man-day can be analyzed by this approach.

Several factors must be taken into account in the utilization of ASV. Copeland et al. [223] have shown that the supporting electrolyte concentration influences both the peak current and the potential at which lead is stripped from the mercury thin film. These effects are due to uncompensated resistance when more resistive solutions are analyzed. [223] Because of the strong dependence of the analytical response on the resistance of the test solution, large errors may result unless sufficient supporting electrolyte is added to swamp the effects of the variable amount of electrolyte naturally present in a sample. While the effect may not be important in the analysis of industrial or sea waters, it is clearly significant when relatively pure waters are to be analyzed.

Rotation of the thin film electrode increases the transport to the electrode during the plating step and strikingly decreases the plating time required. [224] Also, wax impregnation is not required in the preparation of mercury thin film electrodes if the graphite substrate has high density and low porosity. [224] These observations can simplify and improve the use of ASV. A further development of this technique has been the use of a subtractive system to compensate for the deleterious effects of the capacitative current.

Differential Anodic Stripping Voltammetry

The differential modification of the normal approach to ASV was developed by Zirino and Healy. [225] It depends on the fact that hanging mercury drop electrodes can be made such that two electrodes have equivalent characteristics. In use, two electrodes with the same area are placed in the cell, but the sample is electrolyzed only at the test electrode. During the stripping process, the signal from the second electrode is primarily due to the charging effect, and its output can thus be subtracted from that of the test electrode at the summing point of the cell current amplifier. Although a simple amplifier gain circuit must be inserted to adjust (match) the residual outputs of the two electrodes in blank solutions, the approach has proven to be a simple means of compensating for the capacitative current effect on the analytical signal. Clearly, this is better than using current compensation to bias out the effect because it does not rely on the assumption that the residual current varies linearly throughout the potential scan. Martin and Shain [221] developed a similar differential (subtractive) system based on the use of two totally separate electrochemical cells. Differential operation of two such cells is distinctly more complex.

The application of this general approach to thin film electrodes is more difficult because of the electrode uniformity requirement. By fabricating larger graphite electrodes, it appears that uniformity can be obtained for this purpose. [55] The application of the general approach associated with pulse polarography appears to be a more practical alternative.

Differential Pulsed Voltammetric Stripping

Linear scan ASV measures the sum of the Faradaic and charging currents. By superimposing square wave voltage pulses on the linear scan, two advantages can be realized. The charging current can be discriminated against in the same way as in pulse polarography, and greater sensitivity can be realized. Further, an advantage in pulse stripping analysis resides in its repetitive nature. Some of the material stripped from the electrode during a voltage pulse can be redeposited in the thin film during the waiting time between pulses. Analytical signals may consequently be measured several times from the same material, whereas the material is measured only once via the linear scan technique. All of these advantages of pulsed stripping from thin mercury films have been clearly demonstrated, [223, 224] and a quantitative theory explaining the phenomena involved has been developed. [226] The techniques described have been in use at Colorado State University [224] for several months with excellent results. The sensitivity improvements that accrue from the use of this approach are such that most environmental samples can be analyzed by using a pre-electrolysis time of 2 to 5 min. Consequently, 40 to 60 samples can be analyzed per man-day, even when dual standard addition calibrations are carried out on each sample.

Alternating Current Polarography

The Faradaic and non-Faradaic components of current can also be separated by AC polarography and its relative, square-wave polarography. Both alternating potential methods depend on the fact that, for small amplitude excursions, electrochemical processes are reversible, i.e., a current will flow through the electrode in either direction depending upon the direction of the net applied potential. By filtering the total DC plus AC signal to remove the DC portion, the remaining AC current has a known frequency arising from the frequency of the small applied AC perturbation. Frequency-tuned amplifiers can thus be used for large scale amplification of the AC signal. Phase-sensitive detectors, taking advantage of the fact that the Faradaic admittance has a different phase angle from the capacitative admittance, have been used to decrease the charging current component and thereby increase the sensitivity.

Square Wave Polarography

The square-wave analog of AC polarography was pioneered by Barker and Gardner [227] and has proven to have a potential for low detection limits. During each half cycle of a relatively low square-wave frequency, the exponential charging current is allowed to decay to a relatively small magnitude. For the last portion of each cycle, a low-amplitude, high-frequency square-wave is superimposed on this. The high-frequency square-wave current output is rectified and amplified with a frequency-sensitive amplifier.

Buchanan, et al. [228, 229] have used this approach to assay natural water samples for lead in the 2 to 10 ng/ml range. When HCl was used as the supporting electrolyte, a rather large blank was observed. Jennings [230] also reported a shoulder on the cathodic side of the lead peak that may have been due to a change in the double layer capacitance. The blank problem was solved by using a $NaClO_4$-NaF supporting electrolyte that had been purified by exhaustive electrolysis to reduce the blank to 4 ng/ml. [230] Buchanan and McCarten [228] also reported that nitrogen sparging for 5 min. was essential to obtain a stable, reproducible measurement. Such behavior is probably specific to the samples under test, but it illustrates a difficulty with AC electrochemical methods, i.e., the measurements are sensitive to the presence of surface-active substances. It has been reported, [55, 220] for example, that satisfactory measurements cannot be made on some water samples unless they are taken to dryness with $HClO_4$-HNO_3 to destroy organic matter. The percentage of natural water samples that may require this treatment was not stated.

Cathodic Stripping Voltammetry

A unique electrochemical method is cathodic stripping applied to lead in aqueous solution. The method has been previously used only for anion assay by anodic oxidation at a suitable metal electrode and for the oxidation of $Fe(Pb^{2+})$ to $Fe(OH)_3$.

In the method developed by Watkins, [231] (Pb^{2+}) is oxidized at about +1.2 volts versus Ag/AgCl reference electrode for a fixed time and subsequently is cathodically stripped. The deposition reaction is that of the classical electrogravimetric method for lead: $Pb^{2+} + 2H_2O \rightarrow PbO_2 + 4H^+ + 2e^-$. The sensitivity obtained is greatly enhanced over that of the gravimetric method, due to the use of the coulometric measurement during the stripping process. The electrode is conductive SnO_2 deposited on glass. The use of SnO_2 eliminates the blank because the oxide film is reduced during the cathodic cycle.

At present, the method is not competitive with other methods for measuring lead in natural samples for two primary reasons. The stripping process results in a normally shaped initial peak followed by a rather drawn out satellite peak. The integrated sum of both peaks agrees most closely with the theoretical amount of material deposited. The second peak is due to the stripping of lead ion that has been specifically adsorbed on the SnO_2 surface. This material cannot be completely removed with each measurement, leading to an electrode memory; however, overnight storage eliminates the problem. Chloride ions must be minimized to avoid the generation of Cl_2 or reduction of the PbO_2. The phosphate ion forms a basic $Pb^{(4+)}$ phosphate which does not affect the coulometric result, but causes a broadening of the two peaks. To obtain the best results, a current integration method must be used. [233]

Applications of Electrochemical Methods

ASV, pulse anodic stripping voltammetry, square-wave polarography, and pulse polarography have been the most widely used electrochemical techniques for analyses of environmental samples. Although the method of sample preparation may be somewhat unique to each laboratory, the predominant method used was set ashing with nitric-perchloric or nitric-perchloric-sulfuric acid mixtures. A summary of electrochemical applications is given in table 2-6.

TABLE 2-6.–Applications of electrochemical methods

Sample type	Methods[1]	References
Natural waters	ASV	[55]
	PASV	[223]
	SW	[228]
	PP	[232]
Urine	ASV	[55, 223]
	PASV	[223, 224]
Blood	ASV	[55, 223]
	PASV	[223]
Plasma	ASV	[223]
	PASV	[223]
Bone	ASV	[55]
Teeth	ASV	[55]
Biological tissues	ASV	[55]
Air particulates	ASV	[55]

[1] ASV: Anodic stripping voltammetry.
PASV: Pulsed anodic stripping voltammetry.
SW: Square-wave polarography.
PP: Pulse polarography.

NUCLEAR METHODS FOR THE DETERMINATION OF LEAD

The nuclear properties of lead are often used in tracer studies with radioactive ^{210}Pb, but the determination of the element in environmental samples by nuclear methods is difficult. Thermal neutron activation analysis, which gives very low limits of detection for many elements, [234] does relatively poorly for lead because of a combination of low absorption cross sections, short half-lives of activation products, low natural abundances of some isotopes, and the fact that the one isotope produced in reasonably high yield is a pure β^- emitter. The use of fast neutrons has been described, [235] but again the results are anything but encouraging when compared with other methods.

Recently, two different approaches have been used with some success. In the first, ^{207m}Pb is formed by three separate reactions: ^{206}Pb (n, γ) ^{207m}Pb, ^{207}Pb (n,n') ^{207m}Pb, and ^{208}Pb $(n,2n)$ ^{207m}Pb. [236] The gamma energy for the decay of ^{207m}Pb is a favorable 0.57 MeV, but the half-life is only 0.8 sec. Thus very special provision for accurate timing and rapid transfer of the sample to the counter are essential, and chemical separations to remove any interferences are impossible. Because of the very short half-life, this work is most advantageously done in a reactor that can be pulsed at high power levels for short times. Without pulsing, the detection limit is about 4 μg of Pb and with pulsing about 0.4 μg. In spite of the formidable disadvantages and the unimpressive detection limits, there is some current interest in lead determination by this method. [237, 238]

High-energy photons can also induce useful nuclear reactions. A beam of electrons produced by a suitable accelerator strikes a tungsten target, and the bremsstrahlung photons produced by the interaction of the electrons with the target are sufficiently energetic to induce nuclear reactions. The general utility of photon activation analysis has been reviewed by Lutz. [239] Several workers have described its specific application to the determination of lead. [240-242] Recently, Aras et al. [243] described the determination of 14 elements, including lead, on air filters by this method.

The nuclear reaction of interest is ^{204}Pb (γ, n) ^{203}Pb. The half-life of ^{203}Pb is 52 h, and a 0.26 MeV-gamma ray is the primary decay mode. The activity can either be measured on a Ge(Li) detector or, if necessary, the lead activity may be separated and counted on a NaI(Tl) detector. Lutz [240] estimates a detection limit of about 0.5 μg/g by direct instrumental measurement or about 10 ng following a rigorous chemical separation. The detection limit for lead in air is estimated to be 12 ng/m^3, compared with 1 ng/m^3 for flame detection. [243]

Considering the performance of nuclear methods for the determination of lead and the extremely specialized and expensive equipment that is required, it is very difficult to understand why so much work continues in this area. In special installations such as the National Bureau of Standards (NBS) where independent checks on other methods are of paramount importance, the effort is justifiable. However, for general monitoring work, several methods have been described here which, properly used, give superior lead results with equipment that cost orders of magnitude less to build and operate.

TRACING THE ORIGINS OF LEAD IN THE ENVIRONMENT

Contaminants introduced into the environment by man's activities often must be traced in order to assess their impact on the ecosystem. This problem is usually complicated several factors that the concentrations typically undergo diminution during transport, there may be several sources of any single contaminant, and physiochemical changes may occur during movement. Tracing lead from automotive sources, for example, is a complex process due to all of these factors. Consequently, tracing cannot usually be based solely on concentration measurements, but must rely on more specific approaches to measurement and data interpretation. The approaches applicable to lead are discussed in the following sections.

Concentration Ratio Methods

It has been pointed out that the ratios of concentrations of elements in samples can be used more readily to determine generic relationships than the concentrations themselves. [17, 18, 20-23] Tracing methods with concentration ratios are based on the expectation that the ratios will be maintained, while the concentrations themselves undergo drastic changes. Thus they may be used to trace the input of a particular source of contaminants under at least limited circumstances. Clearly, if there are two or more discrete (dissimilar) sources of a particular set of contaminants, the concentration ratios will be changed when the outputs from these sources are mixed. Still, the concentration ratios may often be used to estimate the relative contributions of each source at any particular point of interest. [23] The ability to obtain such estimates obviously depends on the number of sources, the flux of contaminants from each, and the general nature of the system being modeled. Anders [23] has presented some interesting examples of the use of concentration ratio matching in conjunction with a statistical method of interpretation. The correlations developed serve well as indications of the existing generic relationships between environmental samples. [23]

Direct correlation methods are analogous to concentration ratio methods and may be used for tracing purposes. For example, in measuring atmospheric lead, it is appropriate to consider how much of the lead is directly attributable to traffic and how much is due to airborne soil that may contain nominally 10 to 50 μg/g lead. To obtain a general indication of this, it may be assumed that airborne soil will be the primary source of particulate calcium and magnesium collected on air filters, since both are present in soil at relatively high concentrations. Seeley and Skogerboe [175] carried out a correlation analysis on the Ca and Mg concentrations collected on a series of air filters and obtained a correlation coefficient of 0.83. This generally verified the above mentioned assumption. When the lead concentrations were correlated with the parallel calcium concentrations, the correlation coefficient was 0.56. This indicates that the lead contribution from soil was not particularly significant.

Two approaches to differentiating between the natural lead in soil and that originating from automitive sources have been discussed by Seeley et al. [140] Two questions were addressed in examining the atmospheric transport of lead from a highway to the soil. First, at what distance from the highway does significant contamination of soil with airborne lead cease? Second, at what depth does the contamination become insignificant in comparison with the natural lead level of the soil? A direct statistical analysis of lead concentrations determined at several soil depths along a transect to the highway estimated the natural lead content of the soil and its natural variation. Objective statistical tests of these results indicated that the top 6 in. of soil had been contaminated by traffic at least out to 300 ft from the highway. At distances of 100 ft or less, the soil was contaminated to depths of 18 in. [140]

Another means of verifying these results was based on a concentration ratio measurement that used titanium as the "indicator" element. Titanium was selected because it is not used on a wide-spread industrial basis and, where it is used, its release into the environment is probably minimal in comparison with other elements. It is widely distributed in silicate materials at an average terrestial abundance level of 4,400 ppm. 171] Consequently, industrial contamination of the soil with titanium would have to be extensive to cause a significant change (e.g., 5 percent) in its natural concentration level. In effect, the titanium concentration of soil should be relatively constant at any particular location, and the ratio of lead to titanium should also be constant for uncontaminated soils.

Correlation analyses of the Pb/Ti concentration ratio versus the actual lead concentration in each soil sample from a particular location were used to determine the native lead concentration of the soils. The general approach is conceptually illustrated in figure 2-4. If contamination is evident, the approach indicates a region where linear correlation is apparent and another where the points are scattered in an approximately circular pattern. To estimate the native lead concentration, an iterative correlation analysis may be carried out with the apparently linear data. For example, all data above point A (fig. 2-4) may be used in the initial determination of the correlation coefficient. Subsequently, all data above the point B and then above C would be added in the correlation coefficient calculation. When the random scatter due to the natural variations in the native lead is included in the data used for correlation, the value of the correlation coefficient decreases. Thus the iterative correlation analysis detects those points that are randomly and independently distributed, as would be characteristic of the native soil lead. The average lead concentration computed for those samples defines the natural lead concentration, and the standard deviation of those results defines the natural variation characteristic of the soil samples in question. The results obtained by this approach have been extremely useful in differentiating between natural soil lead and contamination levels. [140]

These examples indicate that concentration and concentration ratio methods offer potential for tracing lead in the environment. A primary requisite of such tracing approaches is that objective methods of data interpretation must be used. Certainly, such methods will find further use in environmental tracing studies.

Isotope Ratio Methods

Several studies have clearly shown that lead concentrations in soil and air are enhanced close to a source of lead. [140, 224-247] As indicated above,

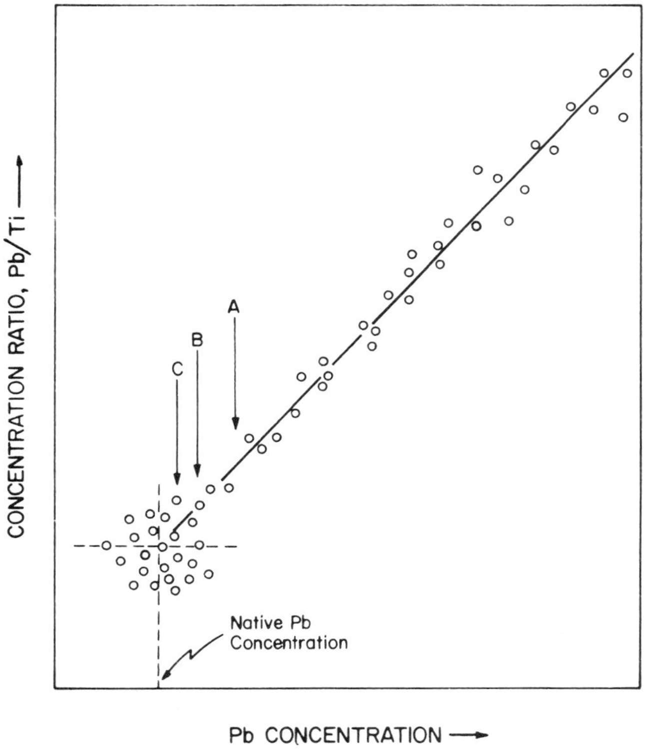

FIGURE 2-4.—Concentration ratio method of tracing lead in the environment.

methods for tracing lead are needed: to define the input from a partucular source, to locate the source, and to differentiate between natural and contamination levels of lead. The problem is complicated by a lack of knowledge of background levels. These are determined by complicated functions of geography, geology, and climate, not only of the local area but over distances of several hundred miles. Sand from the Sahara has been found in air particulate collections in the Bahamas, for example. [248] Considering the diversity of man's activities and the possibility of long-range atmospheric transport for lead, it is obvious that the methods used for tracing the origins of lead must be highly specific. Tracing methods based on isotopic ratio methods appear to offer considerable potential for the solution of such problems.

The domestic and imported lead ores used in the United States come from relatively few mining districts; 93 percent of the U.S. production for 1968 came from the Missouri, Idaho, Utah, and Colorado districts. [249, 250] The isotopic compositions of lead mineral deposits have been thoroughly characterized by several laboratories and have been found to show extensive variations between deposits. [250-253] As a consequence, the source of lead pollutants should be traceable by measuring their isotopic compositions.

Chow [244, 250] has shown a distinct correlation between the lead isotope ratios found in gasolines sold in certain cities and atmospheric particulates or surface soils collected from the same area. On the basis of these studies, Chow has concluded that industrial lead pollutants in the areas studied originate principally from automotive sources. [250]

Isotopic tracer methods have been used by Ault et al. [254] to determine contamination in a number of environmental systems. Their results indicated that topsoil, air, leaves, twigs, and grass along a heavily traveled highway were largely contaminated by automotive lead as opposed to that released from the combustion of coal in the surrounding industrial area. The isotope ratio tracing method also was used to determine the primary distance of transport from the highway. [254] In an associated report, Gast [255] emphasizes that isotopic compositions serve as a valuable complementary means for tracing contamination concentrations. He indicates that further studies are required to fully elucidate the general tracing capabilities that can be derived from the use of such measurement.

A weakness of the stable isotope tracing method has been discussed by Holtzman. [256] Studies to date have shown that the sources of atmospheric lead are not necessarily obvious. While lead in different batches of gasoline may be from different types of ores, the ratios found in air and soil have typically been lower than those measured in the gasoline. [244, 250, 254, 255] This may be due to a variety of factors, but it indicates that lead contamination should be traced by concentration ratio as well as isotopic composition measurements. The results reported to date have been of high interest and suggest that the combined techniques may be very useful. The reports cited above have been based on rather limited sampling, at least partly because of the expense involved in measuring isotopic ratios on large numbers of samples. Had the sampling in each instance been more extensive, it is probable that a better understanding of the capabilities and limitations of the isotope ratio method would have been obtained.

EVALUATION OF THE ACCURACIES OF METHODS FOR LEAD ANALYSIS

In any analysis it is always appropriate to question the accuracy of the determination. Evaluation can be accomplished by: analysis of the same samples by independent analytical methods, analysis of samples (standards) of known composition by using the method under test, collaborative or interlaboratory analyses of the same set of samples, and recovery studies. Any one of these means for checking accuracy is subject to one primary limitation. The actual composition of samples is rarely known and can be established only by careful and time-consuming means. In this regard, NBS efforts to certify a number of biological materials as reference standards for several elements of environmental concern are essential.

A principal difficulty in testing the accuracy of a method originates from the fact that measurements are subject to random (interdeterminate) errors and may at the same time be subject to systematic (determinate) errors. Specifically, an analytical result may include the composite effects of: the random errors, the systematic errors inherent in the analytical method, and the systematic and random errors characteristic of each particular laboratory, set of equipment, or analyst involved in the procedure. [257] Random errors often obscure systematic errors. In effect, the random errors (the imprecision) must be smaller than the systematic errors (the inaccuracy), if the latter are to be detected and their magnitude estimated. When systematic errors are absent from the measurement or obscured by the magnitude of random error, any statements about the accuracy of analysis must be limited by its reproducibility. In the absence of systematic error, a precision estimate implies, and in fact limits, the accuracy estimate of any result. The only possible exception to this occurs when the true concentration of the analytical species is known exactly — an extremely rare case. Because of these factors, methods for the evaluation of accuracy are not as simple and straightforward as one would wish. Various approaches to the problem are illustrated below.

Analysis of Certified Standards

NBS has provided certified standards for analytical programs for many years. Unfortunately, few of these have been of an environmental nature. Recently, certifcation of plant leaf, bovine (liver) tissue, fish tissue, coal, and flyash for elements of environmental interest has been initiated by NBS and will prove extremely useful in environmental research. A limited number of standards of an environmental nature have also occasionally been available on a restricted basis from other agencies such as the American Society for Testing and Materials and the Association of Official Analytical Chemists.

When standards have a general composition comparable to that of the samples of interest, the accuracy of a method may be checked by comparing it with an analysis of the certified standard. It should be noted that certified values have been established by numerous analyses through independent methods and are only accurate within certain specified limits. If the analytical result agrees with the certified value, then the accuracy of the method under test may be considered to lie within the specified limits. Such comparisons provide the simplest and most reliable means for evaluating accuracy. The approach should not be abused, however, by extrapolating comparisons from one sample type to another. Interference effects may be drastically different due to relatively small changes in the general sample composition. For best assurance of reliability, the standard should be run blind, so that it is not given more care and attention than regular samples.

Recovery Studies

The method of adding known amounts of the analyte to samples and carrying out a second analysis to calculate recoveries is widely used. It has the advantage of using the sample matrices of interest. It suffers from the disadvantage that the chemical or physical form of the added analyte may not be even remotely similar to the form in which it occurs in the samples. In essence, this method relies on the assumption that the analyte spike will be affected by the sample preparation and analysis in the same way, and to the same extent, as the analytical species normally present in the sample. Numerous examples can be cited where this assumption has been fallacious. The method should therefore be used with caution.

The spike-recovery method is analogous to the method of standard additions discussed above and is subject to some of the same problems. While one may obtain 100 percent recovery of the spike, there is no guarantee that the determination of the residual analyte in the unspiked sample is accurate. Physical interferences, for example, may affect the residual and spiked samples to the same absolute extent, independent of the analyte concentration; the difference between the two results may be the amount spiked, but the residual determination may still be in error.

Youden [257] has suggested a modification of the spike method that compensates for the fact that the form of the analyte spike may differ from that of the actual analyte. In this method, two or more samples are analyzed by the method in question. Measured portions of each sample are then carefully blended in various ratios to form a set of "new" samples for analysis. The accuracy of the analytical method may then be defined by comparing the results. For example, if samples A and B are analyzed and concentrations C_A and C_B are obtained, equal weights of A and B may be mixed and analyzed to obtain concentration C_{AB}. The inaccuracy, Δ, may then be defined by:

$$\Delta = C_{AB} - \left(\frac{C_A + C_B}{2}\right)$$

The value of Δ must, of course, be larger than the analytical precision associated with the respective concentrations, if the inaccuracy is to be regarded as significant. An additional advantage of the sample mixing approach is that samples of divergent general compositions may be mixed to determine if the method under test is robust, i.e., if it is insensitive to large changes in the matrix. Again, numerous examples can be cited in which this approach will not detect or compensate for inaccuracies due to interference effects; therefore, it should be used with judicious caution.

Whether the sample spiking or the sample mixing method is used, it is good practice to analyze samples and spikes that cover the entire concentration range of interest. If the analysis of interest has a regulatory or diagnostic purpose, the regulatory violation or the diagnostic indicator levels should be bracketed by the recovery measurements.

Collaborative Test Studies

Whether materials of known composition are available or not, collaborative analytical programs may serve well as means for establishing the "true" composition of a material, for comparing the capabilities of analytical methods or techniques, or for evaluating the performance of various laboratories and/or analysts. Whatever the purpose of the collaborative program, it is usually possible to obtain indications in all three of the above areas from the resulting data, if the program is properly designed and adequate interpretive techniques are used. [257]

In the previous discussion of analytical methods for the determination of lead, the comparative statements about methods may be considered to be somewhat subjective. Collaborative test and intralaboratory comparison results are available for many of these methods, allowing more objective comparisons to be made.

An interlaboratory test program on lead involving five laboratories was arranged by Skogerboe [258] in 1971. Three soils, two freeze-dried blood pools, two plant leaf samples, and one freeze-dried liver sample were circulated. Most analyses were by AA, with a few carried out by ASV. Statistical analysis of the results indicated that the intralaboratory analytical precision and agreement were not significantly different at the 95 percent confidence level. Thus the analyses were consistent within the precision of the analytical methods, i.e., the results generally agreed within ± 20 percent or better. While consistency of this order of magnitude is desirable, it serves only as an estimator of the accuracy of analysis. The fact that analysis of the soils by three independent methods produced agreement within ± 20 percent implies that the soil analyses were accurate within those limits. The same observation applies to the pooled blood samples. The plant leaves and the liver samples were standard reference materials obtained from NBS. The comparative results are summarized in table 2-7. The agreement for the plant leaves is within the experimental error, as estimated by the standard deviations given for the round robin results. The poor precision observed for the liver results, coupled with the lack of good agreement with the NBS value, suggests an analytical problem. Reagent contamination during the wet digestion process is suspected as a cause of the high round robin result.

A comprehensive interlaboratory evaluation of the reliability of blood lead analyses has been carried out by Keppler et al. [259] The method most widely used was absorption spectrophotometry with dithizone. Some of the analyses were carried out by AA and emission spectrometry. A critique of the evaluation was published by Weil. [260] The results indicate that the analyses were generally imprecise; the analytical precision estimates ranged from 50 to 100 percent of the amount of lead present. While the program was such that accuracy estimates could not be made, the accuracy cannot be better than the precison. Clearly, these results indicate that the determination of lead in blood is subject to a number of problems that must be defined and alleviated.

A collaborative test program on the determination of lead in milk has been reported by Brandt and Benty. [261] Distressingly large variations were abserved in results obtained on the same samples. In addition, analyses of samples spiked with known amounts of lead exhibited large discrepancies. The results suggest that the inadvertent addition of lead during the sample processing may be a source of the problem. Correction for nonspecific background absorption was not used in the AA analyses, probably causing large positive errors.

A limited collaborative study on the determination of lead in sea water has been reported by Patterson and Settle. [262] Anodic stripping, AA, and isotope dilution mass spectrometry results were accumulated. The stripping and AA results were consistently higher than those of isotope dilution by factors ranging from 3 to nearly 100. The fact that isotope dilution is very nearly an absolute method lends credence to its reliability. This study emphasizes again that the determination of lead is subject to problems

TABLE 2-7.–Comparison of round robin results with NBS certified results [258]

Sample identity	Lead (μg/g sample dry weight) Round robin result	NBS value
Orchard leaves	42.3 ± 1.3	[1]44
Tomato leaves	4.4 ± 0.5	–
Bovine liver	0.66 ± 0.32	[1]0.34 ± 0.08

[1] Tentative value subject to revision.

that must be recognized and eliminated, if conflicting and confusing conclusions in environmental lead studies are to be avoided. It also shows that methods with established reliability on one sample type must be tested again when the sample type is changed.

The U.S. Geological Survey [56] sponsored a collaborative study on two water samples involving 21 laboratories. The lead values ranged from 2 to 160 μg/l for one sample and from 2 to 240 μg/l for othe other. Two high values were rejected from each set of results, leaving means and standard deviations of 16.7 ± 9.0 μg/l and 27.7 ± 11.8 μg/l. While it may be argued that the determination of lead in water is a simple matter, results such as these do not offer verification.

Collaborative test programs do offer a valuable means for checking the reliability of methods, laboratories, and analysts. Such programs should be carefully designed and the results objectively evaluated with approaches such as those suggested by Youden. [257]

Independent Method Comparisons

Laboratories can cross-check results by analyzing the same set of samples with two or more distinct methods. These procedures are subject to different interference effects if they exist, and independent comparisons permit the detection of interferences that may not be recognized when only a single analytical method is used. When independent results do not agree, there is a problem of deciding which method is producing the correct result, or if either result is correct. Once such discrepancies have been detected, the cause often becomes obvious. If not, experiments can usually be designed to locate the cause. Independent methods serve well to keep a laboratory informed as to the quality of its analytical results.

Intralaboratory comparisons of ASV and AA spectrometry results have been reported for blood [263] and for biological tissue analyses. [55] Correlation analyses of the results on tissue indicate that AA tend (on the average) to be slightly higher than those ASV at lower concentrations (approximately 0.1 to 0.2 μg/g) and a little lower at levels above approximately 0.5 μg/g. The opposite effects were observed on the blood lead comparisons, with ASV results slightly higher than those for AA at levels below 30 μg percent and lower at levels above 50 μg percent. [263] While these differences are not extremely large, the trends definitely indicate possible problems for the laboratories involved. Further work will define the extent of these possible problems.

A comparison of X-ray fluorescence and AA analyses of air filters has been carried out by Birks et al. [44] A number of filters used to collect particulates from city incinerators were analyzed by energy dispersive and wavelength dispersive X-ray fluorescence methods and by an AA method. [44] The results are summarized in table 2-8. Agreement among the various X-ray techniques if air, considering that in some cases the samples were not very homogeneously distributed. The AA analyses consistently produced low results compared with the X-ray measurements. Birks, et al. [44] report that this was nearly a universal experience for several elements. They speculate that the particulates collected from incinerators are highly refractory and difficult, if not impossible, to dissolve. Because AA depends on the sample being in solution, the results would be understandably low if this were the case.

Comparison of the energy dispersive (columns A and B, table 2-8) and wavelength dispersive (column C, table 2-8) X-ray results points out some interesting aspects of the two techniques. Major elements can be determined by either of the two techniques while intermediate concentrations can be analyzed by either technique, provided that X-ray tube excitation is used (columns B and C, table 2-8). The lowest concentrations can be measured only through the use of wavelength dispersion because at widely different concentrations the interfering elements affect the signals measured at or near the detection limits. [44] Moreover, it is highly unlikely that computer stripping of the spectra will change this conclusion. It may also be inferred from these observations that nondispersive X-ray analyses of environmental samples for lead will be somewhat limited in application.

Quality Control Procedures

It has been the experience of the authors that many analytical methods are, in fact, quite good. Accuracy problems frequently originate with those who use them. Many researchers have a tendency to use (abuse) methods in ways that were not intended. Such abuses often result from the failure of the user to fully explore the literature relating to that method. Other researchers appear to develop absolute faith in any method available to them and do not consider questioning the results. The comparative results presented above indicate the nature of analytical problems in the determination of lead. For some laboratories these problems are very serious. Certainly, such comparisons point out a need for

TABLE 2-8.—Comparative analyses of filter samples from fixed sources [44]

Sample number[1]	Lead (μg/cm^2)			
	A[2]	B[3]	C[4]	D[5]
I-5	32	46	46	11
I-6	180	230	280	255
I-7	43	56	59	2.1
I-8	130	160	180	159
I-9	ND[6]	ND	ND	0.089
I-3	99	98	104	28
I-6	100	71	116	–
I-1	4	5	4.4	4.8

[1] All samples are from incinerators.

[2] A = 71.5 mCi\\$^{109}$Cd, Si(Li) detector.

[3] B = W X-ray tube, Ni filter, Si(Li) detector.

[4] C = Crystal spectrometer.

[5] D = Atomic absorption analysis.

[6] ND = not detected.

continued efforts to develop more reliable, robust methods of analysis. They also indicate a striking need for laboratories to monitor the quality of the results they obtain. Surprisingly few laboratories formulate or maintain quality control programs. Results from laboratories without such programs should be routinely suspected.

A good quality control program for analytical laboratories should be based on five means for checking results:

(1) Samples should be recycled through the laboratory under blind numbers to eliminate the analyst's bias; the data can indicate the day-to-day repeatability of the analyses. Samples should be selected to cover the general concentration range of interest.
(2) Standard samples, if available, should be periodically submitted under different designations to check both the precision and accuracy over time. Standard samples may be too expensive to re-run frequently. Some laboratories prepare their own "secondary" standards for such repetitive checks with less frequent use of certified standards.
(3) Spike-recovery studies should also be routinely run, but the spiking should be carried out by someone other than the analyst involved.
(4) Analyses of samples by independent methods should also be an integral part of any quality control program.
(5) In addition to these internal check methods, laboratories should participate in collaborative test programs to obtain external evaluation.

All of the methods for assuring precision and accuracy add to the expense and are likely to be carried out only in centralized analytical laboratories. Many researchers prefer to do their own analytical work, but it is difficult to justify the expense of an adequate quality control program unless it can be spread over large numbers of samples. The person supervising the work must be acutely aware of the ways in which the methods can fail and must know alternate ways to get the results when a given approach does fail. A centralized analytical facility, equipped for several different methods, staffed by careful analysts, and supervised by a well-trained analytical chemist, is one of the best ways to assure quality in the results.

CHOICE OF THE MONITORING METHOD

No single set of rules will assure the proper selection of the methods or techniques required to answer a particular set of questions. The primary selection criteria must be determined by the purpose of the research. Sample collection, including selection, numbers of samples, handling, preservation, and storage, must be planned with the methods of analysis and ultimate data interpretation in mind. Conversely, the method(s) of analysis selected will often depend on the number and type of samples to be analyzed, as well as their physical form and other factors. Early planning of the experiments must include input from the analyst as well as other interested parties; several consultations will usually be needed to decide how to achieve the most useful results at minimum cost. Clearly, compromises must be made, but they must be based on objective consideration of previously published work and an honest evaluation of the capabilities of the personnel and equipment involved.

Some general comments can be made, however, on selection of methods for the determination of lead in environmental samples. Some methods are so expensive and specialized that they are unlikely to ever become commonly used. Also, laboratories that have these special capabilities also have highly trained (and expensive) people to supervise the work. Therefore, nuclear methods and the various mass spectrometric methods can be applied only to very important samples or as a check on other, more common procedures. Detailed information on these methods must come from the specialized literature, but some of the capabilities, i.e. the near absolute accuracy of isotope dilution mass spectrometry, should be kept in mind by those working in other analytical areas.

Emission spectroscopy and X-ray fluorescence are moderately expensive and specialized. Both are multi-element methods and lack adequate sensitivity when the primary concern is to measure lead. However, as a qualitative tool the arc emission spectrograph with photographic detection is unequalled. The effort required for calibration and plate reading makes it less attractive for quantitative work. It is unlikely that a laboratory without a spectrograph would want to buy one for lead measurements alone. Direct-reading spectrographs require less effort and are especially attractive (cost effective) when a half dozen or more elements are of interest in large numbers of samples. Developments in the area of stable, flamelike, plasma sources should make them even more valuable. X-ray methods exhibit many of these same characteristics. In addition, they are readily automated and, for samples such as air filters, preparation steps can be virtually eliminated, making the method very attractive in large air monitoring programs.

Most laboratories interested in lead determinations in environmental samples are faced with more modest equipment budgets. In these, the choice is likely to be between spectrophotometric, electrochemical, and AA methods. Only a few hundred dollars are needed to equip a laboratory for spectrophotometric lead determinations; the fact that the method is laborious would be unimportant for small projects. The most severe limitation is that the services of a skilled and patient analyst are an absolute necessity. The average laboratory technician will probably produce results that are often useless or, worse, misleading.

Most laboratories currently use AA to determine lead in environmental samples. There are good reasons for this choice. The equipment is within reach of most budgets, it can be used for many elements, the determinations require only minutes in the most favorable cases, and the results are reliable if a few simple (but extremely important) precautions are observed. The capabilities of the method are readily extended by chemical preconcentration or by the use of carbon furnace atomizers, although in both cases the level of skill and understanding needed by the operator are greatly increased. Limitations are: the sample must be in solution, only one element can be determined at a time, and subtle problems can arise that are likely to go undetected. For the foreseeable future, AA will remain the most generally useful method available for the measurement of lead in environmental samples.

ASV and differential pulsed polarography are the two electrochemical methods that are competitive with AA for trace lead determinations. They offer the advantages of potentially greater sensitivity than even carbon furnace AA and the ability to gain some information on the chemical form of lead. These methods are particularly attractive for water samples in which lead concentrations in the low $\mu g/l$ range are of interest. The disadvantages are: relatively slow analyses, difficulty in maintaining thin film mercury electrodes, and more severe interference problems than those characteristic of AA.

It is the consensus of the authors that several reliable methods for monitoring lead exist among the above choices, but the methods have not always been properly or capably utilized. The people who use them have frequently been less reliable than the methods. This often results when the researcher fails to thoroughly familiarize himself with reports that define the capabilities and limitations of the methods in question. Proponents of the various methods are often over zealous, failing to point out the limitations of a favorite technique with the same care and thoroughness that they give its advantages. As one means of avoiding errors, no laboratory should be satisfied with less than two independent methods for a given determination. One might require more time and effort for routine use, but its importance should not be minimized as a backup method and for cross-checking results. The researcher who does his own analytical work is less likely than a central analytical laboratory to have access to two or more methods for measuring a given analyte.

REMAINING PROBLEMS

Analytical methods that give direct information on the chemical form of lead at trace levels would be very desirable. X-ray diffraction and electron spectroscopy are potentially capable of giving some information, but both lack sensitivity in their current form. Electrochemical methods can differentiate between ionic and bound lead and, in favorable cases, they can show the strength of reversible binding. Considerable information can be gained from indirect evidence, as shown in the chapter on characterization, but direct methods would be a welcome addition.

Methods for total lead that are more tolerant of operator errors and abuse would also be helpful. If they were developed, however, it is likely that the level of abuse in many laboratories would increase a corresponding amount with little change in the final output.

SUMMARY

A great deal of reliable and useful information has been derived from studies involving lead in the environment. At the same time, considerable information has been accumulated based on monitoring methodology that may be objectively questioned. The authors have attempted herein to objectively summarize the state-of-the-art of monitoring methods and techniques, to emphasize the positive and negative aspects of the most widely used sample selection and analytical practices, and to provide the uninitiated reader with a general perspective on the pertinent methods and techniques, together with a core list of references. The authors do not contend that the material or the references cited are complete, nor do they claim infallibility. The authors would hope, however, that the discussion contained herein will stimulate critical evaluation of those monitoring practices subject to question, encourage a more widespread adoption of methods and techniques known to be reliable, and intensify research efforts designed to solve some of the problems mentioned. Given these responses, research dealing with the environmental lead question can be brought to a satisfactory resolution.

REFERENCES

1. G.E.F. Lundell. Ind. Eng. Chem., Anal. Ed. *5*:221 (1933).
2. A. B. Calder. Anal. Chem. *36 (9):* 25A (1964).
3. R. A. Fisher. Statistical Methods for Research Workers, 4th ed. Oliver and Boyd, Edinburgh. (1932).
4. J. Cholak and R. V. Story. J. Opt. Soc. Amer. *31*: 730 (1941).
5. G.E.P. Box and N.R. Draper. Evolutionary Operation. Wiley, New York, NY (1969).
6. W. G. Cochran. Sampling Techniques. Wiley, New York, NY (1963).
7. F. Yates. Sampling Methods for Censuses and Surveys. 3rd ed. Charles Griffin and Co., London (1960).
8. A.P. Altshuller, Anal. Chem. *41*: 1R (1969).
9. P.K. Mueller, E.L. Kothny, L.B. Pierce, T. Belsky, M. Imada, and H. Moore, Anal. Chem. *43*: 1R (1971).
10. A.C. Stern, ed. Air Pollution. Academic Press, New York, NY (1968).
11. B.E. Saltsman, W.A. Cook, B. Dimitriades, E.L. Kothny, L. Levin, P.W. McDaniel, and J.H. Smith, Health Lab. Sci. *6*: 106 (1969).
12. Intersociety Committee. Method of Air Sampling and Analysis. American Public Health Association, Washington, D.C. (1972).
13. B. Gandrud and A.L. Lazrus. Environ. Sci. Technol. *6*: 457 (1972).
14. C.L. Luke, T. Y. Kometani, J.E. Kessler, T.C. Loomis, J.L. Bove, and B. Nathanson. Environ. Sci. Technol. *6*:1105 (1972).

15. T.Y. Kometani, J.L. Bove, B. Nathanson, S. Steinberger, and M. Magyor. Environ. Sci. Technol. *6*:617 (1972).
16. E.R. Reiter. Atmospheric Transport Processes, Part 2. U.S.A.E.C. Division of Technical Information (1971).
17. J.W. Winchester. Tech. Prog. Report No. 3, ORA Project 08903, University of Michigan (May 1970).
18. G.L. Hoffman, R. A. Duce, and W. H. Foller. Environ. Sci. Technol. *3*: 1207 (1969).
19. V. Stenger. Anal. Chem. *43(3)*: 36A (1971).
20. G.V.S. Rayudu, B. Tiefenbach, and R.E. Jervis. Trans. Amer. Nucl. Soc. *11*:81 (1968).
21. R.E. Jervis. Isotop. Radiat. Technol. *6*: 57 (1968).
22. A.K. Perkons and R. E. Jervis. Trans. Amer. Nucl. Soc. *11*: 82 (1968).
23. O.U. Anders. Anal. Chem. *44*: 1930 (1972).
24. R. Dams, J.A. Robbins, K.A. Rahn, and J.W. Winchester. IAEA Symposium on Nuclear Techniques in Environmental Pollution. 139 (1970).
25. ASTM Method D2681-68T. American Society for Testing and Materials. Philadelphia, PA (1971).
26. K.R. Spurný and J. Pich. Collection Czechoslov. Chem. Commun. *30*: 2276 (1965).
27. K.R. Spurný, J.P. Lodge, Jr., E.R. Frank, and D.C. Sheesley. Environ. Sci. Technol. *3*: 453 (1969).
28. A.L. Cohen. Environ. Sci. Technol. *7*:60 (1973).
29. L.J. Snyder and S. R. Henderson. Anal. Chem. *33*: 1175 (1961).
30. L.J. Snyder. Anal. Chem. *39*: 591 (1967).
31. L.J. Purdue, R.E. Enrione, R.J. Thompson, and B.A. Bonfield. Anal. Chem. *45*: 527 (1973).
32. G.A. Jutze and K.E. Foster. J. Air Pollut. Contr. Ass. *17*: 291 (1967).
33. Methods of Air Sampling and Analysis. American Public Health Association, Washington, D.C. (1972).
34. C.P. Rolison and K.E. Foster. J. Amer. Ind. Hyg. Ass. *23*: 404 (1962).
35. J.B. Pate and E.C. Tabor. J. Amer. Ind. Hyg. Ass. *23*: 144 (1962).
36. W.K. Harrison, J.S. Nader, and F.S. Fugman. J. Amer. Ind. Hyg. Ass. *21*: 115 (1960).
37. T.J. Kneip, M. Eisenbud, C.D. Strehow, and P.C. Freudenthal. J. Air Pollut. Contr. Ass. *20*: 3 (1970).
38. R.E. Lee, Jr., S.S. Goranson, R.E. Enrione, and G.B. Morgan. Environ. Sci. Technol. *6*: 1025 (1972).
39. R. Woodriff and J.F. Lech. Anal. Chem. *44*: 1323 (1972).
40. M. Katz. Measurement of Air Pollutants. World Health Organization, Geneva (1969).
41. J.L. Seeley and R.K. Skogerboe. Anal. Chem. *46*: 415 (1974).
42. J.M. Pierrard. Environ. Sci. Technol. *3*: 48 (1969).
43. R. Herrmann and C.T.J. Alkemade. Chemical Analysis by Flame Photometry. 2nd ed. Interscience, New York, NY (1963).
44. L.S. Birks, J.V. Gilfrich, and P.G. Burkhalter. EPA Report R2-72-063 (November 1972).
45. R. K. Skogerboe, Colorado State University, Ft. Collins, CO Unpublished data. (1972).
46. C.D. Burnham, C.E. Moore, E. Kanabrocki, and D.M. Hattori. Environ. Sci. Technol. *3*: 472 (1969)
47. C.D. Burnham, C.E. Moore, T. Kowalski, and J. Krasniewski. Appl. Spectrosc. *24*: 411 (1970).
48. H.E. Allen. In Instrumental Analysis for Water Pollution Control. K.H. Mancy ed. Ann Arbor Science Publishers, Ann Arbor, MI (1971).
49. Standard Methods for the Examination of Water and Wastewater, 13th ed. American Public Health Association, Washington, D.C. (1971).
50. Methods for Chemical Analysis of Water and Wastes. U.S. Environmental Protection Agency, Cincinnati, OH (1971).
51. J.F. Kopp and R.C. Kroner. Trace Metals in Waters of the United States. Federal Water, Pollution Control Administration, Cincinnati, OH (1967).
52. E. Brown, M.W. Skougstad, and M.J. Fishman. U.S.G.S. Techniques for Water Resources Inventory. *5*: Ch. A-1 (1970).
53. F.H. Rainwater and L.L. Thatcher. U.S.G.S. Water Supply Paper No. 1454 (1960).
54. C.J. Velz. Sewage Ind. Wastes. *22*: 666 (1950).
55. Environmental Research Laborotory, University of Illinois. Unpublished data (1973).
56. M. W. Skougstad, U.S.G.S., Denver, CO Personal communication (1971).
57. R.W. Perkins and R.A. Rancetelli. Proc. Amer. Nucl. Soc. Meeting, Columbia, MO (August 1971).
58. D.E. Robertson. Anal. Chim. Acta. *42*:533 (1968).
59. G.G. Eichholz, A.E. Nagel, and R.B. Hughes. Anal. Chem. *37*: 863 (1965).
60. A.W. Struempler. Chadron State College, Chadron, NB Unpublished data. (1973).
61. R.G. Semonin. State Water Survey Report No. AT (11-1)-1199. Urbana, IL (1973).
62. A.L. Lazrus, E. Lorange, and J.P. Lodge, Jr. Environ. Sci. Technol. *4*: 55 (1970).
63. C.W. Francis. G. Chesters, and L.A. Haskin. Environ. Sci. Technol. *4*: 586 (1970).
64. Instrumentation for Environmental Monitoring Vol. II, Water. Lawrence-Berkeley Laboratory, University of California, Berkeley, CA (February 1973).
65. K.M. Mackenthun. The Practice of Water Pollution Biology. Federal Water Pollution Control Administration (1969).
66. V. Sladeckova. Bot. Rev. *28*: 2 (1962).
67. Progress Report. Environmental Pollution by Lead and Other Metals. Institute for Environmental Research, University of Illinois (1972).
68. C.A. Black, ed. Methods of Soil Analysis. American Society of Agronomists, Madison, WI (1965).
69. Soil Testing and Plant Analysis. Soil Science Society of America, Madison, WI (1967).
70. Chemistry of Soil. American Chemical Society Monograph Series No. 160, 2nd ed. Reinhold, New York, NY (1964).
71. M.W.M. Leo. J. Agr. Food Chem. *11*: 432 (1963).
72. S.R. Koirtyohann. University of Missouri, Columbia, MO Unpublished data (1973).
73. W.J.A. Steyn. J. Agr. Food Chem. *7*: 344 (1959).
74. W.H. Smith. Environ. Sci. Technol. *7*: 631 (1973).
75. R.G. Keenan, D.H. Byers, B.E. Saltzman, and F.L. Hyslop. Amer. Ind. Hyg. Ass. J. *24*: 481 (1963).
76. R.O. Farrelly and J. Phybus. Clin. Chem. *15*: 7 (1969).
77. D.G. Mitchell, K.M. Aldous, and F.J. Ryan. Paper presented at the Environmental Health Aspects of Lead Symposium, Amsterdam (1972).
78. K.M. Aldous, NY State Department of Health, Albany, NY Personal communication (1973).
79. D.J. Hammer, J.F. Finklea, R.H. Hendricks, C.M. Shy, and R.J. Horton. Amer. J. Epidemiol. *93*: 84 (1971).
80. L. Kopito, R.K. Byers, and H. Schwachman. New Eng. J. Med. *276*: 949 (1967).
81. R.A. Kyle and G.L. Pease. New Eng. J. Med. *273*: 18 (1966).
82. H.A. Schroeder and A.P. Nason. J. Invest. Dermat. *53*: 71 (1969).
83. L.C. Bate. Intern. J. Appl. Radiat. Isot. *17*: 417 (1966).
84. E.J. Dietrick, E.J. Schlinger, and R. van den Bosale. J. Econ. Entomol. *52*: 1085 (1959).
85. L.R. Taylor. Amer. Appl. Biol. *150*: 405 (1962).
86. R.F. Harwood, Mosquito News. *21*: 35 (1961).
87. P.A. Steward and J.J. Lam. J. Econ. Entomol. *63*: 871 (1970).
88. W.R. Horsfall. J. Econ. Entomol. *55*: 808 (1962).
89. T.T. Gorsuch. The Destruction of Organic Matter. Pergamon Press, New York, NY (1970).
90. G.F. Smith. Anal. Chim. Acta. *8*: 397 (1953).
91. C.A. Evans, Jr., and G.H. Morrison. Anal. Chem. *40*: 869 (1968).
92. G. Tölg. Talanta. *19*: 1489 (1972).
93. C.E. Gleit and W.D. Holland. Anal. Chem. *34*: 1454 (1962).
94. G.I. Ter Haar and M.A. Bayard. Nature. *232*: 553 (1971).
95. E.C. Kuehner, R. Alvarez, P.J. Paulsen, and T.J. Murphy. Anal. Chem. *44*: 2050 (1972).
96. J.M. Mattinson. Anal. Chem. *44*: 1715 (1972).
97. B.D. Stepin, I.G. Gorshteyn, G.Z. Blym, G.M. Kurdyunuv, and J.P. Oglobina. Methods of Producing Superpure Inorganic Substances. J.P.P.S. 53256, U.S. Government Printing Office, Washington, D.C. (1971).

98. J.W. Mitchell. Anal. Chem. *45(6):* 492A (1973).
99. J.H. Yoe and H.J. Koch, Jr. Trace Analyses. Wiley, New York, NY (1957).
100. A. Mizuike. In Trace Analysis - Physical Methods, G. H. Morrison, ed. Interscience, New York, NY (1965).
101. J.K. Taylor, ed. N.B.S. Tech. Note 545, U.S. Government Printing Office, Washington, D.C. (1970).
102. J.E. Allan. Spectrochim. Acta. *17*: 467 (1961).
103. R.R. Brooks, B.J. Presley, and I.R. Kaplan, Talanta. *14*: 809 (1967).
104. S.R. Koirtyohann and J. Wen. Anal. Chem. *45*: 1986 (1973).
105. G. Kisfaludi and M. Lenhof. Anal. Chim. Acta. *54*: 83 (1971).
106. E. Berman, V. Valavanis, and A. Dublin. Clin. Chem. *14*: 239 (1968).
107. J. Jordan. Atom. Absorp. Newslett. *7:* 48 (1968).
108. A.A. Yadav and S.M. Khlpkar. Talanta. *18*: 833 (1971).
109. I. Tsukahara and T. Yamamoto. Anal. Chim. Acta. *61*: 33 (1972).
110. I. Tsukahara and T. Yamamoto. Anal. Chim. Acta. *63*: 646 (1973).
111. M.E. Hofton and D.P. Hubbard. Anal. Chim. Acta. *52*: 425 (1970).
112. R. M. Dagnall, T. S. West, and P. Young. Anal. Chem. *38:* 368 (1966).
113. C.H. Lochmuller, J. Galbraith, R. Walter, and J. Joyce. Anal. Lett. *5*: 943 (1972).
114. T.P. Clark. Paper presented at 7th Conference on Trace Substances in Environmental Health, Columbia, MO (1973).
115. W.L. Hoover, J.C. Reagor, and J.C. Garner. J. Ass. Offic. Anal. Chem. *52*: 708 (1969).
116. L. Kopito and H. Schwachman. J. Lab. Clin. Med. *70*: 326 (1967).
117. N. Zurlo, A.M. Griffini, and G. Colombs. Anal. Chim. Acta. *47*: 203 (1969).
118. Official Methods of Analysis of the AOAC. 10th ed. Association of Official Analytical Chemists, Washington, D.C. (1965).
119. R.L. Mitchell. Technical Communication No. 44., Commonwealth Bureau of Soil Science. Aberdeen, Scotland (1944).
120. T.W. Gilbert. In Treatise on Analytical Chemistry, Part II, Vol. 6. I.M. Kolthoff, P.J. Elving, and E.B. Sandell, eds. Interscience, New York, NY (1964).
121. E.B. Sandell. Colorimetric Determination of Traces of Metals. Interscience, New York, NY (1944).
122. Official Methods of Analysis of the AOAC, 11th ed. Association of Official Analytical Chemists, Washington, D.C. (1970).
123. Official, Standardized and Recommended Methods of Analysis. W. Heffer and Sons, Ltd. Cambridge (1963).
124. Airborne Lead in Perspective. National Academy of Sciences/National Research Council, Washington, D.C. (1971).
125. E.B. Sandell. Colorimetric Metal Analysis, 3rd ed. Interscience, New York, NY (1959).
126. Analysis for Trace Quantities of Lead. Ethyl Corp. Baton Rouge, LA (1972).
127. G.H. Morrison and H. Freiser. Solvent Extraction in Analytical Chemistry. Wiley, New York, NY (1957).
128. T.W. Gilbert. In Treatise on Analytical Chemistry, Part II, Vol. 6. I.M. Kolthoff, P.J. Elving, and E.B. Sandell eds. Interscience, New York, NY (1964).
129. A. Walsh. Spectrochim. Acta. *7*: 108 (1955).
130. C.T.J. Alkemade and J.M.W. Milatz. Appl. Sci. Res., Sect. B. *4*: 289 (1955).
131. C.T.J. Alkemade and J.M.W. Milatz. J. Opt. Soc. Amer. *45*: 583 (1955).
132. L. de Galan and J.D. Winefordner. J. Quant. Spectrosc. Radiat. Transfer. *7*: 251 (1967).
133. S.R. Koirtyohann and E.E. Pickett. Proceedings of XIII Colloquium Spectroscopicum Internationale, Adam Hilger, Ltd., London (1968).
134. L. de Galan and G.F. Samaey. Spectrochim. Acta. *25B*: 245 (1970).
135. J.B. Willis. Spectrochim. Acta. *25B*: 487 (1970).
136. C. Veillon. Handbook of Commercial Scientific Instruments, Vol. 1, Atomic Absorption. Marcel Dekker, New York, NY (1972).
137. W. Slavin. Atomic Absorption Spectroscopy. Interscience, New York, NY (1968).
138. H.L. Kahn. J. Chem. Educ. *43*: A7, A103 (1966).
139. Flame Emission and Atomic Absorption Spectrometry. Vol. 1 and 2. J.A. Dean and T.C. Rains. eds. Marcel Dekker, New York, NY (1969 and 1971).
140. J.L. Seeley, D. Dick, J.H. Arvik, R.L. Zimdahl, and R.K. Skogerboe. Appl. Spectrosc. *26*:456 (1972).
141. S.R. Koirtyohann and E.E. Pickett. Anal. Chem. *37:* 601 (1965).
142. E.F. Dalton, and A.J. Malanoski. J. Ass. Offic. Anal. Chem. *52:* 1035 (1969).
143. G.E. Marks and C.E. Moore. Appl. Spectrosc. *26:* 523 (1971).
144. W.L. Hoover. J. Ass. Offic. Anal. Chem. *55:* 737 (1972).
145. D.L. Dick, S.J. Urtamo, F.E. Lichte, and R.K. Skogerboe. Appl. Spectrosc. *27:* 467 (1973).
146. H.L. Kahn, G.E. Peterson, J.E. Schallis. Atom. Absorp. Newslett. *7:* 35 (1968).
147. H.T. Delves. Analyst. *95:* 431 (1970).
148. D.C. Hilderbrand, S.R. Koirtyohann, and E.E. Pickett. Biochemical Med. *3:* 437 (1970).
149. J.M. Hicks, A.N. Gutierrez, and B.E. Worthy. Clin. Chem. *19:* 322 (1973).
150. B.V. L'Vov. Spectrochim. Acta. *17:* 761 (1961).
151. R. Woodriff and G. Ramelow. Spectrochim. Acta. *23B:* 665 (1968).
152. L. Ebdon, G.F. Kirkbright, and T.S. West. Anal. Chim. Acta. *51:* 365 (1970).
153. K.W. Jackson and T.S. West. Anal. Chim. Acta. *59:* 187 (1972).
154. H. Massmann. Spectrochim. Acta. *23B:* 215 (1968).
155. J.P. Matousek and B.J. Stevens. Clin. Chem. *17:* 363 (1971).
156. E. Norval and L.R.P. Butler. Anal. Chim. Acta. *58:* 47 (1972).
157. J.Y. Hwang, C.K. Mokeler, and P.A. Ulluci. Anal. Chem. *44:* 2014 (1972).
158. H.L. Kahn and D.C. Manning. Amer. Lab. *6:* 51 (1972).
159. J.Y. Hwang, P.A. Ulluci, and C.J. Mokeler. Anal. Chem. *45:* 795 (1973).
160. Y. Talmi and G.H. Morrison. Anal. Chem. *44:* 1455 (1972).
161. J.B. Headridge and D.R. Smith. Talanta. *19:* 833 (1972).
162. H. P. Loftin, C. M. Christian, and J.W. Robinson. Spectrosc. Lett. *3:* 161 (1970).
163. E.E. Pickett and S.R. Koirtyohann. Anal. Chem. *41(14):* 28A (1969).
164. J.M. Mansfield, M.P. Bratzel, Jr., H.O. Norgordon, D.O. Knapp, K.E. Zacha, and J.D. Winefordner. Spectrochim. Acta. *23B:* 389 (1968).
165. R.M. Dagnall, M.R.G. Taylor, and T.S. West. Spectrosc. Lett. *1:* 397 (1968).
166. G. Rossie and N. Omenetto. Talanta. *16:* 263 (1969).
167. R.F. Browner, R.M. Dagnal, and T.S. West. Anal. Chim. Acta. *50:* 375 (1970).
168. S. Greenfield and P.B. Smith. Anal. Chim. Acta. *59:* 341 (1972).
169. G.W. Dickinson and V.A. Fassel. Anal. Chem. *41:* 1021 (1969).
170. F.E. Lichte and R.K. Skogerboe. Anal. Chem. *44:* 1480 (1972).
171. L.H. Ahrens. Spectrochemical Analysis. Addison-Wesley Press, Cambridge, MA (1950).
172. R.L. Mitchell. The Spectrographic Analysis of Soils, Plants, and Related Materials. Technical Communication No. 44. Commonwealth Bureau of Soil Science, Harpenden, England (1948).
173. A. Specht, W. Koch, E. James, and J.E. Resnicky. Soil Sci. *83:* 15 (1957).
174. I.H. Tipton, M.J. Cook, R.L. Steiner, C.A. Boye, H.M. Perry, Jr., and H.S. Schroeder. Health Phys. *9:* 89 (1963).
175. J.L. Seeley and R.K. Skogerboe. Anal. Chem. *46:* 415 (1973).
176. R.D. Sacks and S.W. Brewer. Appl. Spectrosc. Rev. *6:* 313 (1972).
177. M.M. Braverman, F.A. Masciello, and V. Marsh. J. Air Pollut. Contr. Ass. *11:* 408 (1961).

178. M. Slavin. Emission Spectrochemical Analysis. Wiley-Interscience, New York, NY (1971).
179. Methods for Emission Spectrochemical Analysis. American Society for Testing and Materials, Philadelphia, PA (1968).
180. W.A. Gordon and A.K. Gallagher. NASA Tech. Memo. TM X-1220 (April 1966).
181. M. Fred, N.H. Nachtrieb, and F.S. Tomkins. J. Opt. Soc. Amer. *37:* 279 (1947).
182. M. Pierucci and L. Barbanti-Silva. Nuovo Cimento. *17:* 275 (1940).
183. C. Feldman. Anal. Chem. *21:* 1041 (1949).
184. W.A. Gordon. NASA T.N. D2598. Clearinghouse for Federal Scientific and Technical Information, Springfield, VA (1965).
185. W.A. Gordon. NASA T.N. D5236. Clearinghouse for Federal Scientific and Technical Information, Springfield, VA (1967).
186. W.A. Gordon. NASA T.N. D4769. Clearinghouse for Federal Scientific and Technical Information, Springfield, VA (1968).
187. K.M. Hambidge. Anal.Chem. *43:* 103 (1971).
188. L.S. Birks. X-Ray Spectrochemical Analysis, 2nd ed. Wiley, New York, NY (1969).
189. E.P. Bertin. Principles and Practices of X-Ray Spectrometric Analysis. Plenum Press, New York, NY (1970).
190. R.O. Müller. Spectrochemical Analysis by X-Ray Fluorescence. Plenum Press, New York, NY (1972).
191. H.A. Liebhafsky, P.G. Pfeiffer, E.H. Winslow, and P.D. Zemany. X-Rays, Electrons, and Analytical Chemistry. Wiley-Interscience, New York, NY (1972).
192. T.J. Kneip and G.R. Laurer, Anal. Chem. *44(14):* 57A (1972).
193. L.S. Birks. Anal. Chem. *44:* 557R (1972).
194. J.J. Kraushaar, R.A. Ristinen, H. Rudolph, and W.R. Smythe. Bull. Amer. Phys. Soc. *16:* 545 (1971).
195. H. Rudolph, J.K. Kliever, J.J. Kraushaar, R.A. Ristinen, and W.R. Smythe. Analysis Instrum. *10:* 151 (1972).
196. J.A. Cooper. Nucl. Instrum. Methods. *106:* 525 (1973).
197. L.S. Birks, J.V. Gilfrich, and D.J. Nagel. Large-Scale Monitoring of Automobile Exhaust Particles — Methods and Costs. NRL Memorandum Report 2350 (1971).
198. L.S. Birks. Naval Research Laboratory, Washington, D.C. Personal communication (1973).
199. B.R. Kowalski, T.L. Isenhour, and R.E. Sievers. Anal. Chem. *41:* 1705 (1969).
200. J.L. Booker, T.L. Isenhour, and R.E. Sievers. Anal Chem. *41:* 1709 (1969).
201. N.M. Frew, J.J. Leary, and T.L. Isenhour. Anal. Chem. *44:* 665 (1972).
202. N.M. Frew, J.J. Leary, and T.L. Isenhour. Anal. Chem. *44:* 659 (1972).
203. B.C. Wood and R.K. Skogerboe. Appl. Spectrosc. *27:* 10 (1973).
204. R.K. Skogerboe, A.T. Kashuba, and G.H. Morrison. Anal. Chem. *40:* 1096 (1968).
205. C.A. Evans, Jr., and G.H. Morrison. Anal. Chem. *40:* 869 (1968).
206. A.T. Kashuba and G.H. Morrison. Anal. Chem. *41:* 1842 (1969).
207. G.H. Morrison. Anal. Chem. *43:* No. 7, 22A (1971).
208. R.K. Webster. In Advances in Mass Spectrometry, Vol. I. J. D. Waldron, ed. Pergamon, Oxford (1959).
209. F.D. Leipziger. Anal. Chem. *37:* 171 (1965).
210. H. Farror, IV. In Trace Analysis by Mass Spectrometry, A.J. Ahearn, ed. Academic Press, New York, NY (1972).
211. R. Alvarez, P.J. Paulson, and D.E. Kelleher. Anal. Chem. *41:* 955 (1969).
212. P.J. Paulson, R. Alvarez, and C.W. Mueller. Anal. Chem. *42:* 673 (1970).
213. W.H. Latimer. Oxidation Potentials, 2nd ed. Prentice-Hall, Inc., New York, NY (1952).
214. J.J. Lingane. Polargraphy. Interscience, New York, NY (1952).
215. A.E. Martell and M. Calvin. Chemistry of the Metal Chelate Compounds. Prentice-Hall, Inc., New York, NY (1952).
216. A. Lerman, and C.W. Childs. Trace Metals and Metal-Organic Interactions in Natural Water. P.C. Singer, ed. Ann Arbor Science Publishers, Ann Arbor, MI (1973).
217. B.B. Damaskin. Principles of Current Methods for the Study of Electrochemical Reactions. McGraw-Hill, New York, NY (1967).
218. Polarographic Analyzer, Model 174. Princeton Applied Research Corp. Bulletin T-295B-10M-3/73 (1973).
219. I. Shain. In Treatise on Analytical Chemistry, Part I, Vol. 4. I. M. Kolthoff, P.J. Elving, and E. B. Sandell, eds. Wiley, New York, NY (1964).
220. K.H. Mancy. Instrumental Analysis for Water Pollution Control. Ann Arbor Science Publishers, Ann Arbor, MI (1972).
221. K.J. Martin and I. Shain. Anal. Chem. *30:* 1808 (1958).
222. W.R. Matson, D.K. Roe, and D.E. Carritt. Anal. Chem. *37:* 1598 (1965).
223. T.R. Copeland, J.H. Christie, R.K. Skogerboe, and R.A. Osteryoung. Anal. Chem. *45:* 995 (1973).
224. T.R. Copeland, J.H. Christie, R.A. Osteryoung, R.K. Skogerboe. Anal. Chem. *45:* 2171 (1973).
225. A. Zirino and M.L. Healy. Environ. Sci. Technol. *6:* 243 (1972).
226. J.H. Christie and R.A. Osteryoung. Anal. Chem. *46: 351 (1973).*
227. G. Barker and A. W. Gardner. Z. Anal. Chem. *173:* 79 (1960).
228. E.B. Buchanan, Jr., and J.B. McCarten. Anal. Chem. *37:* 29 (1965).
229. E.B. Buchanan, Jr., T.D. Schroeder, and B. Novosel. Anal. Chem. *42:* 370 (1970).
230. V.J. Jennings. Analyst. *87:* 548 (1962).
231. N. Watkins. Ph.D. thesis, University of Illinois (1973).
232. D.J. Myers and J. Osteryoung. Anal. Chem. *45:* 267 (1973).
233. G.D. Christian. J. Electroanal. Chem. *23:* 1 (1969).
234. P. Kruger. Principles of Activation Analysis. Interscience, New York, NY (1971).
235. H.P. Yule, H.R. Lukens, Jr., and V.P. Guinn. Nucl. Instrum. Methods. *33:* 277 (1965).
236. H.R. Lukens. J. Radioanalyt. Chem. *1:* 349 (1968).
237. V.P. Guinn, D.A. Miller, and G.E. Miller. In Irradiation Facilities for Research Reactors. International Atomic Energy Agency, Vienna. 441 (1973).
238. V.P. Guinn. In: Activation Analysis. page 83, International Atomic Energy Agency, Vienna. 83 (1971).
239. G.J. Lutz. Anal. Chem. *43:* 93 (1971).
240. G.J. Lutz. Proceedings of the American Nuclear Topical Meeting on Nuclear Methods in Environmental Research, University of Missouri, Columbia, MO (1971).
241. V.P. Guinn. U.S.A.E.C. Report G.A. 7041. San Diego, CA (1966).
242. R.D. Cooper, D.M. Kinekin, and G.L. Brownell. Nuclear Activation Techniques in the Life Sciences. International Atomic Energy Agency, Vienna. 65 (1967).
243. N.K. Aras, W. H. Zoller, G.E. Gordon, and G.J. Lutz. Anal. Chem. *45:* 1481 (1973).
244. T.J. Chow and J.L. Earl. Science. *169:* 577 (1960).
245. E.A. Schuck and J.K. Locke. Environ. Sci. Technol. *4:* 324 (1970).
246. A. L. Page and T.J. Gonje. Environ. Sci. Technol. *4:* 140 (1970).
247. R.H. Daives, H. Matto, and D.M. Chilko. Environ. Sci. Technol. *4:* 318 (1970).
248. A.C. Delany, D. W. Parkin, E.D. Goldberg, and B.E.F. Reimann. Geochim. Cosmochim. Acta. *31:* 885 (1967).
249. U.S. Bureau of Mines, Minerals Yearbook, 1968. U.S. Government Printing Office, Washington, D.C. (1969).
250. T.J. Chow. Second International Clean Air Congress. 348 (1971).
251. R.D. Russell and R.M. Farquhar. Lead Isotopes in Geology. Interscience, New York, NY (1960).
252. K. Ronkoma. Progress in Isotope Geology. Interscience, New York, NY (1963).
253. J.S. Brown. Econ. Geol. *57:* 673 (1962).
254. W.U. Ault, R.G. Senechal, and W.E. Erlebach. Environ. Sci. Technol. *4:* 305 (1970).
255. P.W. Gast. Environ. Sci. Technol. *4:* 313 (1970).
256. R.B. Holtzman. Environ. Sci. Technol. *4:* 314 (1970).

257. W.J. Youden. Statistical Techniques for Collaborative Tests. Association of Official Analytical Chemists, Washington, D.C. (1969).
258. Impact on Man of Environmental Contamination Caused by Lead. H. W. Edwards, ed. NSF Interim Grant Rept., Colorado State University (1972).
259. J.F. Keppler, M.E. Maxfield, W.D. Moss, G. Tietjen, and A.L. Linch. Amer. Ind. Hyg. Ass. J. *31:* 412 (1970).
260. C.S. Weil. Amer. Ind. Hyg. Ass. J. *32:* 304 (1971).
261. M. Brandt and J.M. Benty. Microchem. J. *16:* 113 (1971).
262. C. Patterson and D. Settle. IDOE Report Study No. 3, California Institute of Technology, Pasadena, CA (1972).
263. R.E. Reiss. Environmental Sciences Associates, Inc. Personal communication (November 1972).

PART II

Transport and Distribution

CHAPTER 3

MODELING ATMOSPHERIC TRANSPORT

Elmar R. Reiter, Teizi Henmi and *Paul C. Katen*
Department of Atmospheric Science
Colorado State University

INTRODUCTION

The transport of lead from automotive sources through the environment follows a network of paths by which the atmosphere, the lithosphere, the hydrosphere, and the biosphere are connected in a complex pattern (figure 3-1). Quantitative estimates of the transport processes taking place along these various paths not only will help provide a better assessment of the problems connected with lead contamination of the environment and in the ultimate control of this contamination; such estimates will also aid in the preparation of studies of the impact of other anthropogenic contaminants on the environment.

This part of the report is mainly concerned with numerical modeling of the atmospheric transport processes involved in the diffusion of gaseous and particulate pollutants from point and line sources. This begins with the well-known solutions to the classical Fickian diffusion equation. It becomes apparent, however, that line sources close to the ground, such as automobile exhaust from a column of vehicles traveling along a highway, are diffusion into a wind field in which wind speed, as well as turbulence intensity, are functions of height above the ground. This functional dependence on height cannot be neglected in the interpretation of fallout measurements downstream from, but relatively close to (nearer than 100 m) a highway. Nor can the gravitational settling of large particles be ignored; they contain a significant portion of the lead that emanates from the exhaust pipes and settles on the road surfaces and in the immediate vicinity of the highways.

Further complications in the application of classical diffusion theory arise from complex terrain configurations in the vicinity of highways and also from the exposure of sampling stations to a network of heavily traveled streets. The interpretation of lead fallout measurements necessitates detailed measurements of the thermal and turbulent structure of the atmosphere, as well as of the mean wind field and its three-dimensional behavior.

No generalized theories are presently available that would allow a theoretical treatment of the turbulent diffusion of pollutants under such complex environmental conditions. Therefore, a semi-empirical approach was developed that incorporates actual measurements of the three-dimensional structure of wind and turbulence into an assessment of the dispersion characteristics of atmospheric contaminants under given source and terrain configurations. Ultimately, this method, it is hoped, will allow the study of diffusion conditions within city street canyons or around obstacles, such as tall buildings. Furthermore, this semi-empirical approach lends itself excellently to a comparison of wind-tunnel data with "real world" data.

The subsequent parts of this chapter contain a brief discussion of the meteorological measurement program that yields parameters through which an interpretation of lead fallout under complicated environmental conditions is possible. A short treatment of simple diffusion equations follows, including an equation that allows for gravitational settling. Finally, a semi-empirical approach is outlined, that allows for the variation of the mean wind field and of turbulence intensity with height.

METEOROLOGICAL MEASUREMENTS AND AIR SAMPLING

The movement of lead from traffic exhaust through the atmosphere can be studied by measuring the space and time dependence of atmospheric lead concentrations near a highway, in conjunction with simultaneous measurement of wind speed and direction, turbulence characteristics, and temperature lapse rate. Concurrent measurement of all atmos-

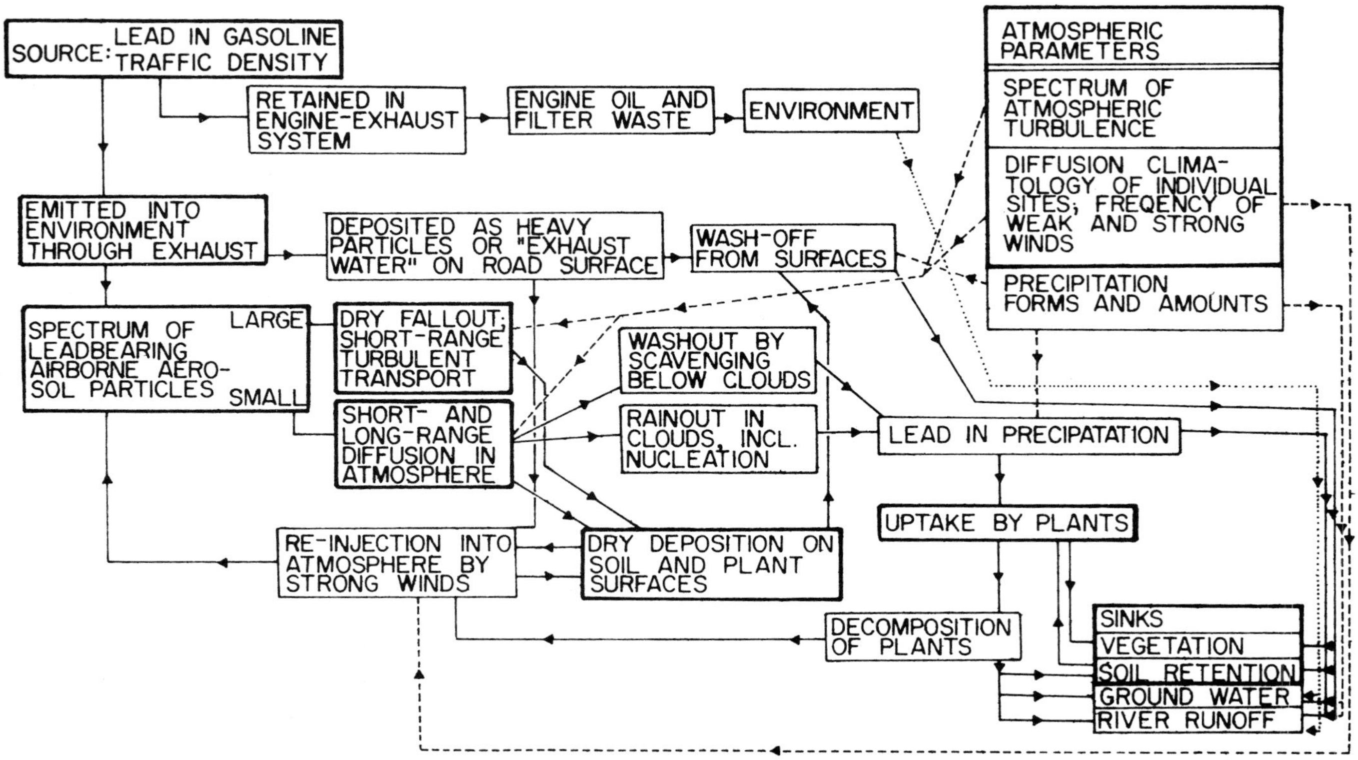

FIGURE 3-1.— Schematic diagram of the influence of atmosphere parameters on lead transport through the environment.

pheric parameters is necessary to apply existing diffusion theory to lead dispersion. This theory, however, takes into account only the component that travels as a gas. Additional measurements of the settling velocity distribution of lead particles as a function of distance from the highway are also needed to explain measured fallout patterns and, hence, measured soil accumulation patterns. Because existing diffusion theory does consider a wide range of atmospheric stability conditions, the measurements of atmospheric parameters are of considerable general utility. Our data, thus, enable us to develop a general approach for describing particulate transport and to test the theoretical model under a variety of atmospheric and meteorological conditions. The meteorological data are also used to couple atmospheric dispersion with soil uptake and accumulation.

The theoretical basis for the meteorological measurements was given first by Sutton [1] and later by Pasquill [2] in a form often used for field measurements. Records of wind direction and velocity as a function of time permit calculation of the variance in each of the wind components. These quantities are then used in the Pasquill diffusion equation which, if the source strength is known, can be used to predict the downstream dilution of a pollutant whose movement can be characterized by gaseous transport equations. The temperature lapse rate strongly affects the atmospheric stability, and its measurement allows the stability conditions to be classified in one of several broad categories. This approach has been used with success for describing atmospheric dispersion of radioactive pollutants. [3, 4]

The field studies established for this program were designed to obtain answers to specific aspects of this complex diffusion problem. Many factors contribute to its complexity:

(1) The exact value of the source strength is unknown.

(2) The initial dilution of the exhaust plume by turbulent automobile wakes is difficult to determine.

(3) The lead-bearing exhaust particulates have a broad spectrum of sizes, some of which settle out of the diffusion plume because of gravitational forces and some of which behave as nonreacting diffusing gases.

(4) The percentage of the total diffusing mass of lead that behaves like a gas, or that falls out rapidly under gravitational settling, is dependent upon the individual vehicle and its operating conditions.

(5) The mean wind distribution and vertical wind fluctuation profiles in the lowest layer of the atmosphere lead to vertical concentration profiles that differ significantly from the normal or Gaussian profiles that elementary models predict, which neglect this region of strongly sheared flow. Typically, simple estimations of lateral and vertical dispersion are made from wind speeds measured at 10 m above ground and from crude estimates of thermal stability. [5] This approach yields usable results for distances greater than 100 m downwind of the source. We are, however, at least initially interested in distances up to 100 m. Therefore,

we felt that the direct measurement of wind fluctuations is a better method of estimating dispersion, for it is the turbulent component of air motions that is responsible for diffusion of the pollutant.

In the initial highway test program, the objective was to obtain measurements of the atmospheric transport of lead from a highway and to arrive at estimates of gravitational settling from the discrepancies between measured lead concentrations downwind from a highway and concentrations predicted with the simplest gaseous diffusion model. (See the discussion later in this chapter.) The application of even a simple model required the estimation of the mean position of the height of the exhaust plume above the roadway. Furthermore, because of initial dilution and subsequent interaction of one exhaust plume with plumes from other vehicles in the same and in adjacent traffic lanes, a "virtual source" was hypothesized at a distance, X', upwind of the first sampler, which is located at the downwind edge of the paved roadway. This virtual line source accounts for the initial plume dilution in the wake of vehicles. Finally, some approximations had to be made for the source strength. This step was accomplished through estimates of average gasoline consumption and average emission rates of consumed lead, both dependent upon vehicle speed. [6]

In the second phase of the highway field program, we attempted to include the effects of particulate settling and of atmospheric vertical nonhomogeneity into a complete dispersion parameterization.

In both cases the field instrument deployments were quite similar. High-volume filter samplers were placed at 1.2 m above the surface (this corresponds to the estimated mean height of exhaust plumes after vehicle wake interaction) at four locations downwind of the roadway up to a distance of 90 m. In addition, bivane anemometers were deployed at two levels (1.67 and 5 m) above the surface (fig. 3-2).

In a further modification of the initial measurement program, two additional samplers were deployed at 2.67 and 12 m above the surface to measure the effects of turbulence and shear flow on the dispersion of lead.

Instrumentation and Meteorological Measurements

Bivane Anemometers: Wind speed and direction are measured by four bivane anemometers manufactured by the R.M. Young Co. Agreement of measured speeds from these four anemometers is within ±0.1 m/sec. when average values over 10-min. periods are considered. Azimuth readings agree within approximately 1°. This is the accuracy to which units may be oriented with reference to a geographic coordinate system. Elevation angles approach an average value of 0° ±1° during 10-min. calibration runs over level terrain.

The bivane instruments are designed for use in mobile field units. Computer programs have been developed to facilitate quick checks on the internal consistency of wind data received from the four units. These data are mean standard deviation values for wind speeds and directions over specified time periods. The computer programs also facilitate correction of wind data with respect to an assigned "true" wind direction. Even though such a correction will normally not be necessary, a possible maladjustment in one or more of the instruments can easily be eliminated. Such maladjustments, if they are present, can be detected from the computer printout previously described. Minor maladjustments of the instruments would normally not be detected until after the field experiments were completed. The correction program, however, allows full use of the data even if such maladjustments exist.

Platinum Resistance Thermometers: To determine the atmospheric lapse rate that affects the turbulence intensity, we measured the temperature profile with platinum resistance thermometers mounted on a 50-ft collapsible mast. When collapsed, the mast can be transported on a small trailer attached to a pickup truck. The thermometer bridge allows the measurement of temperature differences within ±0.01° C between the various levels along the tower and the surface thermometer.

Recording Equipment: Electrical output from sensing instrumentation is recorded on a 20-channel Brush digital magnetic tape recorder. Each channel

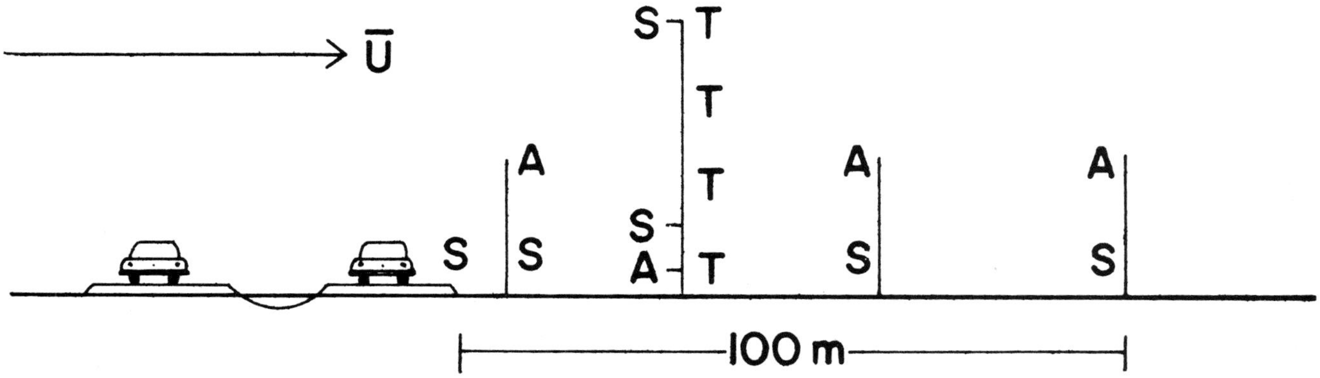

FIGURE 3-2.—Cross-section view through a typical highway sampling site. Anemometers (A), resistance thermometers (T), and high-volume samplers (S) are positioned downwind of the roadway to measure both downwind and vertical dispersion of pollutants.

can be scanned every 0.4 sec. Tapes are wound in self-contained cartridges that facilitate rapid tape change and easy storage. Data from cartridges are transferred onto computer-compatible tape with a Metrodata Systems 620 reader.

Mobile Units: The recorder and display consoles are housed in a pickup truck fitted with a camper. A second truck equipped with a hoist carries the propane-powered generator. This generator is deployed downwind from the sampling site during field measurements as an additional precaution so even the propane fuel used in powering the generator would not contribute to lead contamination. A small trailer carries the main mast and also the three portable wind masts (fig. 3-3).

Air Sampling

The standard technique for measuring atmospheric lead employs a high-volume air sampler, such as the Gelman Instrument Co.'s Hurricane air sampler, which is usually fitted with an 8-in. by 10-in. glass fiber filter. The filter is then analyzed for lead. This is an integrating technique that yields the total amount of lead collected during the sampling period for a given total air throughput. The usual sampling time is 24 h. In the present study because such atmospheric parameters as wind velocity, wind direction, and turbulence are available in considerably greater detail, it was deemed desirable to work with much shorter sampling times for lead concentration measurements. A long sampling time would mask the effect of meteorological parameters, and little useful information would be obtained on the validity of any proposed diffusion model

Furthermore, we found that the standard 8-in. by 10-in. glass fiber filters possess a lead blank, measured by atomic absorption spectrometry, of 39 μg, with a standard deviation of 20 μg. To obtain the required accuracy of 5 percent to be able to resolve the dependence of lead concentrations on distance from the source and on meteorological conditions, it would be necessary to collect a 400 μg sample with this technique. Because the lead content at the highway may vary from 1 to 2 $\mu g/m^3$, it would require a total flow of 400 m^3. With the highest available air sampling rate, the sampling time would be 3 h and 20 min. at 50 ft from the highway, the lead concentration may drop to a quarter of the value measured immediately adjacent to the roadway. The sampling time required would then be over 12 h. These sampling times are far too long for detailed investigations; however, they proved to be satisfactory for routine lead monitoring where the values are to be averaged over long periods.

A sampling technique compatible with the short sampling times required for resolution of meteorological parameters have been devised. The technique involves the use of the standard high-volume air samplers with a 5 μm Gelman triacetate Metricel filter. The measured collection efficiency of this filter is at least as good as the glass fiber, type-A filter normally employed. The efficiency is also comparable to that of 0.01 μm Metricel filters. Our results are consistent with earlier reports that filters of a given pore size can efficiently collect particles considerably smaller than the specified pore size. [7, 8] Impaction and electrostatic effects are probably important in the enhancement of the collection efficiency. Furthermore, the lead blank on the 5 μm Metricel filter averages about 1.5 μg per 8-in. by 10-in. sheet. For a 5 percent accuracy, a total lead sample of 30 μg is required. A standard glass fiber, backup filter is placed behind the 5 μm filter to avoid backstreaming effects and the transfer of lead from the filter screen to the sampling filter. To obtain a 30 μg lead sample from an atmospheric concentration of 1 $\mu g/m^3$, the time required is 30 min. From an atmospheric concentration of 0.25 $\mu g/m^3$ the sampling time is 2.0 h. This time is longer than desired, but it represents the best compromise available for routine measurements. If the backup filter is eliminated, with a possible introduction of a 10 percent error, the sampling time can be reduced to 22 min. from a concentration of 1 $\mu g/m^3$ and to 1.5 h from a concentration of 0.25 $\mu g/m^3$.

The use of a graphite cups as air pollution samplers for metal-containing aerosols is a unique development that affords analytical accuracy on the nanogram level. An emission spectrometer graphite cup electrode is used as the collecting medium in this system. The cups have a diameter of ¼ in. and a length of ¾ in. They are initially doped with a few nanograms of indium, then mounted in a holder with a diameter of 1 in. and length of 2½ in. An air sample is drawn through the cup by a vacuum pump for a period of 5 to 30 min., depending on the ambient air concentrations of the material under investiga-

FIGURE 3-3.—A small trailer carrying the main mast and the three portable wind masts.

tion. Measuring lead from automobile exhaust in the vicinity of a reasonably heavily traveled roadway usually requires air sampling of only 10 to 20 min. Thus time variations of pollutant concentrations under varying environmental wind conditions can be obtained reliably.

To determine atmospheric concentration, it is necessary to know the total volume of air passing through the filter, as well as the total mass of lead collected. Orifice-type flowmeters furnished with high-volume air samples are not satisfactory without further calibration. A precise ionization type flowmeter suitable for high-volume flow rates can be used to calibrate the orifice-meters on the commercial samplers. Flow measurements for the graphite electrode samplers are accomplished with a tube flowmeter.

DISPERSION OF POLLUTANTS FROM LINE SOURCES

Gaseous Pollutants

The average concentration of a pollutant that is not subject to gravitational settling and that emanates from a continuous line source may be expressed by

$$\chi = \frac{2Q}{\sqrt{2\pi}\,\sigma_z \bar{u}} \exp\left(-\frac{h^2}{2\sigma_z^2}\right) \qquad (1)$$

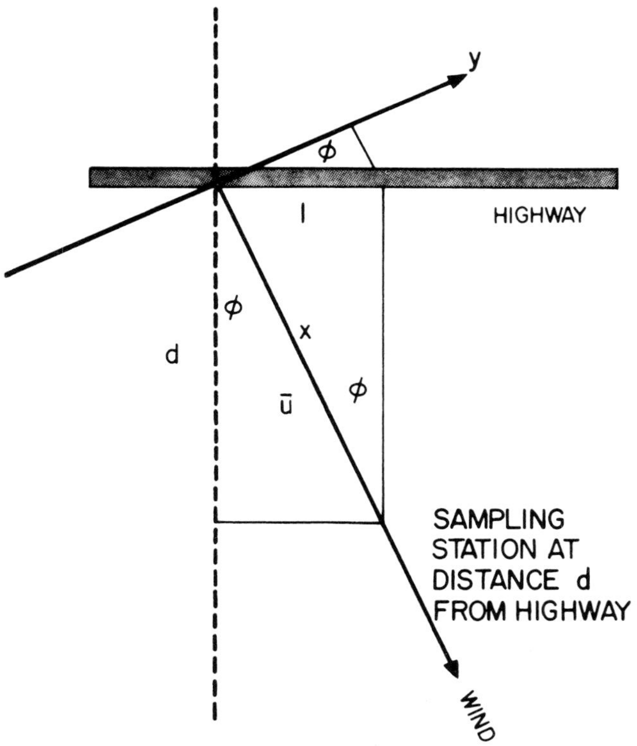

FIGURE 3-4.—Schematic diagram defining sysmbols for air flow at an angle to the highway.

Q is the source strength, σ_z is the standard deviation in the vertical of the location of individual particles with respect to the axis of the diffusing plume, $\bar{u}$ is the average wind speed component perpendicular to the line source, and h is the height of the line source above ground. σ_z is not measured directly. [10] Instead, we are using the approximation

$$\sigma_z = \sigma_w t \qquad (2)$$

which should hold well under unstable thermal stratification. [9] The quantity σ_w is the standard deviation of the fluctuations of the vertical wind component, w. The diffusion time, t, may be obtained from

$$X = \bar{u}t \qquad (3)$$

where X is the distance from the line source.

The initial turbulent mixing of aerosols, caused by the wake of cars moving along the highway, may be taken into account by postulating a ficticious, ideal line source, at a distance $-X'$ upwind from the edge of the highway, that would produce the observed dilution at the first sampling station closest to the roadway.

If the wind direction is not perpendicular to the highway, equation (1) is not strictly applicable. According to figure 3-4, the coordinate x points along the mean wind $\bar{u}$, which now blows at an angle ϕ to a line that extends perpendicular to the highway.

With the equation

$$\frac{\chi}{Q_p} = \frac{1}{\pi\sigma_y\sigma_z\bar{u}} \exp\left[-\left(\frac{y^2}{2\sigma_y^2} + \frac{h^2}{2\sigma_y^2}\right)\right] \qquad (4)$$

for a continuous point source with intensity Q_p, we assume that the highway constitutes an array of such point sources. Instead of simply integrating with respect to y from $-\infty$ to $+\infty$ [which would yield equation (1)], we have to allow for the angle ϕ at which the highway intersects our coordinate system.

We may write

$$x = \frac{d}{\cos\phi} = \bar{u}t$$

$$y = \ell \cos\phi \qquad (5)$$

where d is the perpendicular distance of the sampling site from the line source.

Substitution into (4) yields

$$\frac{\chi}{Q_p} = \frac{1}{\pi\sigma_y\sigma_z \dfrac{d}{t\cos\phi}} \exp\left[-\left(\frac{\ell^2\cos^2\phi}{2\sigma_y^2} + \frac{h^2}{2\sigma_z^2}\right)\right] \qquad (6)$$

We may integrate this equation with respect to ℓ from $-\infty$ to $+\infty$ in order to obtain the summation effect of the array of point sources that constitutes the line source of the highway

$$\int_{-\infty}^{+\infty} \frac{\chi}{Q_p}\, d\ell = \frac{1}{\pi \sigma_y \sigma_z \frac{d}{t \cos \phi}} \exp\left(-\frac{h^2}{2\sigma_z^2}\right) \int_{-\infty}^{+\infty} \exp \frac{-\ell^2 \cos^2 \phi}{2\sigma_y^2}\, d\ell \quad (7)$$

This yields

$$\chi_L = \frac{\sqrt{2}\, Q_L}{\sqrt{\pi}\sigma_z d/t} \exp\left(-\frac{h^2}{2\sigma_z^2}\right) \quad (8)$$

where Q_L is the line source intensity per unit length and χ_L is the concentration representative of this line source.

Using expression (2) we may write

$$\chi_L = \frac{\sqrt{2}\, Q_L}{\sqrt{\pi}\sigma_w d} \quad (9)$$

We have neglected the exponential term; it proved to be close to unity for the relatively low heights above street level at which car exhaust is introduced into the environment.

Computational result of equation (9) will be discussed later.

Particulate Pollutants

When the size of particles is large enough to cause an appreciable settling velocity, diffusion of the particles is affected by this velocity in various ways. As is well known, an immediately obvious effect of the falling of particles is given by a convective term, $W_g \chi$, where W_g is the gravitational settling speed of particles. By taking into account the effect of a finite fall velocity, W_g, equation (4) can be rewritten as

$$\frac{\chi}{Q_p} = \frac{1}{\pi \sigma_y \sigma_z \bar{u}} \exp\left[-\left(\frac{y^2}{2\sigma_y^2} + \frac{(h + W_g x/\bar{u})^2}{2\sigma_z^2}\right)\right] \quad (10)$$

Equation (10) is reduced to (11) following an algebraic manipulation similar to that previously discussed

$$\chi_L = \frac{\sqrt{2}\, Q_L}{\sqrt{\pi}\, \sigma_w d} \exp\left[\frac{-(\bar{u}h + W_g x)^2}{2(\sigma_w x)^2}\right] \quad (11)$$

The exponential term in (11), which has been omitted in equation (9), now becomes important due to the consideration of gravitational settling effects.

Results

The model results based on equations (9) and (11) are presented in figures 3-5, 3-6, and 3-7. In each figure, the solid line represents the best fit for the observed lead particle concentrations, which are given by heavy dots; the dashed line is the predicted concentration from (9); and the smaller black dots are the calculated concentration from (11). The discrepancy between the dashed and the full curve indicates the effect of gravitational settling. Percent values of this settling effect are given numerically in each figure at various distances. [11] The smaller black dots shown in figures 3-5 to 3-7 represent the results, respectively, from three sets of assigned data:

(1) $Q_L = 8.0\,\mu$g/m/sec., $h = 1.5$ m, W_g 0.4 m/sec.

(2) $Q_L = 12.0\,\mu$g/m/sec., $h = 1.5$ m, $W_g = 0.4$ m/sec.

(3) $Q_L = 10.0\,\mu$g/m/sec., $h = 1.5$ m, $W_g = 0.4$ m/sec.

For each individual day, the values for $\bar{u}$ and σ_w obtained from field measurements were used in both equations (9) and (11) for computing the lead concentrations. it is clear that under the chosen values of parameters, the computed magnitudes of χ_L, making use of equation (11), agree very well with the measured data, except in the first 20 m closest to the highway. This discrepancy can be attributed to the assumption of one constant fall velocity value, W_g, regardless of the wide range of the spectrum of particle sizes undoubtedly present. The magnitude of Q_L, even though it appears to be reasonable according to table 3-1, is not based upon actually measured values. More tests are desirable when a better field estimate of Q_L becomes available, in order to establish the general validity of model equation (11).

"Non-Guassian Gaussian" Dispersion

In order to describe the diffusion from an infinite line source in a turbulent shear flow field—where the sources and samplers are not at the same height, and the surface of the earth acts as a barrier to diffusion—the following equation was employed

$$\chi = \left(\frac{Q_L}{\sqrt{2\pi}\sigma_z \bar{u}}\right)\left[\exp\left(-\frac{(z-h)^2}{2\sigma_z^2}\right) + \exp\left(-\frac{(z+h)^2}{2\sigma_z^2}\right)\right] \quad (12)$$

The second exponential factor represents the virtual image source located beneath the surface of the earth, attributed to reflection of the diffusing plume by the surface of the earth.

Since we are usually dealing with a shear flow, we would not expect the vertical distribution of pollution to be described adequately by equation (12), in which $\bar{u}$ is assumed to be constant. Instead, we have attempted to describe the diffusing plume with a

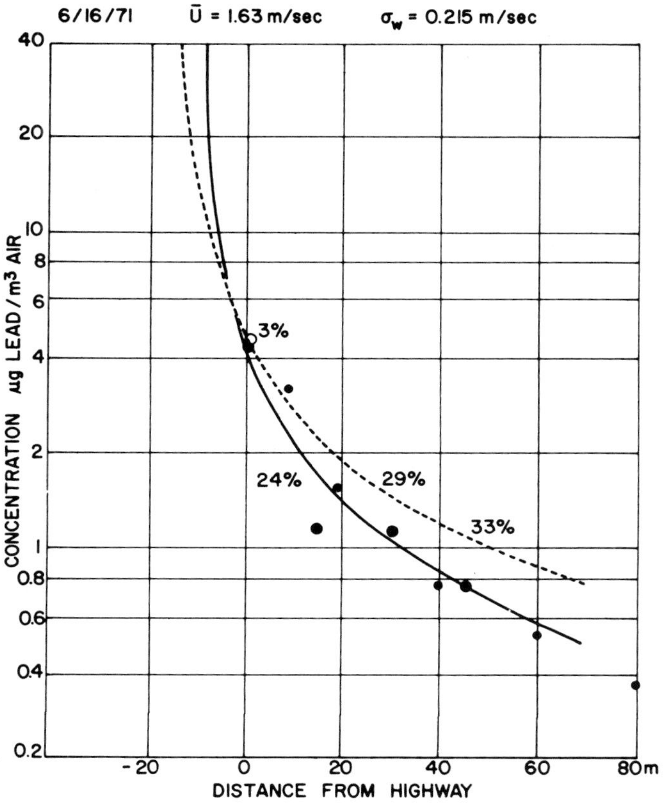

FIGURE 3-5.—Lead concentrations measured by filter samplers as a function of distance from highway (6/16/71). Heavy black dots represent distance from highway. Circle represents concentration values at sampling station closest to the highway, corrected for gravitational fallout. Solid curve gives best fit to observed lead-particle concentration; dashed curve indicates theoretical concentrations if no gravitational fallout had occurred. Percent values give the discrepancy of the observed concentrations as compared with the theoretical ones. Smaller dots represent the calculated concentration from equation (11), utilizing the assigned data: Q_L = 8.0 μg/m/sec., h = 1.5 m, W_g = 0.4 m/sec.

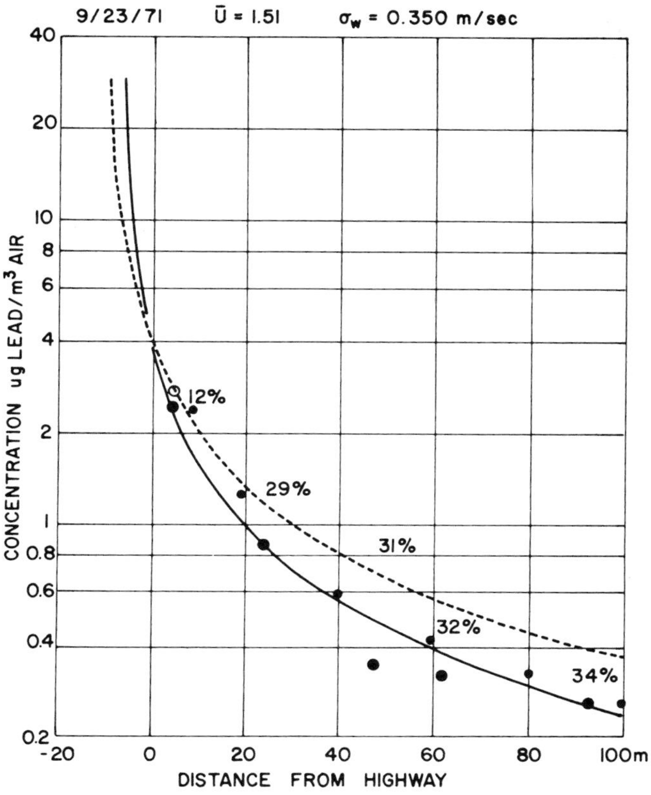

FIGURE 3-7.—Lead concentrations measured by filter samplers as a function of distance from highway (9/23/71). Heavy black dots represent distance from highway. Circle represents concentration values at sampling station closest to the highway, corrected for gravitational fallout. Solid curve gives best fit to observed lead-particle concentration; dashed curve indicates theoretical concentrations if no gravitational fallout had occurred. Percent values give the discrepancy of the observed concentrations as compared with the theoretical ones. Smaller dots represent the calculated concentration from equation (11), utilizing the assigned data: Q_L = 10.0 μg/m/sec., h = 1.5 m, W_g = 0.4 m/sec.

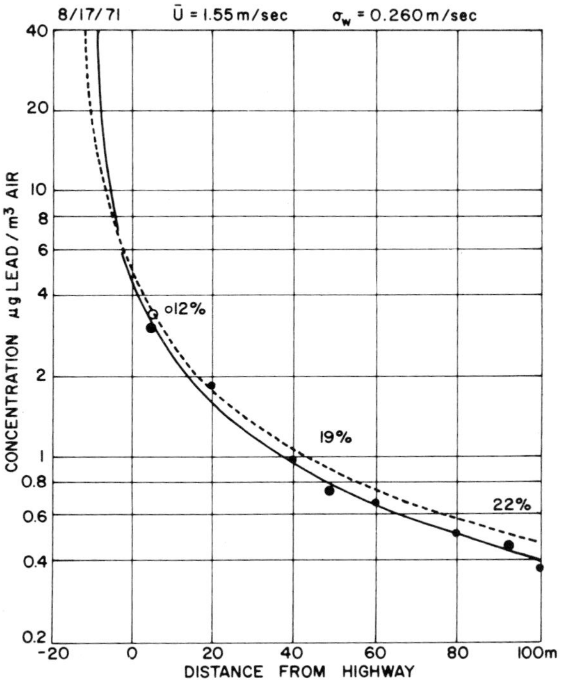

FIGURE 3-6.—Lead concentrations measured by filter samplers as a function of distance from highway (8/17/71). Heavy black dots represent distance from highway. Circle represents concentration values at sampling station closest to the highway, corrected for gravitational fallout. Solid curve gives best fit to observed lead-particle concentration; dashed curve indicates theoretical concentrations if no gravitational fallout had occurred. Percent values give the discrepancy of the observed concentrations as compared with the theoretical ones. Smaller dots represent the calculated concentration from equation (11), utilizing the assigned data: Q_L = 12.0 μg/m/sec., h = 1.5 m, W_g = 0.4 m/sec.

"Non-Gaussian Gaussian" distribution, which takes into account that $\overline{u} = f(z)$, and $\sigma_z = g(x, z)$.

Since previous work has shown that for homogeneous turbulence the dispersion of particles about their center of mass is normally distributed, we can expect this distribution to be normal for short distances about infinitesimal point sources, even in a nonhomogenous shear flow.

Further justification of our description of a non-Gaussian plume with a Gaussian distribution follows from additional physical arguments. In a *homogeneous* flow (not necessarily isotropic) at some distance downstream of their release form an infinite

TABLE 3-1.—*The values of measured (M) and predicted (P) atmospheric concentrations of lead at various stations located at various positions from the roadway and above the surface*[1].

Test / run	Station 1 X_1 = 5.2 m Z = 1.2 m		Station 2 X_2 = 24.5 m Z = 1.2 m		Station 3 X_3 = 47.5 m Z = 2.67 m		Station 4 X_4 = 62.5 m Z = 1.2 m		Station 5 X_5 = 92.5 m Z = 1.2 m	
	M	P	M	P	M	P	M	P	M	P
4/1 (X^1 = 5.6; Q_0 = 14.0; T_0 = 50)	3.65	3.65	–	1.91	0.90 [2]0.57	1.10 [2]0.58	–	0.89	0.58	0.63
5/2 (X^1 = 15.0; Q_0 = 25.0; T_0 = 100)	1.62	2.14	1.30	1.17	0.65 [2]0.29	0.70 [2]0.29	0.41	0.58	0.42	0.41
6/1 (X^1 = 9.0; Q_0 = 8.0; T_0 = 25)	2.46	2.41	0.88	1.01	0.43 [2]0.42	0.60 [2]0.59	0.38	0.48	0.32	0.34
7/1 (X^1 = 8.0; Q_0 = 16.0; T_0 = 25)	4.29	4.38	1.32	1.84	0.95 [2]0.86	1.08 [2]0.95	0.89	0.86	0.62	0.60
8/2 (X^1 = 15.0; Q_0 = 11.0; T_0 = 50)	3.28	2.31	1.35	1.34	0.99 [2]0.74	0.86 [2]0.76	0.41	0.70	0.47	0.51
10/1 (X^1 = 11.0; Q_0 = 23.0; T_0 = 100)	1.85	1.84	1.07	0.87	– [2]0.22	0.51 [2]0.23	–	0.41	0.41	0.29

[1]Stations located at position X meters from the roadway and Z meters above the surface. The value of the intensity of the virtual line source Q_0 [μg/(m-sec.)]; the position of the virtual line source upwind of the near side of the roadway X^1 (m), and the characteristics settling time T_0 (sec.) are given in each case.

[2]Z = 12.0 m.

line source, the fluid particles will have distributed themselves into a Gaussian distribution (fig 3-8). If we now consider this downstream location as the origin of re-release of the same group of particles, at a farther distance downstream the particles (of each subgroup) will have redistributed themselves again into a series of Gaussian distributions; their overall sum, in the limit, is the Gaussian distribution that we would have obtained if we had only considered the diffusion of the particles from the original point of release.

We will now deal with the dispersion of pollutants in a shear flow where the mean velocity and the root-mean-square (rms) values of the vertical velocity fluctuations vary with height, z, above ground (fig. 3-9). If we again consider a point release, we will find a Gaussian distribution at some infinitesimal distance downstream. Over this short distance, in first approximation, the flow characterized by $\overline{u}$ and $\sigma\omega$ may be treated as being homogeneous. If in a second step of approximation, we allow nonhomogeneous effects to influence the re-release of the same particles, at some farther distance downstream we will again have a series of Gaussian distributions. However, since the $\overline{u}$ and $\sigma\omega$ increase with height, the fluid elements at greater height will disperse more rapidly than the particles at low elevations. Therefore, the next set of "Gaussian" distributions will not add up to a normal distribution, but to one that is skewed (fig. 3-10).

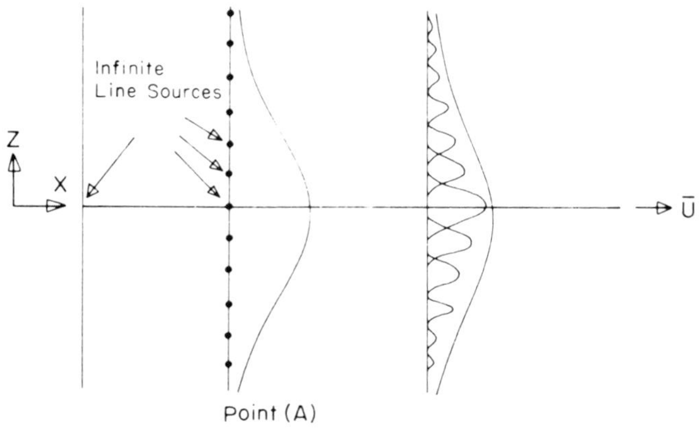

FIGURE 3-8.—Gaussian (normal) distribution of pollutants in a homogeneous flow field.

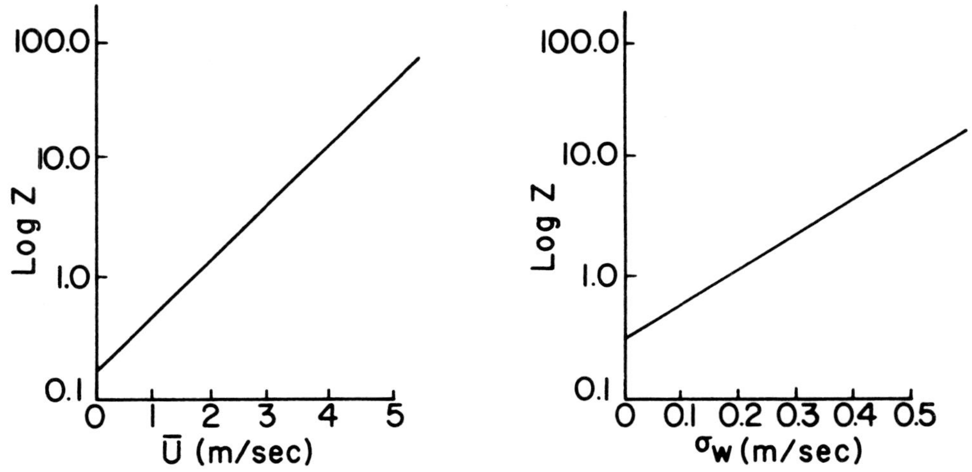

FIGURE 3-9.—Vertical profiles of wind speed and vertical wind fluctuations.

Thus in order to describe the total distribution from the source to the sampling station, we must consider the appropriate values of $\bar{u}$ and σ_z for the point of interest in space. The values to be used are those observed or interpolated at the midpoint of the particle's travel from source to sampler. Linear interpolation is valid only for linear wind profiles that do not adequately describe conditions prevailing near the ground.

In order to replace σ_z in the equation with an easily measurable variable, equation (2) is used. The source strength must be determined before equation (12) can be applied. Earlier studies [11] estimated that 15 percent of the lead emitted from the tailpipes of automobiles would fall out of the diffusing plume within the sampling area. Comparison of measured atmospheric concentrations with estimated values of emission per automobile indicate that 20 percent of the emitted lead settles out within a period of a few minutes. Thereafter, gravitation settling becomes negligibly small, at least over periods of the next few minutes.

From the long-time soil concentrations in the vicinity of the roadway given by Zimdahl, [12] and a combination of settling and dilution rates for the particles, it was possible to estimate the average source strength from the function

$$Q_L = Q_0 \left(1 - 0.20 \left\{ 1 - \exp\left[-\left(\frac{3x}{\bar{u}T_0}\right)\right]\right\}\right) \quad (13)$$

where T_0 is a characteristic time of settling determined by the rms value of the vertical wind fluctuation and $\bar{u}$ is the wind speed. T_0 and $\bar{u}$ are given at the height of the source; x is the distance of the

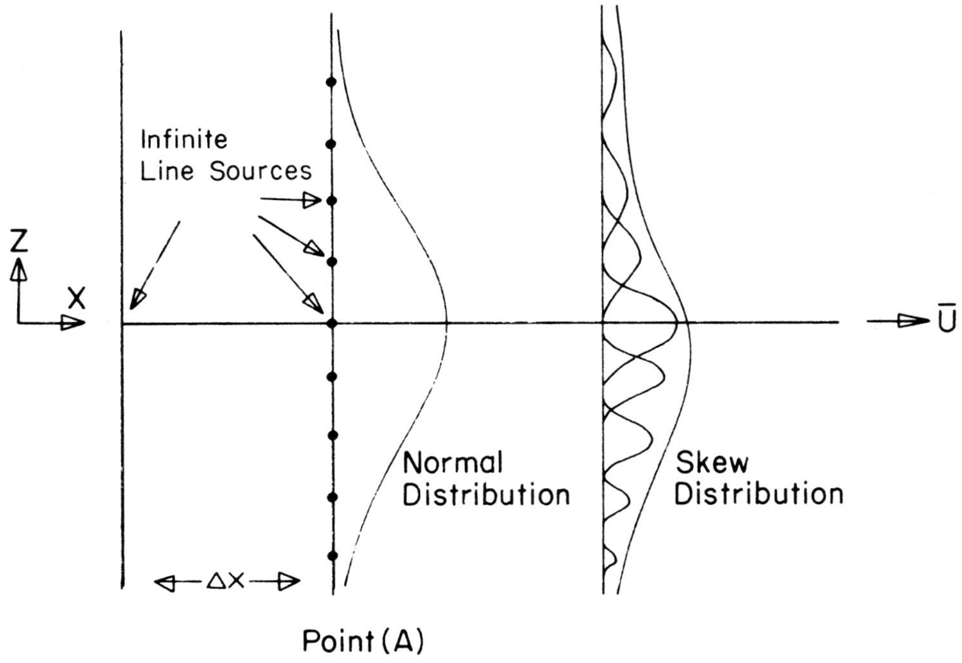

FIGURE 3-10.—Skewed distribution of pollutants in a sheared flow field.

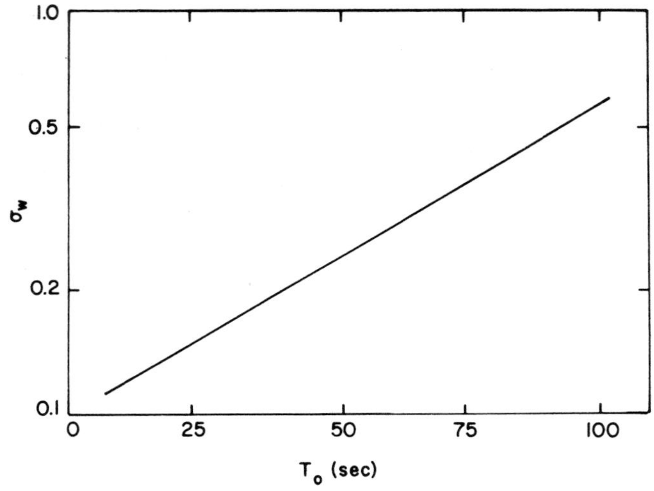

FIGURE 3-11.—Variation of mean characteristic settling time of particulate automobile exhaust as a function of vertical wind fluctuation.

sampling station from the line source. The values for T_0 are given in figure 3-11 as a function of σ_W.

The effective source strength, Q_L in μg/m-sec., which is then used in equation (12) to determine atmospheric concentrations, is weakened by gravitational settling as the distance from the roadway is increased. The actual source strength, Q_0, must be known or estimated from automobile counts and typical emission rates.

Results

Table 3-1 gives the measured concentrations as well as those predicted by equation (12). Of the 36 possible comparisons, measurements were not available for four points (indicated by dashes). The available combinations were plotted on a scatter diagram (fig. 3-12) in order to appraise the accuracy of prediction. The linear correlation coefficient between measurements and model predictions by using equations (12) and (13) was calculated to be $r = 0.97$, indicating good agreement between the two sets of values.

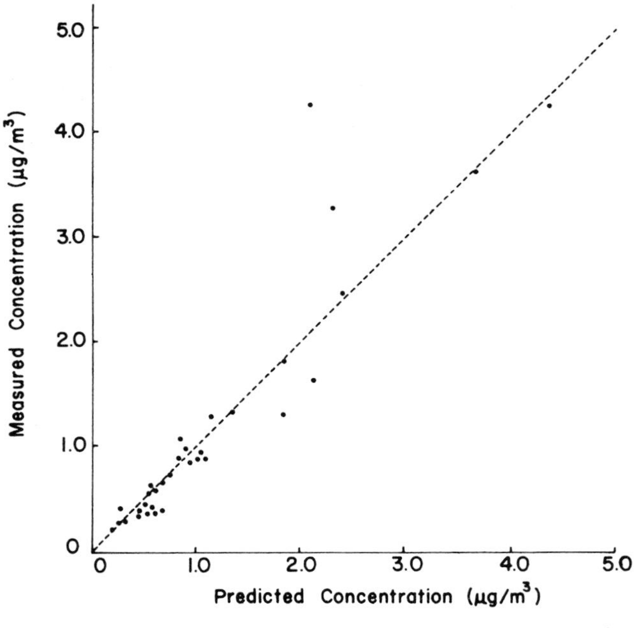

FIGURE 3-12.—Diagram of predicted concentrations as calculated from equations (12) and (13), versus measured concentration of atmospheric lead.

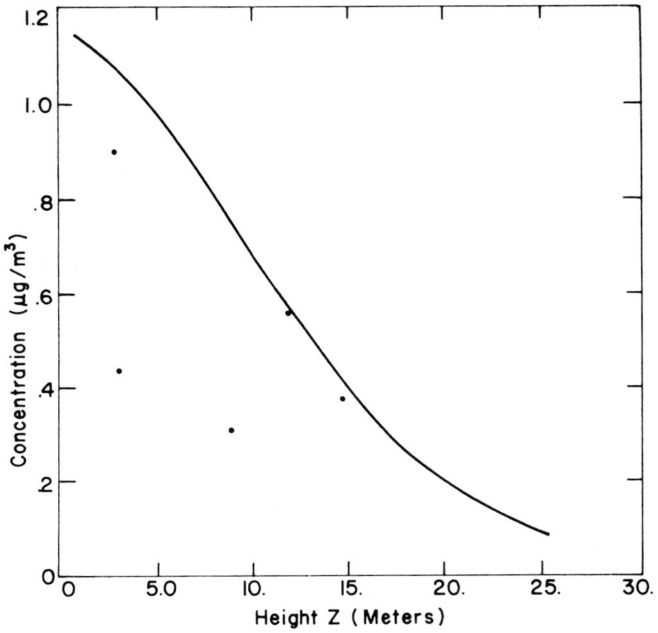

FIGURE 3-13.—Vertical plot of atmospheric concentration of lead as calculated from equations (12) and (13) at a point 52.1 m downstream from a source of strength 14.0 μg/m-sec.

Figures 3-13 and 3-14 show the vertical distribution of lead concentrations in the atmosphere as calculated from equation (12). σ_W in the term $Q_L/(\sqrt{2\pi}\ \sigma_w d)$ is evaluated at the height of the source.

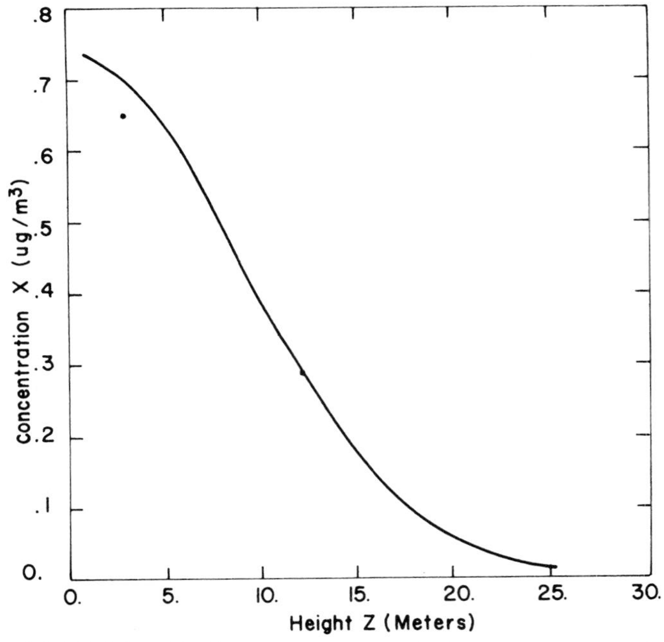

FIGURE 3-14.—Vertical plot of atmospheric concentration of lead as calculated from equations (12) and (13) at a point 62.5 m downstream from a source of strength 25.0 μg/m-sec.

The values of $\bar{u}$ and of σ_w in the exponential terms are evaluated at a point halfway between the point of interest and the real and image sources, respectively. In addition, the measured concentrations have been plotted. It is obvious that this method of computation quite accurately predicts the turbulent diffusion of lead near a highway under conditions of vertical wind shear and vertical variation of turbulence intensity.

DISPERSION OF LEAD IN AN URBAN ENVIRONMENT

The mathematical models of lead dispersion described in the foregoing sections apply to line sources over level and homogeneous terrain. The development of the "Non-Gaussian Gaussian" dispersion model discussed earlier may be applied, however, to complex terrain situations, provided that the three-dimensional distributions of the time-averaged, mean wind field and of the rms values of the wind fluctuations are known. Such knowledge has to be based upon detailed field measurements, augmented by wind tunnel comparisons.

A field program was carried out during February 1973, in downtown Fort Collins, CO, to obtain data to help define the exposure of the population to lead from automotive exhausts in regions with heavy traffic. The object, in part, was to determine the meteorological factors affecting dissipation of air pollution in cities, so that population groups exposed to the greatest hazard could be identified.

Two separate test programs were designed and conducted. The first was in a city street "canyon", where vertical wind profiles and atmospheric lead and carbon monoxide concentrations were measured. In the second test program, discussed in a later section, the daily exposure of residential areas to airborne lead was studied.

In the first experiment, a roadway with approximately uniform buildings on each side (a city street canyon) was instrumented. Ambient meteorological conditions significantly control the pollutant levels in such canyons. The conditions affecting dissipation of the pollutants, and consequently the local population exposure, were determined experimentally.

Anemometers were placed at three levels in the center of the divided roadway and at two levels above the roofs of the buildings. Lead concentrations were measured at three levels in three vertical columns in the canyon and also at two locations above the roofs, for a total of 11 simultaneous samples per test. Carbon monoxide was also measured at several of the lead sampler locations during one-third of the tests. A cross-section through a typical city street canyon as it was instrumented is shown schematically in figure 3-15.

Results of measurements of wind, turbulence, lead, and carbon monoxide are shown in figure 3-16. Figure 3-17 contains results of concentration measurements from a wind-tunnel test for a gaseous material.

A stronger vertical concentration gradient was observed in particulate lead than in gaseous CO. At 2.4 m above the roadway, the Pb/CO ratio was 0.43, while above the rooftop (11.8 m above the roadway) the average ratio was about 0.12. The sample of data shown in figure 3-16 also shows that the Pb/CO ratios decreased dramatically with height. In figure 3-16, the prevailing wind above the rooftops was from the northeast at 2.3 m/sec.; this direction is at an angle of 45° to the city street canyon under study. In this case, the traffic count was 910 vehicles/h. These data indicate surprisingly high lead concentrations. In particular, when compared with concentrations measured near open highways, relatively low traffic densities (< 1,000 vehicles/h) result in quite high atmospheric concentrations (5 to 10 $\mu g/m^3$).

In particular, wind speed, wind direction, and canyon geometry are assumed to be the controlling factors of diffusion. When the mean wind was nearly parallel to the street, the wind tended to advect the pollutants down the street canyon. A slight excess accumulation occurred on the windward side of the buildings. However, when the wind crossed the canyon at a substantial angle, a vortex or helical flow pattern was generated in the street, leading to an accumulation of lead on the leeward side of the buildings.

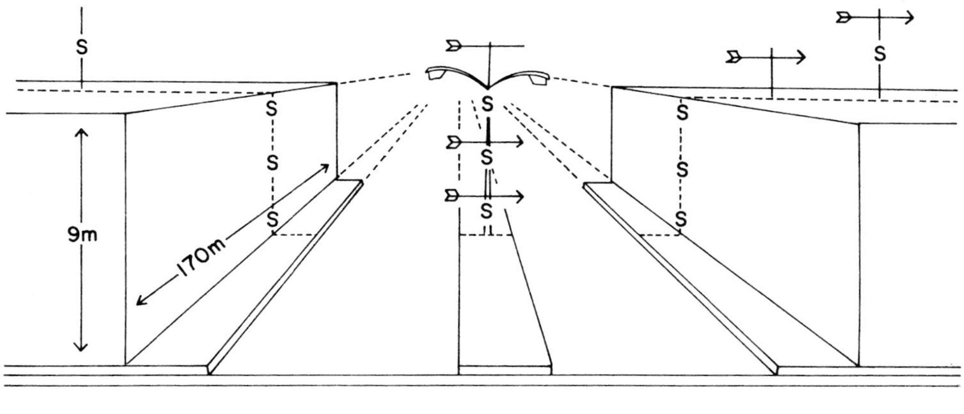

FIGURE 3-15.—Pictorial view of the 100 North block of College Avenue, Fort Collins, CO. Anemometers and air samplers were positioned as shown.

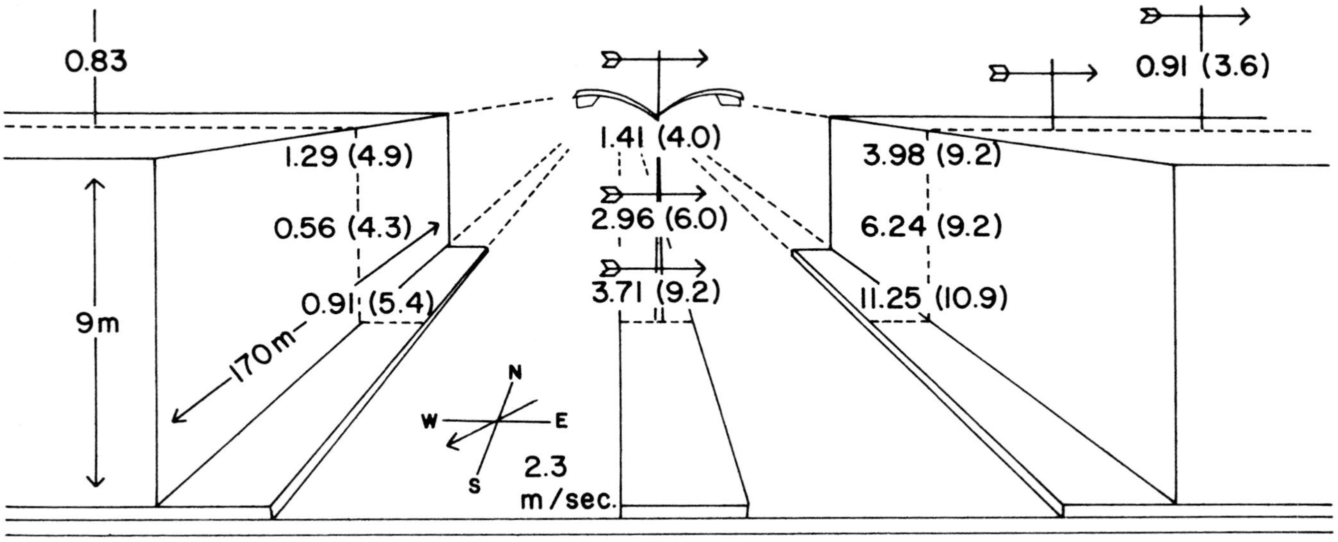

FIGURE 3-16.—Pictorial view of the 100 North block of College Avenue, Fort Collins, CO. Atmospheric lead concentrations are in $\mu g/m^3$. Carbon monoxide values in parenthesis are in parts per million. Wind speed and direction for the highest anemometer are shown.

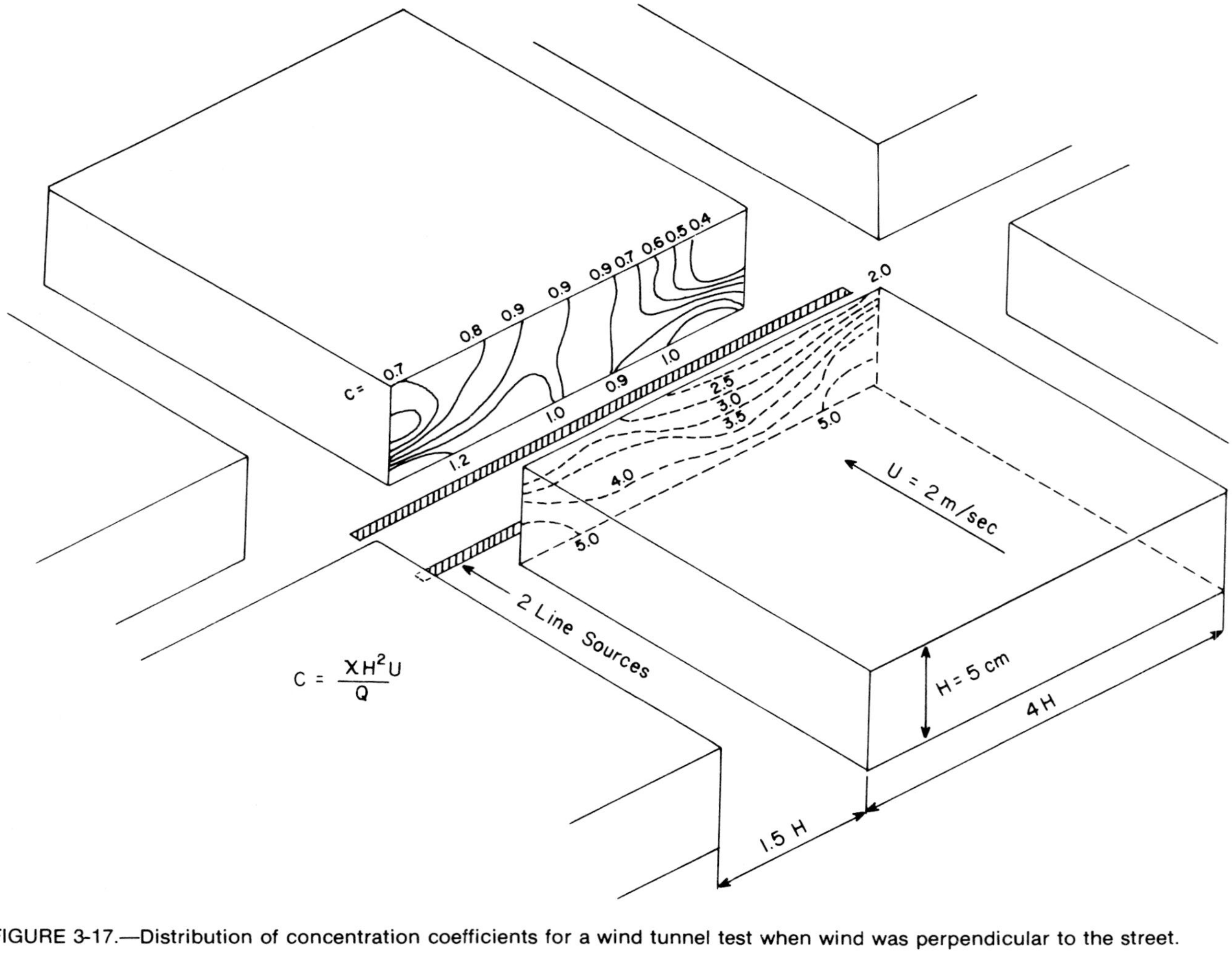

FIGURE 3-17.—Distribution of concentration coefficients for a wind tunnel test when wind was perpendicular to the street.

While the accuracy of the modeling effort is highly dependent upon the accuracy of definition of the flow field in the canyon, which is partially defined by the anemometers, it does not appear that one flow pattern dominated the flow field during the sampling period. That is, even while the mean wind was blowing at a substantial angle to the canyon, the resulting concentration patterns indicated that the component of the wind parallel to the canyon (roadway) could still be distinguished. Similarly, when the mean wind was nearly parallel to the roadway, there was sufficient lateral fluctuation in the wind direction until a circular dispersion pattern could be distinguished, superimposed on the expected longitudinal dispersion pattern. The major discrepancy between these results and those indicated by wind tunnel studies arises because true steady state conditions exist in the wind tunnel, whereas in the field program there was considerable lateral wind fluctuation even over short sampling periods.

As can be seen from figures 3-16 and 3-17, relatively large amounts of lead were deposited within the street canyon and never diffused through the tops of these canyons into the lower troposphere. The degree to which lead and other particulate pollutants remain airborne while in the street canyon depends on the intensity of turbulence which, in turn, is governed by environmental wind conditions and the geometry of buildings and street canyons. Thermal stability is believed to be only a minor factor, in view of the many heat sources that exist in a city.

INTERMEDIATE RANGE TRANSPORT OF LEAD

Sampling Program

The second phase of the urban sampling program discussed earlier considers a broader scale of human exposure to airborne lead. A sampling network of 18 stations was established in and around Fort Collins, CO, covering an area of 125 km^2. One additional station was located outside the network in the mountains northwest of the city to measure background levels of airborne lead. Typically, the total amount of lead collected in a 12-h sampling period at the mountain station was on the order of a few nanograms. This quantity is near the lower limit of detectability of the analytical equipment, with the subsequent result that the background concentrations of airborne lead were negligible during this test program. All other air samplers but one at least 100 m, with some up to 400 m, from major thoroughfares. In general, most were located 20 to 50 m from light-duty residential streets. The object of placing the samplers a considerable distance from the roadway was to sample only that portion of the lead-bearing particles that remained airborne sufficiently long to travel these moderate distances.

The initial problem in this study, as in any atmospheric dispersion modelling, was to establish the pollutant's source strength. Unlike many other automobile exhaust products, there is no satisfactory parameterization of automobile lead emissions at the present time. Several investigations [13,14] have shown that the lead emitted out the tailpipe varies from 10 to 90 percent of the lead consumed, depending upon operating conditions. Recent studies [15] have also shown that the size distribution of the emissions varies substantially depending on several of the other factors in operation. Consequently, a 45 percent emission factor was arrived at as that portion of the consumed lead that was available for intermediate range transport. This figure was based on a representative gas mileage of 13 mpg and the necessary flux of lead at the downwind edge of the city to yield the measured concentration patterns. Average fuel consumption statistics and measured traffic patterns were accumulated, and the actual emissions were determined from average daily vehicle miles pe segment of roadway in the city.

Modeling

The atmospheric dispersion modeling of a group of diffuse area sources is complicated when the source intensities are poorly defined in space and time and the diffusing material exhibits complex behavior. Such is the situation in modeling the atmospheric transport of lead-bearing particulates from automotive sources. Individual automobile lead emissions vary dramatically with operating conditions, exhaust temperature, load, efficiency, and numerous other factors. Thus defining source strengths in a reasonable and practical manner requires consideration of these factors, as well as some actual (but often not reliable) statistics on traffic volume.

A unique model has been developed that allows one to model a series of diffuse sources as an area source and then transforms this area source into an equivalent Gaussian plume model for longer-range diffusion. In this model, sources are portrayed as circular areas whose emission intensity is determined as if all traffic is uniformly distributed over the area. The use of circular sources allows making the transition from a quasi-Lagrangian box model to a Gaussian plume model in a rather smooth step, without the serious surface concentration discontinuities that occur with rectangular area sources, especially when source geometries are fixed and the wind directions are variable. The mixing height in the box is a function of stability and is determined so as to give a concentration within the box that closely approximates that calculated from a finite series of line sources aligned perpendicular to the wind.

Downwind of the area source, a standard Gaussian plume model is used to predict surface concentrations. In order to use the Gaussian plume model, virtual point sources must be utilized; they are located so as to give the vertical and horizontal plume spread corresponding to the radius of the circular area source and the parameterized mixing height.

The model has been designed around the idea of studying intermediate range (10 to 20 km) transport of lead. It does not consider that segment of the large lead-bearing particles that settles out of the atmosphere within 100 m of the roadway, and it does not have a microscale submodel to deal with effects of city geometry and other small-scale factors influencing local urban lead levels. Validating test programs were designed in such a manner that several pertinent questions could be answered. In particular, any systematic discrepancy between measured and predicted patterns could be attributed to other processes. Of concern here is that an additional fraction of the smaller airborne particulates may be removed by a gravitational settling or surface impaction, thus creating a greater concentration gradient than would be predicted. Equally possible is the conversion of organic to inorganic lead forms in these transport scales. Finally, some combination of the previously described processes could be either partially or totally nullifying, and thus undetectable unless a unique tracer experiment is conducted. Sampling equipment capable of measuring organic lead compounds or particle sizes is not always practical or sensitive enough for real-time studies at large distances from roadways. Therefore, we relied on graphite cup samples to measure particles and deduce their overall behavior.

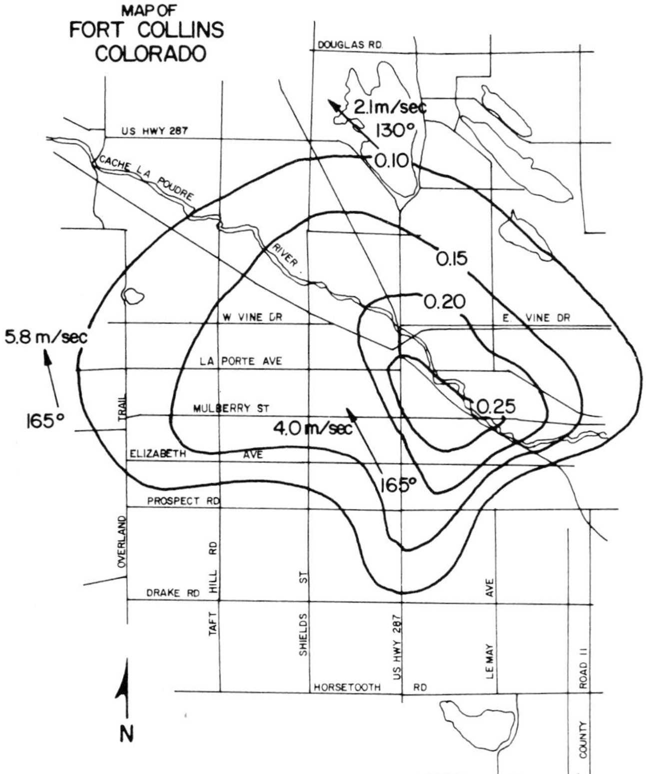

FIGURE 3-18.—Map of Fort Collins, CO showing anemometer stations and isopleths of atmospheric lead concentrations as $\mu g/m^3$. Samples were averaged over 5 1/2 h from 7:00 a.m. to 12:30 p.m. on May 9, 1973.

On the basis of the comparison of test data from the area measurements, figure 3-18, and the model predictions, which show good agreement, we could not isolate any secondary processes from purely diffusional processes. Based on a figure of 13 mpg and traffic density maps, approximately 45 percent of the consumed lead is still airborne at the suburban edge of the city. However, it should be mentioned that these tests were conducted in fair weather. In all likelihood rainout and washout play a significant role in determining the long-range transport and atmospheric budget of lead aerosols.

LONG RANGE TRANSPORT

It is now well-known that lead exhausted from automobiles and industries can be transported for long distances. Annual ice layers from the interior of Greenland show the rapid recent increase in anthropogenic lead particulates in the atmosphere. [16] (See figure 3-19.) It has been noted that a measurable fraction of lead-bearing aerosols are dispersed among small enough particles to be virtually unaffected by gravitational settling. [17] These finely dispersed particulates are available for transport processes on a global scale. In this section, the long-range transport process of lead and the removal mechanisms from the atmosphere are discussed briefly.

Pollutant Release From the Planetary Boundary Layer Into the Free Atmosphere

The lead particulates exhausted from automobiles near the ground undergo dispersion in the atmosphere due to atmospheric turbulence and the turbulent wake generated by automobiles. While undergoing dispersion, the lead particulates are subject to several different processes that lead to substantial depletion within the planetary boundary layer. Among these depletion processes, the most notable is gravitational fallout, the process by which the particulates larger than $1 \mu m$ in diameter are depleted in a short time after exhaustion. However, the particulates unaffected by gravitational settling are dispersed into the planetary boundary layer by turbulent diffusion.

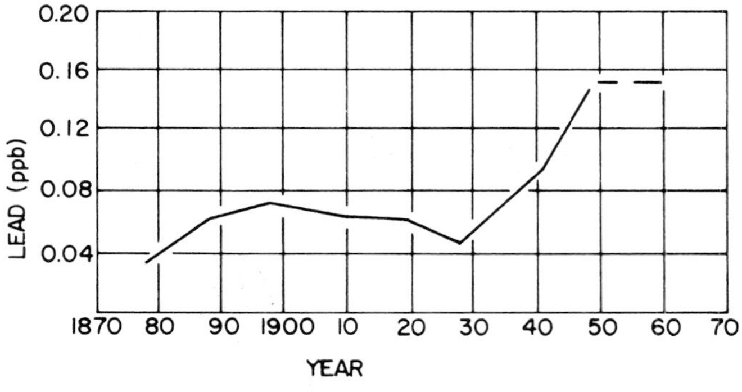

FIGURE 3-19.—Variation of anthropogenic lead dust fall at Camp Century, Greenland, from about 1880 to 1968. (After Murozumi et al. [16])

Pollutants released into the planetary boundary layer are dispersed into the free atmosphere by processes such as small-scale diffusion, convective activity, and the diurnal change of the inversion height that caps the planetary boundary layer. [18-20] The relative effectiveness of these processes depends on the meteorological conditions. Consider, for instance, the situation in which the flux of sensible heat into the planetary boundary layer from a strongly heated surface reaches a maximum in mid-afternoon. As a consequence of this flux, the thickness of the planetary boundary layer also reaches a maximum at that time. Turbulence mixing, mainly by small and mesoscale convective motions, disperses the pollutants throughout the depth of this mixing layer. In the late afternoon, and especially after sunset when the ground cools, a low-level inversion forms and traps the newly generated pollutants underneath. The pollutant contained between the maximum and minimum heights of the inversion is essentially available for long-range transport by geostrophic advection during the nighttime hours. During the next day, far downstream from its original source region, this pollution may again be mixed into the atmosphere layer adjacent to the ground as solar radiation spawns renewed mixing processes during the daytime.

Transport process by convective activity is discussed in the following section.

Long-Range Transport Processes

At the present time, the portion of pollutants escaping into the free atmosphere cannot be estimated easily. The reliable estimation must be based on airplane measurements of the flux of pollutants. However, careful analyses of air trajectories sometimes make it possible to trace the paths of pollution plumes over great distances. Robinson and Robbins [21] were able to describe the observed CO concentrations over the Greenland icecap during the summer of 1967 by such trajectory considerations. Their data showed that on certain days the concentrations of CO increased threefold above background level. Their trajectory analysis, based upon geostrophic wind data at 700 mbar, suggested that the contaminated air had originated in the heavily industrialized areas of the Great Lakes. In a similar manner, Prospero [22] studied the dust concentrations at Barbados that were carried in the northeast trade winds. The aerosol carried by these winds originated in North Africa. Large day-to-day variations in dust loading were observed. There was also a marked annual cycle in dust concentrations, summer values averaging an order of magnitude higher than winter values. The average travel time for a dust cloud to cross the Atlantic was estimated to be 5 days. Smith [23] described the transport of a smoke pall caused by forest fires in northwest Alberta, Canada. The smoke moved across the Atlantic in a westerly flow. Due to the variability of the large-scale wind field and its different character, the smoke reaching Europe dispersed over an area ranging from Scandinavia to Gibraltar. The smoke was injected over Alberta to a height of approximately 3 to 5 km. The atmosphere over the point of origin of the pollution was conditionally unstable and the forest fires provided an intense source of heat. It appeared that the large-scale transport of the smoke was not dependent in any way on conditions in the surface boundary layer.

These examples show that pollutants released into the free atmosphere can be transported over long distances by the large-scale motions of the atmosphere, and that careful trajectory analyses can describe adequately the motion of large pollution plumes. There are, however, several difficulties associated with trajectory analytical techniques. Trajectories computed for different levels in the troposphere over time periods of 3 to 5 days reveal a considerable divergence. [24] Pollution spread over a layer of certain depth in the atmosphere will be diluted due to the effect of vertical shears of the wind vector. A second difficulty arises from the fact that air parcels traveling over long distances in the troposphere move neither along isobaric nor isentropic trajectories. Therefore, for accurate predictions of the path of pollutants in the free atmosphere, it is essential to adopt a statistical treatment of trajectory analyses that can adequately describe the nonisentropic flow and spread of pollutants.

Removal Mechanisms

Precipitation scavenging (rainout) and dry deposition are most likely the dominant processes for the lead particulates injected into the free atmosphere above the planetary boundary layer.

The efficiency of precipitation scavenging depends on microphysical processes in the clouds and on the frequency with which a certain air mass is entrained into a precipitation system. The scavenging of aerosol particles by cloud droplets varies widely with the particle radius. Much work is still required to describe the process quantitatively. The frequency of precipitation events is an important factor in determining the residence time of aerosols in the atmosphere. [25] The frequency and type of precipitation (frontal or convective) are functions of geographical location and of season. Over the United States, for instance, the number of rainy days is especially large along the north Pacific coast and in the region stretching between the Great Lakes and New England. The Southwest is characterized by few precipitation events and by a dry climate. [26] One should expect, therefore, that residence times of aerosols are relatively long in the Southwest and relatively short in the Northeast.

A large number of American and Russian literature references are available showing that for many pollutants wet removal is proportional to the concentration of the pollutant in air. [27] Lodge et al., [28] for instance, showed that the concentration of lead in precipitation is larger in the vicinity of urban areas where the consumption of gasoline is high. Andersson [29] pointed out that the amount of sulfur in rainwater is higher within the perimeter of Uppsala than in the surrounding countryside.

Considerable progress has been made in recent years with regard to improving our knowledge of the dry deposition process on a smooth surface. [30] The mechanism of depletion of a pollutant at the surface of the earth is a microscale phenomenon. The rate depends on two factors: the efficiency with which gas molecules and particles are captured or exchanged at the interface; and the rapidity with which the pollution can be delivered to the interface through the planetary boundary layer, the surface layer, and the viscous sublayer; each of which presents a resistance and results in a vertical gradient of pollution. [24] However, the deposition velocity of lead particulates has not been investigated.

Rainout of Pollutants by Convective Clouds

Convective clouds are an important mechanism not only for transporting air pollutants from the boundary layer into the free atmosphere, but also for cleansing the atmosphere by precipitation processes. Because of the complexity of microphysics, processes for transporting air pollutants by cumulus convection have not been studies extensively. On the other hand, the concentrations of radio-nuclides and of other pollutants in precipitation have been observed by many authors, and the results of these

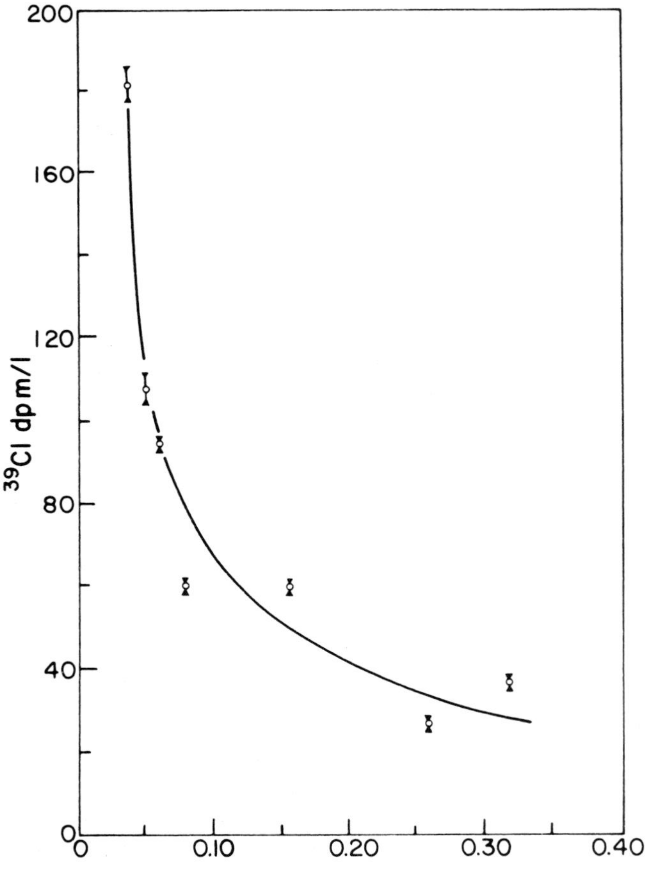

FIGURE 3-20.—^{39}Cl concentration as a function of precipitation rate. (After Engelmann and Perkins [33]).

observations were summarized by Engelmann. [31-33] The tables in Engelmann's papers reveal that the concentration of pollutants in rainwater vary over a wide range, depending not only on the chemical nature of the pollutant, but also on the cloud type from which the rainwater sample is obtained. However, there is a definite inverse relationship between the concentrations in rain and the rainfall rate. A rather striking example of this inverse relationship is shown in figure 3-20.

It is important to study the effects of the physical characteristics of clouds on the concentrations of air pollutants in rainwater. By understanding these effects, the vertical transport processes of pollutants from the planetary boundary layer into the free atmosphere and the cleansing mechanisms of the atmosphere will be better understood.

Model

The cumulus model for this study is one-dimensional and steady-state. (See fig. 3-21.) The assumptions are as follows: the cloud consists of a saturated, cylindrical, continuous vertical current in a steady state. This continuous current of jet is considered to entrain environmental air through its sides. Precipitation effects are not included in the dynamics of the cloud. The details of the equations for the model cloud are found elsewhere. [34-36]

The equation of mixing ratio of pollutant, x, can be obtained from the consideration of the conservation of the quantity x in the portion of the vertical current bounded by the levels z and $z + dz$, as shown in figure 3-22. The conservation is expressed by

$$Mx = (M + dM_e - dM_d)(x + dx) + dM_e x_e - dM_d x - MS\,dz \tag{1}$$

where M is the mass of air that rises vertically through level z within the cumulus cell. x is the mixing ratio of the pollutant in the cumulus cell, x_e is the mixing ratio of the pollutant in the environment, M_e is the mass of air entrained, M_d is the mass of air detrained, and S represents the change of x per unit depth and per unit mass of air due to any source or sink between z and $z + dz$. By neglecting the term $(dMdz)$, the equation can be rewritten as

$$\frac{dx}{dz} + Ex = Ex_e + S \tag{2}$$

where $E = \dfrac{1}{M}\dfrac{dMe}{dz}$

If only the scavenging process by cloud water contributes to the term S, we can write

$$S = \frac{\psi x}{w} \tag{3}$$

where ψ is the scavenging coefficient (sec.$^{-1}$) and w is the updraft velocity at z. The scavenging coeffi-

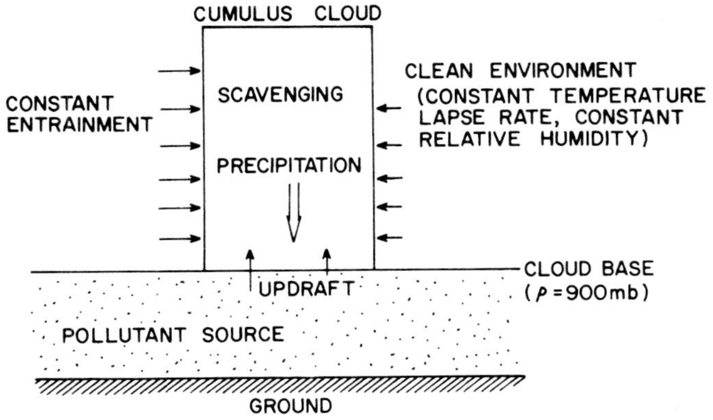

FIGURE 3-21.—Schematic diagram of the model.

cient, ψ, is a function of many physical parameters such as size distribution of cloud droplets and chemical properties of the pollutant. However, for this preliminary study, ψ is assumed to be expressed by

$$\psi = \psi_0 \frac{\rho_a(z)\,Q(z)}{(\rho_a Q) = 1 g/m^3} \tag{4}$$

This means that the scavenging coefficient, ψ, is equal to ψ_0 when the liquid water content is 1 g/m^3 and that ψ varies proportionately to the liquid water content. To simplify the problem, the further assumption is made that the air in the environment of the cumulus cloud contains no air pollutant. Therefore, equation (2) yields

$$\frac{dx}{dz} + Ex = -\psi_0 \frac{x\,\rho_a Q}{w\,(\rho_a Q) = 1\,g/m^3} \tag{5}$$

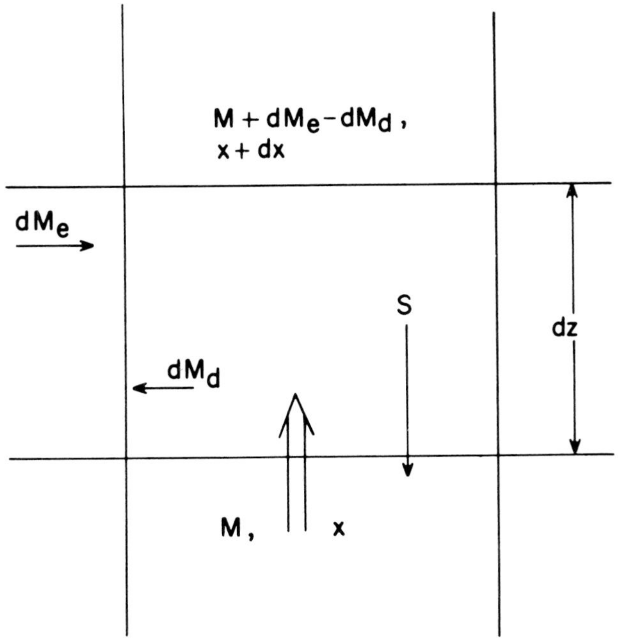

FIGURE 3-22.—Schematic diagram illustrating entrainment, detrainment, and vertical transport of quantity x in a cumulus cloud.

The mixing ratio of the pollutant at the bottom of the cloud is given by x_0. The solution of equation (5) is given by

$$X = \exp\left(-\int_0^z E\,dz\right) \exp\left[\frac{-\psi_0}{(\rho_a Q) = 1 g/m^3} \int \frac{\rho_a Q}{w}\,dz\right] \tag{6}$$

where $X = x/x_0$.

If there were no scavenging process in the cloud, the solution of equation (5) is simply expressed by

$$X' = \exp\left(-\int_0^z E dz\right) \tag{7}$$

Therefore, the amount of the pollutant scavenged in the cloud, T_x, is given by

$$T_x = x_0 \int_0^{z_{top}} \rho_a (X' - X) dz \ \ (\text{g/cm}^2) \tag{8}$$

where z_{top} is the cloud height and ρ_a is the air density. The total water condensed in the cloud is given by

$$T_{water} = \int_0^{z_{top}} \rho_a Q(z) dz \ \ (\text{g/cm}^2) \tag{9}$$

Hence, the average concentration of the pollutant in the cloud liquid water, k, is

$$k = \frac{T_x}{T_{water}}$$

the ratio of the average concentration of the pollutant in liquid water to the concentration of the pollutant in air below the cloud-base is expressed by

$$\left(\frac{k}{x_0}\right)_m = \frac{\int_0^{z_{top}} \rho_a (X' - X) dz}{\int_0^{z_{top}} \rho_a Q(z) dz} \tag{10}$$

At the present stage, the levels within the cloud that contribute mainly to the precipitation are uncertain. The assumption is made, therefore, that by the motion of air the cloud droplets are well mixed so that the concentration of the pollutant in the liquid water is uniform throughout the depth of the cloud. If this assumption is valid, equation (10) gives the value of the ratio of the concentration of the pollutant in rainwater to the concentration of the pollutant in surface air.

In order to have clouds with different physical characteristics, the combinations are composed of physical parameters (temperature lapse rate of the atmosphere, updraft velocity at cloud base, relative humidity of environment, and entrainment rate).

Davis (37) reported that from the observational studies, the scavenging coefficient ψ, for ^{24}Na is larger than 10^{-4} sec.$^{-1}$, and for ^{38}Cl and ^{39}Cl it is larger than 10^{-3} sec.$^{-1}$. Perkins et al. [38] reported that ψ for ^{38}Cl is between 10^{-3} sec^{-1} and 10^{-2} sec.$^{-1}$. For this reason, the following three different values of ψ^0 were chosen: 0.001, 0.10, and 0.05 sec.$^{-1}$.

Results and Discussion

Vertical Distribution of the Pollutant in Cloud Air: The vertical distributions of the pollutant in cloud air are shown in figures 3-23 (*a*) and (*b*) for the two different clouds described previously. In these figures, the scavenging coefficient ψ_0 is taken as an independent parameter. In the case of larger values of ψ_0, the relative concentration of pollutant, x/x_0, decreases rapidly with height. Comparison of the two figures reveals that in the smaller cloud the concentration of the pollutant decreases more rapidly with height. This is due to the slower vertical velocity in smaller clouds. The slower the vertical velocity, the longer the time required for the air to reach the same height above the cloud base. In the case of small values of ψ^0, the concentration of the pollutant in cloud air is significant even in the top portion of a cloud. On the other hand, in the case of larger values of ψ_0, the pollutant is scavenged quickly by cloud water near the lower portion of the cloud. Larger values of ψ_0 may be at least qualitatively regarded as corresponding to the scavenging coefficients for hygroscopic materials, such as sodium chloride. As described previously, the value of ψ is chosen rather arbitrarily. For future work, improved values of ψ should be chosen.

Average Concentrations of Pollutant in Cloud Water as a Function of Cloud Height: The average con-

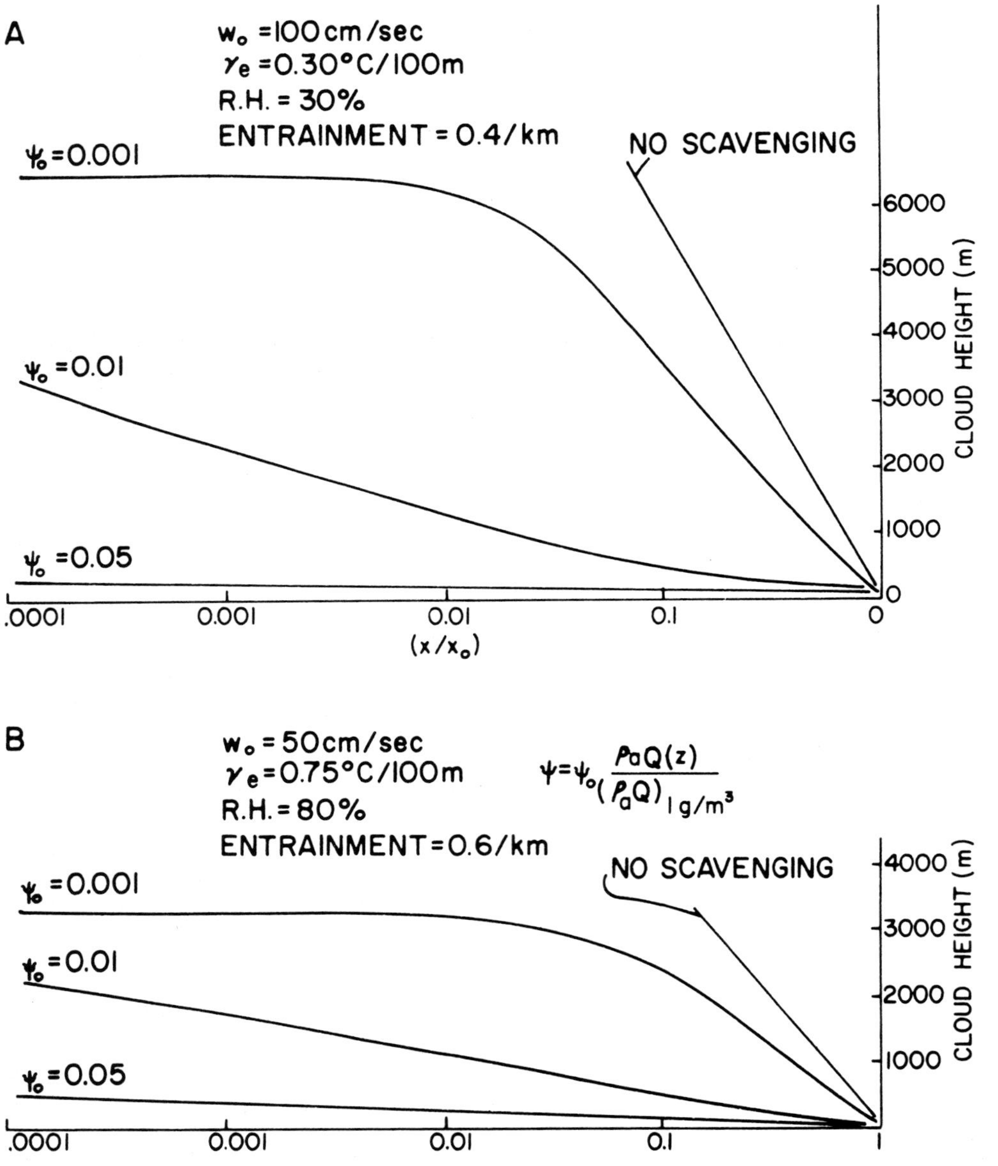

FIGURE 3-23.—Vertical distributions of the pollutant in cloud air for two different clouds.

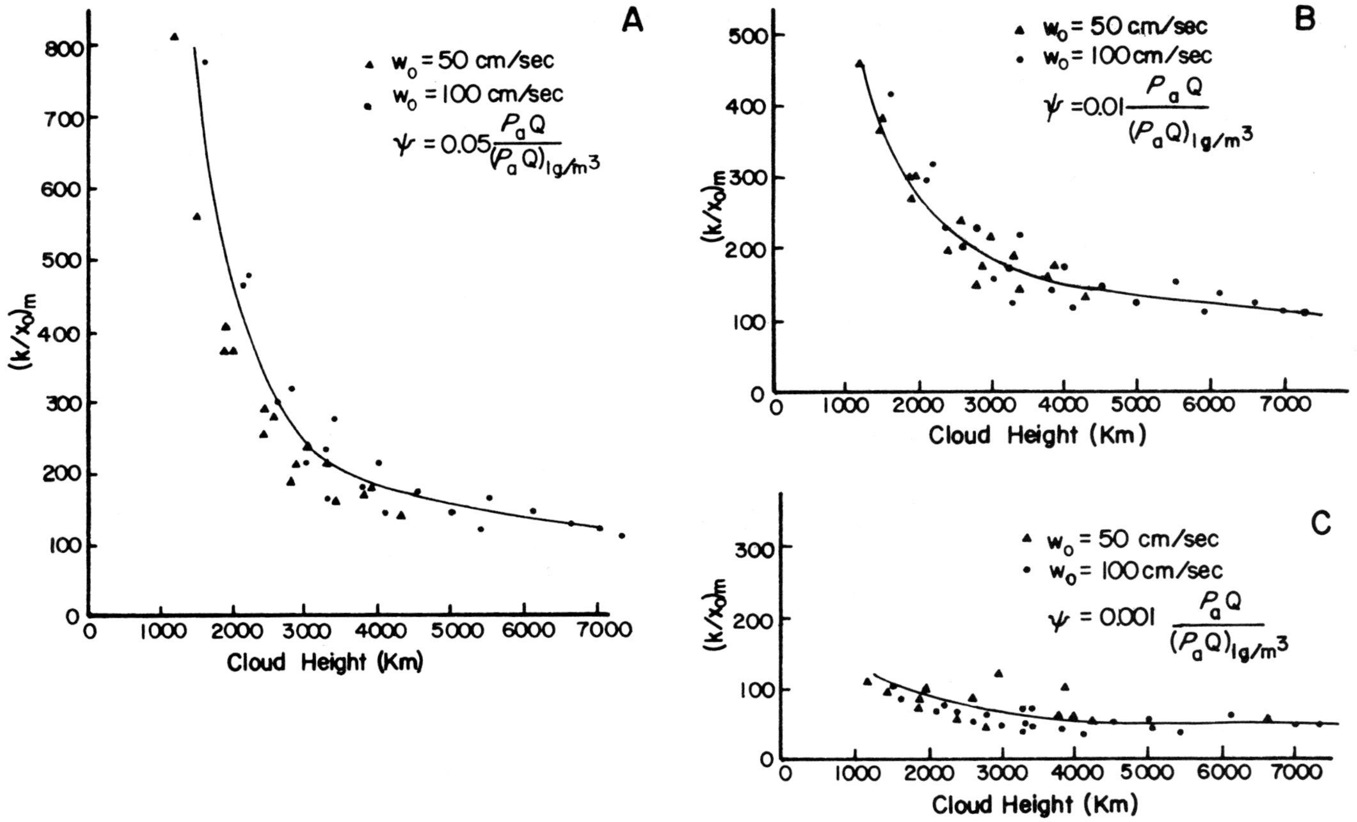

FIGURE 3-24.—Average concentrations of the pollutant in cloud water versus cloud height for different values of the three scavenging coefficients.

centrations of the pollutant in cloud water, which is defined by the equation (10), was calculated for each cloud. Figures 3-24 (*a*), (*b*), and (*c*) show $(k/x_0)_m$ versus cloud height for ψ_0 = 0.05, 0.01, and 0.001 sec.$^{-1}$, respectively. In these figures, the triangular marks and circular marks are for clouds grown under the condition of w_0 = 50 cm/sec., γ_e = 0.75°C per 100 m and of w_0 = 100 cm/sec., γ_e = 80° C per 100 m, respectively. The shape of the curve in figure 3-24 (a) resembles rather strikingly the curve in figure 3-20, which shows the observational results of the concentration of ^{39}Cl in rainwater versus the precipitation rate. It is well known that, in general, the precipitation rate from clouds increases with cloud height. [39]

As is expected, the comparison of figures 3-24 (*a*), (*b*), and (*c*) reveals that the larger the value of ψ_0, the larger the concentration of the pollutant in rainwater. It should be pointed out that the difference of $(k/x_0)_m$ for different ψ_0 is large when the height of the cloud is small, and that for the higher clouds the difference of $(k/x_0)_m$ becomes smaller for different ψ_0. This can be explained as follows: for the chosen value of ψ_0, the scavenging processes in the clouds are so effective that most of the pollutants are scavenged by the liquid water in the lower portion of the cloud. On the other hand, cloud liquid water is fairly uniformly distributed with height throughout the cloud. As a consequence, the difference in the average values of $(k/x_0)_m$ becomes smaller for different ψ_0 in the case of higher clouds. It should also be pointed out that for smaller values of ψ_0, the dependency of $(k/x_0)_m$ on the depth of the cloud becomes smaller. This reflects the fact that the time required for the pollutant to be scavenged by cloud water is longer under inefficient scavenging processes.

The scattering of values in the figure can be explained by the fact that each of the model clouds is grown under different environmental conditions.

Conclusion

If the assumption is valid that cloud water droplets are well mixed vertically as well as horizontally by the motion of cloud air so that the pollutant in the cloud liquid water is uniformly distributed throughout the cloud, the results obtained in this preliminary calculation explain, at least qualitatively, the observation that the concentration of pollutants in rainwater decreases with an increasing rate of rainfall. It is emphasized that the dynamics of clouds are as important a factor as the scavenging coefficient in controlling the concentration of pollutants in rainwater.

The scavenging process involves many physical parameters of the cloud and many microphysical mechanisms. Incorporation of these parameters and mechanisms into a numerical model is a difficult task. An effort should be made to undertake such a task in order to understand the processes of vertical transport of pollutants by clouds and also the eventual removal of certain pollutants from the atmosphere. It should then be possible to estimate the impact of certain pollutants on the global environment by referring back to climatological data on atmospheric transport and precipitation processes.

REFERENCES

1. O.G. Sutton. Atmospheric Turbulence. (1949).
2. F. Pasquill. Atmospheric Diffusion. Van Nostrand Co., London (1962).
3. E. R. Reiter. J. Appl. Meteorol. *2*: 691 (1963).
4. E. R. Reiter. Further Studies on Radioactive Fallout. Colorado State University Atmospheric Science Paper No. 70, Fort Collins, CO (1965).
5. D.B. Turner. Workbook of Atmospheric Dispersion Estimates. Public Health Service Publication No. 999-Ap-26, Cincinnati, OH (1969).
6. K. Hababi. Environ. Sci. Technol. *4*: 239 (1970).
7. C. L. Lindeken, F.K. Petrock, W.A. Philips, and R. D. Taylor. Health Phys. *10:* 495 (1964).
8. W.J. Megaw and R.D. Wiffen. Int. J. Air Water Pollut. *7*: 501 (1963).
9. F.A. Gifford, Jr. An Outline of Theories of Diffusion in the Lower Layers of the Atmosphere. Meteorology and Atomic Energy. TID-24190. U.S. Atomic Energy Commission Office of Information Service (1968).
10. W.L. Godson. Arch. Meteorol. Geophys. Bioklim., Ser. A. *10:* 305 (1958).
11. E.R. Reiter and P.C. Katen. Transport of Lead from Automotive Sources Through the Environment. Presented at AGU Fall Annual Meeting, San Francisco, CA (1971).
12. R.L. Zimdahl. *Impact on Man of Environmental Contamination Caused by Lead.* Interim Report, NSF Grant GI-4, Colorado State University, Fort Collins, CO (1972).
13. K. Habibi. Environ. Sci. Technol. *4*: 223 (1970).
14. K. Habibi, E.S. Jacobs, W. G. Kunz, Jr., and D.L. Pastell. Paper presented to the Air Pollution Control Association Meeting, San Francisco, CA October 8-9. (1970).
15. J.T. Ganley and G.S. Springer. Environ. Sci. Technol. *8*: 340 (1974).
16. M. Murozumi, T.J. Chow, and C. Patterson. Geochim. Cosmochim. Acta. *33*: 1247 (1969).
17. M.L. Corrin. *Impact on Man of Environmental Contamination Caused by Lead.* Interim Report (July 1972 through December 1973), H.W. Edwards, ed. Colorado State University, Fort Collins, CO (1973).
18. G. C. Holzworth. Monthly Weath. Rev. *92*: 235 (1964).
19. G.C. Holzworth. J. Appl. Meteorol. *6*: 1039 (1967).
20. E.R. Reiter. Hydrodynamic Tracers. AEC Review Series, U.S. Atomic Energy Commission (1972).
21. E. Robinson and E.C. Robbins. J. Geophys. Res. *74*: 1968 (1969).
22. J.M. Prospero. Bull. Amer. Meteorol. Soc. *49*: 645 (1968).
23. C.D. Smith. Monthly Weath. Rev. *78*: 180 (1950).
24. R.E. Munn and B. Bolin. Sur. Atmos. Environ. *5*: 363 (1971).
25. H. Rodhe and J. Grandell. Tellus *26*: 442 (1972).
26. G.T. Trewartha. An Introduction to Climate, 4th ed. McGraw-Hill, New York, N.Y. (1968)
27. E.R. Reiter. Atmospheric Transport Processes, Part 4: Hydrodynamic Tracers. AEC Review Series, U. S.Atomic Energy Commission (1975).
28. J.P. Lodge, Jr., et al., Chemistry of United States Precipitation. Final report on the national precipitation sampling network. National Center for Atmospheric Research. (1968).
29. T. Andersson. Tellus. *21*: 685 (1969).
30. G.A. Sehmel. J. Colloid Interface Sci. *37:* 891 (1971).
31. R.J. Engelmann. The Calculation of Precipitation Scavenging. Meteorology and Atomic Energy. TID U.S. Atomic Energy Commission Office of Information Service (1968).
32. R.J. Engelmann. J. Appl. Meteorol. *10*: 493 (1971).
33. R.J. Englemann and R. W. Perkins. J. Atmos. Sci. *28*: 131 (1971).
34. G.J. Haltiner. Tellus. *11*: 4 (1959).
35. R.C. Srivastava. A Model of Convection With Entrainment and Precipitation in Two Studies of Convection. McGill University, Stormy Weather Group Scientific Report MW-38 (1964).
36. T. Henmi and E. R. Reiter. Pollution Transport by Cumulus Clouds. Proceedings of 2nd Annual NSF-RANN Trace Contaminants Conference, Asilomar, CA (1974).
37. W.E. Davis. J. Geophys. Res. *77*: 2159 (1972).
38. R.W. Perkins, C.W. Thomas, and J.A. Young. J. Geophys. Res. *75*: 3076 (1970).
39. P.M. Austin and R. A. Houze, Jr. J. Appl. Meteorol. *11*: 926 (1972).

CHAPTER 4

LEAD IN SOIL

R.L. Zimdahl
Department of Botany and Plant Pathology
Colorado State University

J.J. Hassett
Department of Agronomy
University of Illinois

NATURAL OCCURRENCE OF LEAD

Lead (Pb) is the most abundant of the natural heavy elements, with an atomic number greater than 60. Lead (atomic number 82) occurs in nature as four stable isotopes in varying relative amounts: [1] ^{204}Pb (1.2 to 1.6 percent in almost all samples), ^{206}Pb (20 to 28 percent), ^{207}Pb (20 to 23 percent), and ^{208}Pb (50 to 54 percent). In addition, four short-lived radioactive isotopes: ^{210}Pb, ^{211}Pb, ^{212}Pb, and ^{214}Pb occur in nature as decay products of uranium and thorium.

The insoluble sulfide ore, galena (PbS), is the primary form of lead in the natural state. Lead is also found as plattnerite (PbO_2), cerussite ($PbCO_3$), and anglesite ($PbSO_4$). In its several forms, lead supports an active mining industry. [2, 3] Lead occurs mainly as Pb^{2+}; it is oxidized to Pb^{4+} only under strong oxidizing conditions, and few simple compounds of Pb^{4+} other than PbO_2 are stable. [4]

The accepted average value for the lead content of the earth's crust is 15 μg/g [4-6] or 150,000 tons/mi^3. [7] A range of 0.8 to 500 μg/g has been reported for arable soils. [8, 9] Those sedimentary or alluvial origin often have as little as 0.04 to 0.06 μg/g lead. [10, 11] Other reports [12, 13] list additional values for the Pb content of soils. The highest concentrations of lead occur in the upper horizon of the soil [14, 15] with small additions, leached down to the subsoil.

Parent material influences native lead content, and soils in suspected ore areas have levels up to 45,000 μg/g. [8] Ultrabasic igneous rocks have been reported to contain 0.1 μg/g lead, while carboniferous rocks contain 35 μg/g. [16]

SOURCES OF ANTHROPOGENIC LEAD

Emphasis in the effort to define the role of lead in the environment has centered on automobile fuels to which organic lead (tetraethyl and tetramethyl lead) is added as a means of reducing preignition (knock or ping) problems in high-compression engines. The organic lead is converted to inorganic salts during combustion of the fuel and exits the automobile as halides, hydroxides, and oxides, with smaller amounts as carbonates and sulfates. [17-19]

The relationship of highway traffic to lead in soils has been the subject of intense study in recent years. [2, 3, 20-27] Generally, the consensus of opinion is that lead concentrations decrease rapidly with the distance from the source. Atmospheric lead concentrations stabilize at low levels approximately 100 ft from the road, primarily due to the rapid settling of particles greater than 5 μm in diameter and the downwind traverse of smaller particles entrained in the turbulent atmosphere. [2, 18, 28-31]

Lagerwerff [25] stated that automobile exhaust is currently the dominant source of soil contamination by lead. The quantity of lead discharged into the atmosphere of the United States in 1964 amounted to 100,000 tons, or 10 percent of the total industrial consumption. The average lead content of gasoline is assumed to be 2.1 and 1.5 g/gal. for premium and regular, respectively. Of the lead in gasoline, at least 25 percent is trapped within the automobile [25] and may never reach the atmosphere.

Soils receive about 1 μg/cm^2/year from precipitation, and 0.2 μg/cm^2/year may be deposited in dustfall. [18] Ter Haar et al. [32] estimated that 0.2 percent of the existing natural burden of soil lead is added each year from atmospheric sources. This is only 0.04 to 4 μg Pb/g of soil/year and probably explains the failure to detect accumulation over as long as 3 or more years, even in areas with heavy traffic volume.

The concentration of lead in surface soils in city parks is quite high: [18] Balboa Park, San Diego, 194 μg/g; Golden Gate Park, San Francisco, 560 μg/g; and MacArthur Park, Los Angeles, 3,357 μg/g. Several U.S. studies have reported varying lead concentrations in soils adjacent to highways. Some of the higher values are 540, [26] 160, [27] 403, [23] and 530 μg/g. [33]

The isotopic ratio technique is based on the quantitative differences in the ratios of ^{204}Pb, ^{206}Pb, and ^{207}Pb in lead mined for use in gasoline and the quaternary lead produced by weathering of soil minerals. [34] The technique has also been used by Chow and Johnstone [35] to show that snow on Mt. Lassen, CA, contained lead that originated in gasoline used in Los Angeles. The assumption that an automotive source can be identified by the isotopic ratio is generally accepted, but it has not been established that the isotopic composition of tetraethyl lead differs from lead in other industrial products that may also enter the environment. [18]

Ault et al. [36] found significant differences in lead isotopic ratios ($^{206}Pb/^{204}Pb$) in rock, soil, grass, tree leaves, coal, fly ash, and gasoline. Leaf, grass, and soil samples showed a change in lead isotopic ratios from the highway to 1 mi. windward. The mean value of the 206/204 atomic ratio in topsoil within 500 ft of a major highway was 18.2 +0.2; and beyond it was 18.7 +0.15, which indicates the limited lateral transport of lead. Rabinowitz and Wetherill [37], with a similar technique, found that the major portion of lead near the surface of the soil was isotopically identical to gasoline lead but was different from natural lead. Lead samples from soil profiles in two New Jersey forests showed an increase in the 206/204 atomic ratio with depth ranging from 18.7 at the surface to 19.9 at 30 in. [36, 38] This is similar to the 20.86 value at bedrock reported by Ault et al. [36]

DeTreville [13] stated that no soil samples have been found to be free of lead. Many studies have reported decreases in soil lead levels with distance from the highway, [2, 10, 21-23, 25-27, 29, 33, 38-48], depth, [2, 10, 21, 23, 24, 28, 33, 37-39, 48-50] and changes in traffic volume. [3, 23, 34, 51] In one such study, [2, 33] the lead content of soil adjacent to Interstate Highway 25 north of Denver, CO, was measured as a function of distance from the highway and depth in the profile. Analysis of 0 to 15 cm samples revealed that lead concentrations were highest immediately adjacent to the roadway and rapidly diminished with depth in the profile and distance from the highway. The lead levels of soil samples obtained farther than about 100 ft from the highway were indistinguishable from background. Values up to 1,300 μg/g were found in the top 1 cm adjacent to the highway.

Jenkins and Davies [49] suggested that the enrichment of metals in soils is a function of mineralogy, parent material, texture, organic matter, and pH; soil enrichment due to atmospheric sources is in the order Bi, Pb, Cu > Sn, Zn > Ni > Mo, Be, Ge, Co, Cr > V, Mn. Patterson [52] generally agreed, listing sources of natural lead in the atmosphere as soil dust, volcanic lead halogen aerosols, volcanic silicate smoke, forest fire smoke, aerosols of sea salts, and meteoric smokes. Other sources of atmospheric lead are listed in a National Academy of Sciences report, [18] but no specific data are given as to the amount from each source. Although there is considerable lead in the atmosphere at any given time, it is in the form of very small particles that remain suspended; high concentrations at a source are not causing significant accumulation in soils on a global basis, according to Nathans and Stopps. [53] The same authors state that "no effect is yet discernible on lead concentrations in soils far from the region of injection." They calculate that 1.73×10^{-5} g Pb/cm^2 are deposited annually throughout the world. Most of the lead remains in the northern hemisphere due to meterological phenomena, causing a rise in lead concentrations in Greenland, while no comparable rise occurs in the Antarctic. Patterson [52] suggested that fallout of particulate lead parallels crop production areas, since they are almost always in close proximity to roads. He calculated that 4×10^{-5} g Pb/m^2/day accumulate on crop land in the United States.

Lead may also be added to soil as the insecticide lead arsenate and inadvertently as a fertilizer impurity. This may account for the higher lead levels noted in some agricultural soils. [54]

In the fourth millenium B.C. lead was one of the first metals to be used by man and may have been the first obtained by smelting an ore. In the ancient world, the uses of lead included making glazes, glass, and pigments; and writings were inscribed on lead sheets. By Roman times, lead was used in large quantities for plumbing and other architectural purposes. [4]

Lead pollution of soils from mining and smelting activities is localized, but it may be intense. While lead mining is usually an underground operation, pollution may occur due to windblown ore concentrate from storage dumps, open trucks, railroad cars used for haulage, and trailings ponds. Smelting operations also contribute stack emissions.

Canney [55] investigated the lead content of soils near a lead smelter in the Coeur d'Alene, Idaho, district. He found high ($\geq$ 1,000 μg/g) values of lead in the less-than-80-mesh soil fraction at a distance of 5 miles from the smelter. Most of the lead was in the 0 to 5 cm soil layer. The smelter has been operating for about 40 years. Near two zinc smelters at Palmerton, PA, Buchauer [56] found up to 2,000 μg/g of lead on the lower O_2 horizon (decomposed leaf litter) at a distance of 1 km. The upper A horizon showed between 200 and 1,100 μg/g lead, 90 percent of which was restricted to the upper 15 cm of the soil profile. The study indicated that background values were reached 39 km downwind and 16 km upwind from the smelter. In the Tamar Valley of Cornwall and Devon, England, a district that was intensively mined for nonferrous ores in the 19th century, Davies [57] found a mean lead concentration of 71 μg/g, with a range of 20 to 310 μg/g in the top 15 cm of agricultural soils. Garden soils had a mean concentration of 260 μg/g, with a range of 41 to 522 μg/g. High lead concentrations were also found in the soils of the Avonmouth area of Severnside, England, by Little and Martin, [58] at a distance of at least 10 km from the source. This industrial complex includes one of the largest lead and zinc smelting plants in the world.

Current investigation in the Viburnum Trend or New Lead Belt of southeast Missouri, [59] which is discussed in chapter 7, also shows high lead contents in soils adjacent to a smelter that has been operating from 4 to 7 years. In soils not covered by decaying leaf litter (O_1 or O_2), the lead is contained almost entirely in the top 2.5 cm of the soil. However, most of the area is forested and has a well developed forest floor. The concentration of lead in leaf litter at a distance of 5 miles is 200 to 1,000 μg/g, and increased lead levels have been detected 20 to 25 mi. from the smelter. The distribution of lead on vertical profiles clearly indicates that leaf litter retains most of the lead, in this case up to 98 percent, during the first 4 years of operation.

Lead pollutants from smelters occur mainly in the mineral form. X-ray analyses of baghouse material from the smelter in the Virburnum Trend and from another smelter in southeast Missouri showed the presence of PbS, $PbSO_4$, elemental lead, and a mineral tentatively identified as $PbO \cdot PbSO_4$. The same materials were also identified in the soils near the smelter.

While all investigations of recent smelting activity show that lead is restricted to the top few centimeters of soil, work reported from ancient smelter sites indicates that lead may move downward with time. In the Jadwin Mine area in Burma, Webb [60] found the soil contaminated with lead to a depth of 2.5 m. Strong contamination to a depth of about 50 cm has also been noted at the Bushman Mine in Bechuana Land, an ancient smelting site. According to the investigator, much deeper contamination would have occurred but for the 8.0 to 9.2 pH of the soil.

BEHAVIOR OF LEAD IN SOIL

The chemistry of heavy metals such as lead can qualitatively be described as affected by (1) the specific or exchange adsorption at mineral interfaces, (2) the precipitation of sparingly soluble solid phases of which lead is a constituent, and (3) the formation of relatively stable organic-metal complexes or chelates that result when lead interacts with soil organic matter. [61]

Lead exists in the automobile exhaust primarily in the form of halide salts; [18, 19, 62] but if phosphorous is included in the fuel, it may be in the form of a phosphate-halide compound. [18]. Lead halides are relatively soluble salts and may be carried downward for short distances. With time, these compounds may be converted to less soluble forms. [45]

Nriagu [63-65] investigated the role of indigenous soil phosphate as a potential buffer for lead. His data suggest that lead may be precipitated as a pyromorphite form of $Pb_5(PO_4)_3X$ where X is a halide ion. Santillan-Medrano and Jurinak [61] found under the experimental conditions that $Pb(OH)_2$ appears to regulate Pb^{2+} activity when the pH of the solution is less than 6.6. At pH values greater than 6.6, lead orthophosphate, lead hydroxypyromorphite, and tetraplumbite phosphate are possible solid phases. Miller et al. [66, 67] found a relationship between the uptake of lead by corn and soybeans and soil phosphorus levels. The uptake of lead by corn or soybeans was found to decrease with increasing phosphorus and pH values. They speculated that this was due to precipitations of lead as a phosphate or hydroxide. Zimdahl and Foster [68] also found that applied phosphorus, manure, and lime decreased the lead uptake. Hassett [69, 70] found that the sorption of lead by soil was highly correlated with soil pH, cation exchange capacity (CEC), and phosphorus fertility levels. Hem [71] concluded that in many surface waters the activity of Pb^{2+} is regulated by $PbCO_3$ or $Pb_3(OH)_2(CO_3)_2$.

Keaton [72], Lagerwerff (25), and Dedolph et al. [39] found a high degree of fixation, as compounds with low solubility formed rapidly. For example, 3 days after the addition of 2,784 μg/g lead nitrate, only 17 μg/g soluble lead were found in the soil. In other work by Brewer [73] only 3 μg/g soluble lead was found 3 days after the addition of 6,000 lb of lead nitrate per acre (approximately 3,000 μg/g). Similarly, sorption accounts for a rapid loss of lead from solution. Zimdahl et al. [74] and Hassett [69] found that the sorption of lead by soil could be described by the Langmuir equation.

Studies at Colorado State University have shown that the high degree of lead immobilization by soil may be directly correlated with the soil's cation exchange capacity and inversely with soil pH. [74] The sorption equilibrium for lead is approached quite rapidly. Organic matter was determined to be the primary immobilization agent for ionic lead. Fixation by interaction with clay minerals or by surface adsorption processes appeared to be of less consequence. Stepwise multiple regression analyses with several soil parameters yielded the following regression equation

$$N^* = 4.931 + 0.281 \text{ CEC} + 1.073 \text{ pH}$$

N^* is the theoretical monolayer capacity from the Langmuir model to which the data conform mathematically. The regression coefficient was 0.971, indicating the Langmuir N^* value to be a reasonably good predictor of a soil's capacity to immobilize lead.

Recent work at the University of Illinois [69] with leaching studies and adsorption isotherms has shown that soil pH, cation exchange capacity, organic matter, and the available phosphorus level can affect a soil's capacity to remove lead from solution. A regression equation was determined that predicted the capacity of a soil to sorb lead based on its CEC, pH, and soluble P level.

$$\text{Pb}s = 34.3 + 0.0774 \text{ P}_1 + 5.358 \text{ pH} + 5.337 \text{ CEC}$$

Pbs is the predicted micromoles of lead per gram of soil removed from solution; the coefficient of multiple determination for this equation is 0.997. Results of the regression analyses indicated that soil properties associated with CEC i.e., higher organic matter content, higher surface area, and higher clay content, have a greater effect on lead sorption than soil pH, and that soil pH has a greater effect than soluble P.

Miller et al. [67] grew corn for 4 weeks on lead-amended soils and found that uptake by the corn (Pb_p) depended upon levels in the soil (Pb_t), relative to the capacity of soil to sorb lead.

$$Pb_p = 628 \frac{Pb_t}{Pbs} + 1.61$$

Thus it is the relative amount and not just the total lead that controls uptake.

Mackenzie [75] reported that lead sorbed on montmorillonite was rendered nonexchangeable by drying the clay. Bittell and Miller [76] determined the selectivity coefficients of Pb^{2+}, Cd^{2+}, and Ca^{2+} ions on montmorillonite, illite, and kaolinite. Their data show that in pure clay systems, Pb^{2+} was preferentially adsorbed over Ca^{2+}, whereas the exchange of Cd^{2+} for CA^{2+} had a selectivity coefficient near unity. Lagerwerff and Brower [77] investigated the exchange behavior of Pb^{2+} in small concentrations in kaolinitic, montmorillonitic, and illitic soils pretreated with Al^{3+} or Ca^{2+} and kept at a number of salinity (Cl) levels. A Gapon-type equation was found to describe the reactions. The mean values of the Gapon exchange coefficient describing the distribution of ions between adsorbed and solution phases were 0.31, 0.11, and 0.24 for the Pb-Al system and 4.13, 4.97, and 11.1 for the Pb-Ca system, respectively, in the kaolinitic, illitic, and montmorillonitic soils. Lead precipitated in the Na^+ treated alkalized soils.

Tso [78] showed that limited amounts of ^{210}Pb were removed from soils by leaching with water, and that mild mechanical agitation could increase the rate of removal. Braids and Howe [21] found little movement of lead from the area of application; however, leaching with certain chelating agents successfully displaced the lead.

Lead applied to soil columns as lead nitrate is only very slowly leached by water. [2] Leaching studies carried out at Colorado State University employed two soils amended to soil-Pb concentrations of 460 and 4,300 μg/g. Each column was leached with 3,350 ml of deionized water (equivalent to 105 acre-in., or about seven years of Fort Collins' rainfall) initially containing less than 0.05 μg/g lead; the leaching removed 1.6 percent and 6.2 percent of the lead present on the 460 and 4,300 μg/g columns, respectively. Other studies [79] have shown virtually no leaching in the natural situation. It was determined that the soils could hold from 40 to 72 percent of their cation exchange capacity before leaching occurred. The same workers also found that the soil capacity increased when the deposition was intermittent.

An additional leaching experiment with an equivalent amount of water was carried out with roadside soil collected near Interstate 25 north of Denver, CO. Only 1.4 percent of the total lead was removed. These data suggest that this soil has a capacity for immobilizing substantially more lead than was applied. Theoretical monolayer capacity is from 3.09×10^{-5} to 4.87×10^{-5} moles/g. To conform this effect, the pH of another soil was lowered from 6.8 to 4.0. The N* value was thereby reduced to 4.3×10^{-5} from 6.4×10^{-5} moles/g.

The hydrogen ion is a good competitor for available sorption sites. Most lead salts also increase in solubility as the pH decreases. Thus lead should be more available for plant uptake in acid soils because it is sorbed to a lesser extent and is less likely to be precipitated. John and Van Laerhoven [80] have shown that the formation of lead carbonate (which is unstable at the pH of their experimental soil) does not explain the effect of pH on lead availability. Webb [60] observed that soils with a high pH tended to deter the leaching of lead near a smelter. This observation is in agreement with another [81] that showed more lead immobilized at higher pH values. The mechanism may be an effect of pH on the solubility of lead salts, or a competition between lead and hydrogen ions for available fixation sites; to date, however, it has not been explained.

Goldschmidt [5] proposed that lead is concentrated in the humus or organic fraction of soils in forests because it is taken up slowly by tree roots and transported to the leaves, which fall and decay. The lead then remains near the surface because of adsorption and insolubility. Swaine and Mitchell [9] studied the lead profiles of eight soils in Scotland and found the lead content at 50 in. to be one-half the surface value. They proposed the formation of an insoluble complex in the surface soil, which holds lead in the surface horizons.

Recent results obtained by Stevenson (unpublished data) at Illinois show rather rapid downward movement of lead in a Drummer silt loam soil, where plots established by Baumhardt and Welch [20] had received treatments of lead ranging from 0 to 3,200 kg/ha in 1969. After six years, the maximum lead penetration was 46 to 61 cm for the 800 kg/ha rate; 61 to 76 cm for the 1,600 kg/ha rate; and 76 to 91 cm where 3,200 kg/ha had been added. In addition, both organic matter and total nitrogen increased on lead-amended plots where 20 tons per acre of ground corn cobs were added in 1973 and 1974. These increases appeared due to enhanced preservation of stable humus, perhaps because newly formed humic and fulvic acids were protected from microbial attack by the formation of complexes with the added lead.

Lead is considered by Scalay [82] as an element that is sorbed by humic acids (huminophilic) and becomes concentrated in humus-rich material. A recent study by Baker [83] demonstrates a sharp increase in solubility of lead minerals in 0.1 percent humic acids extracted from soils, when compared with CO_2 saturated water. The investigation also shows that while most of the investigated metal humates displayed "classical insolubility" in water, they are readily mobilized by a 0.1 percent solution of humic acid. These data, as well as recent research discussed by Kirkland [84] on the structure and reactivity of fulvic acids (a relatively low molecular weight humic material), indicate the possibility of aquatic mobilization of otherwise insoluble lead compounds in soil.

Work at Illinois [85-87] with titration curves of soil organic matter (humic acid factor) and lead indicated that at low M^{2+}/humic acid ratios, 2:1 complexes are formed; at high M^{2+}/humic acid ratios, 1:1 complexes are formed. Complexes of Cu^{2+} and Pb^{2+} are considerably more stable than those for Cd^{2+}. Log K_2 values, obtained from the relationship $k_j = b_j / K_i (K_i$ = ionization constant), increased rather dramatically with decreasing salt concentrations and were of the order of those reported in the literature for metal complexes with known biochemical compounds. Differences were slight between humic acids in their ability to bind metal ions, and at least two major sites were involved in the binding.

SUMMARY

The behavior of lead in a soil is dependent to a large extent on the nature of the soil. Soil pH, organic matter, inorganic colloids, iron oxides, and phosphorus fertility, in addition to the amount of Pb in the soil, determine the chemical status of lead and hence its availability to plants.

Lead appears to be tightly bound by most soils, and substantial amounts must accumulate before it represents a hazard to the growth of higher plants. At the present soil levels (due to pollution by tetraethyl lead) and at levels likely in the near future, lead does not appear to have an economic effect on crop growth or quality. The effect of lead due to pollution by mining, milling, and smelting operations is different, as much greater amounts of Pb are being added to the environment in relatively concentrated areas.

The effect of lead on soil microbial populations is as yet unknown, and there has been very little work on this phase of the problem.

REFERENCES

1. W. H. Allaway. Advan. Agron. 20: 235 (1968).
2. H. W. Edwards, M. L. Corrin, L. O. Grant, L. M. Hartman, E. R. Reiter, R. K. Skogerboe, C. G. Wilbur, and R. L. Zimdahl. Interim Report, NSF Grant GI-4, Colorado State University, Fort Collins, CO 169 (1971).
3. A. L. Page and R. J. Gange. Environ. Sci. Technol. *4*: 140 (1970).
4. J. M. Wampler. In *The Encyclopedia of Geochemistry and Environmental Sciences*. R. W. Fairbridge, Van Nostrand Reinhold Co., New York, NY Vol. IVA, 642 (1972).
5. V. M. Goldschmidt. J. Chem. Soc. (London) 655 (1937).
6. A. P. Vinogradov. *The Geochemistry of Rare and Dispersed Elements in Soils*. Consultant Bureau Inc., New York, NY (1959).
7. Distribution of Elements in the Earth's Crust and Sea Water. U.S.G.S. news release (June 1973).
8. D. J. Swaine. Commonwealth Bureau Soil Science Technical Communication No. 48. Herald Printing Works, York, England (1955).
9. D. J. Swaine and R. L. Mitchell. J. Soil Sci. *11*: 347 (1960).
10. K. N. Bagchi, H. D. Ganguly, and J. N. Sudor. Indian J. Med. Res. *28*: 411 (1940).
11. D. Purves. Plant Soil. *26*: 380 (1967).
12. J. Connor. Ph.D. thesis, Rutgers University (1961).
13. R. T. P. DeTreville. Arch. Environ. Health. *8*: 212 (1964).
14. J. R. Wright. Soil Sci. Soc. Amer. Proc. *19*: 340 (1955).
15. J. L. Seeley, D. Dick, J. H. Arvik, R. L. Zimdahl, and R. K. Skogerboe. Appl. Spectros. *26*: 456 (1972).
16. K. K. Turekian and K. H. Wedepohl. Geol. Soc. Amer. Bull. *72*: 175 (1961).
17. K. Habibi. Environ. Sci. Technol. *4*: 239 (1970).
18. Lead: Airborne Lead in Perspective. National Academy of Sciences. A report of the Committee on Biological Effects of Atmospheric Pollution, Division of Medical Sciences, National Research Council. (1972).
19. G. L. Ter Haar and M. A. Bayard. Nature. *232*: 555 (1971).
20. G. R. Baumhardt and L. F. Welch. J. Environ. Qual. *1*: 92 (1972).
21. O. C. Braids and S. W. Howe. Abstr. Amer. Soc. Agron. NY 144 (1971).
22. H. L. Cannon and J. M. Bowles. Science. *137*: 765 (1962).
23. T. J. Chow. Nature. *225*: 295 (1970).
24. M. K. John. Environ. Sci. Technol. *5*: 1199 (1971).
25. J. V. Lagerwerff. In *Agriculture and the Quality of our Environment*, American Association for the Advancement of Science. Publication 85. Washington, D.C. 343 (1967).
26. J. V. Lagerwerff and A. W. Specht. Environ. Sci. Technol. *4*: 583 (1970).
27. H. L. Motto, R. H. Daines, D. M. Chilko, and C. K. Motto. Environ. Sci. Technol. *4*: 321 (1970).
28. I. D. Besner and P. R. Atkins. Tech. Rep. EHE-20-08. Environmental Health. Engineering Laboratory, University of Texas, Austin, TX (1970).
29. J. P. Creason, O. McNulty, L. T. Heiderscheit, D. H. Swanson, and R. W. Bucchley. In *Trace Substances in Environmental Health*, V. D.D. Hemphill, ed. University of Missouri, Columbia, MO 129 (1972).
30. R. H. Daines, H. Motto, and D. M. Chilko. Environ. Sci. Technol. *4*: 318 (1970).
31. E. R. Reiter and P. C. Katen. Proceedings AGU Fall Annual Meeting, San Francisco, CA (1971).
32. G. L. Ter Haar, R. B. Holtzman, and H. F. Lucas, Jr. Nature. *216*: 353 (1967).
33. R. L. Zimdahl and J. H. Arvik. In *Proceedings of Conference on Environmental Chemistry, E. P. Savage ed. Colorado State University, Fort Collins, CO 33 (1972).*
34. *J. V. Lagerwerff. Micronutrients in Agriculture.* Soil Science Society of America Ch. 23 (1972).
35. T. J. Chow and M. S. Johnstone. Science. *147*: 502 (1965).
36. W. V. Ault, R. G. Senechal, and W. E. Erlebach. Environ. Sci. Technol. *4*: 305 (1970).
37. M. B. Rabinowitz and G. W. Wetherill. Environ. Sci. Technol *6*: 705 (1972).
38. J. J. Connor, J. A. Erdman, J. D. Sims, and R. J. Ebans. In *Trace Substances in Environmental Health*, IV. D. D. Hemphill, ed. University of Missouri, Columbia, MO 26 (1970).
39. R. Dedolph, G. L. Ter Haar, R. B. Holtzman, and H. F. Lucas, Jr. Environ. Sci. Technol. *4*: 217 (1970).
40. J. C. Everett, C. L. Day, and D. Reynolds. Food Cosmet. Toxicol. *5*: 29 (1967).
41. R. J. Ganje and A. L. Page. Calif. Agr. *26*: 7 (1972).
42. A. Kleinman. Pest. Monit. J. *1*: 8 (1968).
43. H. O. Leh. Gesunde Pflanzen. *18*: 21 (1966).
44. A. L. Page, T. J. Ganje, and M. S. Joshi. Hilgardia. *41*: 1 (1971).
45. M. J. Singer and L. Hanson. Soil Sci. Soc. Amer. Proc. *33*: 152 (1969).
46. W. H. Smith. For. Sci. *17*: 195 (1971).
47. H. V. Warren and R. E. Delavault. J. Sci. Food Agr. *13*: 96 (1962).
48. W. J. Vandenabeele and O. L. Wood. Chemosphere. *5*: 221 (1972).
49. D. A. Jenkins and R. I. Davies. Nature. *210*: 1296 (1966).
50. T. G. Siccama and E. Porter. Biol. Sci. *22*: 232 (1972).
51. A. Kloke and K. Riebartsch. Naturwissenschaften. *51*: 368 (1964).

52. C. C. Patterson. Arch. Environ. Health. *11*: 344 (1965).
53. M. W. Nathans and G. J. Stopps. Proceedings of 64th Annual Meeting of the Air Pollution Control Association, Atlantic City, NJ (1971).
54. E. A. Schuck and J. K. Locke. Environ. Sci. Technol. *4*: 324 (1970).
55. F. C. Canney. Min. Eng. *11:* 205 (1959).
56. M. J. Buchauer. Environ. Sci. Technol. 7: 131 (1973).
57. B. E. Davies. Oikos. *22:* 366 (1971).
58. P. Little and M. H. Martin. Environ. Pollut. *3*: 241 (1972).
59. Ongoing research, University of Missouri, Rolla, MO.
60. J. S. Webb. International Geology Congress, XX Session. 143 (1958).
61. J. Santillan-Medrano and J. J. Jurinak. Soil Sci. Soc. Amer. Proc. *39:* 851 (1975).
62. D. A. Hirschler, L. F. Gilbert, F. W. Lamb, and L. M. Niebylski. Ind. Eng. Chem. *49:* 1131 (1957).
63. J. O. Nriagu. Inorg. Chem. *11*: 2499 (1972).
64. J. O. Nriagu. Geochim. Cosmochim. Acta. *37*: 367 (1973).
65. J. O. Nriagu. Geochim. Cosmochim. Acta. *37*: 1735 (1973).
66. J. E. Miller, J. J. Hassett, and D. E. Koeppe. Commun. Soil Sci. Plant Anal. *6:* 339 (1975).
67. J. E. Miller, J. J. Hassett, and D. E. Koeppe. Commun. Soil Sci. Plant Anal. *6:* 349 (1975).
68. R. L. Zimdahl and J. M. Foster. J. Environ. Qual. *5*: 31 (1976).
69. J. J. Hassett. Commun. Soil Sci. Plant Anal. *5*: 499 (1974).
70. J. J. Hassett. Commun. Soil Sci. Plant Anal. *7:* 189- (1976).
71. J. D. Hem. J. Amer. Water Works Ass. *65*: 562 (1973).
72. C. M. Keaton. Soil Sci. *43*: 401 (1937).
73. R. F. Brewer. Lead in Diagnostic Criteria for Plants and Soils. University of California Division of Agriculture Science, Riverside, CA 213 (1966).
74. R. L. Zimdahl. In *Impact on Man of Environmental Contamination Caused by Lead.* Interim Report, NSF Grant GI-4 and GI-34813X. (1971).
75. R. C. Mackenzie. Proceedings of International Clay Conference, Stockholm. 183 (1963), *Chem. Abstr. 64:* 8964f.
76. J. E. Bittell and R. J. Miller. J. Environ. Qual. *3*: 250 (1974).
77. J. V. Lagerwerff and D. L. Brower. Soil Sci. Soc. Amer. Proc. *37*: 11 (1973).
78. T. C. Tso. Agron. J. *62*: 663 (1970).
79. S. M. Linnemann. M. S. thesis, University of Missouri-Rolla, MO (1975).
80. M. K. John and C. Van Laerhoven. J. Environ. Qual. *1*: 169 (1972).
81. R. L. Zimdahl. In *Environmental Contamination Caused by Lead.* Interim Report, NSF Grant GI-34813X1 and GI-44423. (1974).
82. A. Scalay. Geochim. Cosmochim. Acta. *28*: 1605 (1964).
83. W. E. Baker. Geochim. Cosmochim. Acta. *37*: 269 (1973).
84. D. F. Kirkland. Chem. Can. 40 (1973).
85. F. J. Stevenson. In *Environmental Geochemistry.* J. O. Nriagu, ed. 95. (1976)
86. F. J. Stevenson. Soil Sci. Soc. Amer. J. (in review).
87. F. J. Stevenson. Soil Sci. (in review).

CHAPTER 5

UPTAKE BY PLANTS

R.L. Zimdahl
Department of Botany and Plant Pathology
Colorado State University

D.E. Koeppe
Department of Agronomy
University of Illinois

INTRODUCTION

When considering ecosystem dynamics, food chains, and the presence of lead, the answers to questions concerning lead uptake by and translocation within plants are important. The lead content of many plant species has been reported to vary with the species and the environment during growth. [1-24] Since all mineral elements in plants come from the growing environment, it is logical to posit this source for lead. Two recognized avenues of uptake are available, one from soil sources through the roots and the other from atmospheric sources through aerial portions of the plant. This discussion deals with the general questions of how lead is taken up by plants, where it is deposited after uptake, and what environmental variables influence this uptake and distribution.

UPTAKE OF LEAD BY PLANT ROOTS

Under most circumstances, ion uptake by roots is from the solution phase of the root growth medium. Ions adsorbed to clay particles or to organic matter are not in solution, and in most instances are not freely available for plant uptake until they are desorbed from the surface into solution. Therefore, when considering the question, "Can plant roots take up lead?", it is logical to first consider experiments done with root growth media without, or with only a few, exchangeable sites that can bind lead. Several such hydroponic experiments have been conducted with a nutrient plus lead medium completely devoid of solid material, or with only sand for root support. They clearly indicate that lead quickly becomes associated with roots after the initiation of treatment, and that some of the lead is absorbed and translocated within the plant. [13, 25, 26] Miller and Koeppe showed that corn (*Zea mays* L.) accumulates large amounts of lead when grown in sand culture. They demonstrated a stunting effect on corn from as little as 24 μg lead per g of sand under phosphate-deficient conditions. The root-to-shoot ratio of absorbed lead decreases as the solution concentration is increased. Root concentration of 32,400, 54,300, and 10,600 respectively, were achieved after 3 days exposure of pinto bean (*Phaseolus* spp.), sugar beet (*Beta vulgaris* L.), and corn to 100 μg/g lead nitrate. [26] Foliar concentrations were 140 and 390 μg/g in pinto bean and corn, respectively.

The binding and exchange capacity of soils are extremely important in determining lead availability to plants. In an experiment by Motto et al., [27] 4.9 times more lead was found in corn and lettuce (*Lactuca* sp.) leaves of plants grown in sand with 4 μg/g lead applied as $Pb(NO_3)_2$, as opposed to soil with 95 μg/g lead from auto exhaust. They hypothesized that lead accumulation in or on leaves in the field, or in roots in the greenhouse, indicated that the primary interaction of lead is with foliage in the field and with roots in the greenhouse. Many investigators have also demonstrated that root absorption is an important source in the field. [14, 17, 22-24, 28-36] However, Menzel [37] stated that soil lead is not absorbed by most plants, while Allaway [38] and Miller et al. [39] reported that different soils produce plants with varying foliar lead content. Other work has shown that although roots do absorb lead, there is limited translocation to shoots; [9, 11, 14, 17, 22, 25, 28, 30, 32, 40] and the failure to wash foliage or incomplete washing prior to analysis may account for some of the root-shoot uptake controversy. However, Lagerwerff [41] stated that studies in which aerial lead was not a factor indicated limited plant response to lead in soil. To support this, he cited studies by Page et al. [15] in which the lead content of strawberries (*Fragaria* spp.) did not change when the extractable lead content of the soil was increased from 8 to 59 μg/g. While the data are valid, the explanation fails to take into account the well-documented limited translocation of lead to edible portions of plants and the rather narrow concentration range employed. On a soil of pH 5.9, a 10-fold increase in extractable soil lead resulted in less than a 2-fold increase in lead uptake by radishes. [42]

Wilson and Cline [43] concluded that soil lead is largely unavailable for plant uptake, and that only 0.003 to 0.005 percent of the total lead in the soil was available. Similarly Motto et al. (27) suggested that plant uptake is probably better related to soluble, rather than total lead in soils. Marten and Hammond [32] showed that an eight-fold increase in total lead content of soil did not increase the lead content of bromegrass (*Bromus inermis* L.). However, Allaway [38] challenged the basis for Brewer's [44] statement that large amounts of soluble lead salts have very little effect on the concentration of lead in plants.

Baumhardt and Welch [1] in an Illinois field study grew corn in Drummer silt loam at pH 5.9; they applied lead acetate as a soil spray to increase the soil lead level to 1,400 μg/g. This treatment increased the lead content of immature whole plants, leaves at tasseling, and mature stover by factors of 15.8, 7.6, and 4.8, respectively, but had no effect on the lead content of grain. These results are at least partially explained in sorption studies by Zimdahl et al. [26] and Miller et al. [39]; they have shown that soils have a very large capacity to immobilize lead, and that rather larger differences in extractable lead are required to detect differences in plant uptake.

MacLean et al. [11] attempted to determine the soil lead available to plants. They were able to correlate 1*N* ammonium acetate extraction with available lead, but only at total lead levels of 1,000 μg/g. Other studies have shown that lead becomes less available with time to extractants such as ammonium acetate. [1,28] There is no method in the literature for determining available lead in a variety of soils over the concentration ranges currently found in natural and contaminated lead environments.

In an interesting alternate hypothesis to uptake from soil, Berger et al. [33] proposed that ^{210}Po, which may act like lead, (Singer and Hanson, [45]), is not taken up from soil directly by plant roots; rather, it is taken up by sorption from dead, moist plant materials at the air-soil plant interface. Because it is known to accumulate near the soil surface, much of the lead in plants undoubtedly comes from the surface horizon.

Uptake of Lead by Plant Foliage

Leaves and stems of grass near roads have been reported to contain as high as 3,000 μg/g lead. [46] The low rate of uptake by roots from soil would *a priori* support the contention that much of this lead is from aerial sources. Such aerial deposition of lead particulate matter on plant surfaces is well documented by Schuck and Locke [47] and in laboratory studies by Wedding *et al.* [48] These studies showed large differences in aerosol deposition that were dependent on characteristics of the leaf surface. While it is probable that much of the aerial lead is deposited only on the surface of plants, two washings with distilled water by Page et al. [15] removed only 60 to 70 percent of the deposited lead.

Leh [49] emphasized the importance of air contamination over soil in a study of rye (*Secale cerale* L.) and potato (*Solanum tuberosum* L.) plants. Other workers have also emphasized foliar transport, [12,18,50,51] but the role of stomata is unclear. Ault et al., [52] Hooper, [53] and Aarkrog and Lippert [54] showed no direct toxic effect from lead sprayed on foliage. Lagerwerff [41] stated that the texture and hairiness of the leaf surface should be considered because leaf hairs may absorb lead selectively, as human hair does. Lead aerosol particles were found to be deposited on rough pubescent leaves in a quantity seven times greater than on smooth waxy leaves. [48] Carlson (personal communication), in work from the same laboratory, found that $PbCl_2$ particles 1 to 3 μm in diameter were not taken into plants from leaf surfaces and were not reentrained into the air with wind speeds of 6.7 m $sec.^{-1}$; however, they were 95 percent removed by simulated rainfall. It is also of interest that large surface depositions of $PbCl_2$ had no effect on leaf photosynthesis.

Mosses are excellent indicators of lead from aerial sources, since they only absorb minerals from precipitation and settled dusts; also, root uptake can be eliminated. Ruhling and Tyler, [51] working in southern Sweden, compared moss samples collected in the last few years with similar species collected and preserved since 1875. Not surprisingly, they found that the lead content of recent samples had risen by a factor of four since the 19th century; they attributed this increase to the combustion of coal in the period 1875-1900 and to the use of leaded gasoline in the period 1950-1968.

Recently, Heichel and Hankin, [55] following an earlier report by Warren and Delavault, [56] found lead associated with chlorine and bromine embedded in tree bark. They noted the similarity of elemental content to compounds emitted in automobile exhaust.

On a road with 30,000 cars/24 h, Dedolph et al. [40] found that a lead concentration of 15 μg/g dry weight (dw) in perennial ryegrass (*Lolium perenne* L.) was correlated with an atmospheric lead concentration of 2.3 μg/m³ at a distance of 40 ft from the highway. At 120 ft, the values were 8.4 μg/g and 1.7 μg/m³.

Much controversy has centered on the question of how much airborne lead is actually getting into the plant under "natural conditions." Lagerwerff [42] has compared the lead content of radishes (*Raphanus* spp.) grown on traffic-contaminated soil at 200 m from a heavily traveled highway (24,000 cars/24 h) and in an environment protected from aerial contamination. Aerial contamination accounted for 40 percent of the lead associated with the tops, but very little of that in the roots. Thus it appears that nearly all of the lead in the roots and 60 percent of that in the tops was accumulated from the soil, and that the aerial contamination was not in the plant. Ter Haar et al. [23] found that 46 percent of the lead content of perennial ryegrass leaf blades and all of that in radish was obtained from soil rather than air, and that there was little or no increased absorption from simulated rainfall. In other work, Ter Haar [21] found that of 10 crops studied, 8 were unaffected by lead in air. The husk, cob, and kernels of sweet corn grown in filtered air contained lead n the ratio 7:3:1, respectively; however, in unfiltered air (1.45 μg Pb/m^3), the ratios were 31:2:1.

However, in the preceding study, as well as in others, actual uptake into foliar cells has not been separated from the possibility that the lead particulates are only present as topical coatings that are embedded in, or fixed to, the waxy cuticle of leaves. A recent study by Arvik [50] addresses this question. In experiments conducted with *Philodendron* leaves, *Malus* fruit, *Lycopersicon* fruit, and *Capsicum* fruit, he found that even under extreme conditions (low pH, high concentration in solution, and long treatment times), less than 2.0×10^{-3} percent of the lead passed the citicular barrier after 144 h of exposure. Dewaxing the cuticle did permit transport, but still not in large quantity.

Although Arvik's study seems definitive, experiments utilizing rinsing techniques [15,18] have failed to remove all doubts about foliar uptake. Rabinowitz [57] has reported foliar absorption and translocation of lead halide aerosols by oats (*Avena sativa* L.) and lettuce grown near a freeway. Similarly, Hemphill [58] has radioautographic evidence of uptake and translocation from lead solutions applied to leaves of sycamore (*Platanus occidentalis* L.) and to lettuce and radish plants. [58,59] However, at maturity, analyses showed absorption and translocation of lead in amounts equal to only 1 percent or less of the amount applied.

A fine structural examination has been carried out by utilizing leaf cells of the bryophyte *Rhytidiadelphus squarrosus* exposed to lead from traffic exhausts. [60] It revealed that detectable amounts of lead had entered the cytoplasm and could be recognized as electron-dense precipitates localized within vesicles or vacuoles, chloroplasts, mitochondria, microbodies, and plasmodesmata.

In summary, it appears that little, if any, aerosol-deposited lead is actually taken in through plant leaves. The deposition of lead on plant surfaces, however, cannot be ignored as a potential hazard in food chains. If lead-contaminated crops are fed to livestock, or if animals graze on contaminated pastures, significant quantities of lead may be ingested. [61]

Distribution of Lead in Plants

There is ample evidence that far less lead is moved within the plant than remains in the roots, and it is generally agreed that lead is not readily translocated to the edible portions of plants. [1,21,22] Kleinman, [10] Warren et al., [24] Page and Ganje, [14,62] MacLean et al., [11] Ganje and Page, [5] and Aarkrog and Lippert [54] have reported the lead content of fruits, vegetables, and grains; in all cases the lead contents were lower than in other vegetative plant parts. Warren and Delavault [56] concluded that oven-dried food products normally contain 0.1 to 1.0 μg/g lead, but the lead content could be up to 10 times higher if the products were grown in areas high in lead. The average value for all plants studied was 10μg/g, with a high of 45μg/g in potato tops. Keaton [28] found less than 3μg/g (dw) in tops, but up to 800 μg/g (dw) in roots of barley grown in soil treated with up to 800 μg/g lead. Liebig et al. [63] found 3μ g/g (dw) in leaves, but 890 μg/g (dw) lead in roots of lemon cuttings grown in solution culture. Gamble [64] reported from 0.3 to 30 μg/g (dw) lead in leaves of various plant species in wooded areas, and Prince [65] found 10 to 25μg/g (dw) in corn leaves. Tso et al. [29,66] reported a much lower level of lead in tobacco (*Nicotiana tabacum)* seed than in the leaf.

Other reports show lead contents of rye chaff and foliage and potato foliage higher by a factor of 2 at 15 ft, as opposed to 300 ft from a highway; however, grain and tuber contents were constant and low at both locations. [49] Privet leaves averaged 45μg/g (dw) at sites remote from highways and 86 g/g (dw) at sites along highways. [67] The normal concentration of lead in hay was 2 to 3μg/g (dw), but values to 284μg/g (dw) were reported near a smelter. [61]

Hevesy [30] proposed that lead is bound in the roots subsequent to root treatment, and that this serves to protect the remaining plant parts from injury. The nature of this binding and its locus have been explained by Malone *et al.* [68] With light and electron microscopic studies of corn grown hydroponically, they showed that roots acquired a surface lead precipitate and slowly accumulated lead crystals in the cell walls. The surface precipitate formed quickly and was independent of plant activity. Two compound forms were postulated but not identified. In contrast, the lead that entered the root was concentrated in an active process in some, but not in all, dictyosome vesicles. After precipitation in the dictyosome vesicle, an unusual and previously unreported sequence of events occurred. Initially, cell wall precursors were added to the vesicle by opposition of vesicles or internal secretion. As the lead crystal grew, more cell wall material was added; eventually, the entire vesicle moved to the periphery of the cell and through the plasmalemma, and then fused within the cell wall. In all instances, lead deposits were concentrated at the cell wall and not with mitochondria or other cellular organelles. Although the sequence of events was observed in the root tips, comparable deposits were observed throughout the plant; it was suggested that a similar deposition process could occur in all plant tissues. It seems likely that this deposition, appearing as a lead-phosphate complex, is rendering the absorbed lead inactive, thus accounting for the presence of high concentrations of lead within plants without noticeable effect on plant growth processes.

Hydroponic and soil studies at Colorado State Universtiy [26,50] suggest that lead uptake may be passive (not requiring expenditure of energy by the plant). This is supported by studies of Goren and Wanner, [6] who examined lead and copper uptake by excised barley (*Hordenum vulgare* L.) roots and found it to be passive. It should be noted, however, that the dictysome deposition of lead in corn roots is most likely not passive, and that both passive and active components were present in these experiments.

In other work to determine accumulation or uptake sites in living organisms, Tornabene and Edwards [69] found lead within the cell wall and on the protoplasmic membrane of the bacterial *Micrococcus* and *Azotobacter* grown in the presence of lead halides; however, thus found very little in the cytoplasm. Brown and Slingsby [70] examined the lichen *Cladonia rangiformis in vitro* and found the lead bound to insoluble anionic sites in exchangeable form and external to the cell membranes. They stated that uptake of lead can take place in the absence of living cells and is a purely physical process. Lead can be replaced by nickel, but replaces potassium in living *Cladonia* cells. Using a histochemical technique whereby lead-containing tissues turn scarlet, Glater and Hernandez [71] were able to identify lead in vascular tissue and on surfaces of contaminated plants.

The relative importance of chelating substances in the movement of lead into and within the plant is hard to evaluate at this time. Chelation of lead with EDTA seems to facilitate lead movement. Wallace and Romney [72] noted that chelating agents modified the uptake of heavy metals by plants from soil. EDTA was used by Marten and Hammond [32] to remove lead from soil, and they found that it increased uptake by plants. Similarly, Hale and Wallace [73] found that DTPA increased uptake of lead by beans regardless of soil type, but that EDDHA and Fe-EDDHA had no effect. Mitchell et al. [74] found that EDTA extraction of soil most closely correlated with plant uptake. EDTA is also used as a wash for plant parts to free them from adsorbed lead prior to analysis. [75]

Tanton and Crowdy [76] found that lead-EDTA entered the plant only in the relatively short root hair zone of the root. The extent of this region is species-specific and dependent upon the previous history of the root. The root endodermis appeared to be a regulator of the passage of lead chelate, with only 3 percent of the lead in the root being translocated to the shoot. Tanton and Crowdy suggested that the translocated lead moved passively with the transpiraton stream, with accumulation at the evaporative surfaces. However, the extent of lead movement within the plant is still an open question, and whether studies with an artificial chelator such as EDTA actually reflect what is happening in nature is open to debate. Studies with compounds that could chelate lead under natural conditions would be a considerable aid in answering this question. Exudates from corn roots do chelate lead [75] and might be important in uptake; however, actual studies assessing the magnitude of this effect are still in progress.

The question of the chemical identity of lead in plants has been only minimally examined. One lead compound, lead pyrophosphate, has been identified in the root or a pinto bean plant in which the lead concentration was 32,400 μg/g on a dry-weight basis. [26] Lead orthophosphate [$Pb_3(PO_4)_2$] has been identified in a soybean (*Glycine max* L.) root. [75] There were, however, differences in the growing conditions of the two plants that may account for the compounds identified.

Deposition, translocation, and uptake studies make it clear that under certain conditions lead is mobile within the plant. However, in general, it seems that there is a 7 to 10 fold decrease in lead concentration when comparing foliage with roots, and a similar decrease between grain and foliage. In other words, if 100 μg/g lead were associated with the roots, it is likely that the grain would contain lead at a level close to 1 μg/g. Because of species differences and varying physiological parameters, such generalities may not always hold true and should be used with caution.

INFLUENCE OF ENVIRONMENT

It is evident from the foregoing discussion that differing soils or species of plants are important determinants of the amount of lead that becomes associated with plants. Wallace and Romney (72) reported that soil pH, soil temperature, calcium availability, heavy metal availability, phosphorus supply, and soluble silicon may be important in the uptake of any heavy metal. Similarly, Warren et al. [24] found variations in plant uptake of lead to be due to pH, soil organic matter content, soil texture, climate, topography, pollution, and geologic background of the soil. In their studies, lead toxicity to corn, beans, lettuce, and radishes was greater under slightly acid soil conditions. MacLean et al. [11] found that the concentration of lead in oats and alfalfa (*Medicago sativa* L.) increased with decreased pH and organic matter, and that the addition of phosphate reduced the uptake of lead.

Working with a series of lead-amended Illinois soils with a range of cation exchange capacities, pH levels, and phosphorus levels, Miller et al. [39] found that lead uptake into foliar portions of corn decreased with an increase in these parameters. Rolfe [77] reported that phosphorus added to three Illinois soils decreased the uptake of lead by six different 2-year-old tree seedlings. In similar studies at Colorado State University, [78] lead uptake by corn was reduced by amending soil with phosphorus. However, large amounts of phosphorus (111 kg/ha with 500 μg/g soil lead) were required to significantly affect uptake. These studies also showed that organic matter (manure) amendments reduced lead uptake by corn. It was suggested that the addition of organic matter should be considered a feasible means of reducing the effects of lead pollution.

In other work, Page and Ganje [14] found that the toxicity of lead to corn, beans, lettuce, and radishes was expressed at lower soil lead concentrations in a slightly acid soil than in calcareous soils. Cox and Rains, [4] John and Van Laerhoven, [8] and Zimdahl [78] have reported a decrease in lead uptake from soil as a result of liming. The importance of liming takes on added significance in the work of Alloway, [79] who reported that apparently healthy radish plants growing on heavily limed soils in Wales contained up to 12,000 μg/g of lead. He suggests a possible counteracting effect of lime on the movement and uptake of lead and also an intercellular antidotal effect of calcium.

John and Van Laerhoven [8] reported little difference in uptake by oats and lettuce when lead was derived from water-insoluble lead carbonate, as opposed to the more water-soluble lead chloride or nitrate. Therefore, the formation of insoluble lead carbonate as a result of liming was not considered to be an explanation of the pH effect. Zimdahl (unpublished data) further suggests the ambiguity in the effects of the form of lead on plant uptake. In soils amended with lead nitrate or sulfate, at concentrations of lead normally found in nature, corn took up more lead into shoots from the sulfate form, while beans and sugar beets took up more from the nitrate form.

The presence of calcium chloride or calcium nitrate was also found to decrease lead uptake by plants grown in solution. [30, 40, 80] Whereas 3 μg/g of lead nitrate killed fescue (*Festuca* spp.) it took 30 μg/g to be lethal in the presence of calcium. [80]

The condition of the plant at the time of lead impact may also be of importance. Young tobacco seedlings accumulated lead more rapidly than older plants, and broad beans absorbed lead more rapidly from 10^{-4} *M* $PbCl_2$ solutions when the cotyledons were removed. [31] Similarly, Hunter [81] found that the lead content of bracken fern (*Pteridium aquilinum* L.) fronds decreased with age. Mitchell and Reith, [82] however, reported a large increase in the lead content of pasture herbage in the fall. This same phenomenon was found in leaves of some deciduous trees at senescence [83] and in wild oats (*Avena fatua* L.). [16]

The location of plants growing under "natural conditions" appears to be an extremely important variable in lead uptake. Because of the large lead input from the automobile, vegetation growing next to highways has been studied to determine the effects of proximity to the source on the lead content of plants. These studies show an inverse correlation with distance from a highway. [3, 5, 14, 15, 19, 20, 27, 35, 43-45, 56, 64, 84, 85] Cannon and Bowles [46] found 100 to 700 μg/g lead in the ash of green samples collected within 5 ft of highways; at 1,000 ft. lead decreasing to less than 5 to 50 μg/g. The distribution of lead seemed to be controlled by traffic volume and prevailing wind direction. Warren and Delavault [56] found higher lead levels in tree stems near traffic than remote from it. Chow [84] reported that grass foliage ranged from 20 to 60 μg/g (dw) along two well-traveled roads. Kloke and Riebartsch [85] found that lead increased in roadside grass foliage from 16.2 μg/g (dw) with 11,000 cars per 12 h to 570 μg/g (dw) with 32,000 cars per 12 h.

SUMMARY

Lead becomes associated with plants via roots and foliage. Movement of lead into higher plants has been convincingly demonstrated through roots, but not so clearly through foliage. Association with foliage is mostly as a topical coating. More lead is deposited on pubescent than on smooth leaf surfaces. Such deposits can be washed off by rainfall, but are reentrained by moderate winds.

Under certain soil conditions (low pH, low cation exchange capacity, low organic matter, and low phosphate levels), large amounts of lead can be taken up by roots. However, the general lack of effect from relatively large lead concentrations in roots has led to the hypothesis that lead is inactivated through deposition in the roots. Electron microscope studies show that lead-phosphate deposits are formed on root surfaces, in peripheral extracellular spaces, and within dictyosome vesicles of root cells outside the endodermis. Dictyosomes containing lead deposits migrate outside the cell itself via reverse pinocytosis, forming extracellular deposits that are most prevalent in roots; however, similar deposits are found throughout corn plants.

Environmental factors, plant age, and plant speciation are important variables in lead uptake by plant roots. For the most part, the alteration of soil parameters to make lead more available in soil solutions increases root uptake. Comparisons of studies done with differing plant age and speciation provide no clearcut possibility for generalization, since the results tend to vary with speciation, but not necessarily with larger plant groupings.

REFERENCES

1. G. R. Baumhardt and L. F. Welch. J. Environ. Qual. *1*: 92 (1972).
2. M. H. Berg. Minn. Acad. Sci. *36*: 96 (1970).
3. J. J. Connor, J.A. Erdman, J. D. Sims, and R. J. Ebans. In *Trace Substances in Environmental Health*. IV, D. E. Hemphill, ed. University of Missouri, Columbia, MO (1970).
4. W. J. Cox and D. W. Rains. J. Environ. Qual. *1*: 167 (1972).
5. T. J. Ganje and A. L. Page. Calif. Agri. *26*: 7 (1972).
6. A. Goren and H. Wanner. Ber. Schweig. Bot. Gaz. *80*: 334 (1971).
7. R. M. Harrison and L. S. Jones. Abstr. Amer. Soc. Agron. NY 145 (1971).
8. M. K. John and C. Van Laerhoven. J. Environ. Qual. *1*: 169 (1972).
9. T. H. Keller and R. Zuber. Forstwissenschaftliches Centralblatt. *40*: 20 (1975
10. A. Kleinman. Pest. Monit. J. *1*: 8 (1968).
11. A. J. MacLean, R. L. Halstead, and B. J. Finn. Can. J. Soil Sci. *49*: 327 (1969).

12. J. D. Martinez, M. Nathany, and V. Dharmarajar. Nature. *233:* 564 (1971).
13. R. J. Miller and D . E. Koeppe. In *Trace Substances in Environmental Health,* IV, D. E. Hemphill, ed. University of Missouri, Columbia, MO 186 (1970).
14. A. L. Page and T. J. Ganje. Agron. Soc. Amer. 88 (1972).
15. A. L. Page, T. J. Ganje, and M. S. Joshi. Hilgardia. *41*: 1 (1971).
16. D. W. Rains. Nature 233: *210* (1971).
17. G. K. Rasmussen and W. H. Henry. Proc. Soil Crop Sci. Fla. *23:* 71 (1963).
18. A. Ruhling and G. Tyler. Bot. Not. *122*: 248 (1969).
19. W. H. Smith. For. Sci. *17*: 195 (1971).
20. W. H. Smith. Science *176*: 1237 (1972).
21. G. L. Ter Haar. Environ. Sci. Technol. *4*: 226 (1970).
22. G. L. Ter Haar. J. Wash. Acad. Sci. *61*: 114 (1971).
23. G. L. Ter Haar, R. R. Dedolph, R. B. Holtzman, and H. F. Lucas, Jr. Environ. Res. *2*: 267 (1969).
24. H. V. Warren, R. E. Delavault, K. Fletcher, and E. Wilks. In *Trace Substances in Environmental Health.* IV, D. E. Hemphill, ed. University of Missouri, Columbia, MO (1970).
25. R. L. Zimdahl and J. H. Arvik. Proceedings of Conference on Environmental Chemistry, Human and Animal Health, E. Savage, ed. Colorado State University. *1*: 33 (1972).
26. R. L. Zimdahl. In *Impact on Man of Environmental Contamination Caused by Lead.* Interim Report, NSF Grant GI-4 and GI-34813X (1972).
27. H. L. Motto, R. H. Daines, D. M. Chilko, and C. K. Motto. Environ. Sci. Technol. *4*: 231 (1970).
28. C. M. Keaton. Soil Sci. *43*: 401 (1937).
29. T. C. Tso, J. M. Carr, E. S. Ferri, and E. J. Baratta. Agron. J. *60:* 647 (1968).
30. G. Hevesy. Biochem. J. *17*: 435 (1923).
31. S. Prat. Amer. J. Bot. *14*: 663 (1927).
32. G. C. Marten and P. B. Hammond. Agron. J. *58*: 555 (1966).
33. K. C. Berger, W. H. Erhardt, and C. W. Francis. Science. *150:* 1738 (1962).
34. T. C. Tso. Agron. J. *62*: 663 (1970).
35. T. C. Tso, N. A. Hallden, and L. T. Alexander. Science. *146*: 1043 (1964).
36. T. C. Tso, N. Harley, and L. T. Alexander. Science *153*: 880 (1966).
37. R. G. Menzel. In *Proceedings of the Hanford Symposium on Radiation and Terrestrial Ecosystems,* Richland, Wash. (May 1965).
38. W. H. Allaway. Advan. Agron. *20:* 235 (1968).
39. J. E. Miller, J. J. Hassett, and D. E. Koeppe. Commun. Soil Sci. Plant Anal. *6:* 349 (1975).
40. R. Dedolph, G. Ter Haar, R. Holtzman, and H. Lucas, Jr. Environ. Sci. Technol. *4*: 217 (1970).
41. J. V. Lagerwerff. In *Micronutrients in Agriculture, Soil Science Society of America. 23. (1972).*
42. J. V. Lagerwerff. Soil Sci. 111: 129 (1971).
43. D. O. Wilson and J. F. Cline. Nature. *209:* 941 (1966).
44. R. F. Brewer. In *Diagnostic Criteria for Plants and Soils.* University of California Division of Agricultural Science. 213 (1966).
45. M. J. Singer and L. Hanson. Soil Sci. Soc. Amer. Proc. *33*: 152 (1969).
46. H. L. Cannon and J. M. Bowles. Science. *137*: 765 (1962).
47. E. A. Schuck and J. F. Locke. Environ. Sci. Technol. *4*: 324 (1970).
48. J. B. Wedding, R. W. Carlson, J. J. Stukel, and F. A. Bazzaz. Environ. Sci. Technol. *9:* 151 (1975).
49. H. O. Leh. Gesunde Pflanzen. *18*: 21 (1966).
50. J. H. Arvik. Ph.D. Thesis, Colorado State University. (1973).
51. A. Ruhling and G. Tyler. Bot. Not. *121:* 321 (1968).
52. W. V. Ault, R. G. Senechal, and W. E. Erleback. Environ. Sci. Technol. *4*: 305 (1970).
53. M. C. Hooper. Ann. Appl. Biol. *24*: 690 (1937).
54. A. Aarkrog and J. Lippert. Radiat. Bot. *11*: 463 (1971).
55. G. H. Heichel and L. Hankin. Environ. Sci. Technol. *6:* 1121 (1972).
56. H. V. Warren and R. E. Delavault. J. Sci. Food Agr. *13*: 96 (1962).
57. M. Rabinowitz. Chemosphere. *4*: 175 (1972).
58. D. D. Hemphill. In *An Interdisciplinary Investigation of Environmental Pollution by Lead and Other Heavy Metals From Industrial Development in the New Lead Belt of Southeastern Missouri,* a Report to the National Science Foundation. B. G. Wixon and J. C. Jennett, eds. (1974).
59. D. D. Hemphill. University of Missouri. Unpublished data (1973).
60. E. M. Ophus and B. M. Gullvag. Cytobios. *10:* 45 (1974).
61. P. B. Hammond and A. L. Aronson. Ann. NY Acad. Sci. *2:* 595 (1964).
62. A. L. Page and T. J. Ganje. Environ. Sci. Technol. *4*: 140 (1970).
63. C. F. Liebig, Jr., A. P. Vanselow, and H. D. Chapman. Soil Sci. *53*: 341 (1942).
64. J. F. Gamble. Final Report No. NYO-10581 to U.S. Atomic Energy Commission, Washington, D.C. (1963).
65. A.L. Prince. Soil Sci. *84*: 413 (1957).
66. T.C. Tso, G. L. Steffens, E.S. Ferri, and E. J. Baratta. Agron. J. *60:* 650 (1968).
67. J.C. Everett, C. L. Day, and D. Reynolds. Food Cosmet. Toxicol. *5*: 29 (1967).
68. C. Malone, R. J. Miller, and D. E. Koeppe. University of Illinois. Unpublished data (1973).
69. T. G. Tornabene and H. W. Edwards. Science *176*: 1334 (1972).
70. O. H. Brown and D. R. Slingsby. New Phytology. *71*: 297 (1972).
71. R. A. B. Glater and L. Hernandez, Jr. J. Air Pollut. Cont. Ass. *22*: 463 (1972).
72. A. Wallace and E. M. Romney. Agron. Abstr. 130 (1970).
73. V. Q. Hale and A. Wallace, Soil Sci. *109:* 262 (1970).
74. R. L. Mitchell, J. W. S. Reith, and I. M. Johnston. J. Sci. Food and Agr. *8:* (suppl. issue) S51 (1957).
75. University of Illinois. An Interdisciplinary Study of Environmental Pollution by Lead and Other Metals. NSF Grant GI-31605, Progress Report (1972).
76. T. W. Tanton and S. H. Crowdy. Pest. Sci. *2:* 211 (1971).
77. G. L. Rolfe. J. Environ. Qual. *2:* 153 (1973).
78. R. L. Zimdahl. In *Environmental Contamination Caused by Lead.* H. W. Edwards et al., ed. Interim Report, NSF Grants GI-34813X1 and GI-44423 (1974).
79. B. Alloway. In *Anomalous Levels of Trace Metals in Welsh Soils.* B. E. Davies, ed. Welsh Soils Discussion Group Report No. 9:87 (1968).
80. *D. A. Wilkens. Nature. 180:* 37 (1957).
81. J. G. Hunter. J. Sci. Food Agr. *4*: 11 (1953).
82. R. L. Mitchell and J. W.S. Reith. J. Sci. Food Agr. *17*: 437 (1966).
83. M. M. Guha and R. L. Mitchell. Plant Soil. *24*: 90 (1966).
84. T. J. Chow. Nature. *225*: 295 (1970).
85. A. Kloke and K. Riebartsch. *Naturwissenschaften. 51*: 368 (1964).

CHAPTER 6

TRANSPORT AND DISTRIBUTION IN A WATERSHED ECOSYSTEM

L. L. Getz
Department of Ecology, Ethology, and Evolution
University of Illinois

A. W. Haney
Department of Botany
University of Illinois

R. W. Larimore and *J. W. McNurney*
Illinois State Natural History Survey

H. V. Leland
U.S. Geological Survey

P. W. Price
Department of Entomology
University of Illinois

G. L. Rolfe
Department of Forestry
University of Illinois

R. L. Wortman
Department of Civil Engineering
University of Connecticut

J. L. Hudson
Department of Chemical Engineering
University of Virginia

R. L. Solomon and *K. A. Reinbold*
Institute for Environmental Studies
University of Illinois

INTRODUCTION

Watersheds are discrete units that provide a logical base for input-output studies. A major advantage is that streamflow, along with dissolved and suspended matter, is routed through a single exit. Thus monitoring for substances that have either soil or water as major sinks is greatly simplified. Because of these advantages, an 86 mi^2 watershed was chosen by the University of Illinois for lead transport and distribution studies. (fig. 6-1) The predominantly agricultural watershed lies primarily to the north of Champaign-Urbana; however, it includes about 90 percent of the metropolitan area with approximately 100,000 inhabitants.

The rural portion of the watershed is drained by the Saline Branch of the Vermillion River, and the urban portion by the Boneyard Creek. The streams join on the east edge of Urbana and flow from the system to the east. The major crops are corn and soybeans. Woodlands, pasture, and wasteland (such as railroad rights-of-way and stream margins) are also included in the watershed area.

The objectives of the study included (1) characterizing the input, accumulation, and output of lead from automobile sources in a typical midwestern ecosystem; 2) understanding the mechanisms controlling fluxes between system components; and (3)

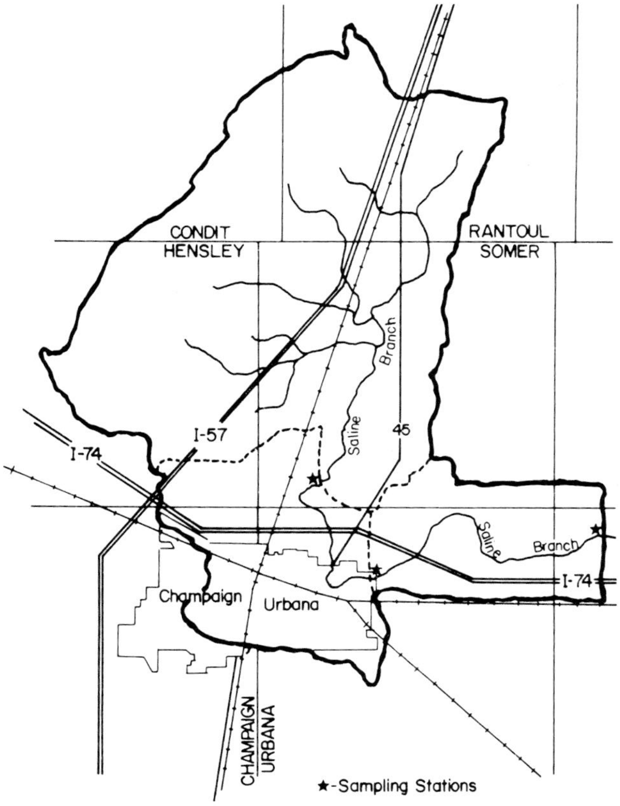

FIGURE 6-1.—Map of Saline Branch Watershed.

evaluating potential effects of lead on ecosystem components other than man. This discussion emphasizes the first objective.

An ecosystem distribution model was developed [1] to identify ecosystem components for intensive study and to provide a frame of reference for the numerous investigators. A diagrammatical representation of the model is shown in figure 6-2. The model includes major components of both the aquatic and terrestrial ecosystems; it represents the system by a network of nodes and branches, where the nodes represent the components of the ecosystem in a general sense, and the branches indicate possible transport mechanisms between nodes.

Source nodes in the model such as auto emissions only have lead exiting, while sink nodes (outflow water) only have branches entering. Three attributes are associated with each node: a mass, a quantity of lead input per unit time for source nodes, and the initial concentration in the node.

Branches represent a variety of transport mechanisms such as diffusion, leaching, biological uptake, and erosion. Self-loops represent the fraction of lead content of a node that remains in that node between time periods. Distribution factors and seasonal factors are used to quantitatively describe the branch flows. The distribution specify the fraction of the lead content of a node that flows in the branches emanating from that node. These distribution factors can be constant or random variables that follow various probability distributions based on the characteristics of the flux in question. Transports may also be affected by the seasons of the year. This is accounted for by modifying the distribution factor of each branch by an appropriate seasonal factor.

The current network for the model contains over 40 nodes and 130 branches to accomodate zoning of

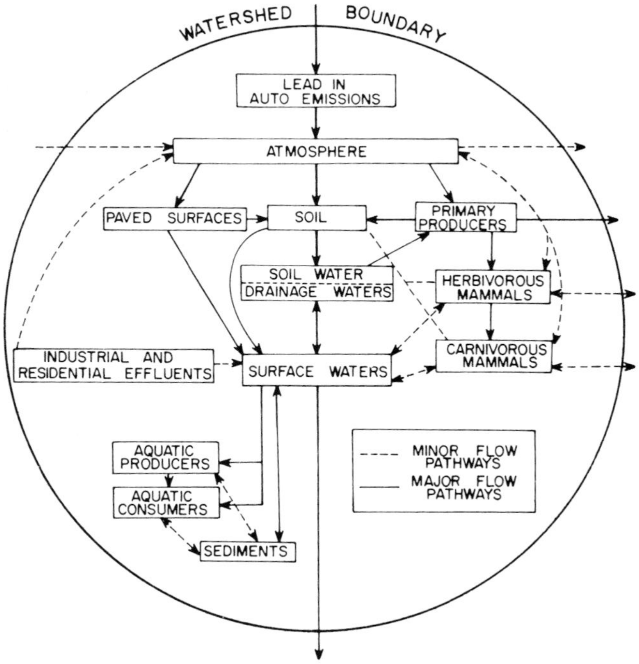

FIGURE 6-2.—Basic ecosystem model.

the system into a range of lead inputs based on traffic volume. Nodes such as primary producers are defined generally in this models but more detailed study of these compartments by species or family can easily be included.

Initial runs of the model provided a sense of direction for the studies in terms of identifying critical geographical areas of the watershed and ecosystem components. The runs also provided individual investigators with an understanding of how their studies and results would fit into the ultimate ecosystem analysis.

SYSTEM INPUTS

Gasoline consumption represents the major source of lead input to the watershed. To determine the effects of lead, the magnitude and distribution of the metal from traffic sources, as well as the spatial and temporal variations of this input, must be determined. Ideally, the amount and distribution of lead emitted to the atmosphere from each automobile in the watershed should be examined, but this would be an impossible task. Rather than sampling emissions, lead input was estimated by relating emissions to gasoline consumption patterns. The gasoline consumption of automobiles has been documented by numerous studies. With these data, if vehicular travel information is available, an estimate of the magnitude can be derived, as well as the spatial and temporal consumption of gasoline for a particular area.

Traffic and transportation engineers have developed techniques for sampling and estimating vehicle travel for use in traffic planning and operations studies. Basically, these involve determining traffic volume on each segment of streets and highways. The traffic volume data and information on street networks can be used to estimate vehicle miles of travel. This was the only feasible method of estimating lead emissions for the watershed area.

Sales data might appear to be an obvious source of gasoline consumption information. However, there is no assurance that the fuel is consumed in the locale where it is sold. For example, gasoline purchased at a location on an interstate highway would not likely be consumed in the immediate area. Sales information has been utilized in this study, but primarily to check the accuracy of estimates obtained from travel information.

Methods used in the current watershed study are as follows. First, the various streets and roads were classified into groups by traffic volume. This classification permitted a rapid examination of the system by looking at those streets and roads that are responsible for most of the lead entering the ecosystem.

Both the rural roads and urban streets were categorized into five classifications by average daily traffic (ADT) volumes. The specifications of each class, as well as the total miles of road in each, are summarized for rural roads (table 6-1) and urban streets. (table 6-2)

TABLE 6-1.—*Classification of rural roads by average daily traffic volume (ADT)*

Class	ADT	Total miles
Class I	over 4,000	17.65
Class II	2,000 to 3,999	0.0
Class III	1,000 to 1,999	6.2
Class IV	250 to 999	15.1
Class V	under 250	64.1

In the second method used in the study, traffic data were analyzed to provide estimates of vehicle miles of travel and, eventually, gasoline consumption. This analysis yielded a spatial distribution of gasoline consumption and, therefore, lead output. Table 6-3 shows the rates of gasoline consumption that were obtained from auto manufacturers to be used in the analysis. Gasoline consumption for the various compartments is summarized in tables 6-3 and 6-4.

TABLE 6-2.—*Classification of urban streets by average daily traffic volume (ADT)*

Class	ADT	Total miles
Class I	over 10,000	16.90
Class II	5,000 to 10,000	20.00
Class III	2,500 to 4,999	15.83
Class IV	1,000 to 2,499	20.20
Class V	under 1,000	70.40

Gasoline consumption by roadway classification is shown in table 6-5 and includes total daily vehicle miles, gallons of gasoline consumed, and the percentage of total gasoline consumption in each street or road class. Note that class I, class II, and Interstate 74 make up 76 percent of all gasoline consumption for the urban streets, while class I makes up 89 percent of the gasoline consumption for the rural roads.

In summary, then, the gasoline consumption for the watershed is:

Urban 46,051 + 2,700 gal./day

Rural 10,524 + 560 gal./day

Total 56,575 + 2,940 gal./day

Gasoline consumption was also compared between a rural subcompartment and an urban subcompartment.

TABLE 6-3.—*Gasoline consumption as a function of speed and classification of streets and roads in the watershed based on average Speeds*[1]

Road type	Speed (mph)	Gas Consumption (mi/gal.)
Local rural	35	16.7
Route 45	60	16
Interstate 57, 74	70	15
Local urban	30	12.5

[1] Based on data from automobile manufacturers.

TABLE 6-4.—*Gasoline consumption within compartments*

Compartment	Total daily vehicle miles	Total daily Gas Consumption (gal.)
Rural	164,318	10,524
Rural subcompartment	1,491	86.5
Urban	575,643	46,051
Urban subcompartment	262,628	21,010

ment. The 5.1 mi^2 rural subcompartment is in the northwest corner of the rural compartment and has 9.4 mi of roads. The urban subcompartment, an area of 4.0 mi^2 in Champaign, has 68.1 mi of streets. Gasoline consumption is about 250 times greater in the urban than in the rural section.

In converting gasoline consumption data to lead emissions in the ecosystem, the following assumptions were made:

(1) Gasoline contains 2.5 g of lead per gal.

(2) In urban areas, due to the nature of the traffic flow, 50 percent of the consumed lead is considered to be emitted from the exhaust system of the automobile.

(3) In rural areas with generally higher vehicle speeds, 80 percent of the consumed lead is emitted.

Total lead input to the watershed on a yearly basis is shown in figure 6-3. The 12 mi^2 urban compartment receives approximately 75 percent of the total lead input, while the 74 mi^2 rural area receives only 25 percent. These emission data, on a per-vehicle volume basis, are 0.101 g lead per vehicle mile for the urban compartment and 0.128 g for the rural compartment. These figures compare favorably with the average lead emission rate for production vehicles of 0.108 g/mi. [2] On the basis of Cantwell's figure, if 0.50 g or 46 percent remains airborne, the total lead input to the ecosystem, with the exception of air, is reduced to 16,000 kg/year. The majority of this deposition occurs in the urban area (10,400 kg) and along major highways (3,900 kg).

If the ecosystem is divided into zones based on traffic volume, the relative concentration factors for lead inputs to each zone can be determined. (table 6-6) Zone 1 is a 50 m strip of land on both sides of highways with traffic volumes greater than 4,000 vehicles/24 h; zone II is the same width strip along

TABLE 6-5.—*Gasoline consumption per day determined by roadway classification*

Class	Vehicle miles	Gallons	Percent of total gallons
	Rural roads		
Class I	144,505	9,335	89
Class II-V	19,813	1,189.4	11
	Urban roads (not including I-74)		
Class I	212,390	16,991	37
Class II	141,685	11,335	25.5
Class III	56,083	4,487	10
Class IV	31,545	2,524	5
Class V	33,968	2,717	6
I-74	80,590	6,447	14
North of I-74	13,982	1,119	2

TABLE 6-6.—*Relative concentrations of lead by zones*

Zone	Vehicles	Percent total area	Relative concentration[1]
I	>4,000	8.5	70
II	1,000-2,000	17	14
III	Rural, Remote	59.5	1
IV	Urban	15	140

[1] The value of 1 is assigned to the rural, remote area.

highways with traffic volumes less than 4,000 vehicles/24 h; zone III represents rural areas remote from highways; and zone IV is the urban area. The relative concentration factors for zones I and IV are very high and indicate areas of concern due to the tremendous lead input per unit area.

The input to the ecosystem of lead in rainwater has also been monitored, and representative data for the urban and rural compartments are shown in figure 6-4. Rainfall input averages 2 percent of the total from automobile emissions. Collections were made by placing 48 acid-washed polyethylene bottles at random locations within the watershed to collect samples during each storm period. The bottles were placed prior to the rainfall and collected afterwards to avoid dustfall contamination. Total rainfall was monitored by recording rain gages.

SYSTEM OUTPUTS

Surface drainage discharge was monitored continuously at five gaging stations located on the Saline Branch of the Vermillion River and Boneyard Creek.(fig. 6-5) The gage locations allow a comparison of the water volume and lead output from the rural portion of the watershed with those from the urban area. This same kind of comparison can be made between a small rural agricultural watershed of approximately 4 mi^2 and a totally urban watershed of similar size, or between the entire rural and urban compartments. An additional gaging station, located 5 mi downstream from the watershed, provides an estimate of stream recovery due to dilution effects of rural drainage waters.

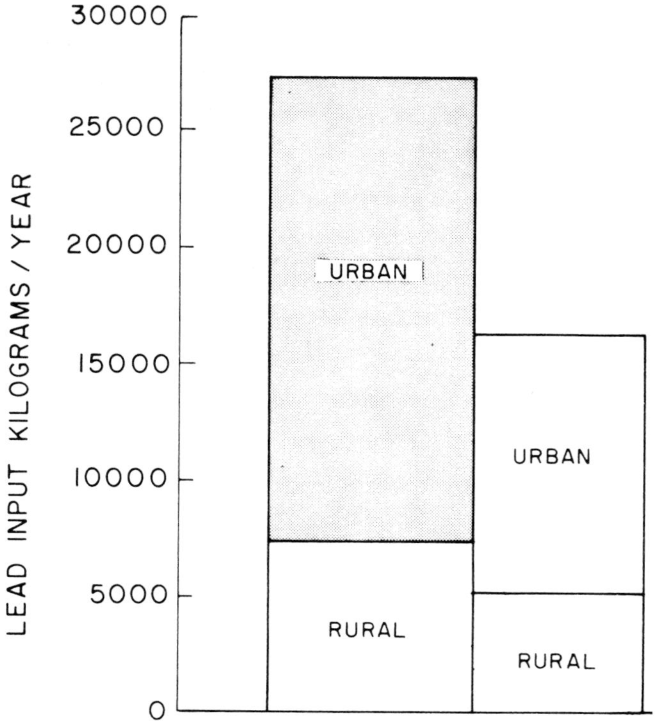

FIGURE 6-3.—Total lead input from automobile emission to the urban and rural compartments.

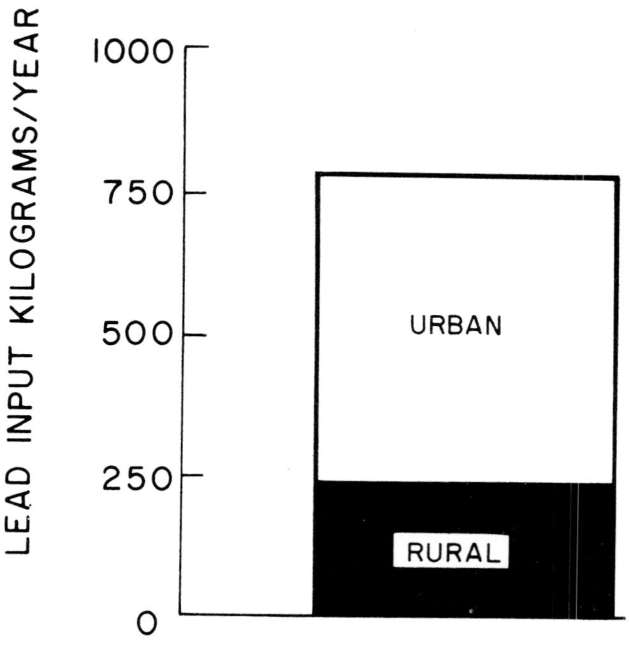

FIGURE 6-4.—Lead input to rural and urban compartments of the watershed from precipitation.

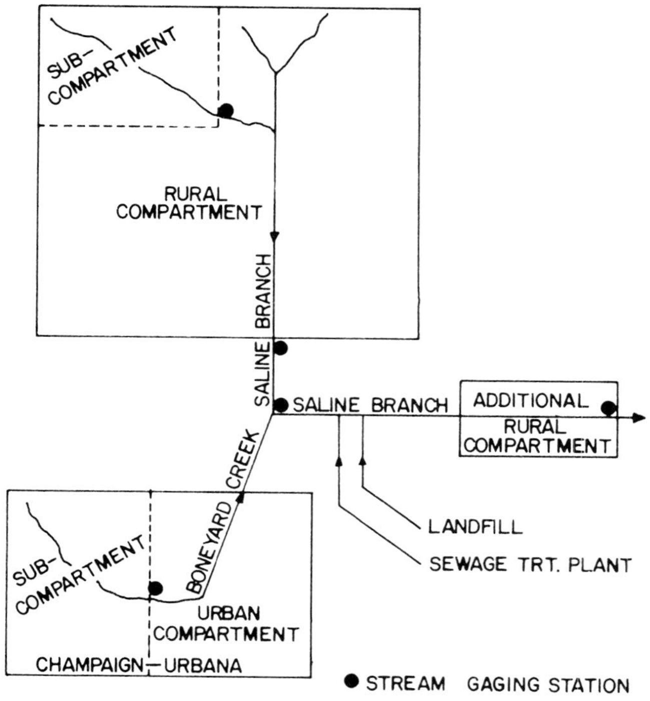

FIGURE 6-5.—Schematic diagram of idealized Saline Branch watershed.

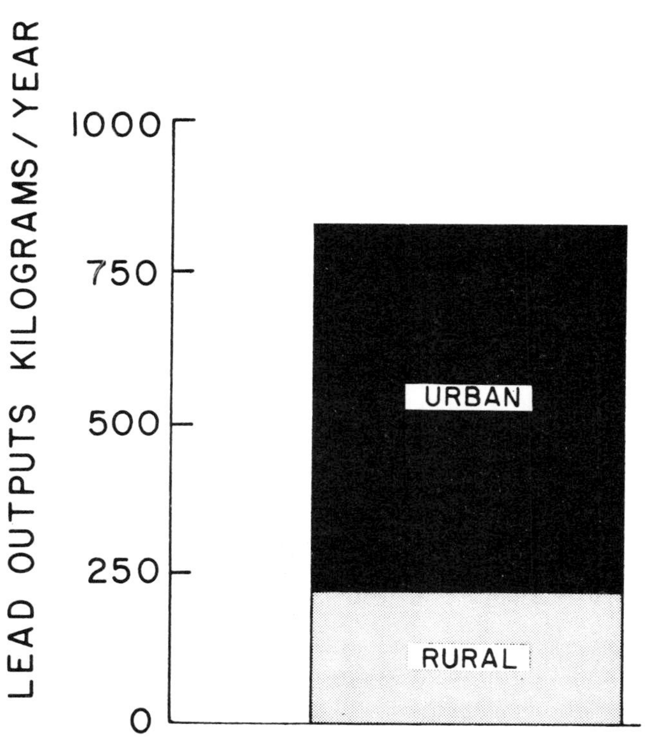

FIGURE 6-6.—Total lead output in streamwater from urban and rural compartments.

Total water volume moving past each gage was determined for each 24 h period. Water samples were taken at 15 min. intervals and composited for each 24 h period at each location by continuous-duty, compositing samplers. During storms, a separate automatic water sampler collected individual 1-l samples at shorter time intervals proportional to the height of the stream.

Water samples were filtered through a 0.45 μm Millipore filter. The filtrate was analyzed by anodic stripping voltammetry, and the suspended solids by atomic absorption spectroscopy. Lead concentrations varied widely, ranging from 0 to 15 μg/l in filtrate and from 15 to 200 μg/l in the suspended solids.

The total output of lead on a yearly basis for the urban and rural compartments is shown in figure 6-6. The 600 kg annual output for the urban compartment is approximately 75 percent of the watershed total. Output from the rural compartment totals 210 kg. A comparison between the total output (fig. 6-6) and the total input (fig. 6-2) indicates that only approximately 2 percent of the total input exits via streamwater, leaving a considerable accumulation of lead in other components of the system.

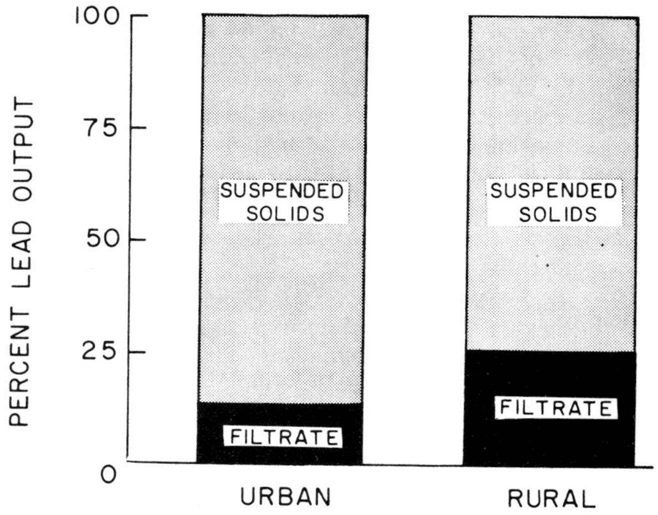

FIGURE 6-7.—Lead distribution between filtrate and suspended solids in urban and rural compartments.

In considering lead in streamwater, it is important to evaluate its distribution in the filtrate and suspended solids. This is shown in figure 6-7 for both urban and rural compartments. As would be expected, most of the lead is associated with suspended solids, with very little dissolved in the filtrate. The ratio of lead in suspended solids to that in filtrates varies from 4:1 in the rural compartment to 27:1 in the urban compartment. The large amount of lead associated with suspended solids in the urban compartment undoubtedly was related to the vast areas of impervious surfaces from which particles are quickly washed away during rainfall.

A comparison of total lead output during selected storm periods, based on the normal 24 h composite samples and the individual 5 min. samples, is shown in table 6-7. The composite sampler underestimated lead output by an average of 30 percent. However, there were no significant differences in sampling system estimates during periods of nonstorm flows.

In 1974 lead output from the urban compartment based on data from 5 min samples from selected storm periods and days without precipitation, amounted to 910 kg, as compared with 600 kg based on composite sampling. Thus the output is 7.5 percent of the estimated 10,400 kg urban input, indicating a large annual storage of lead in the urban ecosystem.

TABLE 6-7.—*Total lead output from the urban subcompartment for several selected storms during 1974*

Storm date	Total precipitation (in.)	Discharge (ft^3)	Lead output, 5 min. estimate (kg)	Lead output, composite estimate (kg)
4/11/74	0.17	30 x 10^4	2.57	1.96
4/29/74	0.28	40 x 10^4	3.41	2.34
5/28/74	0.14	26 x 10^4	2.11	1.80
6/7/74	0.22	63 x 10^4	8.31	5.25
7/8/74	0.18	80 x 10^4	7.96	5.98
8/2/74	0.90	22 x 10^5	8.75	5.69

Air

A number of investigations have been made of atmospheric lead concentrations. [3-8] Most of the studies were conducted in urban areas with high traffic densities and characterized the air quality of an entire city by using only a few sampling sites (at the most six or eight and often only two or three). In some cases, studies covered only 1 or 2 days. Air sampling equipment was often located well above ground level, and many times residential air quality was not well differentiated from that of the central city.

The air monitoring program for the ecosystem study encompassed studies of the distribution of ambient atmospheric lead concentrations in the rural and urban compartments, as well as sample areas with varying traffic densities. This provided an accurate picture of the ecosystem's airborne lead.

Field measurements were obtained with General Metal Works' high-volume samplers. To ensure the most effective use of the samplers, an analytical model was used to predict the distribution of airborne lead concentrations. The samplers were then placed where they would most effectively test the model. Figures 6-8 and 6-9 indicate the locations of the samplers with a letter assigned to each location. Figure 6-8 also shows mean air lead concentrations for the study period.

The samplers were powered by existing electrical outlets or propane-fueled electrical generators at remote locations. Initially, Gelman 8 in. by 10 in., GA-1 Metricel triacetate filters (5 μm pore size) were used as primary filters, with standard 8 in. by 10 in. Type A fiber glass filters (0.8 μm pore size) as secondary filters. Both filters were analyzed for lead content. The single 8 in. by 10 in. fiber glass filter gave results equivalent to those obtained using the dual arrangement and was used in subsequent tests. After completion of a 24 h run (normal procedure), the filter was cut into eight sections, and three random sections were analyzed for lead by atomic absorption spectroscopy.

Rural: From August, 1973 to February, 1974, 11 sampling locations were established throughout the watershed, (fig. 6-8) with eight in the rural and three in the urban compartment. During this period, the rural atmospheric concentrations were intensively sampled, and preliminary information was obtained on urban concentrations. Background levels were assumed to be 88 percent of the measured rural concentrations. [9] In all, 152 measurements were made on 30 different days. Averages over the sampling days appear in figure 6-8 next to each sampling location.

Lead concentrations did not vary greatly within the rural compartment. Lead concentration at remote stations E and I averaged about 0.19 $\mu g/m^3$. However, at stations nearer the urban area but still in the rural compartment, the average was about 0.25 $\mu g/m^3$.

Urban: Although airborne lead samples were taken at three urban stations (D, H, and K) during the intensive rural sampling period, the intensive sampling network for the urban compartment was not established until February, 1974. From May to July, 1974, all 19 samplers were in operation, while two additional sampling stations (C and J) continued to operate in the rural compartment. (fig. 6-9)

The urban compartment was divided into four areas based on use: residential, downtown, university, and suburban-commercial. Nine samplers were in operation in the residential area. Samplers D, L, M, Y, and AA were located on lawns next to streets with traffic volumes less than 1,000 ADT; sites U, V, W, and X were on lawns along main residential streets, with volumes of 4,000 to 5,000 ADT. The downtown areas had three samplers on streets with 12,000 to 20,000 ADT, with one of the samplers (Q) on the roof of a two-story building. Six samplers were placed on streets around the university with greatly varying volumes (2,000 to 20,000 ADT). Station K consisted of a ground-level sampler (K_1) and another sampler (K_2) directly above on the roof of a four-story building. Station N had one sampler (N_1) 17 ft from a high-volume street (20,000 ADT) and another (N_2) 29 ft from the same street. The suburban-commercial area had only one sampling station (H), which was in a light industrial park near the intersection of Interstate highways 57 and 74.

Airborne lead concentrations for each sampling station and traffic volumes of the urban compartment are summarized in table 6-8 for the 3 month sampling period, May through July, 1974. The relative magnitude airborne lead concentrations are in agreement with traffic levels at each station. At station K, the roof-top sampler (K_2) had 40 percent lower concentrations than the ground-level sampler (K_1). This supports the assumption of uniform vertical mixing of pollutants, which is often used in urban "box" models. Measurements from station N indicated a transverse decrease of 14 percent in airborne lead concentrations when the sampling distance was increased from 17 to 29 ft.

Table 6-9 summarizes airborne lead concentrations for ground-level stations at each of the four urban areas along with concurrent rural concentrations.

In conclusion, as shown in tables 6-8 and 6-9, airborne lead concentrations are generally low in the rural and residential areas, while much higher values have been measured in the urban areas with a peak of 3.8 $\mu g/m^3$.

Soils and Plants

The presence of lead in plants, often in concentrations approaching a hazardous level, is well documented in the literature. Numerous investigations have shown that the concentration of lead in soils and plants is significantly higher near highways and drops rapidly along gradients away from this source. [10-16] Although wind direction, speed, and turbulence are important factors in determining the steepness of these lead gradients, Suchdoller [13] report-

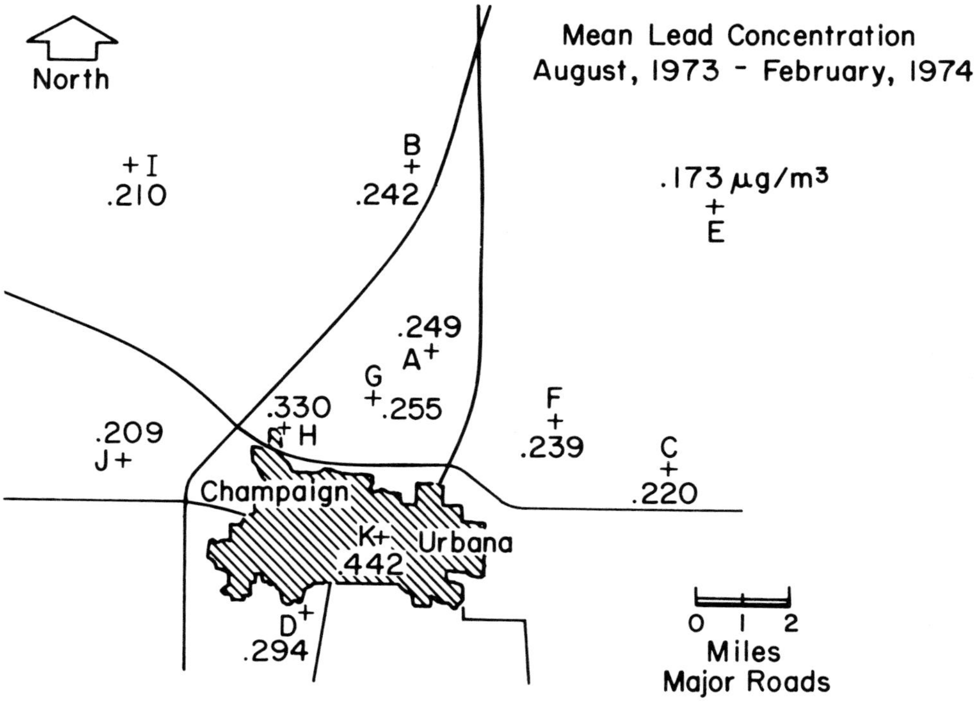

FIGURE 6-8.—Average air lead concentrations in ecosystem.

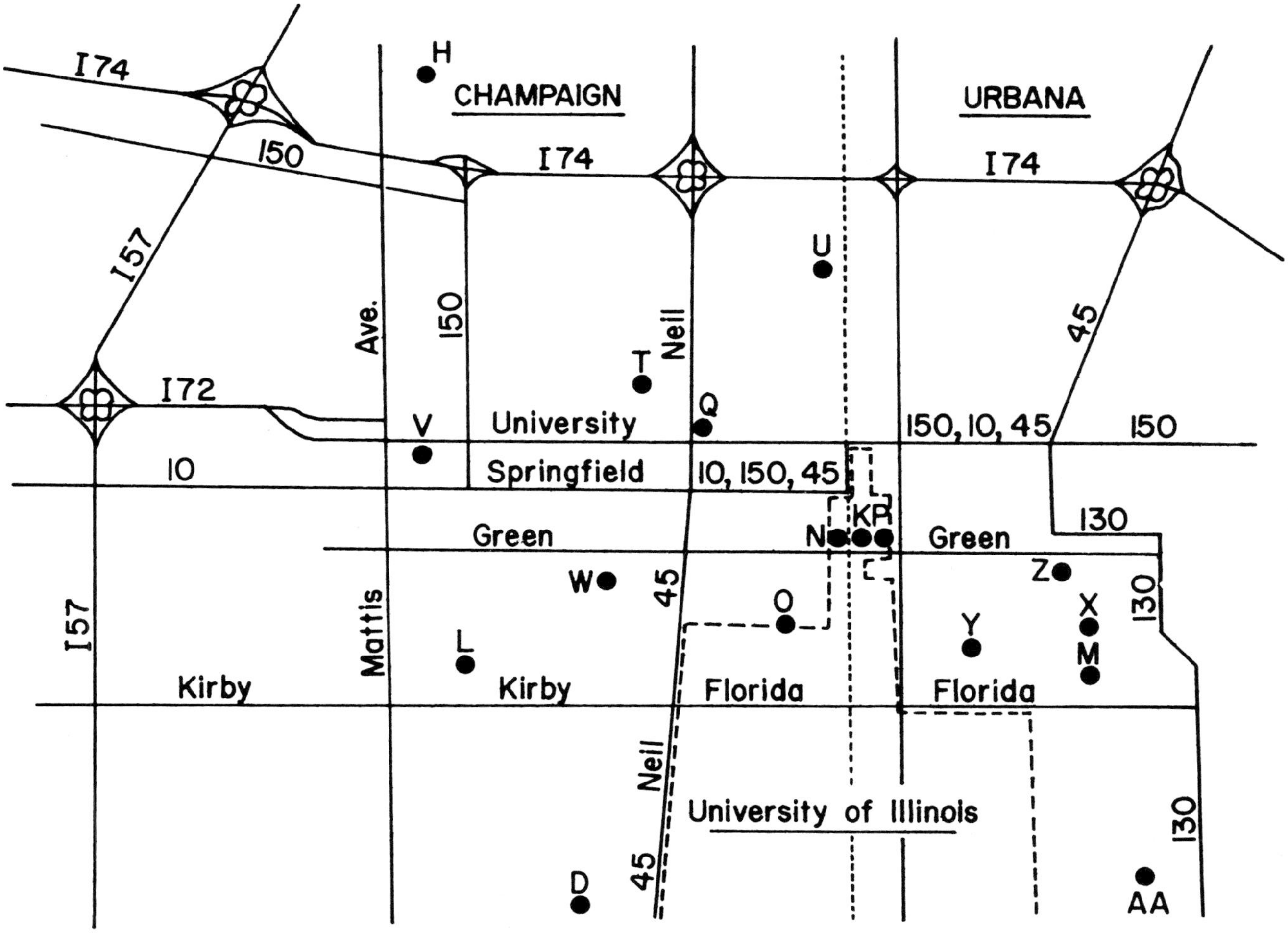

FIGURE 6-9.—Location of urban air samplers in ecosystem.

TABLE 6-8.—*Average urban airborne lead concentrations, May-July, 1974*

Site (station)	ADT	Airborne lead (μg/m³)
Downtown		
Q (roof)	12-20,000 (class I)	0.767
T	12-20,000 (class I)	1.445
Z	12-20,000 (class I)	1.263
University		
O	15,000 (class I)	0.569
P	7,000 (class II)	0.631
K_1 (ground)	2,000 (class IV)	0.425
k_2 (roof)		0.306
N_1 (17 ft)	20,000 (class I)	1.548
N_2 (29 ft)		1.345
Residential		
U	([1])	0.357
V	5,000 (class III)	0.342
W	5,000 (class III)	0.332
X	4,000 (class III)	0.306
D	250 (class V)	0.242
L	700 (class V)	0.250
M	800 (class V)	0.269
Y	800 (class V)	0.284
AA	500 (class V)	0.183
Suburban commercial		
H	([2])	0.244

[1] Difficult to estimate.
[2] Not close to a specific street.

ed that the gradient may be diminished completely within 30 to 40 m of a highway. Motto et al. [11] reported that the lead content of soils and plants sampled along heavily traveled highways increased with traffic volume. They also reported that the major effect of traffic was limited to the surface soil in a narrow zone within 30 m of the road.

Studies of lead in plants and soils were initiated in the summer of 1970 within initial focus on six plant species (including corn, soybeans, bluegrass, clover, dandelions, and fleabane) and the associated soil in transects away from a major highway (approximately 12,000 vehicles per 24 h). [17] In nearly all species tested, lead in or on tissues was a function both of distance from the road and lead in the soil. Analyses of variance were run to determine differences in lead content of taxa and lifeforms (i.e., annual or perennial). No significant differences were found in root lead but the lifeform-family interaction was significant for lead in washed foliage. Table 6-10 shows the interpretation of the significant interaction.

TABLE 6-9.—*Average airborne lead concentrations, ground level stations, May-July, 1974*

Area	Airborne lead (μg/m³)
Downtown	1.352
University	0.747
Residential	0.287
Suburban commercial	0.254
Rural	0.170

TABLE 6-10.—*Analysis of significant lifeform-family interaction of lead in washed foliage by using Tukey's W-procedure at* $P \leqslant 0.05$[1]

Soybeans	Dandelion	Bluegrass	Fleabane	Sweetclover	Corn
11.4	9.6	8.6	6.8	6.0	6.0

[1] Values are mean lead levels as μg/g dry-weight basis.
A common underline indicates no significant difference.

These data suggest that some plant species accumulated more lead than others, and that the accumulation rate was not influenced by whether the species was an annual or perennial. The correlation between distance from the highway and soil lead was significant ($P \leqslant 0.001$) up to 50 m for the 0 to 10 cm samples. The correlation between the surface soil lead content (0 to 10 cm) and the subsurface soil lead content (10 to 20 cm) was also significant ($P \leqslant 0.001$).

Soils in the rural area of the watershed were sampled intensively to further investigate the relationship between traffic volume, distance, and direction from highways. Traffic volume estimates were used to classify roads. Sampling was done in transects to 100 to 200 m from road pavements, with only the surface soils (0 to 10 cm) examined. (fig. 6-10) Each

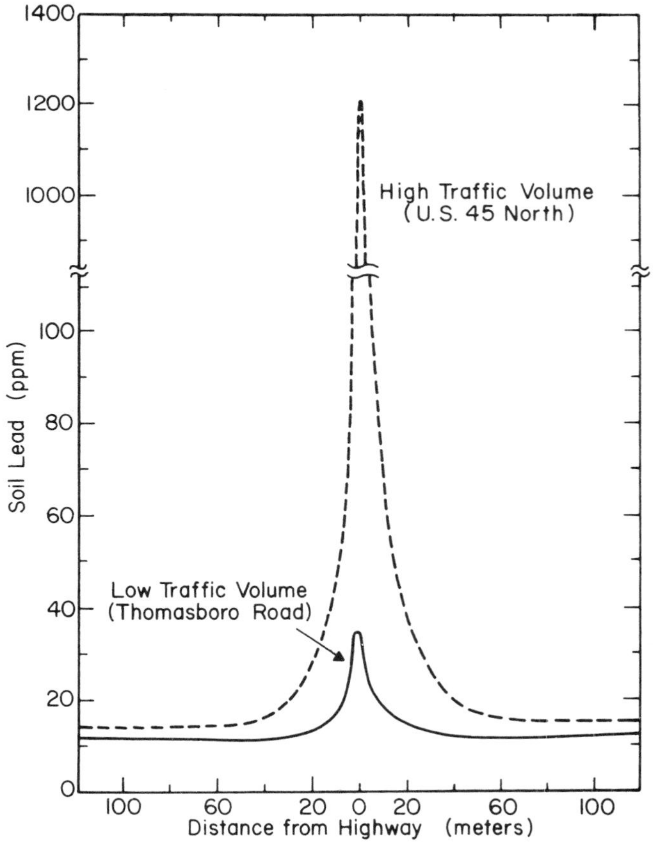

FIGURE 6-10.—Soil lead in relationship to distance from highway.

sample was a composite of six soil probes to a depth of 10 cm, taken at 5 m intervals along the transect. Figure 6-10 represents a composite of 12 transects from high-traffic highways and 12 transects from low-traffic highways; it illustrates the rapid decrease in lead concentration to background. The highest levels were found immediately adjacent to pavements of roads with the highest traffic volumes. On the downwind side (east) of north-south roads, the lead content of soils decreased to background levels within 50 m. On the upwind side, levels immediately adjacent to the pavement were comparable to those on the downwind side, but decreased more rapidly away from the road, reaching background levels within 20 m. Roads with low traffic volumes (less than 2,000 vehicles per 24 h) had associated soil lead gradients to background levels within no more than 5 to 10 m of the pavements on both the upwind and downwind sides. Thus soil lead was seldom influenced beyond the right-of-way of roads with less than 2,000 vehicles per 24 h. East-west roads with high traffic volumes had soil lead gradients that were essentially comparable on the north and south sides and intermediate to the east and west gradients on north-south roads. Lead in soil nearest the pavements was much the same as for the north-south roads, but the lead gradient decreased to background within 30 m of the pavement.

The results indicate that the higher the traffic volume, the higher the soil lead content, and the farther the lead gradient extends from the highway. Even on the most heavily traveled roads, the strong enhancement of soil lead extended only 50 to 75 m downwind and 25 to 40 m upwind. Extremely high lead levels (> 50 μg/g) were seldom found in this study beyond 20 m from pavements, except in unusual circumstances such as ditches draining road surfaces.

Lead accumulation in urban and rural soil profiles is shown in figure 6-11. The concentrations are averages of 10 profiles. The urban profiles were taken within 10 m of a city street (12,000 vehicles per 24 h). Figure 6-12 shows an urban profile sampled within 1 m of a heavily traveled city street (15,000 to 17,000 vehicles per 24 h), with a breakdown of the top 3 cm into 1 cm segments. The rural profiles were taken in an agricultural field remote from a highway. Maximum concentrations were found in the upper 10 cm, with a sharp decline in concentration between 10 and 20 cm. Below 20 to 30 cm, the lead concentration was relatively uniform. Considerably higher concentrations were evident in the urban profiles, which again showed a strong correlation with traffic volume.

Intensive studies of lead in the vegetation of the ecosystem, during the summers of 1971-74 showed a similar influence of the proximity and direction from highways as that observed for soils. Sampling was completed when the plants were fully mature in the field. Along the heaviest traveled roads (12,000+ vehicles per 24 h), there was significantly more lead in or on corn and soybeans within 20 m of the pavement; by 30 m from the pavement, however, lead in

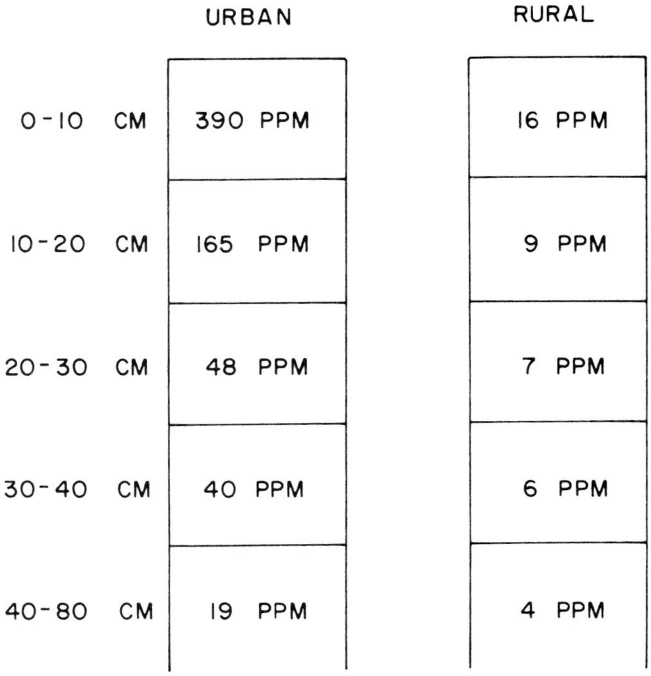

FIGURE 6-11.—Lead concentrations in urban and rural soil profiles. (Values shown are the mean of 10 profiles in each location.)

or on crops was not different from the average for the field. On lesser traveled roads, there was no observable influence of the highway on crop lead. Along the heavily traveled roads, corn and soybeans growing within 20 m of the pavement averaged about 30 μg/g lead, whereas plants farther away from the pavement and along secondary roads averaged 8 μg/g. Analyses of kernels of corn and soybeans showed a consistently low lead content, averaging less than 2 μg/g. Figure 6-13 shows a composite of 12 transects from high traffic highways and illustrates the influence of traffic on lead in vegetation and soils.

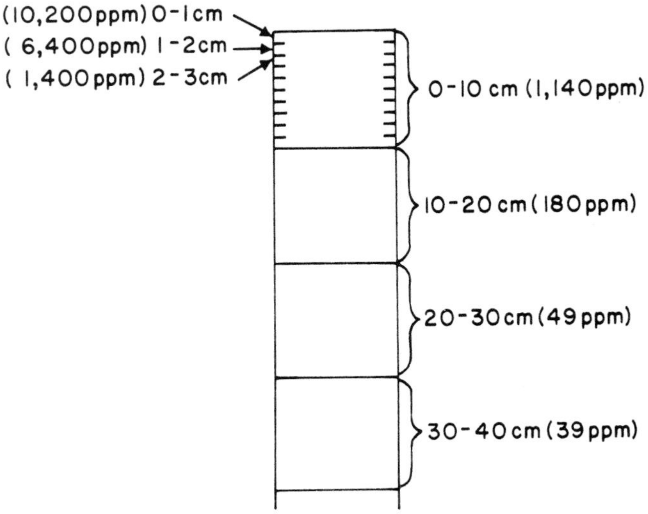

FIGURE 6-12.—Urban soil profile with top three 1 cm segments analyzed individually.

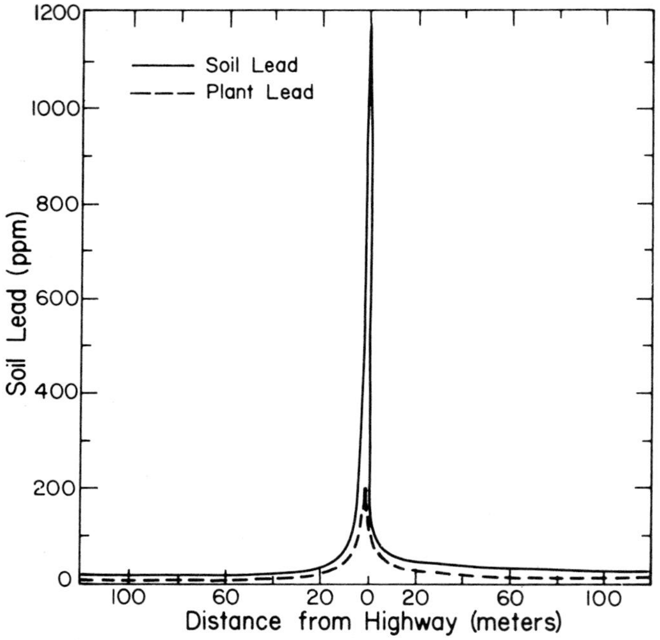

FIGURE 6-13.—Soil and plant lead as related to distance from highway.

Lead content in the vegetation of the ecosystem is summarized in table 6-11. The size (223 km²) and diversity of the watershed precluded a routine approach to sampling. Consequently, the watershed was divided into rural and urban compartments. The rural compartment was subdivided into vegetation habitats. A description of the divisions is given below:

A. Urban: All areas within the watershed developed for residential or industrial use.
   1. Pervious: All nonpaved surfaces.
   2. Impervious: All paved surfaces, including roofs.

B. Rural:
   1. Fields
      a. Cultivated fields: All crop fields that were tilled within the last year.
      b. Sod fields: All crop fields and/or pastures that are tillable.
   2. Roads (See section on traffic sampling.)
      a. Pavement.
      b. Roadsides: Up to and including bordering fencerows.
   3. Railroads
      a. Track pavement.

TABLE 6-11.—*Summary of lead content of vegetation and habitat areas in a Champaign County watershed*

Habitat	Total area (km²)	Percent of watershed	Average Pb Content of vegetation (μg/g)	Average dry biomass of vegetation (g/m²)	Total Pb in vegetation of habitats (g)
A. Urban					
1. Pervious	14.5	6.4	67	287	278,820
2. Impervious	17.2	7.6	–	–	–
B. Rural					
1. Fields					
a. Cultivated	166.4	74.6	8.0	1,400	1,863,680
b. Sod	3.8	1.7	4.1	800	124,640
2. Roads					
a. Class I					
Pavement	0.8	0.3	–	–	–
Roadside	1.6	0.7	34.0	159	8,650
b. Class II					
Pavement	0.1	0.1	–	–	–
Roadside	0.1	0.1	21.0	155	326
c. Class III					
Pavement	0.1	0.1	–	–	–
Roadside	0.2	0.1	15.6	194	605
d. Class IV					
Pavement	0.1	0.1	–	–	–
Roadside	0.3	0.1	9.0	288	778
e. Class V					
Pavement	1.1	0.5	–	–	–
Roadside	2.2	0.9	7.5	262	4,323
3. Railroad					
a. Track pavement	0.5	0.2	–	–	–
b. Right-of-way	0.5	0.2	6.7	520	1,742
4. Homestead	5.9	2.7	6.1	112	4,031
5. Streams	3.1	1.4	4.7	[1] 623	9,077
6. Fencerows	3.0	1.3	5.8	691	12,023
7. Waste areas	3.0	1.3	6.5	[1] 664	12,948
8. Woods	1.5	0.7	8.6	[1] 127	1,638
9. Pond (surface water)	0.3	0.1	–	–	–
Total	226.2 or (87.3 mi²)			Total	2,211,105 or 2,211.1 kg

[1] Herb layer only.

b. Right-of-way: Up to and including fencerows.
4. Homesteads (See urban description).
5. Streams: All vegetation bordering streams that was not utilized for crops or pasture. In waste areas, this included only the stream bank vegetation.
6. Fencerows: All untilled areas between fields, whether a fence was present or not.
7. Waste areas: Miscellaneous areas not identified in other categories. Included were refuse areas, dry borrow pits, permanently fallowed fields that lacked complete arborescent cover, rural cemeteries, and unused areas around isolated outbuildings.
8. Woods: Forested areas with all or nearly all arborescent cover.
9. Ponds: Generally borrow pits. Surface water only was included.

The total area for each compartment and subdivision was determined from aerial photographs. Ground checks were made in all cases where vegetation or land use was unclear. All compartments and subcompartments were randomly sampled with ¼ m^2 circular frames at peak standing crop. Transects were used to determine the lead content relative to distance and direction from roads, but, with the exceptions noted previously, traffic flow had little influence on the lead content. Because sampling was random within each subcompartment, the observed biomass (averaged) was multiplied by the total area to arrive at the total biomass. The average lead content of the vegetation in a subcompartment was then used to calculate total lead for a subcompartment. (table 6-11)

An extensive sampling of urban vegetation and soils was completed in 1974. (table 6-11) In the transect studies, samples were taken at several locations within the urban area depending on traffic volume; the results show a similar pattern to the rural area. Somewhat different results were obtained on one particular set of four transects (street to house) on streets with low traffic volume (200 to 400 vehicles per 24 h) and four on streets with high traffic volume (12,000+ vehicles per 24 h), each of which was sampled at four different times during the year. (fig. 6-14) The concentration of lead in both soils and plants along these transects, especially in the high traffic areas, increased as the houses were approached. These houses were all brick, unguttered but with painted trim. It is quite likely that lead particulates deposited on the roofs of these houses were washed off by rainfall and accumulated in soils and plants along the drip zone. However, a portion of this lead accumulation may be a result of leaching from paint on the house trim.

Temporal distribution of lead is shown in figure 6-15 from urban and rural tree-ring samples. The urban samples were collected from trees within 10 m of heavily traveled city streets (greater than 10,000 vehicles per 24 h). The rural samples were from trees within 10 m of a country road (1,800 vehicles

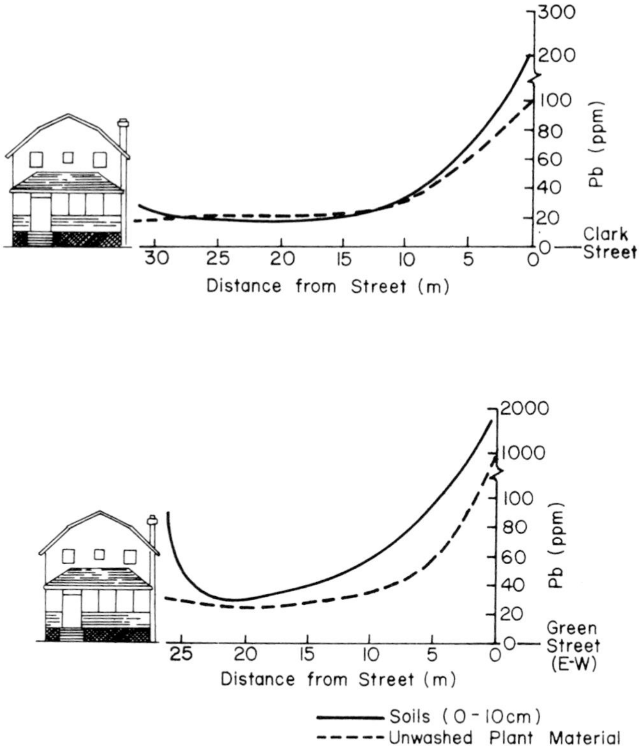

FIGURE 6-14.—Soil and plant lead concentrations along transects from low (Clark Street) and high (Green Street) traffic volume city streets.

per 24 h). The values are averages of 20 samples and show a significant increase between 50-year-old tree rings and current rings. This graphically illustrates the increased insult of lead to the environment during the past half century. [18]

Urban Dust

Two important points concerning the distribution of lead were discussed earlier in this chapter. First, an estimated 75 percent of the lead emitted from automobile exhausts was deposited in the urban compartment of the ecosystem. Second, the highest concentrations of lead in surface soils were within

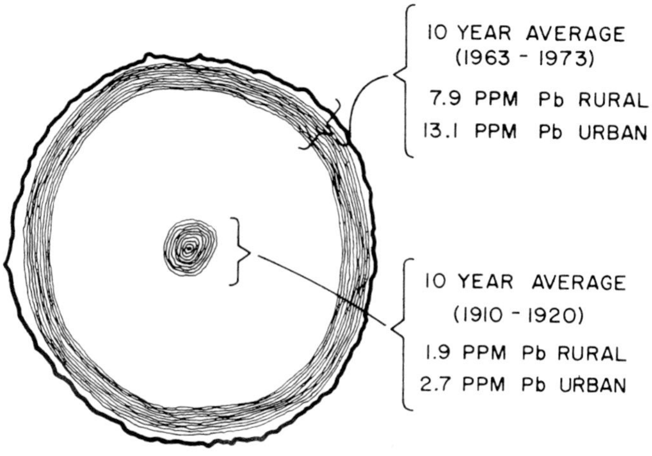

FIGURE 6-15.—Tree ring analysis of urban and rural trees.

20 m or less of traffic thoroughfares. In urban areas, this means that a high percentage of the lead is deposited in, on, or around residences, commercial establishments, and other buildings. Further distribution and concentration are likely as lead associated with dust is blown around by the wind or tracked into buildings from streets, sidewalks, and other outside areas. Thus potential human exposure is much greater in urban than rural areas, and the total body burden is likely to increase where people and automotive traffic are concentrated.

An intensive study of lead in urban dusts and soils has been conducted in the urban compartment (Champaign-Urbana) of the ecosystem. Cadmium was also considered, but to a lesser extent than lead. Results from about 1,400 samples have shown unusually high concentrations in homes, schools, business establishments, etc., as well as on streets and surrounding areas.

It should be emphasized that the data reported here for urban soils are from different locations than those discussed in the preceding section. Because dust and soil are so intimately connected, it was essential that both be collected at the various sampling sites. Thus there is no duplication in the two discussions of urban soils.

Dust samples were collected by a vacuuming technique with the specially designed equipment shown in figure 6-16. On carpeted areas, a vacuum inlet pressure of 7.5 psi was used, while a full pressure of 20 psi was applied to other sufaces. Samples were taken within a 0.5 m by 0.5 m area. A single pass with carefully overlapping, single-direction strokes was sufficient to remove 90 percent of the lead from sampling surfaces.

Dust collected in the nylon chamber, (fig. 6-16) along with the glass fiber filter, was placed in tared glassene bags and weighed. Prior to use, both the filters and glassene bags were temperature and humidity equilibrated for a period of 24 h. Weighings were made rapidly to minimize hygroscopic effects. Plastic gloves were worn in operations where contamination from bare hands might be a problem.

In very dusty areas (city streets, for example) or if stones were present, samples were taken with a standard wet-dry industrial-type vacuum cleaner. The cleaner was modified by fitting two glass fiber filters back-to-back in place of the foam rubber filter that protects the motor. The vacuum reservoir was lined with a plastic garbage bag that was cut to fit and sealed tightly around the vacuum inlet. Fine dust that failed to drop into the reservoir was trapped by the filters. Final weighings included the filters, plastic bag, and the dust sample. All samples were dried and passed through a 30 mesh (ASTM) stainless steel screen prior to analysis. Lead and cadmium were determined by atomic absorption spectroscopy.

Soil samples were taken from the surface 2.5 cm with a standard auger. The samples were crushed, passed through a 2 mm screen, and dried for 4 h at 105° C prior to analysis.

Interior Residential: A total of 239 floor dust samples were taken in 12 homes, four of which were

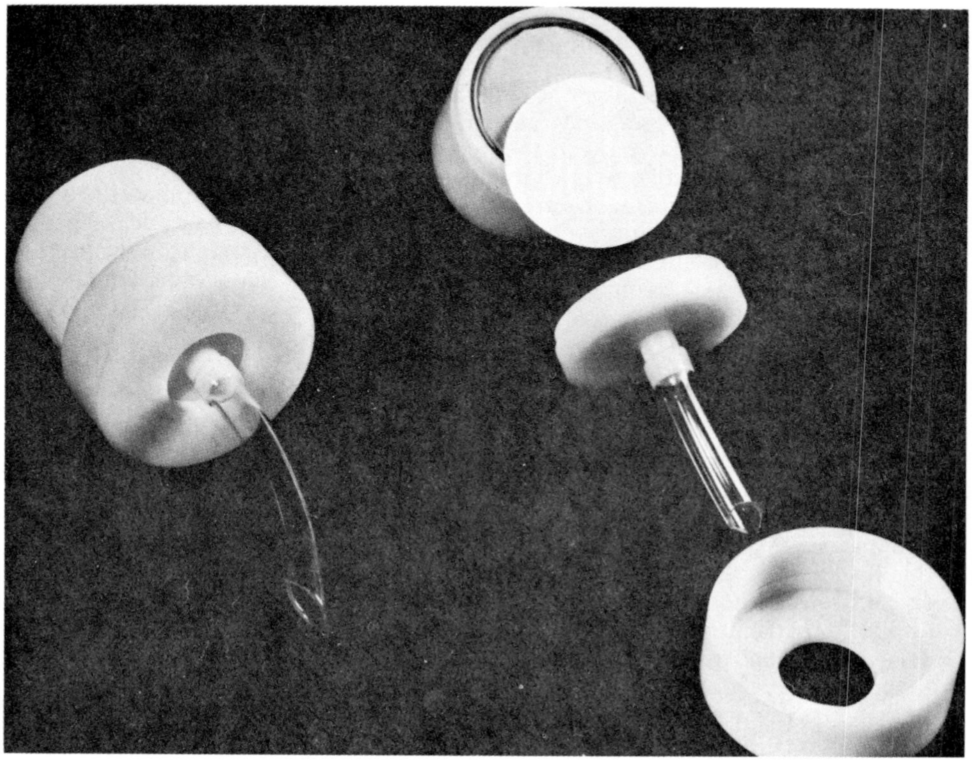

FIGURE 6-16.—Vacuuming techniques used in urban dust studies. (A) Equipment

sampled twice. Ten of the homes were single-family detached dwellings,and two (sites F and K) were located in multi-story apartment buildings. All of the homes were upper-middle class residences and in excellent condition. They were painted with nonlead or low-lead paints and located in relatively low traffic density areas. Except for site L, all homes were heated with forced air, natural gas furnaces. All dust samples were taken in the central area of rooms, well away from walls, windows, and doors.

Eight of the sites (A, B, C, D, E, G, I, L) were located where the lead level of outside air was also being monitored. The remaining four sites (F, H, K, L) were fairly close to monitoring stations.

Detailed results for rug- and nonrug-covered surfaces are shown for both lead and cadmium in table 6-12. These data are summarized in table 6-13. Average dust-leadlevels were 600 μg/g 680 g/m2 for the 12 homes. Lead concentrations were higher on wood and tile than on rugs—950 and 450 g/m2, respectively. However, the total amount of settled lead was higher on rugs when expressed on an area basis. This is because rugs accumulate more dust than bare floors. The lead levels in dust were similar to those reported by Kreuger [19] for suburban Boston.

In general, the amount of indoor lead in dust was proportional to outside air concentrations, even though the latter were relatively low (mean, 0.28 μg/m3; range, 0.18 to 0.34 μg/m3). Site H was an exception where lower than expected amounts of lead were found. The apartment locations (F and K) were also lower than anticipated, based on the traffic levels in the area. Airborne lead was not measured near the apartment sites, and little is known concerning such inside apartment buildings.

TABLE 6-12. –*Mean lead and cadmium levels in settled floor dusts at residential sites*

Site code	Lead		Cadmium	
	μg/g	μg/m²	μg/g	μg/m²
A	1,440	1,180	28	26
B	690	710	19	14
C	940	1,700	10	31
D	760	580	10	11
E	900	1,590	20	37
F	350	760	48	132
G	240	100	26	7
H	500	70	25	3
I	170	120	7	4
J	270	400	9	15
K	430	650	12	15
L	460	220	12	6
Average (12 sites)	600	680	18	25

Cadmium levels as high as 105 μg/g and 219 μg/m2 were measured in some homes. Extremely large amounts, averaging 44 μg/m2, were found in dust from rugs. A major source of this cadmium appears to be the rubber backing on many carpets. A random sample of this material assayed 2,000 μg/g cadmium, and abrasion might well cause its presence in dust.

Interior Nonresidential: For a more complete evaluation of human exposure, it is necessary to establish lead and cadmium levels in places where people

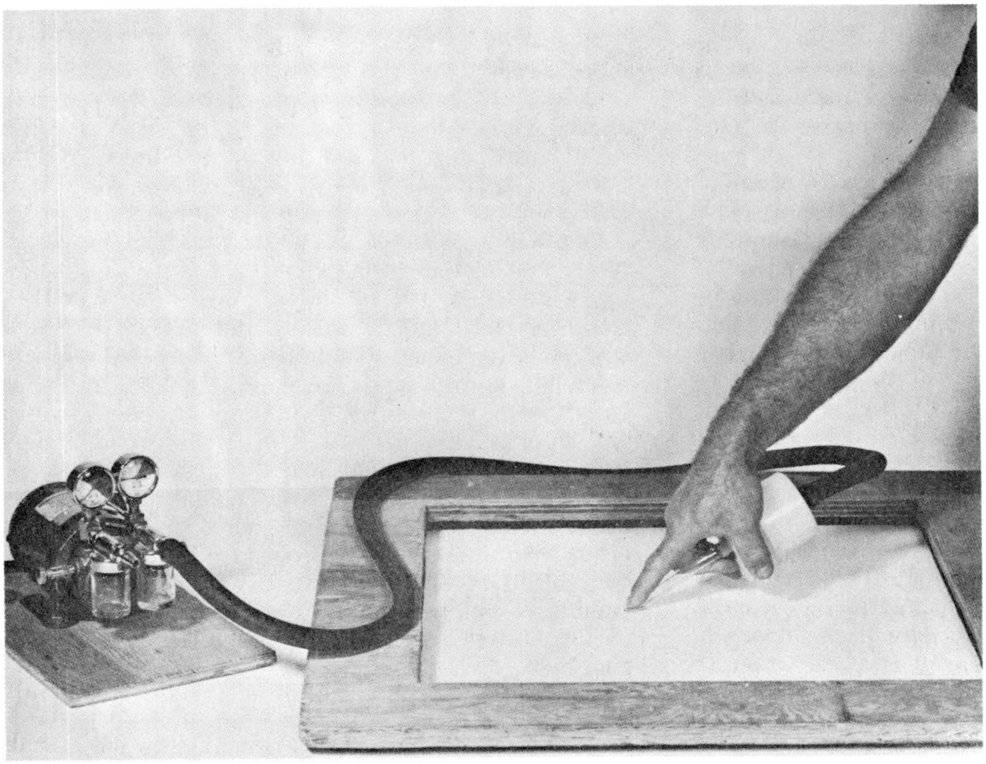

FIGURE 6-16.—(B) Collection technique.

TABLE 6-13.–*Lead and cadmium content of floor dust in homes (rug-covered versus nonrug-covered surfaces)*

Site code		Number of samples	Lead µg/g	Lead µg/m²	Cadmium µg/g	Cadmium µg/m²
A	Rugs	7	830	1,490	24	43
	Nonrugs	13	1,780	1,020	29	17
B	Rugs	7	550	590	16	18
	Nonrugs	6	1,440	1,980	13	18
B (repeat)	Rugs	5	420	460	17	20
	Nonrugs	10	480	140	25	6
C	Rugs	14	640	2,710	14	58
	Nonrugs	9	770	360	7	4
C (repeat)	Rugs	14	520	1,940	10	34
	Nonrugs	7	2,580	930	7	5
D	Rugs	5	380	860	14	31
	Nonrugs	17	870	510	9	5
E	Rugs	8	340	3,060	16	75
	Nonrugs	11	900	970	14	13
E (repeat)	Rugs	8	590	1,110	12	20
	Nonrugs	9	1,690	1,490	36	46
F	Rugs	7	470	1,320	64	219
	Nonrugs	1	250	40	2	4
F (repeat)	Rugs	6	240	350	66	71
	Nonrugs	1	290	60	105	22
G	Rugs	11	220	110	14	7
	Nonrugs	3	320	70	71	8
H	Rugs	1	300	170	11	6
	Nonrugs	4	550	40	28	2
I	Rugs	13	210	160	7	6
	Nonrugs	6	100	20	6	1
J	Rugs	7	370	800	13	30
	Nonrugs	8	190	40	4	1
K	Rugs	4	580	1,160	13	26
	Nonrugs	4	270	140	10	4
L	Rugs	3	440	640	13	20
	Nonrugs	10	470	90	12	2
	All rugs	120	450	1,220	19	44
	All nonrugs	119	950	600	18	10

work, shop, or visit outside the home. In addition to direct exposure in these areas, some of the dust may be inadvertently carried back to the home on clothing or shoes.

More than 350 samples were collected in schools, hospitals, food markets, and university laboratories, offices, and classrooms. As in the homes, samples were taken in the central areas, away from walls. All areas sampled were in good physical condition, with no evidence of peeling of chipping paint. Detailed results are shown in table 6-14 and are further summarized in table 6-15.

Lead and cadmium were found, often in very high concentrations, in all buildings sampled. Extreme concentrations were found in chemistry laboratories at the university. Since levels of both metals were much less in adjacent offices, the source of such high levels was probably chemicals being used in the laboratories. For this reason the data from chemistry laboratories were not included in the average values shown in table 6-15. Had these data been included, the average values of 1,400 µg/g and 2,040 µg/m² would have been almost doubled.

In general, the levels of lead and cadmium were about twice as great as the residential averages. This, especially in the case of lead, undoubtedly reflected airborne lead levels, which were from two to five times greater in nonresidential than in residential areas. Heavier foot traffic may also be a factor in tcking dust and soil into the buildings from the outside. Again, high levels of cadmium were associated with rubber-backed rugs or floor mats.

Outdoor Residential: Airborne lead that moves directly into houses with air currents is perhaps a relatively minor source of interior dust-related concentrations. The major source is more likely related to lead emissions that accumulate in dust and soil surrounding houses and are later tracked inside or blown in as reentrained dust through open windows and doors. The magnitude of such accumulations was determined around the 10 single-family dwellings used for the indoor residential study. Eight of these homes were of frame construction, and the other two had painted wood trim. The exteriors were all in excellent condition, with no indication of peeling or chipping paint. Analyses of paint from 8 of the houses indicated 1 percent lead and 50 µg/g cadmium.

Data from 288 samples (all analyzed for lead, 177 analyzed for cadmium) have been grouped by areas around a typical house and are shown in figure 6-17. Both median and range values are shown to compen-

TABLE 6-14.–*Lead and cadmium in indoor floor dusts at nonresidential sites*

Sampling site	Number of samples	Lead µg/g	Lead µg/m²	Cadmium µg/g	Cadmium µg/m²
Halls/corridors					
Roger Adams Laboratory	31	2,720	690	14	6
Mechanical engineering building	36	3,380	3,010	9	8
Krannert court	11	500	1,190	7	21
Federal buildings					
Rugs	1	1,710	18,000	23	240
Nonrugs	2	4,680	360	369	28
Krannert garage	10	1,830	8,650	5	25
Columbia School	4	1,010	1,640	9	15
Gregory Hall	3	740	230	5	1
Rooms in university buildings					
University laboratories					
299 RAL	35	11,400	3,390	185	57
221 RAL	18	6,280	5,770	86	74
57-59 RAL	8	910	120	33	4
136-138 RAL	7	5,740	3,560	69	34
11 RAL	9	8,060	26,560	79	302
University offices					
105, 115, 116, 202, 222 RAL	11	1,450	640	13	7
University classrooms					
Various – Gregory Hall	7	590	190	24	3
Various – Lincoln Hall	8	930	155	1	0
Rooms in other buildings					
Federal building offices					
Rugs	6	2,320	11,780	1,033	2,943
Nonrugs	1	2,960	560	1,060	200
Columbia School classes					
Rugs/mats	10	730	3,990	29	122
Nonrugs	17	650	200	7	4
All floors	27	680	1,600	15	48
Dr. Howard School classrooms					
Rugs/mats	12	410	2,360	30	115
Nonrugs	26	430	210	19	5
All floors	38	420	890	23	71
Carle Hospital complex					
Entry areas (carpet)	11	620	7,940	20	390
Hospital corridors/rooms	11	360	180	22	4
Food markets					
Plain floors	34	490	100	9	2
Matted areas	2	730	18,000	24	290

sate for variations in data related to differences in house locations, orientation, etc. The term "near" on the figure is defined as 1 m.

Lead levels were high at all locations surrounding the homes. Soil lead was somewhat greater near the road and the house, but it did not vary greatly from one location to another. In contrast, concentrations in dust varied considerably. Amounts ranged from 240 to 6,640 µg/g away from the house and from 130 to 11,760 µg/g near the house. On an area basis, amounts were as follows (values expressed as µg/m²): (1) street gutters—median 21,300, range 1,270 to 211,000; (2) driveways—median 5,820, range 1,050 to 86,000; (3) walks—median 1,590, range 720 to 15,200; (4) near house—median 1,840, range 200 to 9,100.

Cadmium levels were somewhat higher in soils and dusts near the house (1 µg/g in soil, 6 to 8 µg/g in dust). Total cadmium levels in dusts were relatively uniform, ranging from 7 to 20 µg/m²..

The houses used in this study represent relatively ideal conditions: low traffic density in the vicinity; use of low-lead or unleaded paints, and excellent maintenance of both interiors and exteriors. Soil and dust lead levels increased dramatically around houses that did not meet these standards. This was

TABLE 6-15.–*Mean lead and cadmium levels in settled floor dusts at nonresidential sites*

Surface	Number of samples	Lead µg/g	Lead µg/m²	Cadmium µg/g	Cadmium µg/m²
Rugs/mats	42	860	6,670	70	604
Nonrugs/mats	212	1,500	1,120	19	12
All floors	254	1,400	2,040	44	110

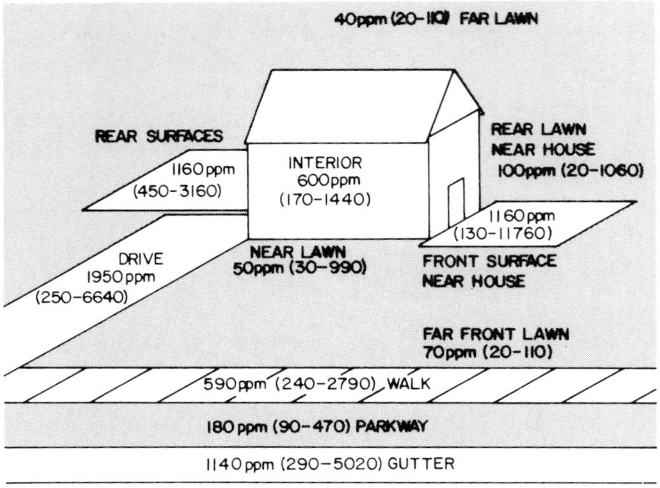

FIGURE 6-17.—Composite diagram showing lead in soil (median and range values) in residential areas.

shown in a special study of areas surrounding 7 additional frame structures. Each was painted with an identical type of leaded paint (27 percent Pb by weight). They were located in a higher traffic density area where airborne lead contents were two to three times greater than in the previously mentioned residential area. These houses are not now used as personal residences, although they were in the past.

Soil lead levels are shown in figure 6-18 as the broken line for painted frame houses. Soil lead levels were much higher around these 7 houses as compared with the 10 homes prevously discussed. The largest increases were in soil adjacent to the curb and near the house. However, these maximum values were of the same order of magnitude as those in settled dusts on walks and driveways of the 10 houses painted with low-lead paints.

Outdoor Nonresidential: Lead and cadmium concentrations were determined in soil and dust samples around several of the nonresidential buildings used for the indoor study. Dusts were sampled at 7 sites (195 samples) and soil at 20 locations (183 samples). All samples were analyzed for lead as well as 70 percent for cadmium.

All of the buildings were of brick or stone construction. Traffic density was heavier than at either of the two locations discussed above, with volume amounting to 5,000 to 20,000 vehicles/24 h. The amount of lead in settled dusts, shown in figure 6-19 as a function of distance from the street, reflected the higher traffic volume. Average lead concentrations were about twice as great as those found in the residential zone where traffic was much lighter. Mean lead accumulations as great as 190,000 $\mu g/m^2$ were found on sidewalks, but the concentrations decreased with distance from the street. There was no increase in lead values near buildings. However, undisturbed dust samples were obtained adjacent to buildings at only two locations, and the results were inconculsive.

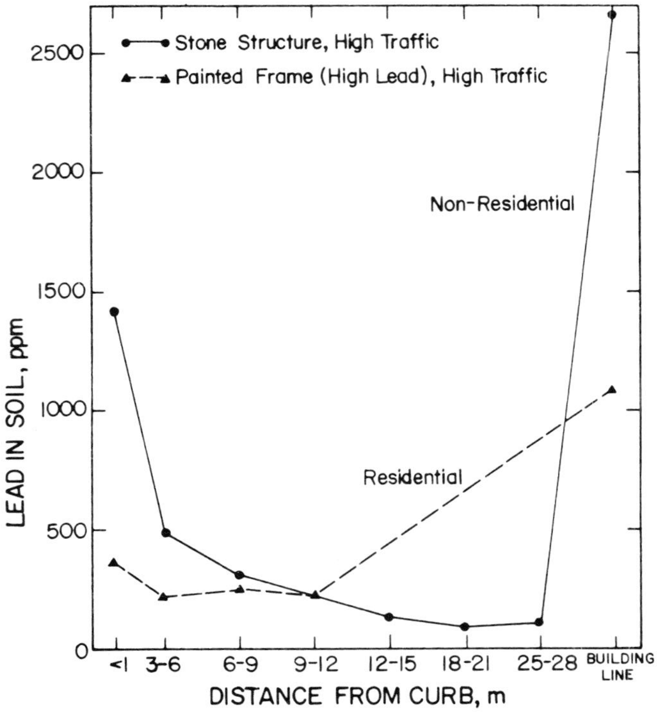

FIGURE 6-18.—Mean lead concentration in soils around stone and painted frame buildings in high-traffic areas.

In figure 6-18 soil lead levels around stone structures are designated by a solid line, as a function of distance from the curb. Levels were highest at the building line, then dropped rapidly and increased again nearer the street.

In the nonresidential area soil lead contents were several times greater than the residential values. Also, concentrations at the curb were much greater than those found in the special study of 7 houses painted with leaded paint. (figure 6-19) In both cas-

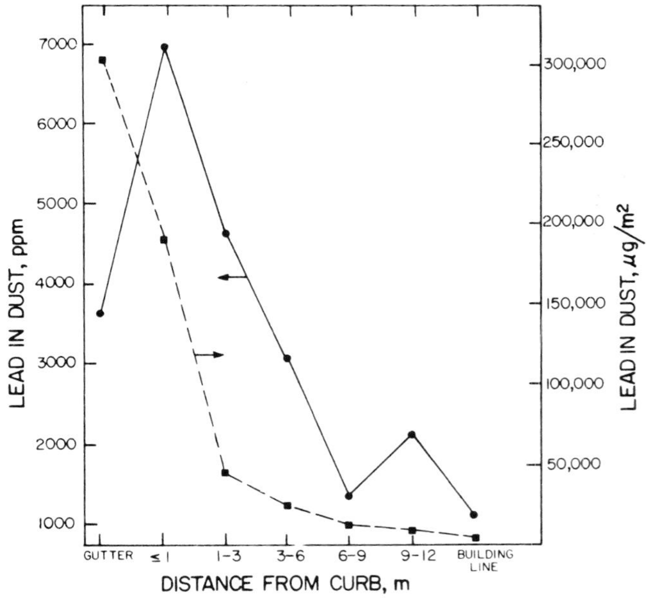

FIGURE 6-19.—Mean lead concentration in dusts from nonresidential sites.

es, the higher lead values reflected the greater traffic density in nonresidential areas.

Cadmium levels averaged 1 $\mu g/g$ in soils and 2 to 6 $\mu g/g$ in dusts with no well-defined trends. Total cadmium in dust decreased from 134 $\mu g/m^2$ at the street to 1 $\mu g/m^2$ near the building line.

Sources of Lead: The accumulation of lead in dusts and soils around buildings appears to be directly related to the volume of traffic on neighborhood streets and roads. Similar increases in the lead content of air as a function of traffic density suggest that the major source is exhaust emissions following the combustion of leaded gasoline in automobiles. Lead deposited on streets, sidewalks, and driveways is subject to further distribution by the physical forces of wind and water. Windblown particles associated with dust are apt to be redeposited within the urban environment because of the extremely complex wind currents caused by buildings and street canyons. As a consequence, the rather large buildup of lead in soil and dusts around buildings should be expected.

The degree to which leaded paint contributes to lead concentrations in soil near buildings is not so clear. Leaching from leaded paint is obviously a source, as shown by the higher concentrations of soil lead near the 7 frame houses painted with leaded paint, as compared with those near the frame dwellings where low-lead paints were used. However, a further comparison of the same 7 frame houses wth the brick and stone buildings (fig. 6-18) shows twice the concentration of soil lead near the building line of the latter. Although the brick and stone buildings had painted wood trim, it is highly improbable that this relatively small amount of painted surface could account for the difference. Lead washed from unguttered roofs, ledges, and walls is a more likely major source. Lead subject to washoff could be particulates deposited directly from the air or associated with windblown dust that collects on buildings.

Concentrations of lead in house dust may come directly from airborne sources or indirectly from exterior soil and dust. The airborne contribution would be greatest during warm weather when windows and doors are open for ventilation. In contrast, dust and soil are tracked into houses during all seasons and are likely to be the major source of lead. Of the two, dust is perhaps the most important. Overall lead values are higher in dust, and it is continually being blown around by the wind. Soil is generally more stable, especially around established buildings where it is usually covered with vegetation and less susceptible than dust to wind transport.

Health Implications: Obviously, the large quantities of lead and cadmium in urban environments raise questions about the possible effects of increased exposure on human health, especially that of small children. In the absence of clinical evidence, it is only possible to conjecture these effects. Because small children commonly play on the floor and the ground, and also put their hands (or almost any available object) in their mouths, the potential for increased ingestion of these toxic substances most certainly exists. It is significant that this potential exists in well-maintained, better homes and not just in the rundown tenement, with its peeling and chipping paint, that is sually associated with the critical ingestion of lead by children. The mere fact that such conditions exist in heretofore unsuspected neighborhoods should be recognized by medical diagnosticians and clinicians.

SMALL MAMMALS

Samples of small mammals were obtained by snap-trapping from the following situations: within 10 m of a high-use road (>12,000 vehicles/24 h); within 5 m of medium-use roads (2,000 to 6,000 vehicles/24 h); and within 5 m of low-use roads (4,000 vehicles/24 h). Control samples were taken from situations at least 5 m away from any road (normally the closest road was one with low traffic volume). Trapping was done seasonally (November, March, June, and August).

All specimens were washed thoroughly in glass-distilled water to remove as much particulate lead from the fur and skin as possible; the specimens were then freeze-dried. The entire animal (including the digestive tract) was processed when the dried specimen weighed less than 10 g. Specimens heavier than 10 g were ground in a blender and 1 g subsamples were processed for lead analyses. Individuals were also selected from each trapping for tissue analysis. All lead concentrations were recorded in $\mu g/g$ (dω).

All species except the white-footed mouse, *Peromyscus leucopus,* had higher lead concentrations in habitats adjacent to high-traffic roads. (table 6-16). Since the home range of the white-footed mouse averages more than 50 m in diameter, even those individuals caught nearest the highway were undoubtedly spending a considerable amount of time much further removed. This probably accounts for the low level of lead in this species.

Lead concentrations in small mammals taken along medium-use roads were usually intermediate between those taken near low- and high-traffic roads. There was no significant difference in lead concentrations in small mammals along low-traffic roads and those taken more than 50 m from any road.

Along high traffic roads there was a correlation between habitat requirements and lead concentrations in small mammals. Species that require dense vegetation and do not range out into cultivated fields (*Microtus ochrogaster,* prairie vole; *Blarina brevicauda,* short-tailed shrew; *Cryptotis parva,* least shrew; and *Reithrodontomys megalotis,* white harvest mouse) had higher lead concentrations than did those that do extend their ranges into cultivated areas and thus out more than 50 m from roads; (*Peromyscus mainculatus,* white-footed mouse; *Mus musculus,* house mouse). There was no correlations between lead concentration in the small mammals and the season of the year.

Along heavy-use roads there was a general correlation between feeding habits and lead concentra-

TABLE 6-16. –*Lead content of small mammals*[1]

Species	Lead concentration (µg/g dry weight)				
	Traffic levels[2]				
	High (12,000)	Medium (2 to 6,000)	Low (<400)	Fields[3]	Urban
Blarina brevicauda	[4]15.2 (46)	[4]6.5 (71)	[4]3.9 (49)	3.6 (16)	31.7 (37)
Cryptotis parva	12.3 (13)	5.4 (3)	3.2 (3)	7.4 (7)	–
Peromyscus maniculatus	[4]5.5 (57)	3.7 (103)	[4]2.6 (63)	2.8 (99)	12.1 (8)
Peromyscus leucopus	2.6 (16)	2.6 (23)	–	2.7 (4)	–
Reithrodontomys megalotis	10.8 (32)	3.8 (23)	3.1 (6)	3.1 (20)	–
Microtus ochrogaster	[4]8.2 (40)	[4]4.3 (60)	[4]2.6 (50)	3.3 (13)	11.2 (8)
Mus musculus	6.9 (51)	3.5 (150)	4.2 (59)	4.6 (104)	13.7 (4)
Spermophilus tridecemlineatus	–	8.2 (13)	–	4.3 (17)	16.0 (1)

[1] Sample size in parentheses.
[2] Vehicles/24 h. Animals taken within 10 m of road.
[3] Animals taken more than 50 m from a road.
[4] Difference between lead concentrations of samples of a given species from the indicated sites significant at P < .01.

tions in the body tissues. Insectivores (the shrews) had the highest concentrations of lead; herbivores (prairie voles) had intermediate levels; while granivores (deer, white-footed and house mice) had the lowest levels of lead recorded. The white harvest mouse (a granivore) did not fit this pattern; likewise, *Spermophilus tridecemlineatus*, 13-lined ground squirrel (also a granivore), showed higher levels of lead than other species along medium-use roads. The reasons for the differences in these two species could not be determined; the small sample size may have been at least partially responsible in regard to the 13-lined ground squirrel.

The higher lead levels in the shrews may have resulted from food chain concentration. Insects, the major diet of the shrew, taken from areas adjacent to high-traffic roads, also showed relatively high concentrations of lead. Carnivorous insects at these sites displayed higher lead concentrations than did herbivorous or granivorous species.

Lead concentrations in the organs and tissues of the small mammals itemized in table 6-17 were generally highest in the bones; other organs and tissues did not have lead concentrations significantly higher than that of the total body. The lead levels in bone were generally low in terms of absolute amounts, averaging 109 µg/g for three specimens of the white harvest mouse and 67 µg/g for 49 specimens of the short-tailed shrew. There was no indication of lead concentrations in any organ that could cause pathological problems in the individual.

Due to their size, only selected organs and tissues of the larger mammals were analyzed for lead content. The organs were removed from the animal, freeze-dried, and then analyzed. The results are presented in table 6-18.

Although the absolute values were low for the large mammals in both the rural and urban areas, the animals captured within the urban compartment (it was assumed they spent most of their lives therein)

TABLE 6-17. –*Mean lead concentrations in organs and tissues of small mammals*

Species	Exposure	Lead concentration (µg/g dry weight)							
		Total body	Gut	Spleen	Liver	Lung	Kidney	Bone[1]	Muscle[2]
Blarina brevicauda	High	18.4	24.0	4.5	4.6	16.9	12.4	67.1	9.7
	Medium	6.7	7.0	3.6	2.0	5.6	5.8	19.9	5.7
	Low	5.7	3.1	2.3	1.0	7.8	3.9	12.2	5.4
Microtus ochrogaster	High	5.1	11.0	5.3	1.6	2.8	8.1	16.6	8.2
	Medium	5.9	18.4	2.2	1.2	1.8	7.6	23.2	3.0
	Low	1.9	2.8	2.4	1.0	1.3	2.8	4.6	2.0
Peromyscus maniculatus	High	6.3	19.2	19.4	3.5	6.4	7.9	24.6	6.8
	Medium	4.3	6.0	3.0	1.7	2.4	9.0	8.0	7.4
	Low	3.3	4.5	6.5	1.8	6.1	3.0	6.4	1.8
	Control[3]	3.1	4.3	3.7	1.1	1.5	1.8	5.7	2.1
Mus musculus	High	6.8	18.6	12.1	2.9	2.8	8.1	19.2	5.9
	Medium	6.0	8.8	3.1	1.6	3.4	6.6	21.0	3.9
	Low	6.7	4.8	5.1	1.6	1.7	3.1	23.5	3.4
	Control	2.0	2.7	2.1	1.9	3.4	3.4	9.3	3.8
Riethrodontomys megalotis	High	12.3	17.8	14.5	4.7	20.9	–	109.5	27.5
	Medium	3.1	6.2	9.2	1.1	4.2	2.1	–	4.4
	Low	2.7	3.5	5.6	2.3	4.7	4.8	18.4	–

[1] Femur [2] Thigh [3] Fields more than 50 m from a road.

TABLE 6-18.–*Lead concentration (µg/g dry weight) in organs and tissues of large mammals*

Area	Number	Lead concentration (µg/g dry weight)					
		Spleen	Liver	Lung	Kidney	Bone[1]	Muscle[2]
Urban	[3]6	2.4	7.9	3.1	8.8	21.8	0.9
Rural	[4]6	0.4	1.4	0.6	3.7	3.0	0.4

[1] Femur.
[2] Thigh.
[3] Gray squirrel, 3; opossum, 2; raccoon, 1.
[4] Gray squirrel, 1; fox squirrel, 1; raccoon, 2; rabbit, 1; gray fox, 1.

had higher lead concentrations than did those from the rural compartment.

Total body lead concentrations appeared, in general, to have been relatively low in all species of mammals studied. There was no information as to whether lead concentrations such as those found in this study would have influenced the various population parameters. It was also not known what effect, if any, such lead concentrations would have on the reproduction and survival of individuals living adjacent to high-use roads.

However, even if lead had a deleterious effect, it probably would not be a significant factor affecting populations of the species studied. Habitats within 50 m of heavy-use roads constituted only 12.5 percent of the total available habitat for the prairie vole, white harvest mouse, and least shrew; 11 percent of the available habitat for the short-tailed shrew; less than 1.5 percent of that for the deer and house mouse; approximately 10 percent of that for the 13-lined ground squirrel; and less than 1 percent of the total available habitat for the white-footed mouse. Any increased mortality or reduced reproduction resulting from the lead taken up by small mammals would affect only a minor segment of the total population of any species in the watershed. Such losses would quickly be replaced by dispersal from adjacent populations.

It is highly doubtful that the very low concentrations of lead in the larger mammals could have a significant impact on the population dynamics of these animals.

Birds

Samples of the bird population were obtained by shooting individual specimens with a pelletgun or shotgun with copper shot. Several factors were considered in obtaining specimens from areas of high environmental lead. First, few birds nest close to roads with heavy traffic. In addition, the territories of birds are sufficiently large that individuals nesting adjacent to a road would spend a major part of their time outside the zone of high lead concentration. Thus birds from the urban compartment would be more representative of those living in high-lead situations, and most of the samples for the high-lead area were taken there. However, 4 specimens of breeding red-winged blackbirds were taken from a marsh 5 m from an interstate highway. Control specimens for each species were taken from sites more than 50 m from any road.

In almost all cases, the birds from the urban compartment and the red-winged blackbirds obtained along the interstate highway had higher lead concentrations in the organs and tissues sampled than did individuals from rural areas. (table 6-19) The feathers and bones (and in starlings, the kidneys) had the highest lead concentrations of any tissue sampled from terrestrial vertebrates. Lead concentrations were no higher in the robin and red-winged blackbird, which feed extensively on invertebrates, as compared with seed feeders (i.e., house sparrow) or omnivores (the remaining species). Although sample sizes were too small for statistical analysis, adult house sparrows, starlings, grackles, and robins had somewhat higher lead concentrations in most tissues than did immature individuals.

The highest concentrations occurred in the feathers and bones of urban birds, relatively "inactive"

TABLE 6.-19.–*Lead concentration in bird organs and tissues from areas of high-lead and low-lead environments*[1]

Species	Lead level	Lead concentration (µg/g dry weight)						
		Feathers	Gut	Liver	Lung	Kidney	Bone[2]	Muscle[3]
Red-winged blackbird	Low (10)	26.5	[3]2.1	5.8	0.4	[3]2.1	6.9	[3]0.8
	High (4)[5]	66.8	[3]2.6	1.2	4.1	[3]4.1	9.1	[3]0.6
House sparrow	Low (16)	27.0	2.3	0.6	0.9	3.5	16.9	0.9
	High (11)	158.3	26.2	12.0	6.9	33.9	130.4	2.1
Starling	Low (11)	6.4	1.3	4.0	[3]2.8	3.6	12.8	[3]0.8
	High (13)	225.1	6.0	16.1	[3]5.2	98.5	213.0	[3]2.4
Grackle	Low (10)	36.0	1.4	2.5	[3]2.3	3.5	21.5	0.8
	High (11)	81.4	10.2	12.1	2.7	13.5	62.8	1.4
Robin	Low (10)	25.3	3.2	2.4	2.2	7.3	41.3	[3]1.0
	High (10)	79.7	24.5	10.5	10.3	25.0	133.7	[3]1.2

[1] Sample size in parentheses.
[2] Femur.
[3] Pectoral.
[4] Differences between low and high lead environments not significant at the $P < .05$ level; all others significant at least at the $P < .05$ level.
[5] Samples taken 10 m from interstate highway.

tissues. However, it is highly doubtful that the lead concentrations found in even the more critical organs were great enough to affect the survival of birds.

In contrast to some observations of insectivorous small mammals, there was no indication of food-chain concentration of lead in birds in this study. The prominent insectivorous species had no higher concentrations than did the seed, fruit, or omnivorous feeders. Although the birds were obtained during the spring when many were feeding on fruits and berries, the young birds should have had diets consisting primarily of invertebrates. However, in most instances where there was a difference, the young had lower levels than the adults. It would appear that age is more important in the accumulation of lead than actual feeding habits.

Total Lead in Terrestrial Vertebrates

Estimates of the total amount of lead in small mammals in the rural compartment were as follows:

(1) If all species were at a maximum population density at the same time, the total amount of lead stored would be 47.2 g.

(2) If all species were at their lowest population density at the same time, they would contain a total of 2.2 g.

(3) The best estimate of the average amount of lead in the mammal's body at any given time was 23.7 g.

Thus the amount of lead in the total standing biomass of small mammals in the ecosystem at any given time was very small, representing an insignificant factor in the flux of lead through the ecosystem.

In conclusion, terrestrial vertebrates do not appear to be significantly affected by current environmental concentrations of lead from vehicle exhaust. Neither do these animal groups appear to be significant factors in the lead flux in either the urban or rural compartments of the ecosystem.

Insects

Insect communities in the ecosystem were sampled on a weekly basis during the 1973 and 1974 growing seasons. Sample sites were located in various compartments of the ecosystem primarily in relation to potential lead input. High-lead-input areas were close to an interstate highway bordered by mowed grass (>12,000 vehicles/24 h); intermediate areas were along secondary roads (2,000 vehicles/24 h); and low-input areas were remote to traffic or less than 500 vehicles/24 h. Samples were also taken in fields of corn and soybeans, both adjacent to the interstate highway and remote from heavy traffic, and in remote areas of natural vegetation. Samples were collected with a portable suction trap, sorted, dried for 48 h at 85° C, and weighed. Mass collections of insects for lead analyses were made by walking at a set distance from the pavement for about 10 min. Additional collections, used to estimate insect biomass per unit area, were taken by placing the suction trap nozzle diameter 34 cm) vertically onto the ground and extracting the insects from this known area (0.0855 m^2) of substrate. Insects were categorized according to their feeding habits: sucking plant juices, chewing plant parts, or preying upon other insects. Samples from a given site within the 3 categories were composited until about 1 g (dw) of insect material was collected. This material was then analyzed for lead content by atomic absorption spectroscopy.

No significant differences in the lead content of insects were observed between samples taken at different sites with the same traffic density and vegetation type, nor between samples taken at different times of the year. Therefore, the data were grouped

TABLE 6-20.–*Mean lead content (μg/g) in the urban and rural compartments of the ecosystem*

Sample site	Leaf-hoppers *(Cicadellidae)*	Plant bugs *(Miridae)*	Long-horned grasshoppers *(Tettigoniidae)*	Short-horned grasshoppers *(Acrididae)*	Flower flies *(Syrphidae)*	Ants *(Formicidae)*	Lady beetles *(Coccinellidae)*
				1973			
Urban							
High traffic volume	15.0	29.5	63.2	57.5	39.6	99.1	–
Intermediate	–	6.2	77.4	–	24.9	54.6	–
Low traffic volume	5.7	–	–	–	6.7	66.8	–
Rural							
High traffic volume	11.6	11.0	49.4	13.4	12.9	52.3	–
Low traffic volume	10.5	7.5	44.5	10.8	15.8	21.5	–
Remote from traffic	7.8	5.4	4.8	–	5.5	10.4	–
				1974			
Urban							
High traffic volume	4.1	9.8	–	–	15.5	45.1	11.3
Intermediate	3.1	3.0	–	–	3.0	13.3	8.3
Low traffic volume	3.4	3.4	–	–	4.8	20.1	10.7
Private yard	12.9	–	–	–	8.4	27.0	3.3
Rural							
Low traffic volume	–	–	–	–	–	2.4	14.0
Remote from traffic	3.5	3.9	–	–	–	3.8	3.7

and analyzed for differences between insect feeding types and between each of the major traffic density and vegetation categories.

Lead content in insects was positively correlated with lead-emission levels, decreasing from areas of high-traffic volume to areas remote from roads. (table 6-20) A comparison of families sampled during 1973 and 1974 shows that the lead concentrations in 1974 were about 50 to 60 percent lower than in 1973. The reasons for observed populations shifts and lower lead levels are unknown.

In high-lead-emission areas, there was a trend of increasing lead content from sucking, to chewing, to predatory insects; no such trend was evident in low-emission areas. (fig. 6-20) [20] Chewing insects probably ingested more lead from surface deposits on leaves than did insects sucking liquids from the internal vascular tissues of plants. Data on predatory insects that feed on lead-containing herbivores suggested that lead was selectively retained in the body, leading to biological concentration in this two-trophic-level system. There was no evidence of lead concentration by herbivorous insects.

There was also a decrease in the lead content of insects with increasing distance from a high-traffic-volume highway. (table 6-21) This pattern was similar to that for lead in soils and plants.

The biomass of insects, estimated by suction-trapping a known area, differed at each locality. Near the interstate highway, from 0 to 7 m from the pavement, the mean biomass was 0.324 g/m^2; the mean lead content of 25.3 μg/g, resulting in a lead concentration of 8.21 μg/m^2. From 13 to 20 ms from the interstate, the biomass was 0.408 g/m^2; the lead content was 10.9 μg/g, resulting in a concentration of 4.43 μg/m^2. At Brownfield meadow next to a low-use road, the biomass was higher: 0.500 g/m^2 with a mean lead content of 10.2 ppm, yielding a concentration of 5.07 μg/m^2. The levels of lead per unit area were much lower in sites remote from roads and in agricultural crops, even though crops adjacent to the interstate (i.e., about 30 m from the pavement) were sampled. Natural vegetation habitats had a mean insect biomass of 0.378 g/m^2, while soybean and corn fields had an insect biomass of only 0.146 g/m^2. Lead concentrations were 1.58 μg/m^2 and 0.61 μg/m^2, respectively. The low insect biomass and resulting low lead concentrations in cropped areas may be attributed to the low diversity of habitats in cultivated fields, as well as to the use of insecticides.

TABLE 6-21.—*Lead concentration of insects at two distances from a high-traffic-volume road (interstate highway)*

Insect feeding type	Lead concentration (μg/g) 0-7 m from pavement	13-20 m from pavement
Sucking	15.7	9.8
Chewing	27.3	10.4
Predatory	31.0	20.0
Mean	24.7	13.4

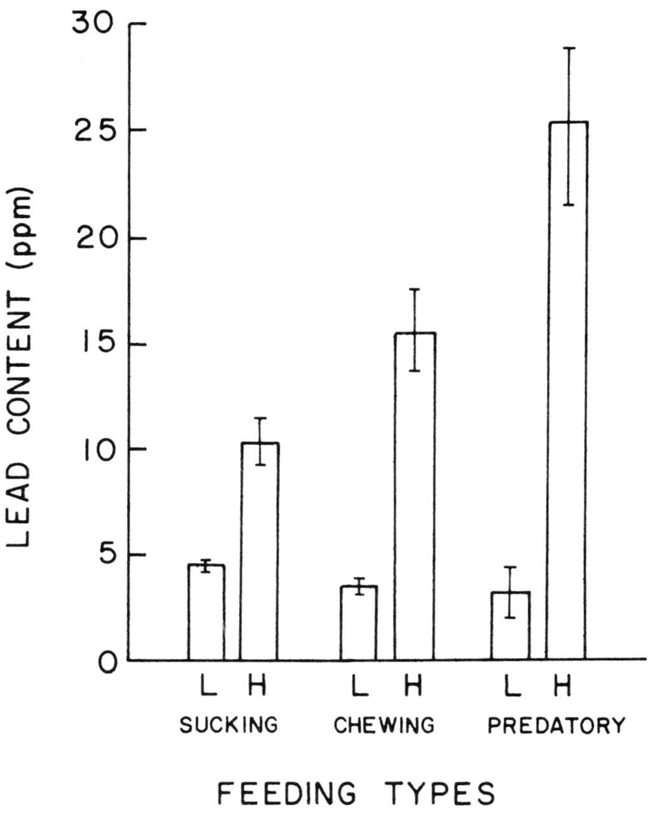

FIGURE 6-20.—Lead content of various insect trophic levels for low (L) and high (H) traffic areas.

Because about 75 percent of the ecosystem consisted of cultivated land areas with low insect biomass and because the levels of lead in the existing insect population were relatively low, it seems that the insect component of the ecosystem did not act as a reservoir for large quantities of lead. However, insects did contribute significantly to the flow of energy and pollutants up the food chain, and lead concentration near high-emission areas could result in poisoning of insect predators.

Conclusions

Lead in the terrestrial portion of the system was concentrated in the soils, plants, animals, and insects in the urban area or near high-traffic-volume highways. Samples taken in locations remote from traffic sources had relatively low lead concentrations, which may be considered to represent the background level. However, because of the considerable enhancement of lead concentrations in soils and organisms collected near high-traffic areas and because of the increase in the environmental lead load with time, as shown by tree-ring studies, it is important that areas with high-traffic volumes be closely monitored in the future to ensure that environmental lead concentrations do not reach toxic levels. Additional studies in these high-lead environments are currently in progress to evaluate the possible loss of plant or crop productivity. It is critical that threshold as well as toxic concentrations be determined, especially in the case of plant productivity, to allow a more thorough evaluation of the lead problem.

LEAD IN THE AQUATIC ECOSYSTEM

From the terrestrial part of the ecosystem, lead enters the aquatic system primarily in surface runoff as suspended particles or adsorbed to soil particles. Atmospheric fallout, aerosols scrubbed from the air by precipitation, and direct contact with products containing lead are other sources of lead to the aquatic system. Since many aquatic systems have no major exports of lead, they must be considered potential sinks for lead pollutants.

The sediments, plants, invertebrates, and fish in the streams were examined to determine the distribution and accumulation of lead by higher trophic levels, and to measure the physical transport of lead by aquatic invertebrates and fish.

The Saline Ditch drainage above the village of Mayview (which includes all of the ecosystem under study) was divided into four sectors—rural, urban, marginal, and combined. (fig. 6-21) The 14 sampling stations assigned to the various sectors included a variety of habitats and localities. (fig. 6-21)

Essential measurements of stream size and character were made and updated throughout the study period. Width, depth, current velocity, and substrate type were measured on transects at 400 m intervals along the stream length. Discharge, bottom composition, and stream area for each station were calculated from these data.

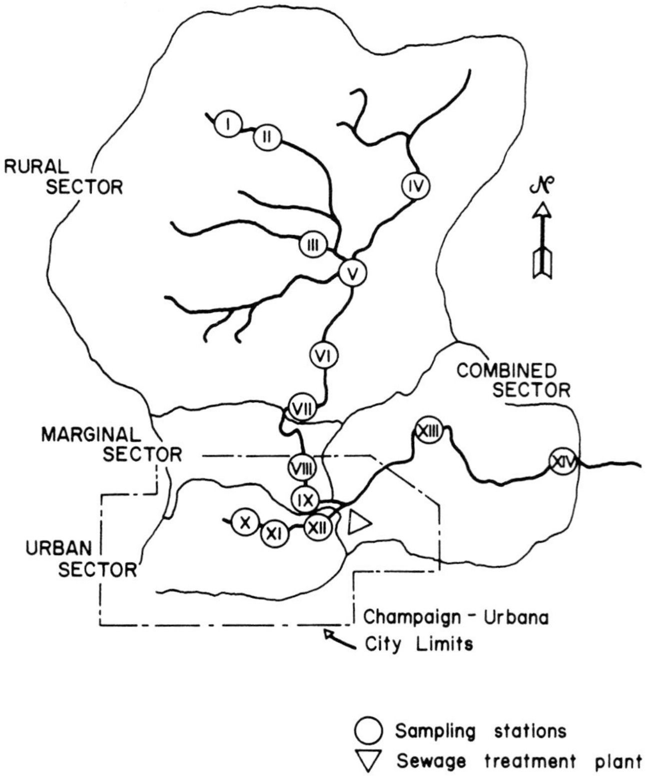

FIGURE 6-21.—The Saline Ditch drainage basin, study area divisions, and sampling stations.

TABLE 6-22.—*Substrate composition of the benthos sampling stations in various parts of the Saline Ditch drainage*

Division-station	Sand	Silt	Gravel (percent)	Rubble	Concrete
Rural tributaries					
I	26	50	24		
II	5	95			
III	4	96			
IV	2	98			
Rural mainstream					
V	54	28	18		
VI	82	7	11		
VII	35	40	25		
Marginal					
VIII	44	21	25	10	
IX	50	11	24	15	
Urban					
X	96	2		2	
XI	57	8		9	26
XII	13		31	36	20
Combined					
XIII	34	8	44	14	
XIV	52	7	41		

The rural sector included 136.3 km^2 of predominantly agricultural land along the main stream of the Saline Ditch and its headwater tributaries north of Champaign-Urbana. Most of the 47.5 km of rural streams have been channelized to facilitate drainage. Four sampling stations (I to IV) were established on the primary and secondary tributaries and three (V to VII) on the main stream. The average discharge of the rural sector was about 20 m^3/min. The predominant substrate of the tributaries was silt, with sand and gravel in the main stream. (table 6-22)

The marginal sector, a 5.5 km stream section with a drainage basin of 22.9 km^2, is an extension of the rural main stream and was considered a transitional area between the rural and urban sectors. Due to the complexity of external influences, principally of urban origin, minimal effort was expended on the two stations (VIII and IX) in this sector. The discharge and substrate composition were very similar to those of the main stream in the rural sector.

The urban sector included 16.1 km^2 of the Boneyard Creek drainage basin. The creek, a 4.8 km tributary of the Saline Ditch, is located within the city limits of Champaign and Urbana. The average discharge is about 5.0 m^3/min. Almost 20 percent of Boneyard Creek is routed through channels of poured concrete, segments of which have bottom sediments entirely composed of sand and gravel. (table 6-22) High total dissolved solids and periodic occurrences of ammonia characterized the chemical nature of the tributary.

The combined division was comprised of the discharges from the Saline Ditch headwaters, Boneyard Creek, and the urban sewage treatment plant. Two stations (XIII and XIV) were located along the 10.6 km length. The discharge from the total ecosystem watershed of 60 m^3/min. was measured at station XIII downstream of the effluent outlet for the urban sewage treatment plant. Urban and rural lands, as well as a sanitary landfill, were within the 37.7 km^2 drainage basin of this sector. The substrate was almost entirely sand and gravel, (table 6-22) and the water was generally high in the nutrients.

Sediments

Sediment samples were taken with a 1/300 m^2 position coring device to a depth of 10 cm because most of the penetration of benthic organisms and water circulation was judged to occur only in that portion of the sediment. Samples were screened to remove material larger than 2.0 mm in diameter, because preliminary work indicated that more than 90 percent of the lead was associated with smaller particles. Samples were dried and thoroughly mixed before subsamples of 1 to 2 g, obtained by a rifle-type sample splitter, were analyzed for lead.

The lead concentration of sediment samples differed significantly in various sections of the Saline Ditch drainage. (table 6-23) The lead concentrations in Boneyard Creek sediments of the urban sector reached a maximum of 1,100 μg/g near station XII. Sediments at station X, the headwaters of Boneyard Creek, had a mean lead content of 584 μg/g. Other urban sediment sampling points showed a wide variation in lead content, but were well below the mean value of 387.5 μg/g. However, the urban concentration was more than 10 times greater than that of the combined sector. Main stream rural sediments had lead concentrations less than those of the combined sector. There was no significant difference in sediment lead concentrations between stations on the rural tributaries and those in the marginal sector, which receives more lead due to some urban runoff. The relatively high concentration of lead in the sediments of the rural tributaries resulted from the ability of these very fine highly organic sediments to bind available lead more successfully than the sand that dominates stream beds in the marginal and other sectors. The main stream rural sections (V to VII) had the lowest significant concentrations.

Lead content of sediments decreased with depth as shown by separate analyses of 1 cm segments of 10 cm core samples taken from 6 rural and 6 urban points. Concentrations were exceedingly high in the top 1 cm from both areas. (fig. 6-22) These relatively high concentrations were probably due to accumulation of new lead particles through constant mixing at the sediment-water interface.

Lead storage in stream sediments, especially urban areas, was substantial. (table 6-23) Storage in the upper 10 cm of sediment varied from 1.5 g/m^2 in the rural main stream to 5.7 g/m^2 in the combined sector; a maximum was reached in the urban sector, where the mean was nearly 60 g/m^2. These values were more than 6 orders of magnitude larger than the values for storage found in the aquatic biological components. (table 6-24)

Sediment profiles were also examined from silt substrates up to a depth of 30 cm in the urban stream and the rural main stream. At quarterly intervals during the year, 24 samples were taken and averaged with each stream segment. Lead concentrations decreased with increasing depth in

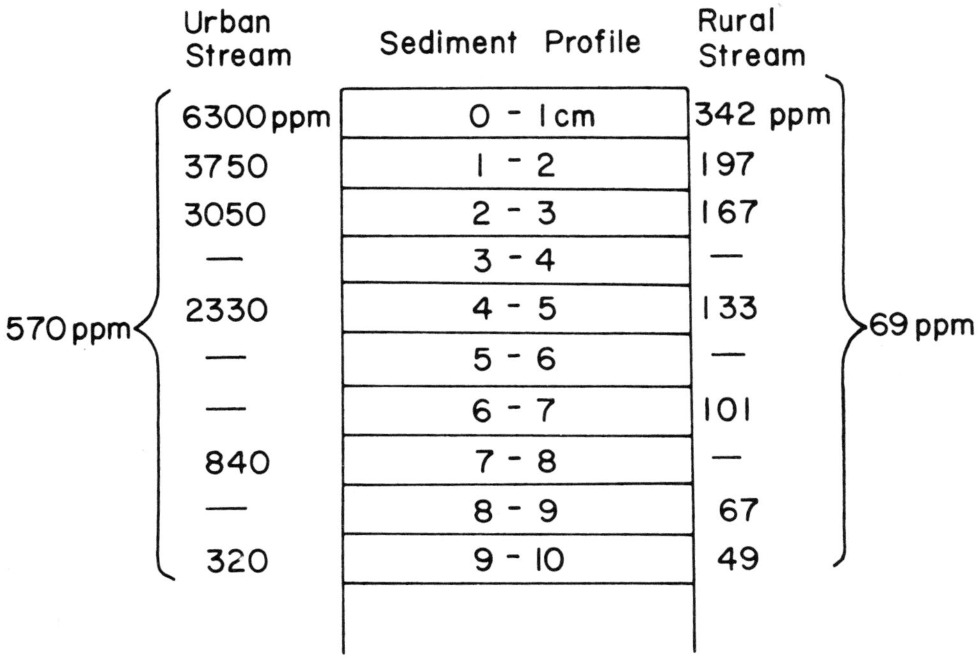

FIGURE 6-22.—A comparison of the lead concentration in 1 cm segments of the top 10 cm of urban and rural sediments.

TABLE 6-23.—*Mean lead concentrations and storage in sediments, to a depth of 10 cm, from the various compartments of Saline Ditch and Boneyard Creek*

	Urban	Combined	Tributaries	Marginal	Main rural
Station	X-XII	XIII-XIV	I-IV	VIII-IX	V-VII
Number of samples	104	68	75	15	47
Lead storage (g/m^2)	58.1	5.7	2.4	2.1	1.5
Pb level ($\mu g/g$ dry weight)	387.5	37.8	15.8	13.8	10.2
Significance[1]	______	______	______	______	______

[1] Common underscore indicates means that are not significantly different at the 5 percent level.

both the rural and urban profiles. (fig. 6-23) However, the values were much higher for the urban stream. The lead distribution pattern of stream sediments resembled that of terrestrial soil profiles.

Biota

The biomass and lead concentrations were determined for aquatic organisms in the drainage system and used to calculate lead storage in the biota. The biomass in each sector was calculated from the biomass density at each sampling station. Lead concentrations for each sector were determined from samples accumulated from all stations in a particular sector. The mass (g) of a taxonomic group was multielied by the specific lead concentration ($\mu g/g$) to find the lead storage. This method permitted the lead accumulation of a taxonomic group to be weighted for the amount of habitat suitable for that group.

Plants: Aquatic plants occurred in all sectors, but were abundant only during the summer. Plants and invertebrates were sampled together with substrate samplers of three sizes and kinds: ¼ m^2 open cylinder, 1/25 m^2 grab sampler, or 1/300 m^2 corer. The size used depended on the density of organisms. Samples were washed on a No. 30 mesh sieve, and the plants were separated from each other and from invertebrates. Aquatic plants were washed with water, dried, and ground to a powder prior to analysis.

The plant biomass during the warm months was highly variable in all sectors, ranging from 0 to 300 g/m^2 (wet weight). The alga *Cladophora* occurred throughout the system and had highly variable lead concentrations, ranging from means of 20.1 $\mu g/g$ (dry weight) in the rural sector up to 347 $\mu g/g$ (dry weight) in Boneyard Creek. (table 6-25) These values reflected the relative amount of lead input of the two sectors. *Potamogeton* was common in rural streams (30.0 $\mu g/g$ lead), while *Elodea* was an important constituent of the combined sector (89.9 $\mu g/g$ lead).

The total lead stored by plants was highly variable due to fluctuations in biomass. (table 6-24)

Invertebrates: Invertebrate samples were obtained and separated in the same manner as aquatic plants. They were washed in 95 percent ethyl alcohol, dried, and ground to a powder for analysis.

The distribution and biomass of benthic invertebrates were highly variable. (table 6-26) The biomass density in the combined sector was consistently higher than in the rural or urban sectors, primarily because of the increased nutrients available from

TABLE 6-24.—*Seasonal lead storage in biota of Saline Ditch and Boneyard Creek*

	Lead concentration ($\mu g/m^2$)			
Compartment/organism	Summer	Fall	Winter	Spring
Rural				
Fish	2.36	3.52	2.81	2.60
Benthic invertebrates	24.07	27.45	60.48	40.90
Macrophytes	[1] 298.91			
Total	324.34	30.97	63.29	43.50
Urban				
Fish				
Benthic invertebrates	112.22	874.50	4,497.41	839.13
Macrophytes	[1] 2,608.70			
Total	2,720.92	874.50	4,497.41	839.13
Combined				
Fish				
Benthic invertebrates	339.44	899.68	5,304.96	312.85
Macrophytes	[1] 109.75			
Total	449.19	899.68	5,304.96	312.85

[1] Highly variable due to fluctuations in biomass.

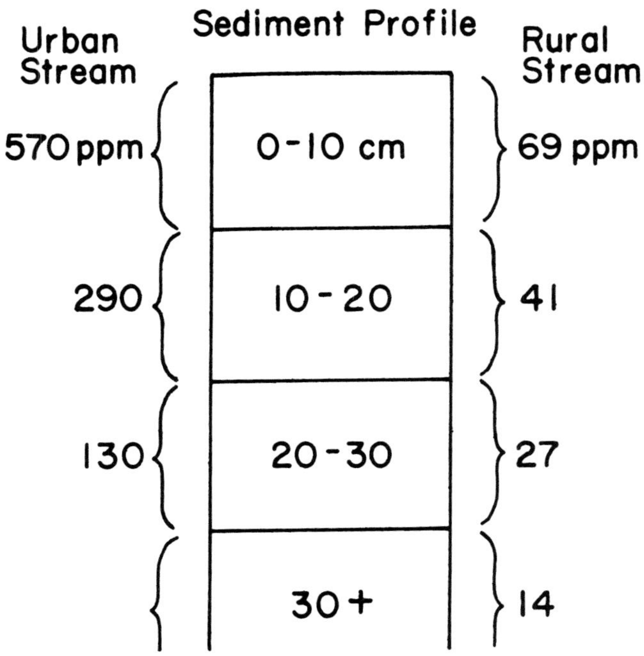

FIGURE 6-23.—Mean lead concentrations (μg/g) in urban and rural stream profiles (silt substrate).

agricultural runoff and septic tanks in this sector. The weighted average biomass densities between the rural and urban sectors were similar. During the spring and fall, periods of relatively high oxygen concentrations, benthic populations may well have been limited by substrate instability. In all sectors and in all seasons, the biomass density was consistently highest on silt substrates. (table 6-26) In the rural sector, nearly 70 percent of the invertebrates were found on the 25 percent of stream bottom covered with silt. In most instances, these were areas of deposition protected from the current and least subject to disturbance. The instability of these soft substrates, when exposed to increased currents, limited invertebrate production in all sectors. Thus substrate stability and stream oxygen content were limiting factors for benthic invertebrate biomass, which in turn affected the biological lead storage and flux in all sectors.

In general, the benthic populations were highest in winter and lowest in summer, regardless of substrate. (table 6-26) These seasons were, respectively, the periods of highest and lowest oxygen availability. While the small seasonal differences in the biomass of the rural sector might be explained by natural benthic cycles based on oxygen availability, the population changes in the urban and combined sectors cannot be explained this way. Here the substantial seasonal changes in biomass occurred mostly among tubificids, organisms with nonseasonal cycles. The Tubificidae and Chironomidae, which together comprise more than 95 percent of the benthic invertebrates of the urban and combined sectors, have respiratory pigments to improve their oxygen exchange efficiency, enabling them to survive summers when the oxygen concentration is frequently low.

Seasonal changes in lead concentrations were determined only for the Tubificidae from the urban sector. No significant difference was found in mean seasonal lead concentrations for spring, summer, and fall. However, the lead concentration in tubificids increased from a mean of 322 μg/g (spring through fall) to 484 μg/g in winter. This peak value corresponds with a period of high surface runoff because the ground was relatively impermeable due to freezing.

Lead storage in the rural and combined sectors (table 6-24) varied as a result of changes in biomass, but storage in the urban sector reflected changes in both biomass and lead concentrations.

The only measurable biological transport of lead in the Saline Ditch drainage was due to the drift of aquatic invertebrates. Plankton is essentially nonexistent, and fish movements are minimal. (See table 6-27.) Samples of drifting invertebrates were taken in late summer when peak values should have been encountered. A 1 ft^2 net of No. 20 mesh Saran cloth was placed in the stream for between 30 min. and 1 h. The samples were taken during the day within 1 h after sunset and within 1 h before sunrise. Measurements of flow rate through the mouth of the net at the beginning and end of each sampling period permitted quantification of drift biomass. Samples were analyzed for lead concentrations in the drift were expressed on a daily basis. (table 6-27) Lead fluxes in the urban and rural sectors were similar because the low stream flow of the urban sector was offset by its relatively high lead concentrations in the biota. The amount of lead transported during peak periods of stream drift (late summer) was several orders of magnitude less than the amount of lead moved in solution or adsorbed to suspended material. Thus drift organisms do not constitute a significant lead transport.

Fish: Fish were sampled by confining them in a measured portion of stream with 20 ft seines of 3/16 in. mesh. The upstream seine was slowly moved downstream, driving the fish into a bag seine at the lower end of the blocked-off section. A series of samples from an isolated stream section indicated that this technique effectively removed 50 percent of the fish population without specific selection, except that large carp *(Cyprinus carpio)* successfully avoided capture. The uniformity of the banks and bottom of the Saline Ditch, due to earlier channelization, permitted effective use of this technique. Fish were sorted after capture, oven dried without a preservative, and ground to a powder for lead analysis.

The biomass distribution of fish was easily determined because they occurred in measurable numbers only in the rural sector and portions of the marginal sector. The fish population was consistently near 3 g/m^3 (wet weight) in the rural sector.

Lead concentrations in fish were lower than those in aquatic plants or invertebrates. (table 6-25)

TABLE 6-25.–*Lead concentrations of predominant organisms from rural, urban, and combined compartments of the Saline Ditch*

Compartment/organism		Lead concentration (μg/g dry weight) Arithmetic mean	Number of samples	Standard deviation
Rural				
Plants	*Cladophora*	20.1	11	5.1
	Potamogeton	30.0	15	5.4
Invertebrates	*Hirudinea*	12.6	12	14.8
	Oligochaeta	13.2	7	7.6
	Tubificidae	16.0	13	11.8
	Sphaeriidae	5.5	11	3.1
	Lirceus fontinalis	7.6	5	3.4
	Hexagenia limbata	10.4	15	9.7
	Anisoptera	6.6	6	2.4
	Chironomidae	20.1	12	18.2
	Decapoda	4.7	23	5.2
Fish	*Catostomus commersoni* (white sucker)	2.4	19	1.6
	Etheostoma nigrum (johnny darter)	4.1	9	2.4
	Ericymba buccata (silverjaw minnow)	1.8	29	0.9
	Notropis umbratilus (redfin shiner)	1.9	29	0.6
	Pimephales notatus (bluntnose minnow)	2.7	83	2.7
	Semotilus atromaculatus (creekchub)	1.4	35	0.6
Urban				
Plants	*Cladophora*	347	6	139
Invertebrates	*Tubificidae*	367	29	373
	Chironomidae	153	5	106
	Decapoda	11	4	
Combined				
Plants	*Elodea*	89.9	22	93.3
	Cladophora	34.9	7	7.5
Invertebrates	*Tubificidae*	48.6	71	33.9
	Chironomidae	42.7	12	16.4
	Physa	41.7	10	25.5
	Psychodidae	32.3	4	

TABLE 6-26.–*Biomass density of benthic invertebrates from the Saline Ditch and Boneyard Creek*

Compartment	Substrate	Biomass density (g/m^2 wet weight) Summer	Fall	Winter	Spring
Rural					
	Silt	9.2	16.9	20.5	18.9
	Sand	1.7	1.8	3.9	1.9
	Gravel	0.6	1.3	2.3	0.7
	Weighted average	6.5	12.4	13.8	12.2
Urban					
	Silt	34.2	5.0	64.2	63.7
	Sand	9.8	12.8	80.1	19.1
	Gravel	3.7	51.8	–	–
	Weighted average	6.0	11.2	41.8	15.8
Combined					
	Silt	177.5	587.6	1,251.3	313.9
	Sand	28.2	32.5	272.8	5.0
	Gravel	47.1	23.9	–	–
	Weighted average	56.1	73.1	299.4	49.1

TABLE 6-27.–*Peak period drift from the Saline Ditch and Boneyard Creek and the amount of lead transported by drifting organisms*

Organism	Peak period drift		
	Wet (g/100 m³)	Dry (g/day)	Lead (mg/day)
Rural compartment			
Chironomidae	0.042	1.09	0.01
Ephemeroptera	0.006	0.14	0.01
Coleoptera	0.011	0.50	0.02
Hemiptera	0.030	1.43	0.10
Tirchoptera	0.002	0.07	–
Physa	0.019	0.90	0.01
Nematamorpha	0.013	0.40	0.01
Zygoptera	0.001	0.05	–
Total	0.124	4.57	0.16
Urban compartment			
Chironomidae	0.559	1.26	0.13
Psychodidae	0.067	0.15	0.02
Culicidae	0.002	–	–
Amphipoda	0.003	–	0.00
Tipulidae	0.002	–	0.00
Terrestrial	0.151	0.38	0.00
Total	0.784	1.80	0.15
Combined compartment			
Chironomidae	1.780	388.8	17.7
Psychodidae	0.156	35.5	1.0
Coleoptera	0.010	3.9	0.2
Corixidae	0.010	4.0	0.2
Gastropoda	0.128	50.8	2.8
Trichoptera	0.007	1.7	0.1
Tubificidae	0.004	1.0	–
Nematamorpha	0.019	4.8	0.2
Terrestrial	0.016	3.9	–
Total	2.130	494.4	22.2

Seasonal lead storage in fish was essentially nonexistent. (table 6-24)

Interactions

In stream sediments and three selected groups of organisms, (fig. 6-24) the lead concentrations showed progressive increases that correlated with the relative input received by a particular stream sector. Values of lead concentrations from the rural main stream and tributaries (considered baseline data) were exceeded, successively, by values from the marginal, combined, and urban sectors. The highest lead concentrations, in all cases much greater than any others, were found in the urban division.

Perhaps the most striking result was the variability of lead concentrations n different species from the same sector. In the rural sector, where the only complex community structure existed, lead concentrations of dry animal tissue ranged from 1.4 to 16.0 μg/g. (fig. 6-25) The organisms in the rural streams were divided into four groups based on their food and habitat requirements.

The first group—composed of the mayfly (*Hexagenia limbata*), the Tubificidae, and the oligchaetes,—burrow into and ingest the substrate. [21] These organisms showed a definite preference for silt substrate where they were continually exposed, both internally and externally, to high lead sediments.

The second group was composed of the following benthic organisms that inhabit the sediment-water interface: crayfish (*Orconectes virilis*), which feeds on vegetation and detritus; [22] the pelecypods, which are filter feeders that utilize suspended detritus; and the johnny darter (*Etheostoma nigrum*), a benthic, carnivorous fish that occasionally ingests detritus and silt. [23] These organisms occasionally contacted lead-rich material in their food source and environment.

The remaining two groups consisted of fishes that had very little close body contact with settled or suspended sediment material. These include the white sucker *(Catostomus commersoni)* and the bluntnose minnow *(Pimephales notatus)*, which are bottom feeding omnivores that often ingest detritus. [24-26] The creekchub (*Semotilus atromaculatus*), silverjaw minnow *(Ericymba buccata)*, and redfin shiner *(Notropis umbratilus)*, which comprise the fourth group, are principally carnivores. [26-30] Their most likely lead source is their food. The creekchub's mean lead concentration was significantly lower than those for all other species. It tends to be a piscivore, while the silverjaw minnow and redfin shiner are usually restricted to eating invertebrates.

The lack of lead accumulation through trophic levels, particularly the low concentrations in the carnivores, (fig. 6-25) clearly indicated that lead accumulation by aquatic organisms through ingestion was of minimal importance. Thus in contrast to some terrestrial food chains, there appeared to be no biological magnification of lead through trophic levels in the aquatic system.

Conclusions

In the rural sector, the highest lead concentrations in stream sediments were only slightly higher than in the surrounding field soils, while sediments in the urban sector were many times higher. Stream sediments acted as a sink where lead accumulated to high levels. The highest concentrations were in the urban sector.

Lead levels of sediments and organisms reflected the relative amount of urban runoff entering the stream. In the urban sector, a dramatic increase in lead accumulation by Tubificidae corresponded to a period of increased high lead runoff.

Lead concentrations of aquatic organisms varied substantially within and between sectors. Lead concentrations in the same kind of organisms were generally 2 to 4 times higher in the combined sector than in the rural sector. Similar organisms in the urban sector had lead concentrations 10 to 20 times higher than those in the rural compartment. The mean concentrations for each species within a sector were markedly different. Organisms at low trophic

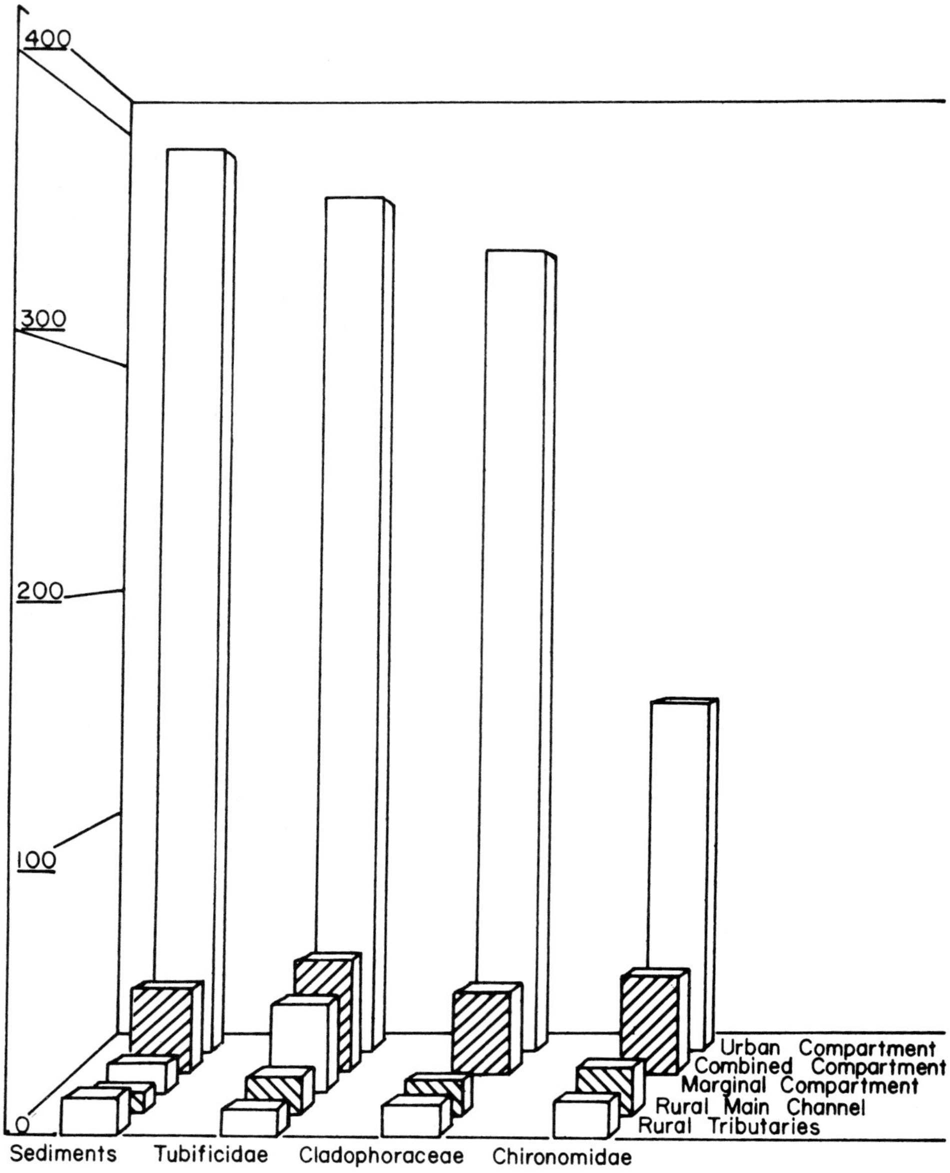

FIGURE 6-24.—Mean lead level of representative organisms and sediments of compartments of compartments in the Saline Ditch.

levels had the highest concentrations, while those at the top trophic level had the lowest lead concentrations. Thus there was no biological magnification of lead through the aquatic trophic structure.

The biological lead concentrations were apparently influenced by the amount of contact an organism had with the substrate. This phenomenon was regulated by habitat preferences and feeding habits. Thus the actual amount of lead stored by the biota of a stream segment dependends on the types and amount of habitat available, the community composition, and the lead concentration in the sediments. Fluctuations in biomass due to changes in habitat or community structure were common and caused significant changes in the amount of lead stored. However, lead storage variations in the urban sector were a function of biomass and biota lead concentrations.

The amount of lead transported by drifting stream invertebrates was insignificant in this ecosystem.

SUMMARY AND CONCLUSIONS

An 86 mi^2 watershed was studied to determine the movement of lead into and through the ecosystem, as well as the effects of lead on various biotic components. Gasoline consumption was the major source of lead input to the watershed. It was estimated that the total input to the ecosystem, with the exception of air, was 16,000 kg/yr. The 12 mi^2 urban compartment received 75 percent of the total lead input.

The majority of the lead exiting from the ecosystem was carried in the drainage water, primarily in association with suspended solids. It was estimated that 910 kg of lead exited from the ecosystem via stream water annually. Of this, 75 percent was from the urban compartment.

With 16,000 kg of lead per year entering the watershed and only 910 kg per year exiting via stream water, 14,800 kg per year, or 92.5 percent of the lead input, remained within the ecosystem.

Airborne lead concentrations were low in the rural area (0.17 $\mu g/m^3$), and in the urban area concentrations varied from 1.352 $\mu g/m^3$ on high-use downtown streets to 0.287 $\mu g/m^3$ on low-use residential streets.

In both the urban and rural compartments, lead concentrations were highest in the top 10 cm of soil in areas immediately adjacent to streets and roads with high traffic volumes. A study of lead in dusts and soils in the urban compartment of the ecosystem showed high concentrations inside buildings. Lead levels in dusts within homes averaged 660 $\mu g/g$ and 680 $\mu g/m^2$. In nonresidential interiors, dusts averaged 1400 $\mu g/g$ lead and 2,040 $\mu g/m^2$. The concentrations of lead in dusts and soils around buildings and in dusts inside buildings appeared to be directly related to the volume of traffic on neighborhood streets and roads.

Concentrations of lead associated with vegetation, in small mammals and other vertebrates, and in insects were also highest in areas adjacent to high-use roads and streets. Because the vegetation biomass was large, it was estimated that 2,211 kg of lead were associated with the vegetation of the ecosystem.

For all animals, the total body lead concentrations were relatively low. The highest concentration for an urban-dwelling mammal was 31.7 ppm. Although there may be some food-chain magnification in insectivores that feed near roads with high traffic volumes, terrestrial vertebrates do not appear to be significantly affected by current environmental lead concentrations, nor do these terrestrial vertebrates appear to be significant factors in the lead flux in either the urban or rural compartments of the ecosystem.

In contrast to the case with vertebrates, there is an increase in lead concentrations in insect food chains from sucking to chewing to predatory insects in high-lead areas next to heavily traveled roads. However, the low biomass and low lead concentrations indicate that insects are not a major reservoir of lead in this ecosystem.

In the aquatic system, lead concentrations in the top 10 cm of stream sediments were high in the urban stream (387.5 $\mu g/g$), while rural stream lead concentrations were many times lower. The high urban concentrations were related to lead particulates washed from the large impervious areas of the urban compartment during storms.

Aquatic biota had lead concentrations many times higher in the urban stream than in the rural stream. There was no biological magnification of lead in aquatic food chains. Lead in aquatic organisms was related to the amount of contact with substrates containing high lead concentrations. The stream drift of invertebrates was insignificant in transporting lead from the ecosystem.

In conclusion, most of the lead that entered the watershed remained within it. The major lead reservoirs were the soils, plants, and stream sediments.

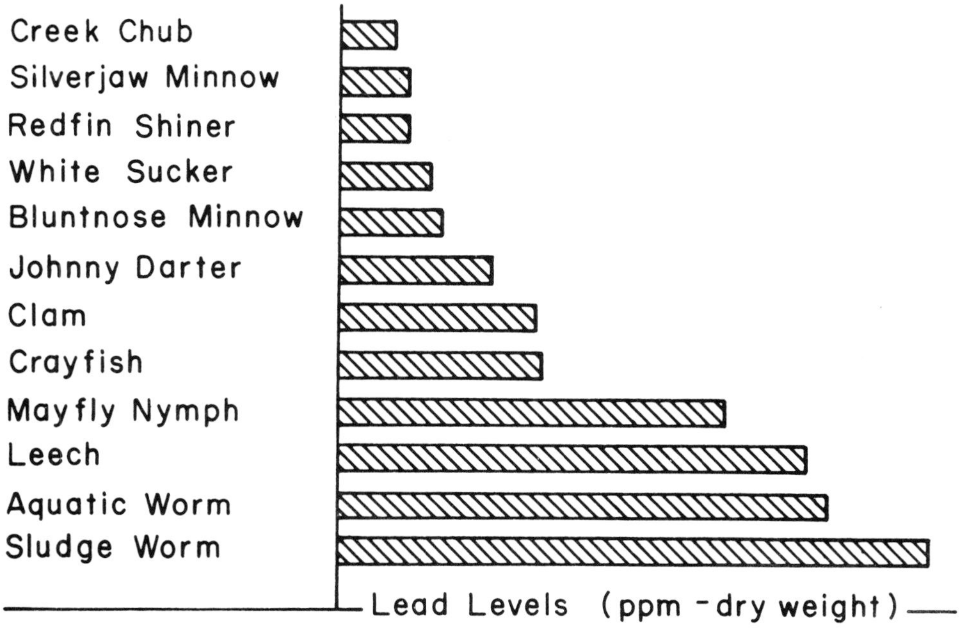

FIGURE 6-25.—Relative lead concentrations of organisms from the rural compartment, moving upward in the tropic structure.

The highest concentrations and a major portion of the stored lead were associated with the urban compartment and ares adjacent to high-use rural roads. These areas should be monitored in the future to detect changes in lead storage and concentrations with time and with additional lead input, to determine if toxic levels are reached.

Insect food chains and insectivorous birds should also be closely monitored in high-lead-input areas. Aquatic systems should receive closer examination in the future, as sediments are major metal reservoirs.

REFERENCES

1. G. L. Rolfe, A. Chaker, J. Melin, and B. B. Ewing. J. Environ. Sys. *2(4):* 339 (1972).
2. E. N. Cantwell, E. S. Jacobs, W. G. Lunz, Jr., and V. E. Liberi. Cycling and Control of Metals. 95 (1972).
3. U. S. Public Health Service. Publication No. 999-AP-12 (1965).
4. L. B. Tepper and L. S. Levin. University of Cincinnati, College of Medicine, Department of Environmental Health (1972).
5. T. J. Kneip, M. Eisenbud, C. D. Strehlow, and P. C. Freudenthal. J. Air Pollut. Contr. Ass. *20:* 144 (1970).
6. J. M. Colucci, C. R. Begeman, and K. Kumler. J. Air Pollut. Contr. Ass. *19:* 255 (1969).
7. H. R. Bowman, J. G. Conway, and F. Asara. Environ. Sci. Technol. *6:* 558 (1972).
8. P. R. Harrison, W. R. Matson, and J. W. Winchester. Atmos. Environ. *5:* 613 (1971).
9. J. L. Hudson, J. J. Stukel, and R. L. Solomon. Atmos. Environ. (1975 in press).
10. A. L. Page, T. J. Ganje, and M. S. Joshi. Hilgardia. *41:* 1 (1971).
11. H. S. Motto, R. H. Baines, D. M. Chilko, and C. K, Motto. Environ. Sci. Technol. *4:* 231 (1970).
12. A. Ruhling and G. Tyler. Bot. Notiser. *121:* 321 (1968).
13. A. Suchdoller. Ber. der Schweizerischen Botanischen Gesellschaft. *77:* 266 (1967).
14. G. C. Marten and P. B. Hammond. Agron. J. *58:* 553 (1966).
15. A. Kolke and K. Riebartsch. Naturwissenschaften. *51:* 367 (1964).
16. H. L. Cannon and J. M. Bowles. Science. *137:* 765 (1972).
17. A. Haney, J.A. Carlson, G.L. Rolfe. Ill. Acad. Sci. *67* (3): 323 (1974).
18. G.L. Rolfe. For. Sci. *20*(3): 283 (1974).
19. H.W. Kreuger. N. W. Project 72-3. Kreuger Enterprises, Cambridge, MA (1972).
20. P.W. Price, B.J. Rathcke, and D.A. Gentry. Environ. Entomol. *3* (3): 370 (1974).
21. B. P. Hunt. Mich. Dept. Conserv. Bull. Inst. Fish. Res. *4:* 1 (1953).
22. M. J. Caldwell and R. V. Bovbjerg. Proc. Iowa Acad. Sci. *76:* 463 (1969).
23. C. L. Turner. Ohio J. Sci. *22:* 41 (1921).
24. N. H. Stewart. Bull. U. S. Bur. Fish. *42:* 147 (1926).
25. W. C. Kraatz. Ohio J. Sci. *28* (2): 86 (1928).
26. S. A. Forbes and R. E. Richardson. Illinois Laboratory of National History Survey, 2nd ed. (1920).
27. C. L. Hubbs and G. P. Cooper. Cranbrook Inst. Sci. Bull. *8:* 1 (1936).
28. J. S. Dinsmore. Proc. Iowa Acad. Sci. *69:* 296 (1962).
29. R. D. Hoyt. Amer. Midl. Nat. *84* (1): 226 (1970).
30. S. A. Forbes. Bull. Ill. St. Lab. Nat. Hist. *1* (3): 18 (1880).

CHAPTER 7

TRANSPORT AND DISTRIBUTION FROM MINING, MILLING, AND SMELTING OPERATIONS IN A FOREST ECOSYSTEM

*J. Charles Jennett**, *Bobby G. Wixson*, *Ivon H. Lowsley*, and *Krishnier Purushothaman*
Department of Civil Engineering
University of Missouri-Rolla

Ernst Bolter
Department of Geology
University of Missouri-Rolla

Delbert D. Hemphill
Program Director - Environmental Trace Substances Research Center
University of Missouri-Columbia

Nord L. Gale
Division of Life Science
University of Missouri-Rolla

William H. Tranter
Department of Electrical Engineering
University of Missouri-Rolla

HISTORY AND DEVELOPMENT

Lead was first discovered in Missouri by French developers in 1701 and mined from shallow surface deposits as early as 1720. [1] Since that time production has increased gradually, with most of it coming from the Old Lead Belt in Madison and St. Francis counties, both in southeast Missouri. Explorations by the St. Joe Lead Co. (now St. Joe Minerals Corp.) discovered new detoits in Washington County and led to development of the Indian Creek Division Mine in 1953. (These mines are just north of, and not considered to be part of, the Viburnum Trend.) In 1955 St. Joe Co. verified initial lead ore discoveries near Viburnum. This rich lead-zinc deposit extends approximately 40 mi due south from Viburnum and is more or less continuously mineralized from a width of a few hundred to a few thousand feet. The lead-zinc ore varies from low to an extraordinarily high grade and has proven to be of much higher quality than was forecast in the initial exploration program. Both lead and zinc are primarily recovered as coproducts of the lead ore. [2] This new lead-rich area was named the Viburnum Trend or New Lead Belt of Missouri. The geographical location is shown in figure 7-1 along with the location of mines and one smelter. Most of this major mineral discovery is on federal land (Clark National Forest), which presented many problems in the development of major mining industries within the multiple-use program of the U.S. Forest Service.

Since 1970 the Viburnum Trend or New Lead Belt has ranked first in the world in lead ore production. [3-5] Zinc has also increased so that Missouri ranks second in the nation, passing both Tennessee and Colorado, and is challenging New York for first place. Missouri also ranks fifth and seventh, respectively, in the United States for the production of silver and copper.

The hilly topography of the area confines discharges from the mining and smeltering operations in separate stream systems; hence pollution effects can be determined for individual mines, mills, and lead smelters. Streams that flow into the area from the eastern edge of the drainage basin (Black River watershed) are not affected by either the mining development or populated regions and so can be used as controls. [6] The forested Ozarkian terrain

*Presently chairman, Department of Civil Engineering, Syracuse University.

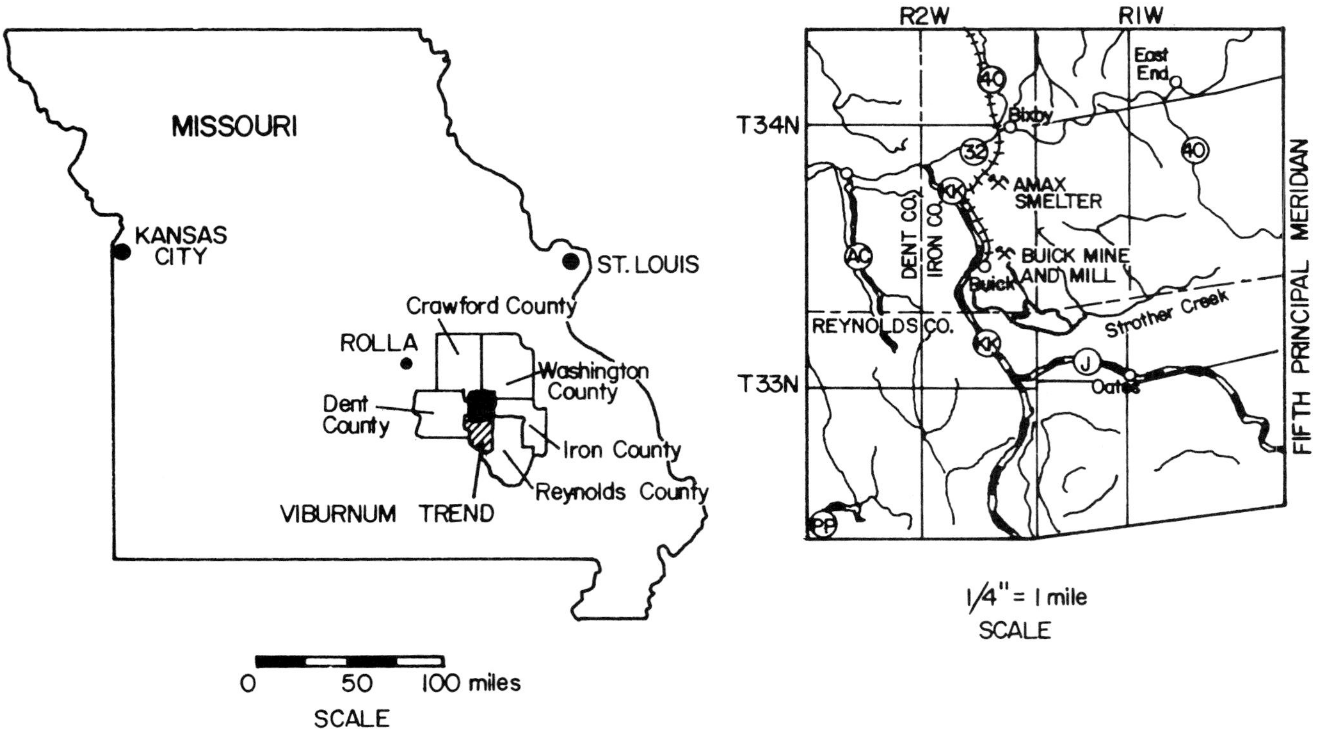

FIGURE 7-1.—Location of the Viburnum Trend in the New Lead Belt of southeast Missouri.

has a decided effect on wind direction, which affects the movement of stack emissions and other possible industrial and domestic sources of air pollution. While the prevailing wind direction is from the west and south, the rolling terrain makes it difficult to determine air pollution effects on the forested or cleared areas near the mining and smelting operations.

Galena (PbS) is the principal ore mined, with lesser quantities of sphalerite (ZnS), pyrite (CuFeS), and silver recovered as economic coproducts. Geologically, the lead is disseminated throughout the Cambrian-age Bonneterre formation, mostly a dolomite at depths ranging from 700 to 1,200 ft. Since the producing formation also serves as a major aquifer, most mines must employ constant pumping (ranging from 3,000 to 6,000 gal/min.) to prevent flooding. This inflow of water has reduced mine dust as a potential problem, but the mine water pumped to the surface contains appreciable amounts of fine galena and associated trace metals as well as spillage from underground repair and equipment maintenance operations.

Generally, the ore is crushed underground and hoisted to the surface in skip loaders. Water is pumped from the mines to the surface, and part of it is used in the milling procedures. Chemical reagents may be added to separate lead, zinc, or copper minerals from the finely ground rock by flotation. The remaining mine water is treated in large sedimentation or tailings ponds.

The lead and zinc (and sometimes copper) minerals from the flotation process are pumped into thickeners where the concentrate settles. The water, excess flotation reagents, and colloidal and supercolloidal minerals are then discharged as a milling effluent. The concentrate is vacuum dried and transported to smelters for final conversion to metal. A typical mine-mill production flow chart is illustrated in figure 7-2.

The water and tailings waste from the mining and milling operations are discharged into settlement and treatment lagoons, generally known as tailings ponds. Dams for the ponds are usually constructed from tailings (generally 200 mesh in size), which are centrifuged into fine and coarse fractions during the ore concentration process. After the dams are constructed, the tailings and other liquid wastes are introduced upstream until the lagoon is effectively filled. The lagoons serve as settling basins for the residual tailings to biologically degrade any spent organic reagents discharged from the milling process. Figure 7-3 is a production flow chart depicting a typical modern smelting operation with various types of emission controls.

Pollution Sources

Figure 7-4 summarizes potential sources for environmental contamination by heavy metals associated with the development of a major lead mining region.

The most obvious source of lead emissions is the smelter owned by the AMAX Lead Co. The 200 ft main stack discharges a baghouse effluent containing 28.5 lb/h of total particulates under normal operating conditions; 5.1 lb/h is particulate lead. Less obvious sources of lead near the smelter are the so-called fugitive sources such as truck and railroad transport of ore concentrates and windblown dust from metal concentrate piles (containing up to 70

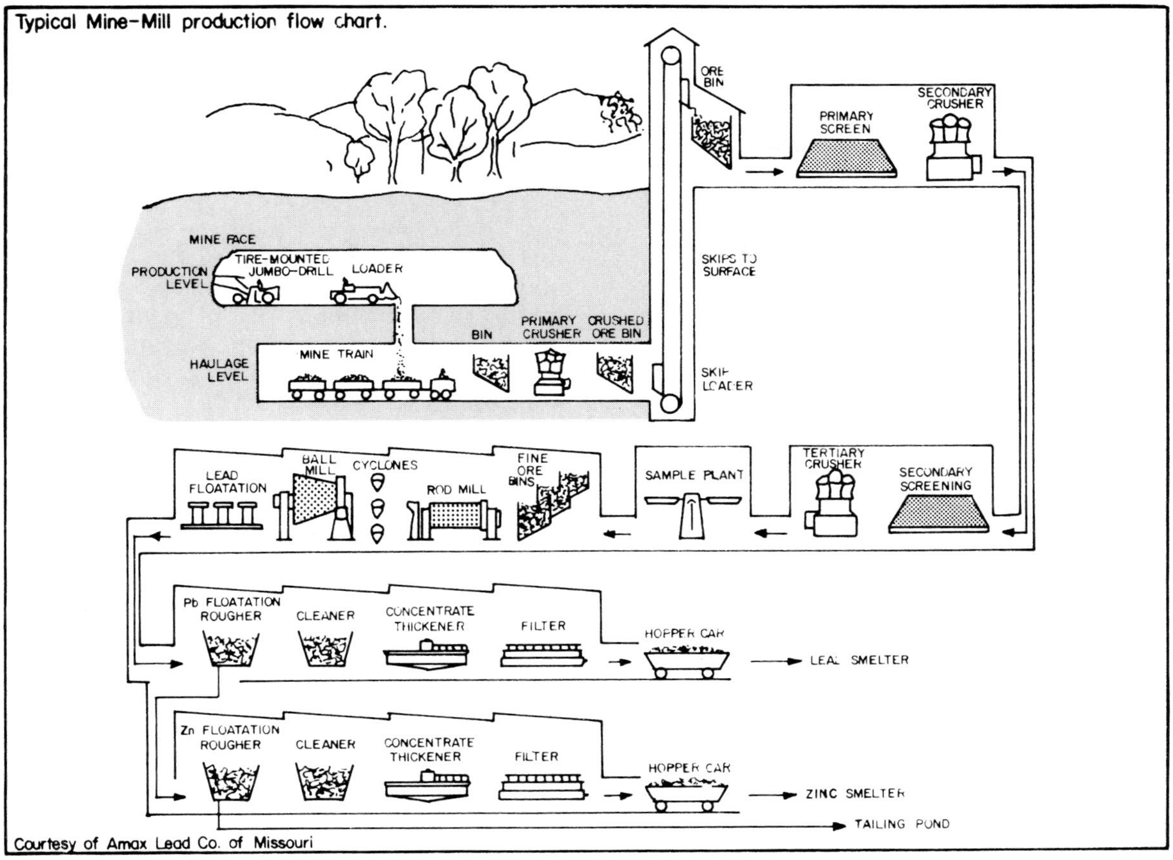

FIGURE 7-2.—Typical mine-mill production flow chart. (Courtesy of AMAX Lead Co.)

Smelter-refinery production flow chart.

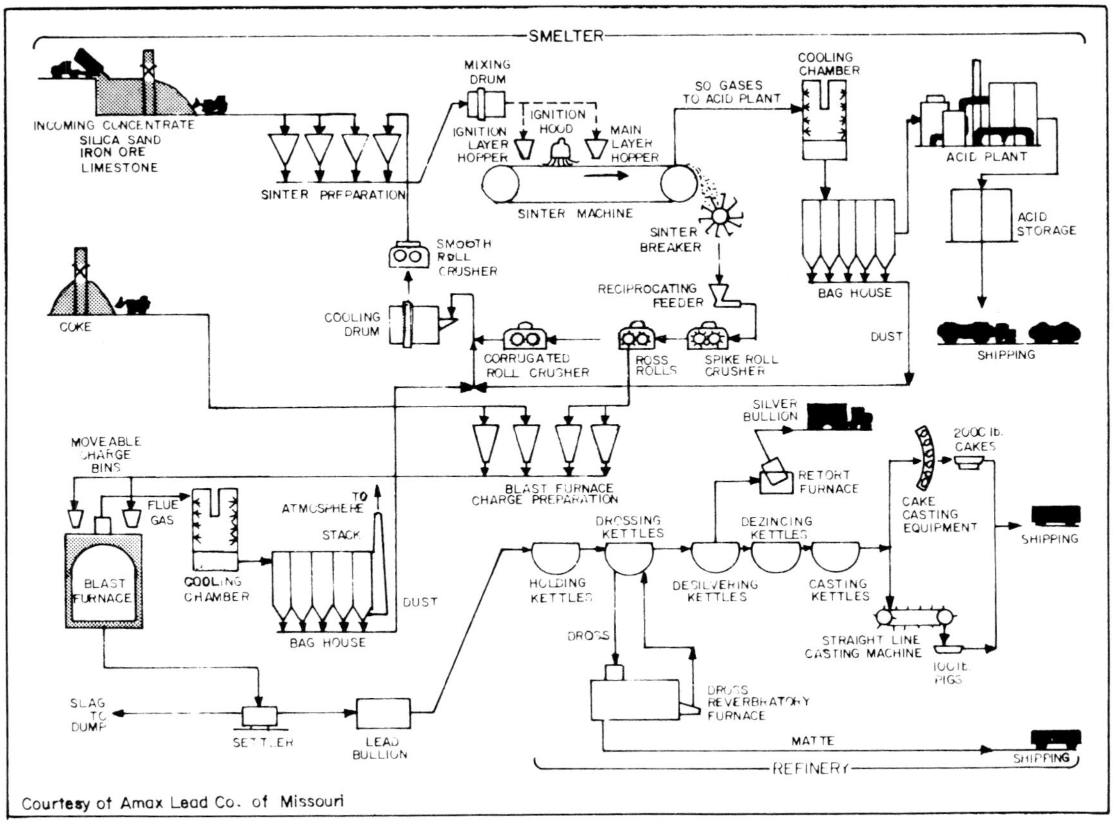

FIGURE 7-3.—Smelter-refinery production flow chart. (Courtesy of AMAX Lead Co.)

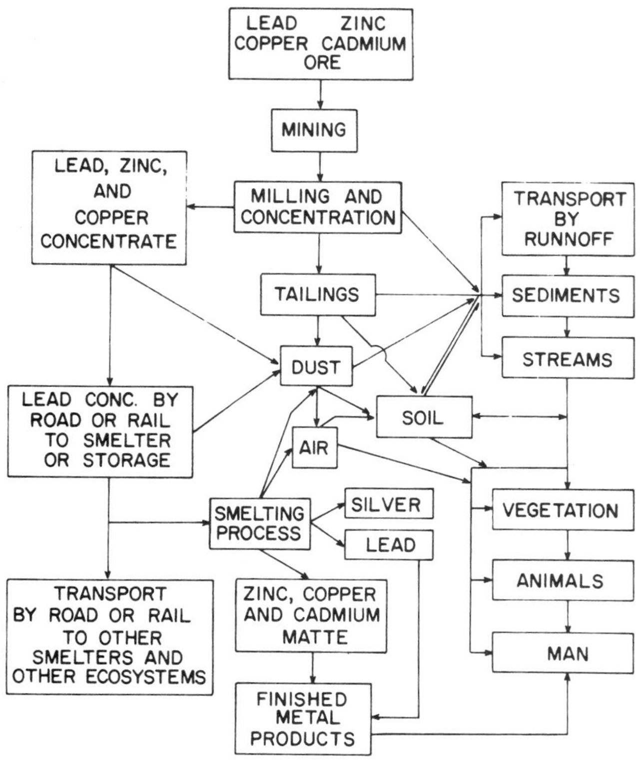

FIGURE 7-4.—Sources of trace metals in the environment of the New Lead Belt of southeastern Missouri

percent lead by weight). By comparing particle size distributions, Purushothaman [7] has theorized that these are the major sources of lead deposited within about 3 mi of the smelter. At greater distances, the stack emissions are the prime contributors of lead to the soil.

Other sources of lead and trace metals released to the environment include: [8]

(1) Mine water. Most mines must remove from the operational levels water that is generally rich in dissolved CO_2, carbonates, minor hydraulic and oil pills, unused blasting materials, and finely ground rock. Some formations (not in the New Lead Belt) may have pyrites in the ore; depending on the amount present, they may produce acid conditions and concomitant high levels of dissolved metals.

(2) Mill waters. These waters include finely ground gangue (rock containing no ore), unrecovered minerals, excess organic milling reagents, and dissolved and suspended heavy metals.

(3) Tailings. This gangue material represents the largest volume of solid wastes and is used to construct the tailings dams; it contains widely varying amounts of lead, zinc, etc. The tailings are finely ground (less than 200 mesh) and may blow onto the soil or wash into streams.

(4) Concentrates. These materials, containing approximately 70 percent lead by weight, not only blow onto soil and plants, but may also be washed away to the soil and streams. Transport in open vehicles and railroad cars aids in disseminating these materials.

(5) Smelter emissions. In addition to the stack and fugitive emissions from blowing concentrates and dusts from the sintering operations, liquid wastes from lagoons and acid neutralization systems may inadvertently contain heavy metals. In the New Lead Belt, the smelter operations are unusual in that both liquid and airborne wastes frequently contain anomalous amounts of cadmium. This element, normally present in the ore in minute quantities, is apparently concentrated during the smelting process.

This rapid and continuing industrial development in a rural forest ecosystem has presented a unique opportunity to studying the transport and distribution of lead and associated trace metals from mining, milling, and smelting operations. [9-10] Research into the various ecosystem compartments as related to lead in air, soil, vegetation, and water has determined background values, established natural baselines, and evaluated the lead mining and smelting industries as sources of trace metals in the environment. [11]

LEAD IN THE AIR

The distribution of heavy-metal bearing particles emitted into the atmosphere from lead mining and smelting operations forms a basis for estimating the transport of heavy metals to other components of the total ecosystem. Investigations of the atmospheric transport and distribution of lead in the New Lead Belt involved three specific objectives: (1) to determine the strengths of the stationary and mobile emission sources of particulate lead and other heavy metals in the smelter area; (2) to characterize the particulate lead emissions from the stack, particulate lead in the air, and windblown particulate lead; and (3) to determine the short-range distribution of particulate heavy metals in air and on the ground.

Source Strength and Characterization

The sources of particulate lead and other heavy metals in the AMAX smelter area fall into two categories. First, the process-oriented stationary sources—in this case, the 200 ft main stack receiving effluent gases from a baghouse attached to the blast furnace. In addition to the blast furnace off-gases, the baghouse received the weak off-gases from the sinter machine. The second category is the industrial sources (mobile) such as trucking and railroad operations, exposed piles of process materials, dust on the ground, and spillage around the smelter.

A sampling study to determine particulate loading and size distribution in the stack gas involved three tests run under normal smelter operations. The Results are shown in table 7-1 in terms of particulate loading and emission rates. The average mass emission rate of heavy-metal bearing particulates was 28.45 lb/h; particulate lead amounted to 4.66 lb/h or

TABLE 7-1.—*Stack particulate emissions from AMAX lead smelter under normal operating conditions*[1]

Test date	Particulates		Emission rate[2]							
			Lead		Zinc		Copper		Cadmium	
	Gr/DSCF[3]	lb/h	Gr/DSCF	lb/h	Gr/DSCF	lb/h	Gr/DSCF	lb/h	Gr/DSCF	lb/h
2/6/73	0.0122	20.87	0.0029	4.95	0.00027	0.47	$8.2x10^{-6}$	0.0140	0.00011	0.193
2/7/73	0.0240	38.69	0.0026	4.18	0.00016	0.339	$8.3x10^{-6}$	0.0134	0.00010	0.157
2/11/73	0.0111	25.80	0.0021	4.85	0.00021	0.495	$5.2x10^{-6}$	0.0120	0.00012	0.281
Average	0.0158	28.45	0.0025	4.66	0.00021	0.435	$7.2x10^{-6}$	0.0131	0.00011	0.210

[1] Process: sinter machine and blast furnace operations.
[2] Test point 13 baghouse outlet flue. Baghouse received sinter machine and blast furnace off-gases only. The sinter machine strong gases are taken to an acid plant for SO_2 recovery.
[3] Grains per dry standard ft^3.

16.3 percent by weight of particulates. At the same time, particulate zinc, copper, and cadmium were discharged at rates of 0.435, 0.013 and 0.21 lb/h, respectively. Although the ambient levels and deposition rates of settleable particulate cadmium were constantly lower than those of particulate copper, the emission rate was almost 20 times higher than that of copper. There are two reasons for the high emission rate of particulate cadmium. First, lead ore concentrate from the New Lead Belt contains a very high concentration (930 μg/g) of cadmium; [12] in most concentrates the Zn to Cd ratio runs around 200, but the New Lead Belt ratio is closer to 100. Secondly, copper was found to be associated with particles of 7 μm mass median diameter (MMD). Such particles are more efficiently removed in the baghouse than cadmium, which was attached to particles of 1 μm MMD. The relative amounts of zinc, copper, and cadmium in the stack emissions were approximately 1/10, 1/400, and 1/20 of lead, respectively.

The particulate emissions from the stack were characterized with an Anderson in-stack cascade impactor. (table 7-2) Approximtely 25 percent of the particulate lead, zinc, and cadmium was associated with 0.74 μm particles; in the case of particulate copper, however, about one-half of the amount was concentrated in particles greater than 8 μm. This suggests that a greater percentage of the particulate copper would be removed by the baghouse, while more of the particulate cadmium would escape through the stack.

The size distribution of the heavy metal-bearing particulates is shown in figure 7-5. Approximately 90 percent of the particulate lead, zinc, and cadmium was less than 10 to 20 μm, with a MMD of 1 to 1.5 μm. However, the particulate copper had a MMD of 7 μm, with 25 percent larger than 12 μm.

Generally, particles with an equivalent size of 10 to 20 μm, depending upon their density, are considered to have such small settling velocities that they tend to remain suspended in the atmosphere for long periods of time and hence are subject to long-range rather than short-range transport and dispersion.

Assuming the average density of lead-bearing particles to be 10 g/cm^3, then settling velocity would be affected by the particle diameter under a given set of conditions. In short-range transport of lead, it can be assumed that gravitational sedimentation will be negligible for particles with settling velocities $\leq$ 1 cm/sec. [13] Hage's data [14] on the settling velocity of smooth spherical particles with a density of 5 g/cm^3 are given by Vander Hoven [13] as a function of particle diameter and altitude. Particle size was reported to be the dominant factor affecting its velocity. Accordingly, particles with diameters of less than 10 μm are not affected significantly by gravitational settling.

Windblown dust in the smelter area was characterized with an Anderson nonviable sampler and aluminum foil collection plates at a location exposed to trucking activities. The sampling was carried out for 4.5 h at 1 ft^3/min. The results are summarized in table 7-3.

TABLE 7-2.—*Particle sizing of heavy metal-bearing particulate emissions from a primary lead smelter stack*

Stage	ECD[1] (μm)	Amount collected (μg)				Percent of total mass			
		Lead[2]	Zinc	Cadmium	Copper	Lead[2]	Zinc	Cadmium	Copper
0	12.7	540	35	14	4.5	10.8	7.4	7.4	23
1	8.0	505	39	15	4.5	10.1	8.3	8	23
2	5.3	365	27	12	3.7	7.3	5.7	6.4	18.9
3	3.75	265	26	9	3	5.3	5.5	4.9	15.3
4	2.35	335	30	11	0.7	6.6	6.4	5.9	3.6
5	1.2	720	64	21	1	14.4	13.5	11.2	5.1
6	0.74	1,235	108	47	1.2	24.6	22.9	25	6
7	0.49	665	72	33	1	13.3	15.3	17.6	5.1
Backup filter		380	71	26	0	7.6	15	13.6	0

[1] Corrected effective cutoff diameter at 50 percent impaction efficiency.
[2] Ref. [14].

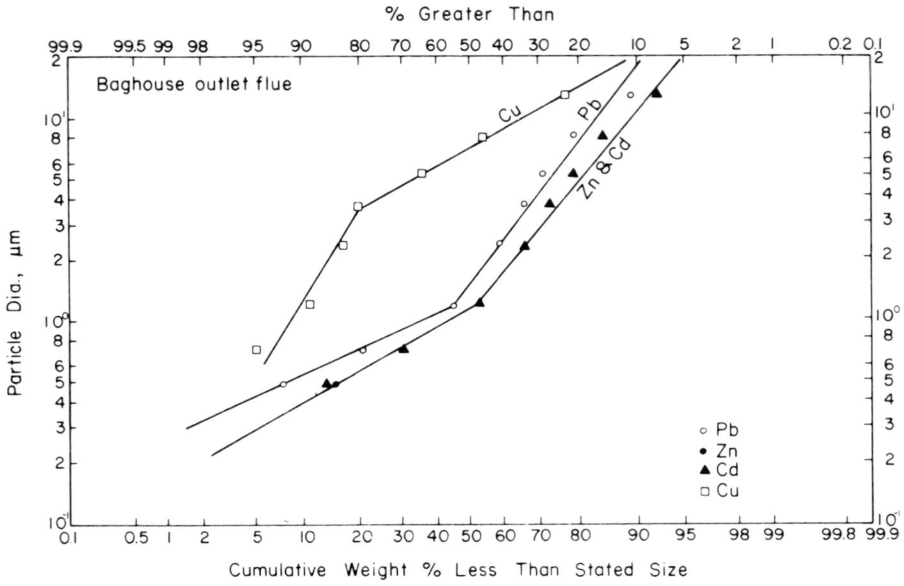

FIGURE 7-5.—Size distribution of particulate trace metals from a primary lead smelter stack.

More than 90 percent of the windblown particulate lead was associated with particles greater than 3.3 μm, while almost equal amounts of particulate copper were concentrated in each particle cut-off size ranging from 0.43 to 11 μm. The particulate lead concentration decreased from 40.3 to 0.8 μg/m³; particulate copper levels had two peaks of 3.8 μg/m³. Apparently, the dusts generated by the trucking activities, as well as that resuspended by the wind, are significant contributors to the high levels of lead and other heavy metals observed in the surrounding area.

The particle size distribution of lead and copper in the fugitive emissions is presented in figure 7-6. Only two-thirds of the particulate lead (with a MMD of 9 μm) were less than 11 μm in diameter; however, approximately 85 percent of the particulate copper (MMD = 3.5 μm) could be attributed to particles of 11 μm diameter or less.

Transport and Distribution of Settleable Particulate Heavy Metals

The transport and distribution of settleable particulate lead and other heavy metals were determined by field measurements with twenty dustfall stations in an area 3 mi by 3 mi surrounding the smelter. (fig. 7-7) As expected, the deposition rate for settleable particulate lead decreased with distance in each direction from the stack. The total deposition rate in any direction was a function of the pevailing wind.

The transport of settleable particulate zinc, copper, and cadmium followed the same trend as the particulate lead except that the deposition rates of Zn, Cu, and Cd were approximately 10, 20, and 100 times, respectively, lower than that of lead. This gives a cadmium-to-copper ratio of about 0.20. This compares with the cadmium-to-copper ratio in stack emissions of about 20; thus the cadmium-to-copper ratio in the settleable particulates is several orders

TABLE 7-3.—*Characterization of windblown (fugitive) particulate heavy metals in the vicinity of a primary lead smelter*[1]

Stage	ECD[2] (μm)	Amount collected (μg)		Percent of total mass		Concentration (μg/m³)	
		Lead	Copper	Lead	Copper	Lead	Copper
0	11	308	29	33.5	13.3	40.3	3.8
1	7	268	29	29.2	13.3	35	3.8
2	4.7	148	29	16.1	13.3	19.5	3.8
3	3.3	128	24	13.9	11	16.7	3.1
4	2.1	46	24	5	11	6	3.1
5	0.65	6	29	0.7	13.3	0.8	3.8
6	0.43	0	29	0	13.3	0	3.8
7		0	19	0	8.7	0	2.5
Backup filter		15	6	1.6	2.8	1.9	0.8

[1] Zinc and cadmium values were low and hence are not reported.
[2] Corrected effective cutoff diameter at 50 percent impaction efficiency.

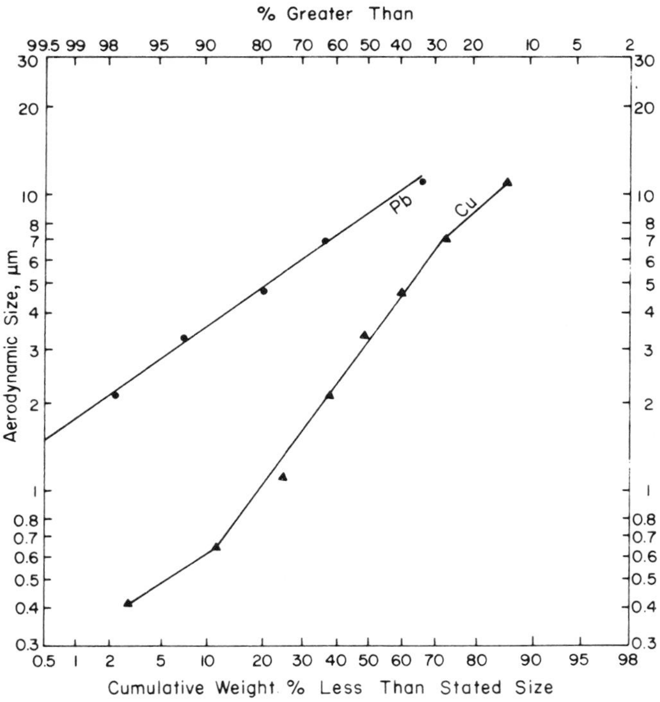

FIGURE 7-6.—Size distribution of windblown particulate lead and copper around a primary lead smelter.

of magnitude lower, suggesting that fugitive sources are the major contributors rather than stack emissions.

The directional distribution of settleable particulate zinc, copper, and cadmium indicated again that prevailing winds were responsible for higher deposition rates at the south, east, and northwest stations as compared with the west stations. The proximity of the Magmont Mine tailings pond did not increase the deposition rates at the 1 and 1.25 mi stations.

The settleable particulate lead, zinc, copper, and cadmium reached background values of about 55, 9, 3, and 0.3 mg/m2-mo, respectively, at a distance of 1.5 mi from the smelter.

The radial distribution of settleable particulate lead around the AMAX Smelter is shown as isopleths in figure 7-8. The isopleths were obtained by using the geometric mean values of the deposition rate at each station. [15]

In summary, transportation activities inherent to the smelting industry, combined with meteorological conditions that cause intermittent and unpredictable amounts of fugitive dusts are major factors in the accumulation of heavy metal particulates within the smelter area. Beyond the smelter boundary, the loading of settleable particulates drops to low yet significant values.

Transport of Ambient, Suspended, Particulate, Heavy Metals

The transport of ambient, particulate lead, and other heavy metals, may best be considered as both short- and long-range. Short-range transport is due to the rapid settling of larger particulates, while smaller particles remain suspended in the atmosphere and are transported over longer distances.

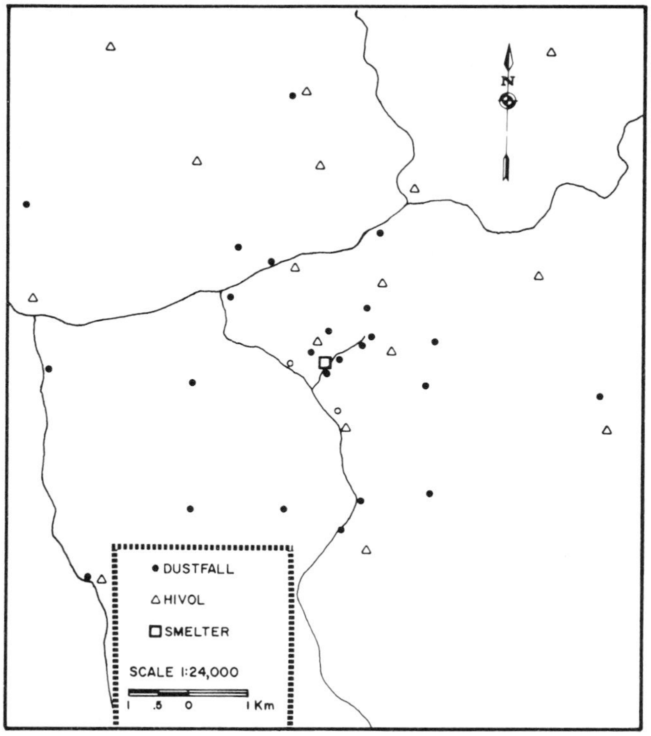

FIGURE 7-7.—Location of air monitoring stations around a primary lead smelter.

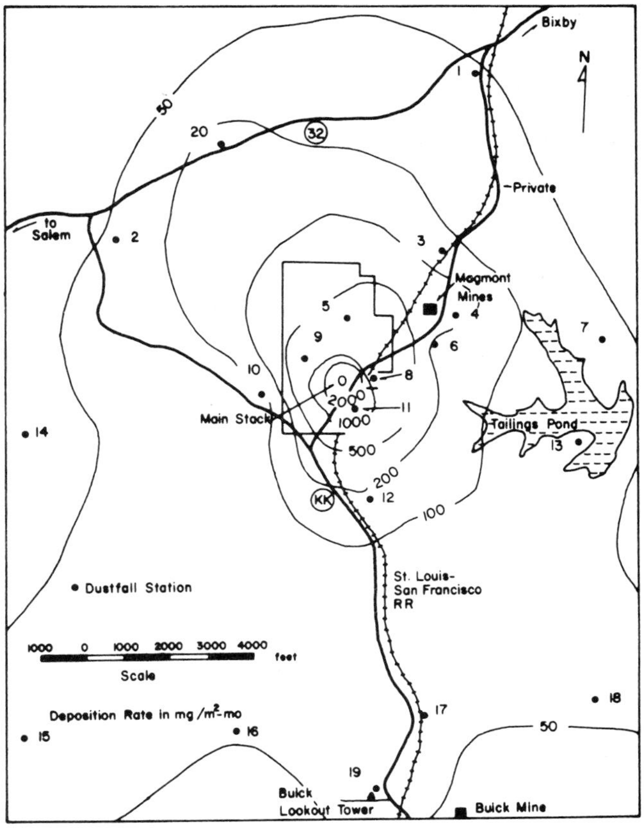

FIGURE 7-8.—Radial distribution of settleable particulate lead around a primary lead smelter.

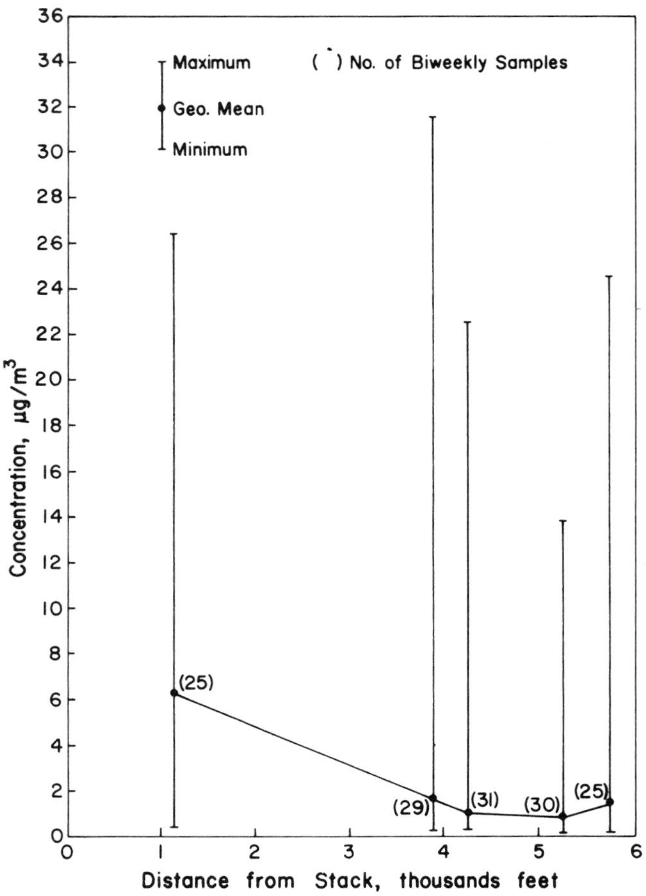

FIGURE 7-9.—Short range transport of ambient lead.

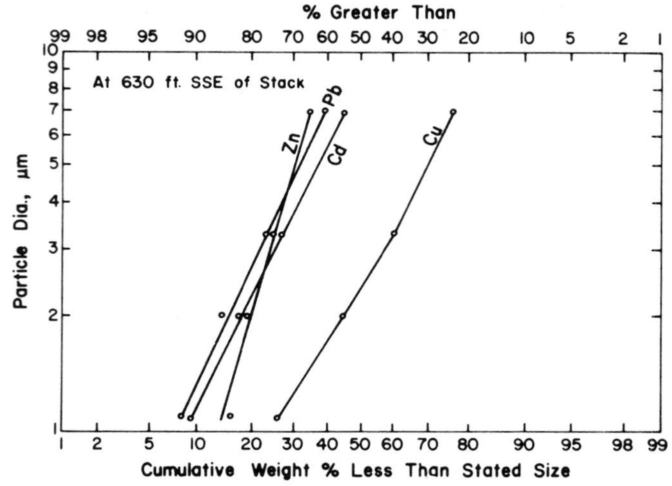

FIGURE 7-10.—Size distribution of ambient particulate heavy metals near a primary lead smelter, 630 ft SSE of stack.

The short-range transport of the ambient lead was monitored with five high-volume stations within 6,000 ft of the smelter stack as shown in figure 7-9. Bi-weekly samples were collected with 8 in. by 19 in., Type A, glass fiber filters from June 1972 to June 1974. High-volume cascade sampling was also done to evaluate the characteristics of the particulate lead and other heavy metals.

The concentration of ambient particulate zinc was often equal to or less than the blank value of the glass fiber filter, in spite of the fact that samples were taken for a 24 h period. [15] This was corrected by using Whatman 41 cellulose filters with a low zinc content.

The short-range transport of ambient suspended particulate lead indicated that the lead level decreased with distance, ranging from 6.4 $\mu g/m^3$ at 1,140 feet to 1 $\mu g/m^3$ at 1 m. from the smelter stack. However, the maximum levels varied from 31.5 to 13.8 $\mu g/m^3$ depending upon wind speed and direction.

The short-range transport of ambient suspended particulate cadmium and copper is shown in table 7-4. The lead-to-copper ratio was about 25 at the nearest location from the stack and about 30 at the farthest point. The lead-to-cadmium ratio was about 60 closer to the stack and about 40 at the more distant location.

The physical characterization of ambient particulate lead and other heavy metals was achieved with a high-volume particle sizing head and a 12 in. diameter glass fiber collection disc at 20 ft³/min. Samples were collected for 48 to 72 h to obtain sufficient amounts of heavy metals for analysis. Figures 7-10 and 7-11 show the size distribution of ambient heavy metal-bearing particulates at 630 and 1,425 feet SSE and ENE, respectively, of the AMAX smelter stack. The MMD of the lead-bearing aerosols varied from 7 to 11 μm, and that of particulate copper ranged be-

TABLE 7-4.—*Short-range transport of ambient suspended particulate copper and cadmium around a primary lead smelter*

Distance from stack (ft)	Concentration ($\mu g/m^3$)					
	Copper			Cadmium		
	Geo Mean	Max	Min	Geo Mean	Max	Min
1,140	0.25	2.3	0.02	0.11	6.68	0.01
3,875	0.07	0.75	0.01	0.02	0.34	0.01
4,250	0.03	0.5	0.01	0.03	1.68	0.01
5,250	0.03	0.18	0.01	0.02	2.29	0.01
5,750	0.05	0.4	0.01	0.04	0.7	0.01

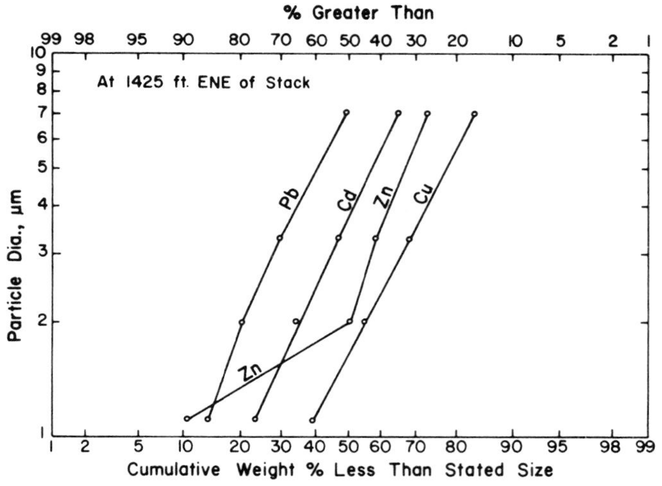

Figure 7-11.—Size distribution of ambient particulate heavy metals near a primary lead smelter, 1,425 ft ENE of stack.

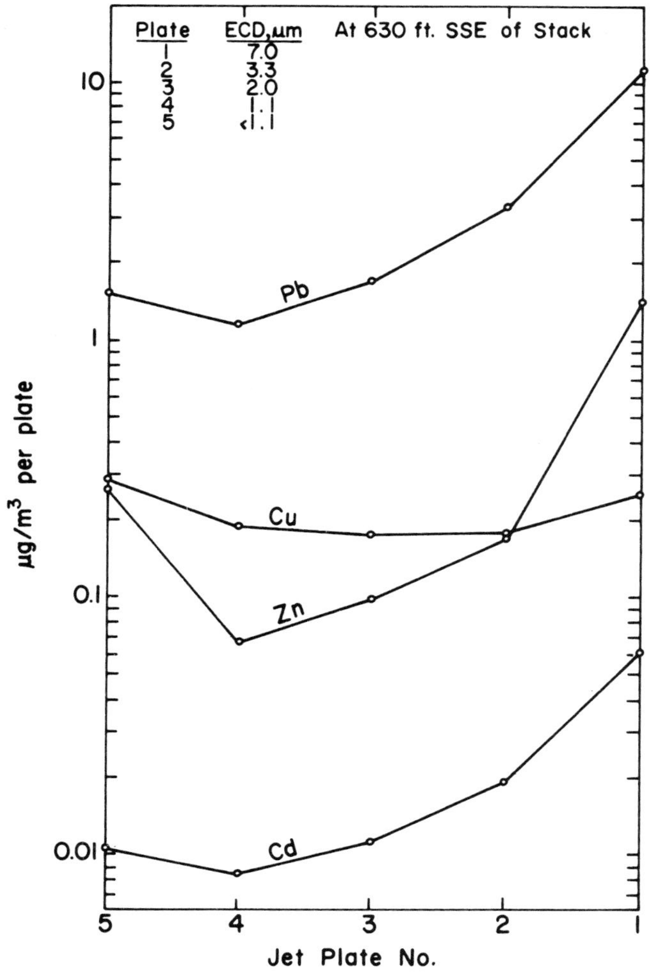

FIGURE 7-12.—Relative concentration of ambient particulate heavy metals as a function of particle size, 630 ft SSE of stack.

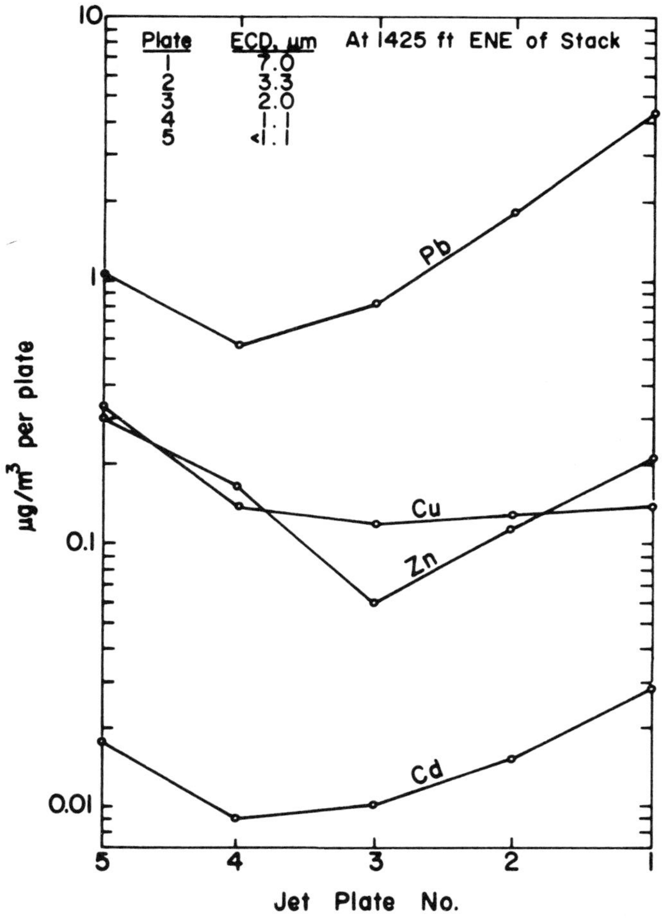

FIGURE 7-13.—Relative concentration of ambient particulate heavy metals as a function of particle size, 1,425 ft ENE of stack.

tween 1.6 and 2.5 µm. These sizes correlate with those for windblown dust shown in figure 7-6, confirming that the heavy metals are of windblown origin. In the case of ambient zinc and cadmium, about 25 to 65 percent of the aerosols were larger than 7 µm, with a MMD of 2 to 4 µm.

The size class distribution of heavy metals was evaluated in terms of relative concentration as a function of particle size. (figs. 7-12 and 7-13) This type of representation brings out some interesting points that are not evident from the particle size distribution curves shown in figures 7-9 and 7-10. The ambient lead, zinc, and cadmium were concentrated in 7 µm and less than 1.1 µm particles; however, the ambient copper was mostly associated with particles less than 1.1 µm in diameter.

The ambient concentrations of heavy-metal particulates determined with the high-volume impactor closer to the smelter stack are presented in table 7-5. The concentration of heavy settleable particulates and the decrease in heavy metal concentration with increasing distance could be attributed to the change in the wind direction over a period of 48 h.

CONCLUSIONS

Based on data presented and discussed above, several conclusions can be drawn concerning the distribution of settleable particulates and suspended particles of lead and other heavy metals due to emissions from lead mining and smelting operations in the New Lead Belt area. It should be emphasized that the lead industry has utilized these data to change or improve the various operations and con-

TABLE 7-5.–*Ambient particulate heavy metal levels as obtained by high-volume impactor*

Sampling period	From stack		Concentration (µg/m³)			
	Distance (ft)	Direction	Lead	Zinc	Copper	Cadmium
July 12-15, 1973	630	SSE	18.68	1.74	1.1	0.11
July 9-11, 1973	1,425	ENE	8.63	0.64	0.86	0.08

trol many of the problems noted in the research findings. The conclusions are:

(1) During the period of study, the major sources of lead and other heavy metals within 1 mi of the AMAX smelter were the fugitive dusts; these caused the deposition of settleable particulate lead in excess of 2 g/m^2-mo in the smelter area. Numerous changes have been made to alleviate these conditions.

(2) The particulate emissions from the smelter stack were of significance in the long-range (greater than 1 mi) heavy-metal deposition.

(3) The short-range ambient suspended particulate lead and cadmium concentrations reached levels of 32 and 6.7 μg/m^3 in ENE and NNW directions, respectively.

LEAD IN SOIL

Soil and geochemical studies were carried out to (1) determine the extent and characteristics of soil pollution patterns produced by the various smelting and mining activities, (2) determine the chemical and mineralogical characteristics of the pollutants emitted from various sources, (3) assess the relative importance of the individual pollution sources, and (4) determine the geochemical behavior and mobility of polluting heavy metals.

The seven active lead mines and their associated mills in The New Lead Belt are potential pollution sources through windblown material from stored ore concentrates and, to a lesser extent, tailings. One lead smelter of the AMAX Lead Co. of Missouri is located in the district, with a second smelter, ASARCO, about 28 mi to the southeast. Both smelters started operation in 1968.

Sampling and Analysis

Most of the study area is heavily forested with oak and local stands of pine. Since early data [16] indicated that heavy metals accumulate in the decaying leaf litter with only minor amounts reaching the underlying mineral soil, emphasis was placed on the collection and analysis of the organic horizons. The samples of leaf litter from soil profiles were described according to a proposed international system for the description of soil horizons. [17] The following abbreviations are used for the various samples:

W. Oak = leaves taken directly from white oak trees
Of = partly decomposed leaf litter
Oh = well-decomposed leaf litter, usually more than 1 year old
O = undistinguishable leaf litter

Soil samples were classified according to their depth.

Grass samples were collected from scattered pastures in the vicinity of the smelters. These samples were subdivided into "standing grass" and "matted grass." The vegetation was dried at 100°C and then ground in a Wiley mill. The soil samples were also dried at 100°C and passed through an 80 mesh (0.171 m) sieve to remove pebbles. Most of the samples were analyzed for lead, zinc, copper, and cadmium by atomic absorption spectroscopy.

Considerable variation was found in the heavy-metal concentrations of replicate samples from heavily contaminated areas. (table 7-6) This was due to the nature of the leaf materials rather than the analytical techniques, and it emphasizes the need for a relatively large number of samples. It is quite likely that the heavy-metal particulates are not uniformly distributed within the litter layers. It is also difficult to sample specific organic horizons without some mixing from adjacent layers. In addition, heavy-metal particles may be partially removed or concentrated in certain parts of samples during collection and preparation for analysis. These same factors were also evident in grass samples.

Background Concentrations

Soil samples from 125 stations covering most of the New Lead Belt do not indicate the natural occurrence of abnormal heavy-metal concentrations in the soils of the district. High concentrations can always

TABLE 7-6.—*Variation in heavy metal content of duplicate leaf litter (Of layer) and soil samples (0-1 in. depth)*[1]

	Pb	Zn	Cu
	Of-leaf litter (μg/g)		
	85,000	1,460	1,400
	81,000	1,400	1,280
	34,000	1,800	1,500
	74,000	1,530	1,120
	62,000	1,240	920
	76,000	1,280	1,180
	67,000	1,400	1,040
	81,000	1,360	1,220
Average	70,000	1,434	1,208
Range	34,000-85,000	1,240-1,800	920-1,500
Standard deviation	16,423	174.4	187.9
	Soil, 0 to 1 in. depth (μg/g)		
	3,200	200	80
	1,260	125	40
	2,160	160	60
	1,020	130	30
	1,280	100	30
	580	130	20
	1,940	150	50
	1,860	145	50
Average	1,663	143	45
Range	580-3,200	100-200	20-80
Standard deviation	812.4	29.5	19.3

1. Sample collected within 1 m^2 located 0.25 mi northeast of smelter.

TABLE 7-7.–*Estimated background concentrations of heavy metals 25 to 30 mi from AMAX smelter*

	Pb	Zn	Cu (μg/g)	Cd
Leaf litter Of	40-70	35-50	6-8	0.7-1
Leaf litter Oh	80-100	50-70	8-11	0.9-1.5
Standing grass	35-50	12-17	3-6	<0.5
Matted grass	50-100	25-60	5-7	<0.5-1.6
Soil (minus 80 mesh)	15-20	10-20	1.5-4	<0.5

be related to potential pollution sources. The background concentrations reported in table 7-7 are based on samples collected at a distance of 25 to 30 mi from the AMAX smelter.

Lead Distribution in Soil Profiles

Analysis of soil profiles from oak forests in the vicinity of the smelters, which have been operating from 4 to 7 years, indicates that most of the lead and other heavy metals are still in the leaf litter. Only the top 1 in. of the underlying mineral soils shows elevated levels in areas of heavy contamination. The concentrations of lead usually approach or reach background at depths of 2 to 3 in. (tables 7-8 and 7-9) Thus if data from soil samples below the leaf litter were used alone, both the intensity and the extent of the pollution patterns would be seriously underestimated. In contrast, elevated levels were found to a depth of several inches in soil samples taken from profiles without a cover of leaf litter. (table 7-10) [18]

TABLE 7-8.–*Heavy metals in soil profiles 0.2 mi from AMAX smelter*

	Pb	Zn (μg/g)	Cu
Of (0.25 in., pine needles, oak leaf litter)	130,000	1,850	2,200
Oh (0.25 in., soil + oak leaf litter)	39,000	1,350	760
0-0.25 in. soil	3,200	150	100
0-25-0.5 in. soil	840	80	35
0.5-1 in. soil	330	60	20
1-2 in. soil	20	50	<10
2-3 in. soil	<20	50	<10
3-4 in. soil	<20	50	<10

Distribution Patterns of Lead Near Smelters

Since the distribution patterns of lead around the two smelters differ in several respects, the results are discussed separately.

TABLE 7-9.–*Heavy metals in soil profiles 0.3 mi from AMAX smelter*

	Pb	Zn (μg/g)	Cu
Leaf litter (0.5 in.)	7,500	330	150
0-0.25 in. soil	380	100	30
0.25-0.5 in. soil	80	60	10
0.5-1.0 in. soil	<20	50	<10
1-2 in. soil	<20	30	<10

AMAX Smelter: Soil samples were collected along six sampling lines radiating for a distance of 25 to 30 mi from the smelter. These lines generally followed highways, but the samples were collected at least 1,000 ft from paved and well-traveled roads. The area showing elevated lead concentrations in the leaf litter is slightly elongated to the north and south from the smelter, which agrees with the prevailing wind directions. At a distance of about 1/4 mi from the smelter the lead concentration in the Oh layer is between 30,000 and 60,000 μg/g, with some samples as high as 130,000 μg/g. At a distance of 0.5 mi, the lead concentration is in the range of 20,000 to 45,000 μg/g and then drops sharply for about 5 mi. Figure 7-14 shows the distribution of lead, after 5 to 6 years of smelter operation, for a distance of 30 mi. The oldest well-decomposed leaf litter (Oh) contains the highest lead concentrations. However, either the recent leaf litter (Of) or leaves collected from trees would be suitable to establish the extent of the pollution pattern, especially if leaves are collected just prior to annual defoliation, thus allowing maximum time for deposition.

It is interesting in this context to compare the distribution patterns of zinc, copper, and cadmium with lead. Cadmium and copper reach background concentrations at 15 and 10 mi, respectively, from the smelter, while zinc rarely shows elevated values in leaf litter beyond 5 mi from the source. Thus neither copper, cadmium, nor especially zinc could be used as indicators of lead pollution patterns. The differences in extent and intensity of the distribution patterns of the four heavy metals are only partially due to their relative concentrations in the source (stack emission). For instance, zinc is emitted from the smelter in higher concentrations than either cadmium or copper. It appears that the solubilities of the heavy-metal particles and their subsequent geochemical mobilities are of greater importance.

TABLE 7-10.–*Heavy metals in soil profiles without a cover of leaf litter located 0.25 mi from AMAX smelter*

	Pb	Zn (μg/g)	Cu
0-0.25 in. soil	4,500	600	140
0.25-0.5 in. soil	500	70	10
0.5-1 in. soil	330	50	10
1-2 in. soil	100	50	<10
2-3 in. soil	140	50	<10
3-4 in. soil	160	50	<10

Figure 7-15 shows the distribution of lead in pasture grass north of the smelter. Background concentrations seem to be reached at a distance of about 10 mi. As expected, the highest lead concentrations are in the matted grass rather tn in the standing grass.

ASARCO Smelter: The distribution pattern of lead in leaf litter around this smelter is strongly elongated to the north and south. Background concentrations to the north and south are reached at 8 to 10 mi from the smelter and at about 5 mi to the east and west. The pronounced elongation of the polluted area

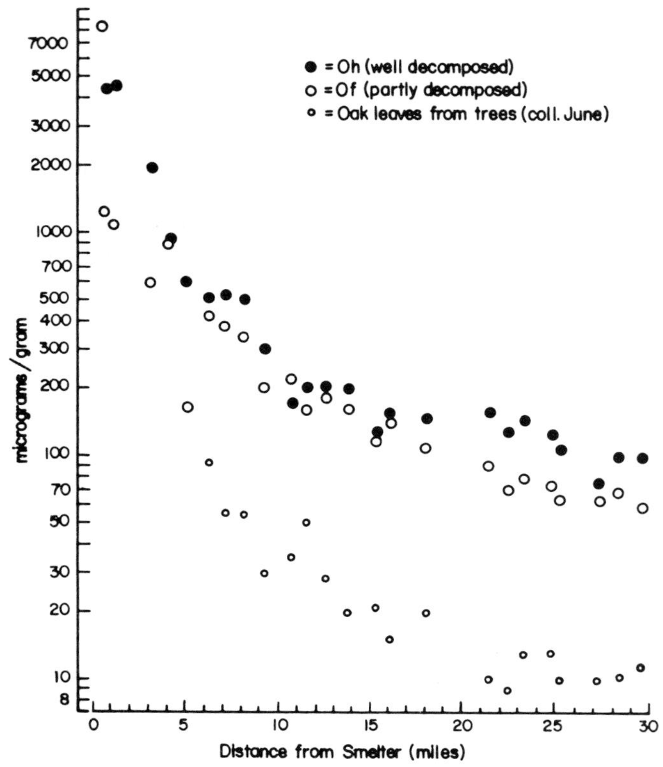

FIGURE 7-14.—Lead in Oak Leaf Litter NNW of AMAX Smelter.

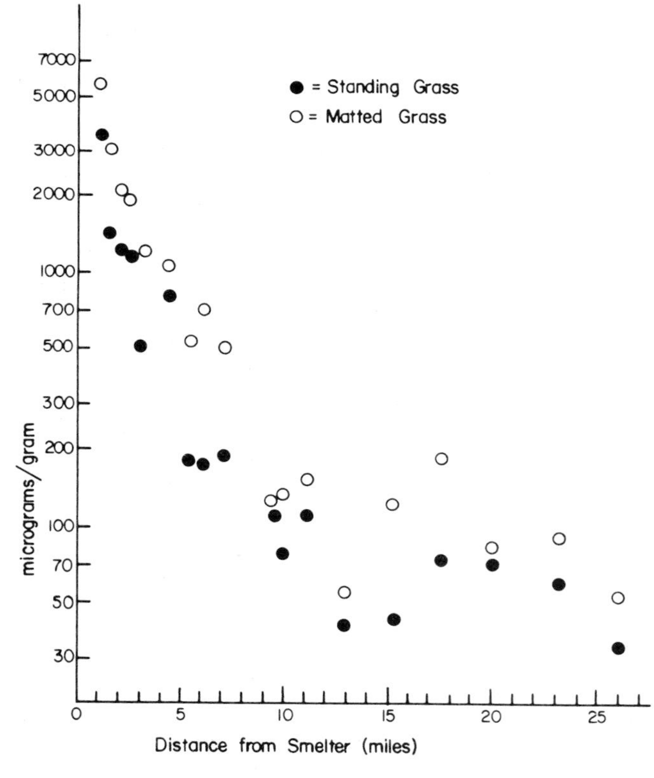

FIGURE 7-15.—Lead in Grass Samples Collected North of AMAX Smelter in April 1974.

is probably caused by the location of the smelter in a deep north-south trending valley, as well as by the prevailing wind directions. In comparison, the distribution pattern for cadmium is similar in extent to lead, while those of copper and especially zinc are much less extensive.

The distribution of lead to the north of the smelter is shown in figure 7-16. Contrary to the distribution of lead around the AMAX smelter, the concentrations of lead near the ASARCO smelter is generally higher in the more recent, partly decomposed leaf litter (Of layer) than in the older Oh horizon.

Pollution by Mining and Transport Activities

Pollution of leaf litter by mining activities (essentially handling of ore concentrate and storage of tailings) does not appear to be as serious as that from smelter emissions. Samples taken along three sampling lines for a distance of about 4 mi from a mine indicate that some leaf litter samples within 1 mile may contain lead concentrations of several hundred thousand micrograms/gram. At greater distances, as well as in many samples close to the mine, the concentrations are at background levels.

Transport of ore concentrate in open trucks and railroad cars does contribute high lead concentrations to the environment for a distance of up to 500 ft. Data for leaf litter near a highway are shown in figure 7-17. Major pollution (in excess of 500 μg/g) is generally restricted to a distance of 200 ft from the transport routes. Railroad transport appears to be a less significant source than highway trucking. This aspect of lead pollution is discussed in more detail elsewhere in the chapter.

Nature and Mobility of Metal Pollutants Near Smelters

Lead smelters have two potential pollution sources: emission through the stacks and fugitive sources, such as stored ore concentrate and dust blown from contaminated soil in the smelter area. Identification of the relative contributions from these sources is of major interest since the ultimate objective of pollution research is to help reduce or eliminate the causes. For instance, reducing stack emission at great expense would obviously not fulfill this goal, should the major source be fugitive dusts. The problem of source identification appears rather simple and straightforward since the two sources consist of materials that differ to some extent in their chemical and mineralogical compositions. However, experience has shown that the problem is less easily solved than might be expected.

Particulate matter that leaves the stacks is difficult and expensive to collect. However, baghouse dust is probably similar in mineralogy and chemical composition, an assumption supported by the comparisons of small amounts of particulates collected from the two sources.

Lead compounds in baghouse dust have been identified by X-ray diffraction as $PbSO_4$, $PbO \cdot PbSO_4$, and metallic lead. In addition, the dust

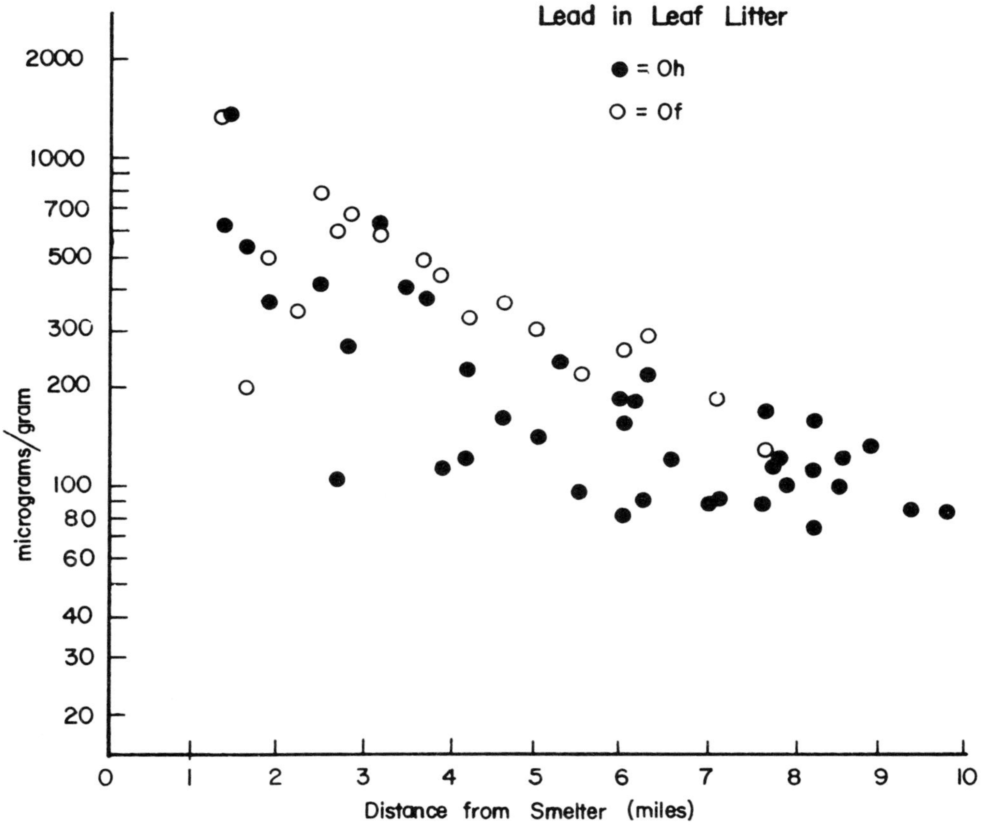

FIGURE 7-16.—Lead In Leaf Litter North of ASARCO Smelter.

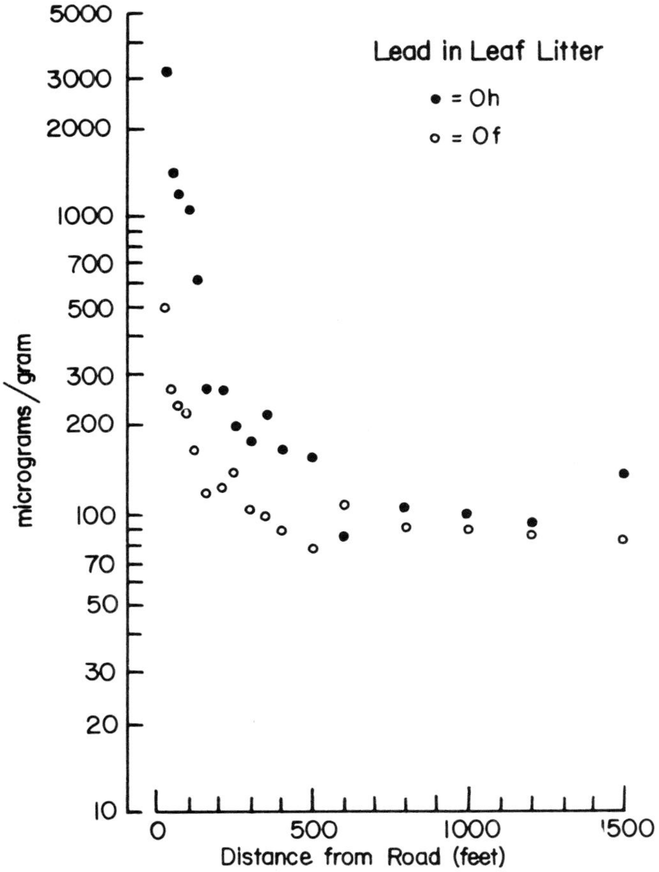

FIGURE 7-17.—Lead in Leaf Litter near Highway 72, an Ore Trucking Route East of Reynolds.

of the AMAX smelter contains approximately 30 percent PbS as the predominant lead compound. The amount of PbS in the baghouse dust of ASARCO is less than 5 percent. The chemical composition of baghouse dust is typically in the range of 50 to 60 percent lead, 1 to 13 percent zinc, 0.25 to 13 percent copper, and 0.5 to 6 percent cadmium.

Fugitive sources are essentially lead-ore concentrates with PbS as the major mineral and a composition of about 70 to 75 percent lead and 0.03 to 1 percent cadmium. Fugitive sources, therefore, consist predominantly of PbS with a very high Pb:Cd ratio (about 1,000:1), while the Pb:Cd ratio in smelter emission is much lower (about 20:1). Also, the baghouse dust, especially at the ASARCO smelter, contains very little PbS. It might be expected, therefore, that areas with a high PbS content and a high Pb:Cd ratio receive fallout from fugitive sources, while a low PbS and low Pb:Cd ratio would indicate stack emissions as the major source. Unfortunately, the available results are less than conclusive on this point. Concentrating lead compounds from soil or leaf litter for quantitive X-ray analysis is difficult. In addition, this method is not very suitable near the AMAX smelter because of the high PbS content of the baghouse dust. Use of the Pb:Cd ratio along a detailed sampling line north of the AMAX smelter was of limited success. The Pb:Cd ratio ranges from 175 to 275 for a distance of about 0.5 mi from the smelter. From 0.5 to 8 mi the ratio ranged from 100 to 175; from 8 to 30 mi, the range was from 60 to 100. Based on these results, a tentative conclusion would be that fugitive sources are a significant contributor for a distance of at least 3,000 ft. This conclusion si supported by the occasional microscopic observation in soils of this area of material that appears to be slag.

The reason that the Pb:Cd ratio does not clearly delineate the relative significance of the different sources becomes more obvious when the solubilities of the heavy metal compounds are considered. For example, solubility studies of baghouse dust (table 7-11) indicate that cadmium is more soluble than the other three heavy metals, especially lead and copper. This could increase the Pb:Cd ratio in the leaf litter. Ongoing studies also demonstrate (table 7-12) that the natural water-soluble organic (humic) acids present in the decaying leaf litter significantly increase the solubility of lead, zinc, and copper compounds. In addition, the various lead compounds have different solubilities, PbS being only slightly soluble while $PbSO_4$, $PbO{\cdot}PbSO_4$, and metallic Pb are much more soluble. (table 7-12) Experiments with ion-sensitive electrodes indicate that lead and copper become strongly complexed and zinc weakly complexed by the organic acids. At a low pH (about 4.5), cadmium remains primarily in the ionic state. In their complexed state, the heavy metals are less easily adsorbed on soils or precipitated, thus becoming both more mobile and easily removed. [19] The pathways of dissolved heavy metals are not yet fully understood. However, considerable amounts are removed by runoff during heavy rains, a point discussed elsewhere in this chapter. Analyses of runoff water after a period of dryness revealed up to 500 μg/l lead, 4,800 μg/l zinc, and 194 μg/l cadmium after filtration through a 0.45 μm millipore filter. As the result of these different solubilities and probable subsequent mobilities, the Pb:Cd ratio in the leaf litter changes, and PbS is enriched compared with other lead compounds.

In this context, the differences in pollution patterns of the AMAX and ASARCO smelters are of interest. The much lesser extent and intensity of pollution around the latter is partially caused by such factors as lower production and differences in the treatment of ore concentrate and baghouse dust. However, differences in the mineralogy of the two stack sources may also have a major influence. Stack emissions at AMAX contain a high percentage of rather insoluble PbS, thus permitting a buildup of lead a considerable distance (10 to 20 mi) from the smelter where the deposition rate is small. In contrast, emissions from ASARCO consist largely of the more soluble lead compounds, and 10 mi north of the smelter the rate of deposition may be exceeded by the rate of solution and subsequent removal of lead. If these conjectures are correct, then the area around the smelter that receives heavy metal fallout is actually larger than that currently showing elevated soil lead values. Also, changing operational procedures in smelters to eliminate emission of insoluble lead compounds such as PbS would perhaps result in less extensive and intensive soil pollution. At the same time, release of more soluble lead compounds would probably result in higher heavy metal pollution of streams during periods of runoff.

TABLE 7-11.–*Solubility of lead compounds in water (H_2O) and natural organic acids (OA)*[1]

		Pb	Zn	Cu	Cd
			(percent soluble)		
AMAX	OA	16.6	61.4	4.2	38.8
(main baghouse)	H_2O	4.8	15.6	<3.3	33.8
ASARCO	OA	30.0	38.6	12.9	57.8
(blast furnace)	H_2O	4.2	3.1	<5.0	62.6
ASARCO	OA	17.6	30.0	2.8	71.8
(sinter plant)	H_2O	10.5	18.1	<1.0	68.8

[1] Conditions: pH 3.8, 50 mg dust in 100 ml, 40 h, OA = 0.01 percent.

TABLE 7-12.–*Solubility of lead compounds in water (H_2O) and natural organic acids (OA)*[1]

	Lead (percent soluble)	
	H_2O	OA
$PbSO_4$	15.5	24.0
$PbO{\cdot}PbSO_4$	3.0	4.5
PbS	0.05	0.1
Pb	0.75	0.5

[1] Conditions: 0.02 g sample, 100 mi OA (0.01 percent) or H_2O, pH 5.5, 24 h.

Effects of Metal Mobility on Leaf Litter and Biomass

The effects of the increased mobility of heavy metals, as a function of humic acids in the leaf litter and soil, are being jointly investigated by the University of Missouri-Rolla and the Oak Ridge National Laboratory. [20] This work was being done on the Crooked Creek watershed, located near the center of the Viburnum Trend in the New Lead Belt. This watershed also forms the basis for the development and verification of a unified model of the transport of elements by atmospheric and hydrologic processes. [21]

The high solubility of cadmium with respect to lead is of particular concern since there are indications that Cd may be moving into the mineral soil horizons more rapidly than Pb. Cadmium is more toxic to plants and microbial organisms than lead, and its ultimate effect may be a disruption of some normal ecosystem functions.

Watson [22] has characterized the litter horizons at selected sites on Crooked Creek according to biomass, macronutrient status, and heavy metal concentrations. She found a significant decrease in the biomass of the O_2 litter horizon with increasing distance from the smelter. Over the same rnge of sites, the O_1 layer remains relatively constant. Concentrations of Cd, Pb, Cu, and Zn follow expected logarithmic decreases in concentration with increasing distance from the smelter.

The effect on the O_2 horizon biomass appears to be due to a reduction in the rate of decomposition by microbial and arthropod activity. These decomposing organisms are completely missing within ½ mi of the smelter. There has also been some alteration in the turnover rates of macronutrients, some being retained in the litter while other have leached from the system. The cause of this has not been determined, but the long-time consequence could be a depletion of essential nutrients, with a corresponding decrease in forest productivity.

SUMMARY

Data from an intensive sampling program indicate that heavy pollution of soils occurs from lead-smelting activities and to a lesser extent from mining and milling operations. This has occurred during the 5 to 7 years of an estimated 60 year period of operation in the New Lead Belt.

The organic horizons are the most heavily contaminated parts of soil profiles in the surrounding forest ecosystem. Except where bare soil is exposed to fallout from emissions, only the top 1 in. of the mineral soil shows elevated levels of lead and other heavy metals. Thus analyses of soil samples alone, rather than leaf letter, would have resulted in a serious underestimation of the problem.

Humic acids present in the leaf litter and soil enhance the solubility of heavy metals in the decending order of Cd, Zn, Pb, and Cu. Lead compounds are also differentially soluble in these acids with $PbSO_4$ the most soluble and PbS the least. These differential solubilities influence both the extent and intensity of pollution patterns.

A major effect on the biomass of the O_2 litter horizon occurs within ½ mi of the smelter stack and appears related to an absence of decomposing organisms within the immediately affected area. Turnover rates and patterns of macronutrients have also been observed. While the long-term effects of these changes are currently a matter of conjecture, they could result in nutrient depletion and a decline in productivity of the forest.

LEAD IN VEGETATION

Plants are often excellent indicators of certain environmental contaminants due to their rapid phytotoxic response to sulfur dioxide, ozone, herbicides, etc. Their response to other contaminants, such as toxic heavy metals, may not be readily visible. However, analyses of their leaves, roots, or other tissues may serve as indicators of their geochemical environment. Plants, for instance, require 16 elements for normal growth, 13 of which come from the soil. Plant tissue analyses may show the presence of many times this number of elements.

Vegetation forms an important node and link in the transport and distribution system of lead entering the environment from mining, milling, and smelting processes. The residence time may be quite short or relatively long. Lead deposited on the leaves of trees moves to the soil system with annual leaf fall or may be washed to the ground by rainfall. Leaves may be eaten by herbivores, thus introducing attendant lead into natural food chains. When deposited on forage crops, lead may be ingested by domesticated animals. Deposits on leafy vegetables may be directly consumed by humans. Plant roots also absorb lead from the soil solution, and it may be transported and stored in other plant parts. Lead incorporated in the woody tissues of a tree might remain for many years. In contrast, that in the edible portion of a radish would be transferred to another node in a matter of weeks.

Lead accumulations on or in plants growing near mines, mills, and smelters have caused problems in many countries. [23-25] In Missouri, lead poisoning was diagnosed as the cause of death of several horses pastured in the vicinity of a lead smelter at Glover. There is, therefore, an obvious need to monitor vegetation for the possible accumulations of lead and other heavy metals, especially in the vicinity of industrial operations.

The distribution of lead in vegetation of the New Lead Belt has been investigated in three general areas:

(1) Lead deposited on the leaves of natural forest vegetation near mines, mills, and smelters.
(2) Lead deposited on vegetation along main haul roads and transport routes.
(3) Lead accumulated by vegetable crops grown in areas where potential contamination is high.

Lead Accumulated by Vegetation Near Mines, Mills, and Smelters

Industrial operations in the New Lead Belt are surrounded by forests. The accumulation of lead by some of the important overstory and understory species was investigated during the summers of 1971-1973. These included white oak (*Quercus alba* L.), post oak (*Quercus stellata* Wangenh.), short-leaf pine (*Pinus enchinata* Mill.), and dryland blueberry (*Vaccinim pallidum* Ait.).

Post oak and pine foliage were sampled intensively in the vicinity of the AMAX smelter, while oak and blueberry leaves were collected systematically over a 12 mi by 25 mi area that included most of the Viburnum Trend in the New Lead Belt. Soil samples were also taken at the same location as the white oak and blueberry foliage. All samples were analyzed without prior washing for lead, copper, cadmium, zinc, and manganese by atomic absorption spectroscopy. Thus unless otherwise stated, the results represent internal as well as surface accumulation.

Unusually high levels of lead were found in and on leaves of post oak and short-leaf pine near (0.5 mi or less) the AMAK smelter. (table 7-13) The maximum values were 8,125 and 11,750 μg/g for post oak and pine, respectively. Elevated levels of lead were detected at distances greater than 2 mi.

Lead accumulated by white oak and blueberry foliage from the 12 mi by 25 mi sampling area is shown in figures 7-18 and 7-19 and tables 7-14 and 7-

TABLE 7-13.—*Concentration of lead in leaves of post oak* (Quercus stellata) *and needles of short-leaf pine* (Pinus echinata) *in vicinity of lead smelter operations*

	Post oak			Short-leaf pine		
	Lead concentration (μg/g dry weight)			Lead concentration (μg/g dry weight)		
Distance from smelter (mi)	Range of values	Mean	Standard deviation	Range of values	Mean	Standard deviation
0-0.5	230-8,125	3,776.70	4,032.2	420-11,750	3,546.36	3,322.83
0.5-1.0	71-3,800	771.40	777.36	101-1,475	497.37	366.58
1.0-1.5	50-1,580	250.00	273.2	52-1,050	273.56	216.27
1.5-2.0	45-640	192.80	145.8	62-412	142.85	178.43
More than 2.0	18-1,360	168.97	222.8	22-661	123.29	125.43

TABLE 7-14.—*Concentration of lead in leaves of white oak* (Quercus alba) *and leaves of blueberry* (Vaccinium pallidum) *in vicinity of lead smelter operations*

	White Oak			Blueberry		
	Lead concentration (μg/g dry weight)			Lead concentration (μg/g dry weight)		
Distance from smelter (mi)	Range of values	Mean	*n*[1]	Range of values	Mean	*n*
1-2	75.8-1,221.0	574.3	3	141.0-874.0	495.3	3
2-3	122.0- 557.0	299.0	7	93.0-338.0	203.9	10
3-4	39.8- 155.0	80.2	4	21.0-146.0	76.1	5
4-5	28.2- 276.0	93.0	10	34.2-181.0	68.6	11
5-6	23.5- 228.0	75.6	12	24.0-155.0	64.3	10
6-7	26.0- 121.0	54.5	9	29.0-101.0	41.6	11

[1] *n* = number of samples.

15. In the northern half of the study area, the smelter was found to be the primary source of contamination, although the influence of mines and mills is evident. (figs. 7-18 and 7-19) Additional vegetation and soil contamination undoubtedly resulted from ore haulage by truck and railroad.

The Fletcher Mine and mill complex was perhaps the primary source of lead contamination in the southern half of the study area.

Leaded motor fuel and truck transportation of lead ore were also contributing factors. The accumulation of lead by vegetation was not as elevated in the vicinity of mines and mills as around the smelter. The distances at which anomalous levels of lead occur were also less than in the vicinity of a smelter.

The amount of lead accumulated by white oak leaves was slightly higher than that accumulated by blueberry foliage. This slight difference is perhaps due to the fact that the blueberry is an understory species subject to less aerial deposition than the overstory oaks.

Elevated levels of Cd, Cu, and Zn (figs. 7-20 and 7-21) were associated with leaves in the vicinity of the industrial operations and were correlated with lead levels. Manganese levels were high (over 3,000 μg/g in many instances), but were not correlated with the contamination by Pb and other heavy metals. Manganese is readily available for uptake by plants in the acid soils of the area. Cadmium levels ranged from <0.50 to 4.0 μg/g for the 516 white oak and blueberry leaf samples. High levels of Cd were found in the samples of leaves containing high levels of Pb. In these high lead samples, Cd ranged from 0.25 to 2.4 μg/g, with only a few samples greater than 1.0 μg/g.

Lead Contamination Along Roadways

Numerous studies have shown that soil and vegetation near highways contain elevated levels of lead

TABLE 7-15.—*Concentration of lead in leaves of white oak* (Quercus alba) *and leaves of blueberry* (Vaccinium pallidum) *in vicinity of Fletcher mine and mill*

	White Oak			Blueberry		
	Lead concentration (μg/g dry weight)			Lead concentration (μg/g dry weight)		
Distance from road (mi)	Range of values	Mean	*n*[1]	Range of values	Mean	*n*
0-1	39.0	39.0	1	39.0-132.0	85.5	2
1-2	16.5- 70.0	35.9	6	16.5- 40.0	25.8	5
2-3	20.0- 26.5	22.6	6	16.5- 27.5	20.6	5
3-4	10.8- 57.7	20.7	10	11.5- 42.5	20.6	12
4-5	12.5-124.0	29.5	15	5.8- 73.9	27.1	17
5-6	12.5- 36.5	22.6	11	5.9- 24.0	17.7	11

[1] *n*=number of samples.

that decrease with increasing distance from the highway. [26-32] Some of the elevated levels have been assumed to result from the use of leaded motor fuels.

A new source for roadside lead deposits has been added along the highway or railroad system in the New Lead Belt, where large quantities of lead, zinc, and sometimes copper concentrate are hauled by rail or truck to the smelters. Although some of the lead concentrate is smelted near the mines and mills in northwest Iron County, a major portion is either shipped by rail to a smelter near St. Louis or hauled by truck and smeltered in southeastern Iron County. The haulage trucks may or may not be covered, and the combination of open trucks, rough roads, and wind results in considerable blowing and deposition of the ore concentrate on roadside soils and vegetation concern about the hazard from open-truck hauling was heightened when unusually high concentration of lead, copper, zinc, and cadmium were report-

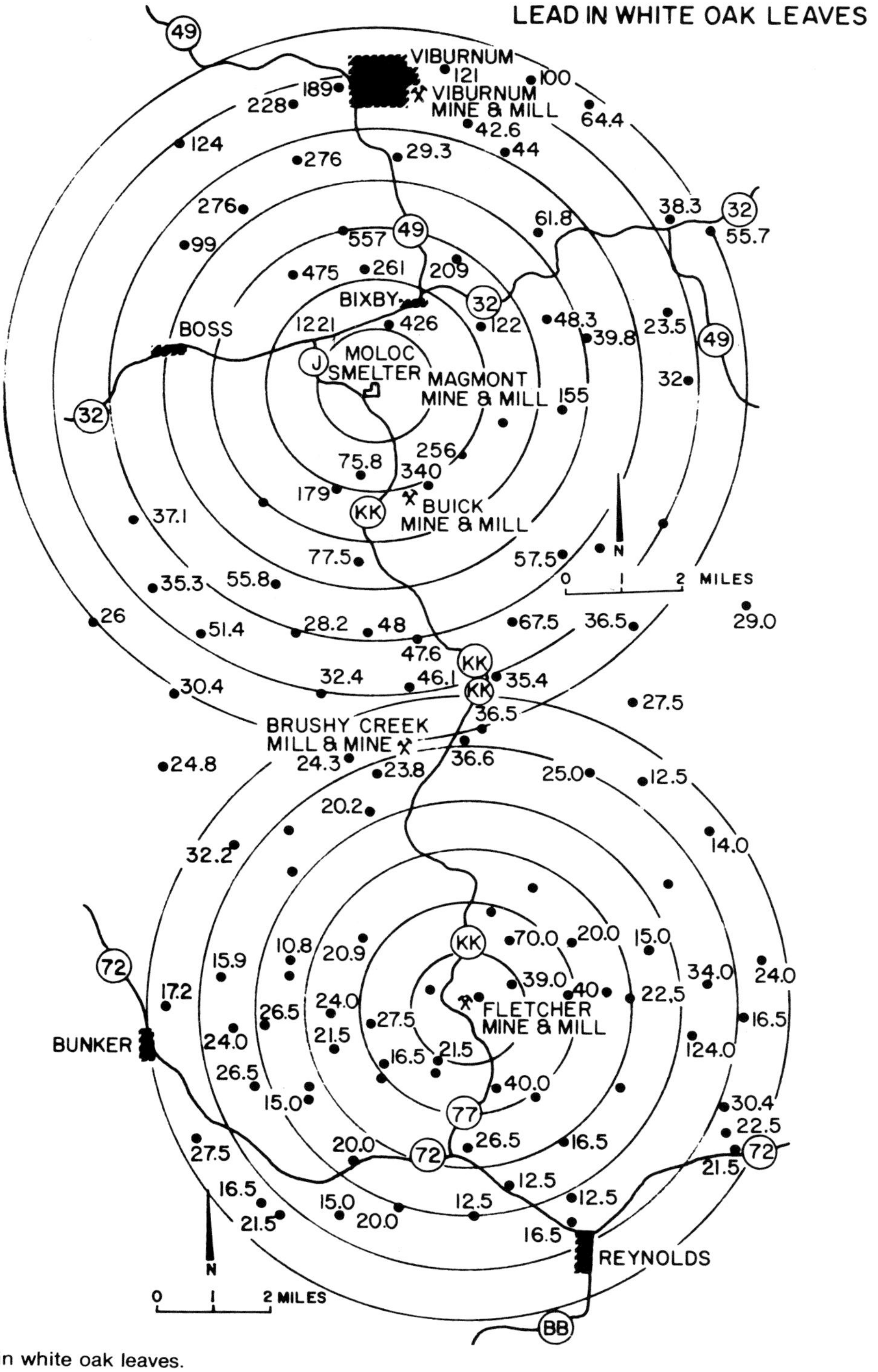

FIGURE 7-18.—Lead in white oak leaves.

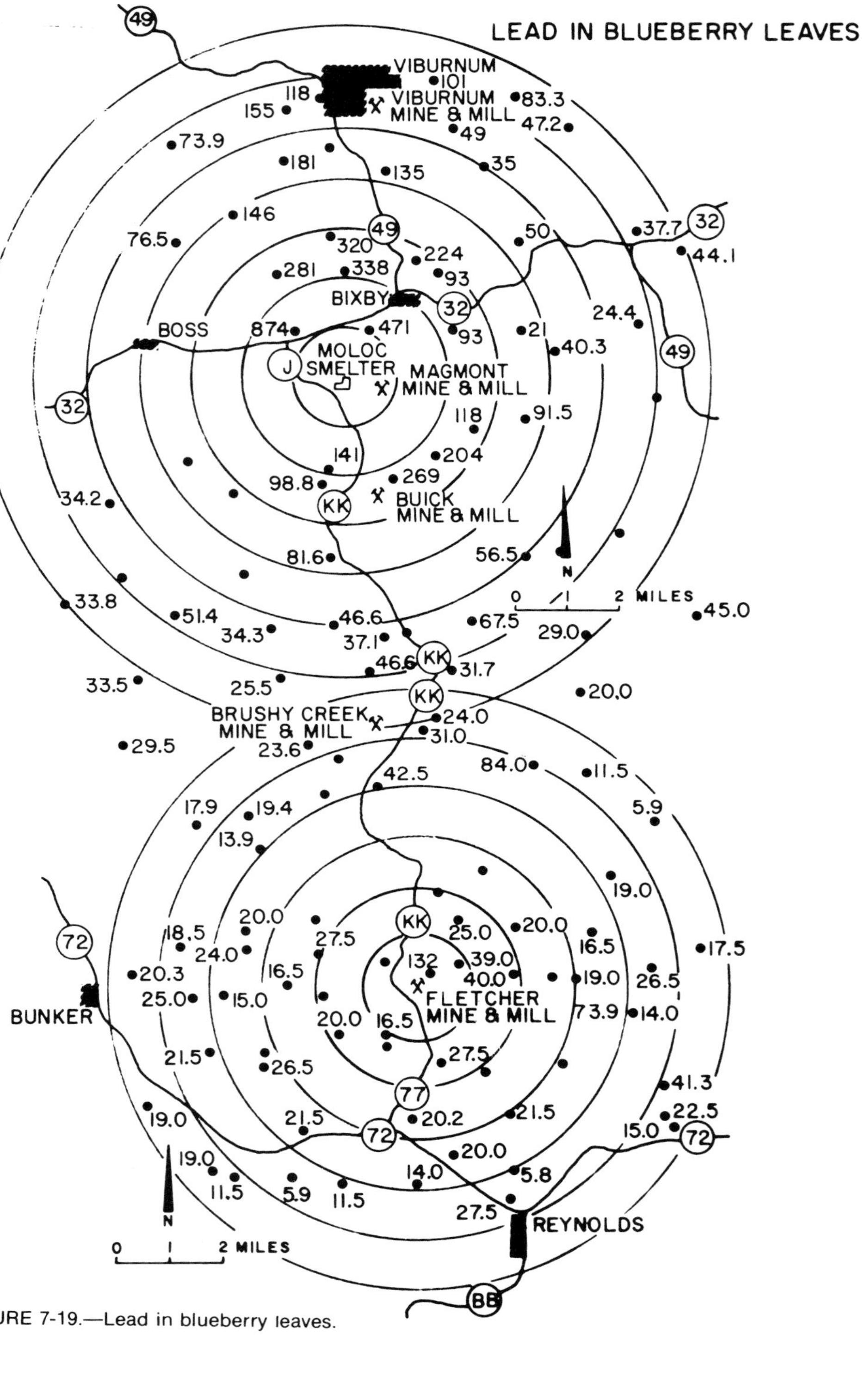

FIGURE 7-19.—Lead in blueberry leaves.

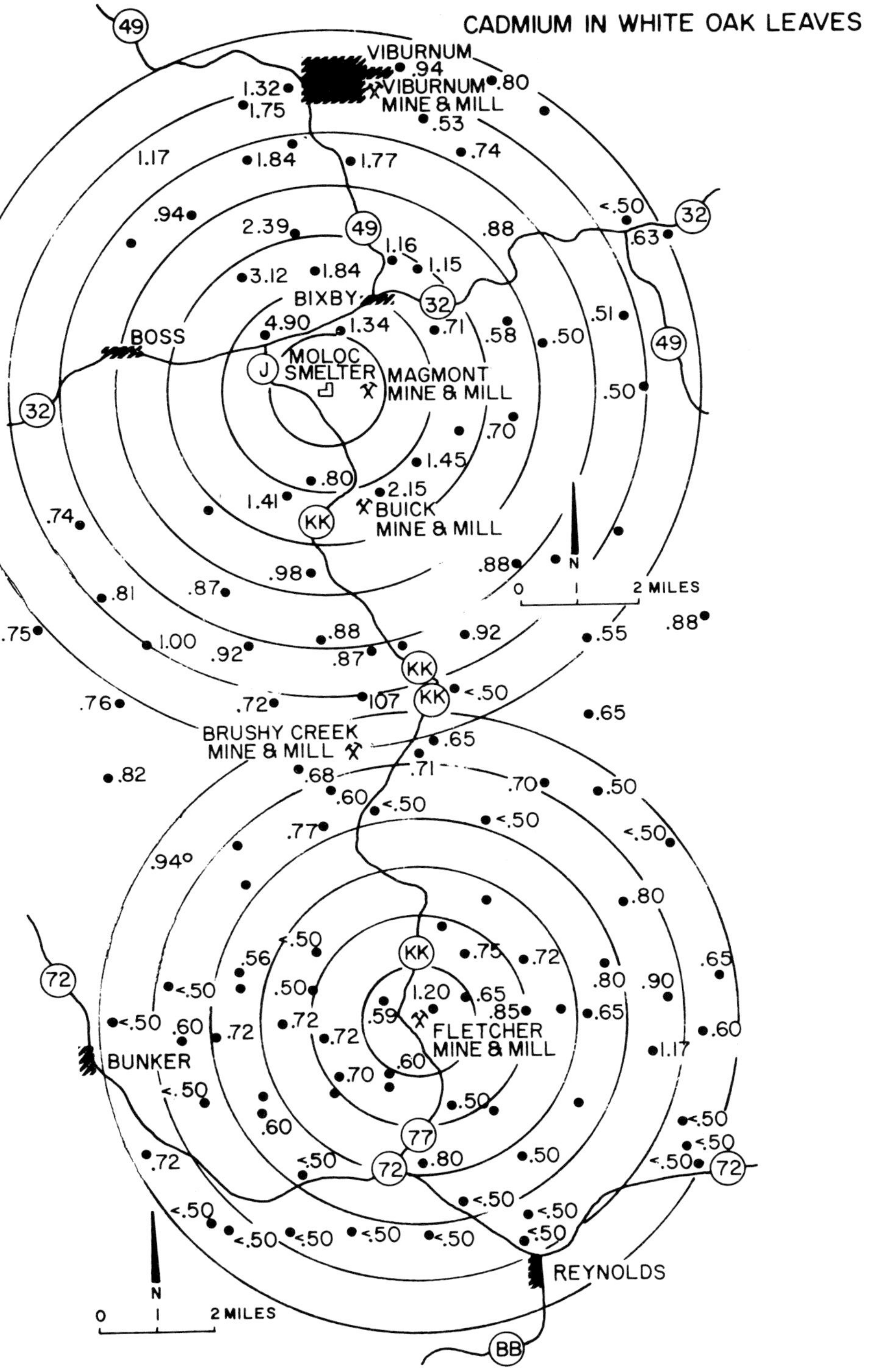

FIGURE 7-20.—Cadmium in white oak leaves.

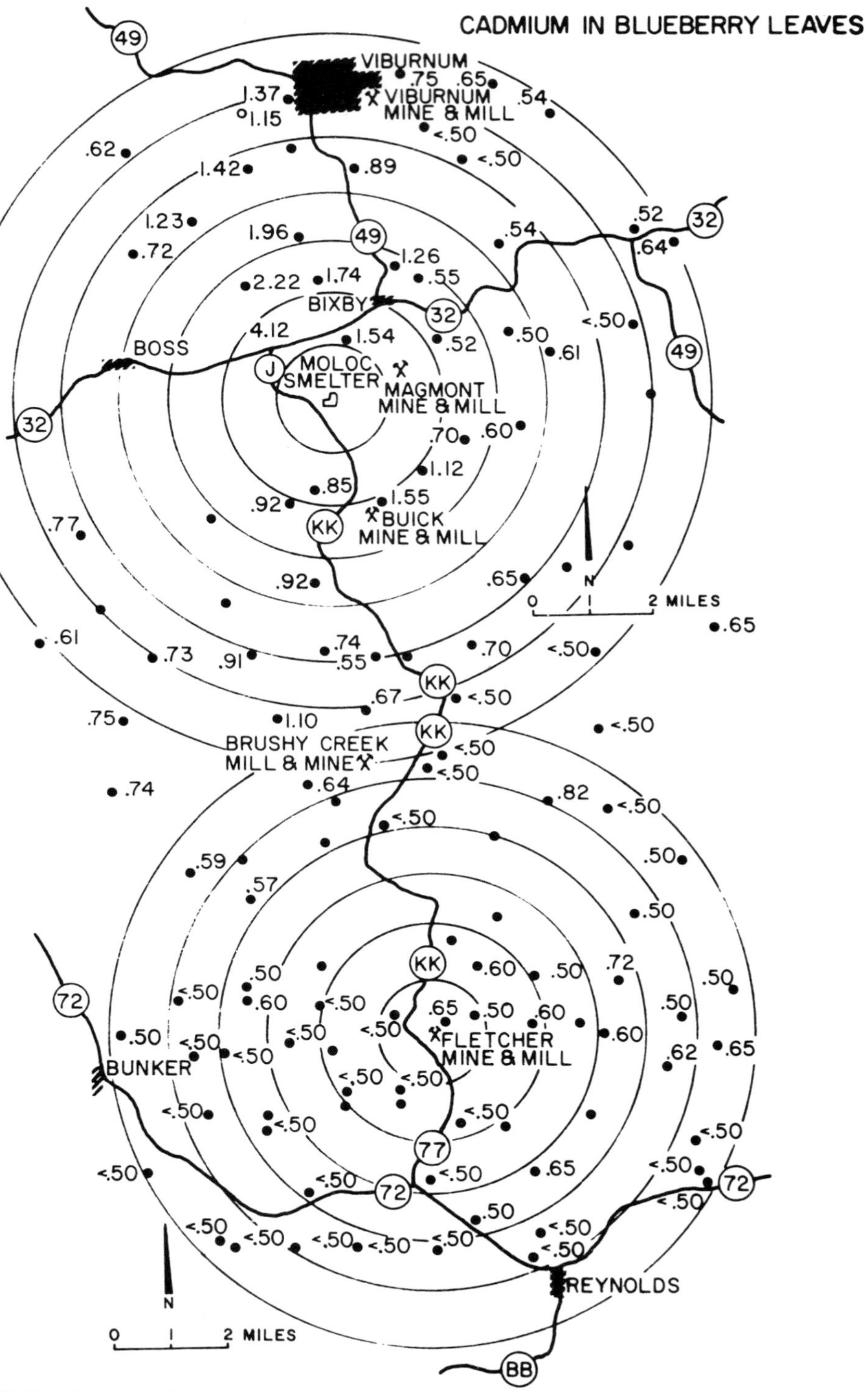

FIGURE 7-21.—Cadmium in blueberry leaves.

ed in red cedar (*Juniperus virginiana,* L.) foliage along haulage routes near Centerville. [33] These were thought to reflect vehicular transport of lead-bearing ores, rather than mineralized rock. An intensive sampling of both soil and vegetation was initiated to determine the extent to which areas adjacent to the highways were being contaminated by the transport of lead concentrate from mines and mills to the smelter.

The main haulage routes are shown in figure 7-22. Soil and vegetation samples were taken at approximately 3 mi intervals beginning at Glover. Control samples were collected southward (samples 73 to 80) of the southern-most mine of the Viburnum Trend in Reynolds County, westward into Dent County (samples 127 to 134) from the smelter and mine near Bixby (northwest Iron County), and north of the main truck routes (samples 135 to 142) into Washington County. The area traversed by the trucks varies from relatively open cultivated or pasture lands to dense oak-hickory-pine forests.

In open areas, samples of soil and vegetation were collected at the roadside and at distances of 100, 200, 300, and 400 yd from the roadway. In densely wooded areas, sampling was limited to the road right-of-way and, where possible, at least 100 yd into the forest. Since no one vegetation species was present at all sites, some were collected during July

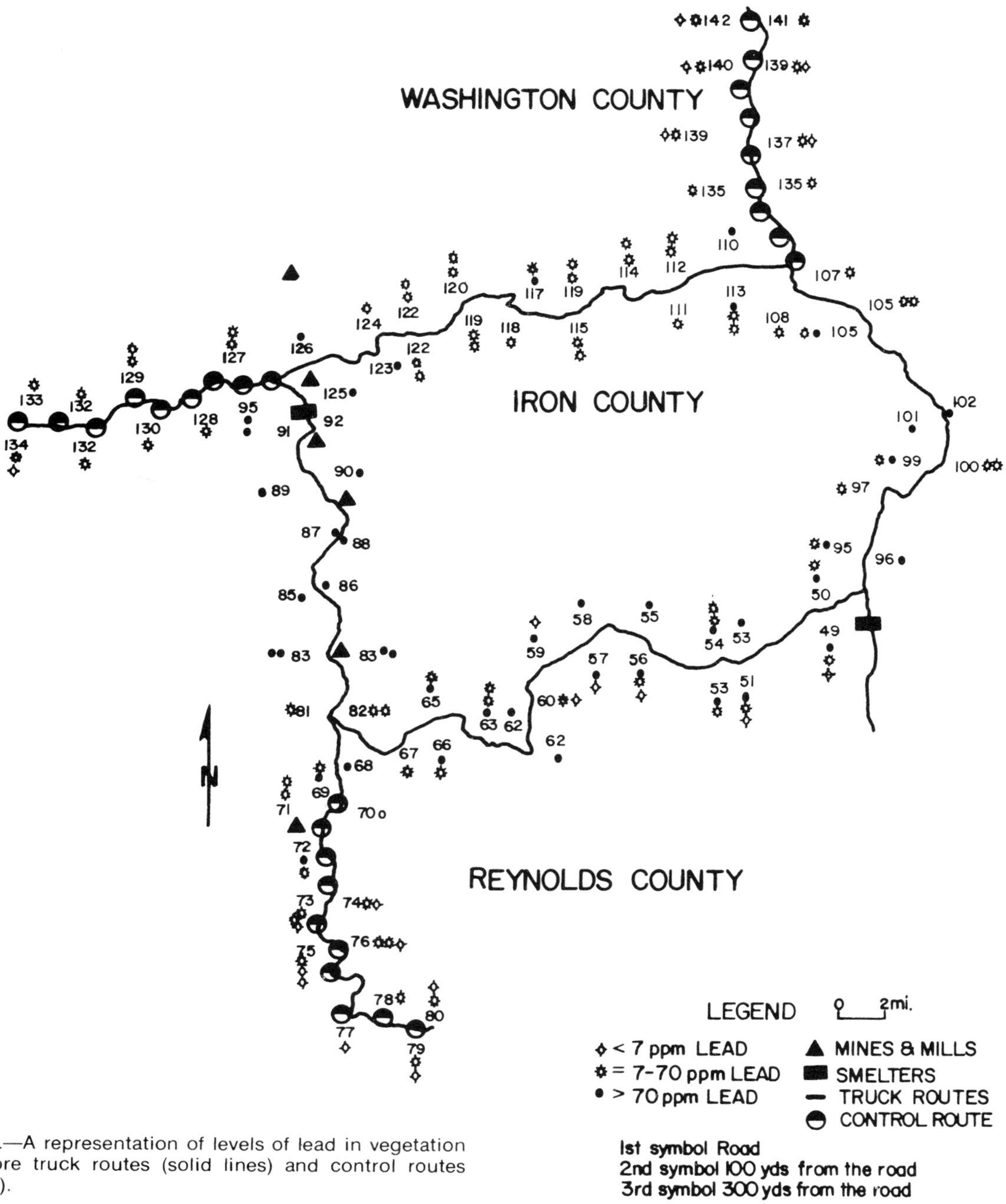

FIGURE 7-22.—A representation of levels of lead in vegetation along lead ore truck routes (solid lines) and control routes (broken lines).

and August of 1970: fall fescue *(Festuca arundinacea schreb)*, a domesticated forage species that is also extensively seeded on highway rights-of-way; purple top *(Tridens flavus L. Hitch)*, a native grass species; and dryland blueberry *(Vaccinium pallidum)*. Purple top was the most common species. Additional samples were collected during the summer and early fall of 1971 along a series of transects at distances of 25 to 1,000 ft from the main roads. Lead ore has been transported over most, but not all, of these routes.

Grass plants were severed approximately 1 in. above the soil line, and all visible soil particles were brushed from the base of the plant prior to analysis. Mature leaves of blueberries were taken from the center one-third of the current season's growth, while oak leaves were collected from branches approximately 6 ft above the ground. The vegetation samples were not washed prior to analyses.

Soil samples were collected to a 6 in. depth with a soil auger. Soil samples were extracted with 0.2 percent acetic acid for 2 h to determine available lead. Total lead values were determined by digestion with concentrated nitric acid.

Analyses of vegetation samples from the truck and control routes are given in table 7-16 and 7-17. Lead concentrations associated with plants along the haul roads varied markedly among individual sampling sites. The highest levels were in samples collected near smelters, mines, and mills and along rough areas in the highway surface that caused the truck transports to bounce. Lead content decreased rapidly as the distance from the roadway increased.

TABLE 7-16.—*Concentration of lead in vegetation along ore truck routes*

Distance from road right-of-way (yards)[1]	Lead concentration (μg/g dry weight)			
	Low	Mean	High	n[1]
	All Species			
0	1.0	279.7	4,350.0	80
100	1.9	34.2	537.0	58
200	2.4	11.6	35.0	16
300	4.8	8.5	15.0	8
400	4.0	6.5	9.1	6
	Blueberry			
0	7.0	98.0	537.0	21
100	10.4	45.0	172.0	22
	Fescue			
0	20.0	341.0	915.0	3
100	4.9	9.6	14.3	8
200	7.3	10.6	19.0	5
300	6.5	10.0	15.0	3
400	6.9	8.0	9.1	2
	Purple top			
0	1.0	344.0	4,350.0	56
100	1.9	32.0	537.0	28
200	2.4	12.0	35.0	11
300	4.8	7.6	10.1	5
400	4.0	5.8	8.3	4

[1] n=number of samples.

TABLE 7-17.—*Concentration of lead in vegetation along control routes*

Distance from road right-of-way (yards)	Lead concentration (μg/g dry weight)			
	Low	Mean	High	n[1]
	All species			
0	2.4	18.1	49.0	27
100	0.15	8.1	20.0	15
200	4.0	7.2	12.0	9
300	2.5	3.9	5.8	5
400	2.3	3.9	5.8	5
	Blueberry			
0	10.0	17.3	49.0	24
100	0.15	10.7	18.0	11
	Purple top			
0	2.4	18.2	49.0	24
100	4.0	7.4	18.0	11
200	4.0	7.2	12.0	9
300	2.5	3.9	5.8	5
400	2.3	3.9	5.8	5

[1] n=number of samples.

Lead levels were several times higher along ore transport routes than on control roads.

Elevated levels near control highways probably reflected contamination from leaded gasoline. Also, the Missouri Highway Department occasionally spread slag from the smelters on highway shoulders, but this practice has been discontinued. Levels in plants approached background levels beyond 200 yd from the roadway.

Lead added to soil in a soluble form soon changes to unavailable or fixed forms. It was desirable to determine whether there was a better correlation between the plant levels and soluble lead, as idicated by extraction with 0.2 percent acetic acid for 2 h, or between plant levels and total lead determined by digestion of the soil with concentrated nitric acid. The concentration of soluble or theoretically available lead in soils along the ore-truck routes is shown in table 7-18 and along control routes in table 7-19. Marked differences were evident between the truck and control routes near the road, and a several-fold difference at greater distances from the road.

The total lead contents of the soil are shown in table 7-20 for ore-truck routes and for control routes in table 7-21. A 10-fold difference was evident in

TABLE 7-18.—*Concentration of available lead in soils along ore truck routes*[1]

Distance from road right-of-way (yards)	Lead concentration (μg/g dry weight)			
	Low	Mean	High	n[2]
0	0.09	17.54	135.0	56
100	0.01	0.45	4.5	46
200	0.03	0.75	6.6	17
300	0.02	0.56	2.9	9
400	0.04	0.85	3.8	6

[1] Soil samples collected to 6 in. depth.
[2] n=number of samples.

TABLE 7-19.—*Concentration of soluble available lead in soils along control routes*[1]

Distance from road right-of-way (yards)	Lead concentration (μg/g dry weight)			
	Low	Mean	High	n[2]
0	0.08	0.58	3.80	24
100	0.19	0.46	1.40	16
200	0.04	0.27	0.50	9
300	0.15	0.25	0.40	5
400	0.15	0.28	0.50	5

[1] Soils collected to 6 in. depth.
[2] n=number of samples.

samples collected on the road right-of-way. Later studies by Bolter et al. [34] have shown that lead concentrations are greatest in the organic layers or the top few millimeters of mineral horizons in uncultivated soils. Much higher values were attained once these fractions were analyzed.

The accumulation of lead by all plant species is shown in figure 7-21. Vegetation containing more than 70 μg/g lead (dry weight) denoted by the bold black circles was located only on road right-of-ways along truck routes. Near mines, mills or smelters,

TABLE 7-20.—*Total lead in soils along truck routes*[1]

Distance from road right-of-way (yards)	Lead concentration (μg/g dry weight)			
	Low	Mean	High	n[2]
0	16.7	809.6	3,792.0	56
100	12.7	32.5	80.0	28
200	20.0	36.0	48.0	11
300	33.0	52.0	55.0	5
400	36.0	55.0	176.0	4

[1] Soil samples collected to 6 in. depth.
[2] n=number of samples.

levels were more elevated at 100 yd or more from the road-right-of-ways.

Concentrations of lead in white oak and blueberry leaves along ore-haulage and control roads are shown in tables 7-22 and 7-23. There is good agreement between values obtained for the two species. Lead levels decreased with increasing distance from the road, but were still elevated at 1,000 ft from ore-haulage routes. Some lead may also blow onto the roadside vegetation from smelters, mines, and mills. Thus the absolute amount of contamination due to ore haulage cannot be determined.

TABLE 7-21.—*Total lead in soils along control routes*[1]

Distance from road right-of-way (yards)	Lead concentration (μg/g dry weight)			
	Low	Mean	High	n[2]
0	10.7	75.6	384.0	24
100	9.3	23.2	49.0	16
200	8.3	17.3	27.0	9
300	13.3	17.3	20.0	5
400	13.3	22.0	33.0	5

[1] Soil samples collected to 6 in. depth.
[2] n=number of samples.

The importance of ore transport as a source of contamination is illustrated by data from the pasture where problems with horses were encountered. Samples of forage grasses, collected in July approximately ½ mi north of the smelter, had a lead content of 115 μg/g dry weight nearest the smelter, 344 μg/g at 100 yd from the southern edge, and 1,140 μg/g at the northern edge, which was immediately adjacent to the highway used to transport lead ore to the smelter.

Lead Accumulation by Vegetable Crops

Although lead mining, milling, and smelting operations have been carried on for over 100 years in seven southeastern Missouri counties, only recently has concern arisen over possible contamination of fruits and vegetables grown in the area. Elevated lead levels found in miscellaneous fruit and vegetable samples in June and July 1970 emphasized the need for a more thorough investigation of toxic metals in crops used for human and animal consumption. In response to this need, a study was initiated to determine the accumulation of lead in leaf lettuce, radishes, and green beans grown near sources of contamination. [35] These three crops were chosen because they are commonly grown, and each represents a different edible plant part. Control samples were collected in three north central Missouri counties, well removed from any known lead deposits or associated industrial operations, and away from heavily traveled roads. All samples were obtained from home gardens.

The vegetables were harvested at the edible stage, and samples were immediately washed in distilled water. For analytical purposes, lettuce was divided into roots and tops, radishes into edible roots and tops, and greenbeans into pods and roots. The results of the analyses are shown in tables 7-24, 7-25, 7-26, and 7-27, respectively, for control counties, Old Lead Belt counties, New Lead Belt counties, and a small town located within 2 mi of a smelter that has operated since the 1890's.

Mean lead levels were generally much greater for plants grown in gardens subject to contamination than in the controls. Considering only the edible plant parts, accumulations were greater for lettuce leaves than for radish roots or greenbean pods. Variations in lead values for individual samples were great and were related to the degree of exposure.

In the Old Lead Belt, samples of lettuce and radishes with very high lead contents were grown near old mines or on soils where dolomitic limestone mine waste had been applied to gardens. In the New Lead Belt, lead values depended on the proximity of gardens to industrial operations or haul roads with the highest levels near smelters.

Exposure to smelter emissions in the small-town gardens produced the highest lead levels recorded for lettuce and greenbeans. The values for lettuce ran as high as 1,324 μg/g, about two times greater than elsewhere. Lead values in greenbean pods ranged up to five times higher than did values in the

TABLE 7-22.–*Concentration of lead in leaves of white oak (Quercus alba) and leaves of blueberry (Vaccinium pallidum) from traverse samples along roads in Missouri's New Lead Belt*

Distance from road (ft)	white oak: Lead concentration (μg/g dry weight)			blueberry: Lead concentration (μg/g dry weight)		
	Range of values	Mean	n [1]	Range of values	Mean	n [1]
0- 50	69.0-604.0	209.0	15	47.0-524.0	200.7	16
51- 100	32.2-495.0	149.0	21	21.1-419.0	176.8	23
101- 200	21.7-636.0	115.3	30	14.6-322.0	103.0	28
201- 400	12.0-286.0	81.6	26	11.9-351.0	92.9	27
401- 600	13.0-293.0	78.9	23	17.5-259.0	79.1	20
601- 800	11.0-113.0	42.3	11	5.0-137.0	46.3	12
801-1,000	9.8- 66.5	53.4	11	5.0-111.0	34.9	11

[1] n=number of samples.

TABLE 7-23.–*Concentration of lead in leaves of white oak (Quercus alba) and leaves of blueberry (Vaccinium pallidum) from traverse samples along roads without lead ore haulage in Missouri's New Lead Belt*

Distance from road (ft)	white oak: Lead concentration (μg/g dry weight)			blueberry: Lead concentration (μg/ g dry weight)		
	Range of values	Mean	n [1]	Range of values	Mean	n [1]
0- 50	52.4-104.0	81.0	3	47.0-100.0	69.2	3
51- 100	34.5- 45.8	41.3	3	21.1- 44.7	29.2	6
101- 200	21.7- 26.6	24.5	4	19.5- 43.7	32.4	4
201- 400	12.0- 19.8	16.3	4	11.9- 30.0	20.2	4
401- 600	17.5	17.5	1	19.1	19.1	1
601- 800	11.6	11.6	1	12.1	12.1	1
801-1,000	9.8- 14.9	12.0	4	12.8- 22.2	16.7	4

[1] n=number of samples.

TABLE 7-24.–*Lead content of vegetable crops in control counties*

Crop	Lead concentration (μg/g dry weight)[1]: Low	High	Mean	n [2]
Lettuce-root	10.0	50.0	20.3	12
Lettuce-leaf	6.9	33.9	20.6	12
Radish-root	5.0	11.0	7.7	12
Radish-top	5.0	32.0	14.0	12
Greenbean-root	5.0	20.0	9.1	12
Greenbean-pod	<5.0	<5.0	<5.0	12

[1] All samples washed in distilled water immediately after harvest.
[2] n=number of samples.

TABLE 7-25.–*Lead content of vegetable crops in Old Lead Belt counties*

Crop	Lead concentration (μg/g dry weight)[1]: Low	High	Mean	n [2]
Lettuce-root	11.7	492.0	68.8	30
Lettuce-leaf	10.3	742.0	83.8	30
Radish-root	5.0	518.0	33.4	30
Radish-top	5.0	117.0	76.7	30
Greenbean-root	5.0	67.0	8.6	30
Greenbean-pod	5.0	10.1	5.4	30

[1] All samples washed in distilled water immediately after harvest.
[2] n=number of samples.

TABLE 7-26.–*Lead content of vegetable crops in New Lead Belt counties*

Crop	Lead concentration (μg/g dry weight)[1]: Low	High	Mean	n [2]
Lettuce-root	10.0	660.0	89.7	28
Lettuce-leaf	8.9	923.0	114.0	28
Radish-root	4.4	107.0	22.3	28
Radish-top	13.3	578.0	94.4	28
Greenbean-root	5.0	53.0	9.9	28
Greenbean-pod	5.0	40.0	8.8	28

[1] All samples washed in distilled water immediately after harvest.
[2] n=number of samples.

TABLE 7-27.–*Lead content of vegetable crops in a small town with a lead smelter*

Crop	Lead concentration (μg/g dry weight)[1]: Low	High	Mean	n [2]
Lettuce-root	25.0	3,244.0	636.2	8
Lettuce-leaf	47.0	1,324.0	284.0	9
Radish-root	7.4	50.0	22.3	8
Radish-top	32.0	385.0	119.5	8
Greenbean-root	10.0	704.0	119.3	8
Greenbean-pod	5.0	136.0	25.1	8

[1] All samples washed in distilled water immediately after harvest.
[2] n=number of samples.

Old Lead Belt and three times higher than those in the New Lead Belt. [36]

Lettuce roots had higher mean lead concentrations than either radishes or greenbeans. The highest values were in the smelter town, where concentrations exceeded 3,000 $\mu g/g$ and averaged 636 $\mu g/g$.

The effects of these high lead levels on human health have not been determined. However, the vegetables analyzed do not constitute a major part of the diet for people in the area. Thus the possibility of chronic effects is somewhat lessened.

LEAD IN SURFACE WATERS

Research studies were performed to determine the importance of surface waters and aquatic systems in the New Lead Belt area as a possible sink, a transport mechanism, a habitat for fish and other aquatic organisms and a recreational resource. [37]

The topography and resulting drainage pattern of the New Lead Belt are uniquely adapted to monitoring and tracing the effect of industrial pollutants. First, the mining, milling, and smelting effluents from each operation are channeled into separate stream systems, so that any pollution effects can be individually determined. Equally important are the many streams without mining development that serve as controls. Five of the streams receiving mine discharge eventually flow into the Black River basin and then into Clearwater Lake (approximately 70 mi away). This is the only lake in the system and potentially a major long-term sink for heavy metals entering the system. A map of the basin, emphasizing the overall drainage pattern, is shown in figure 7-23.

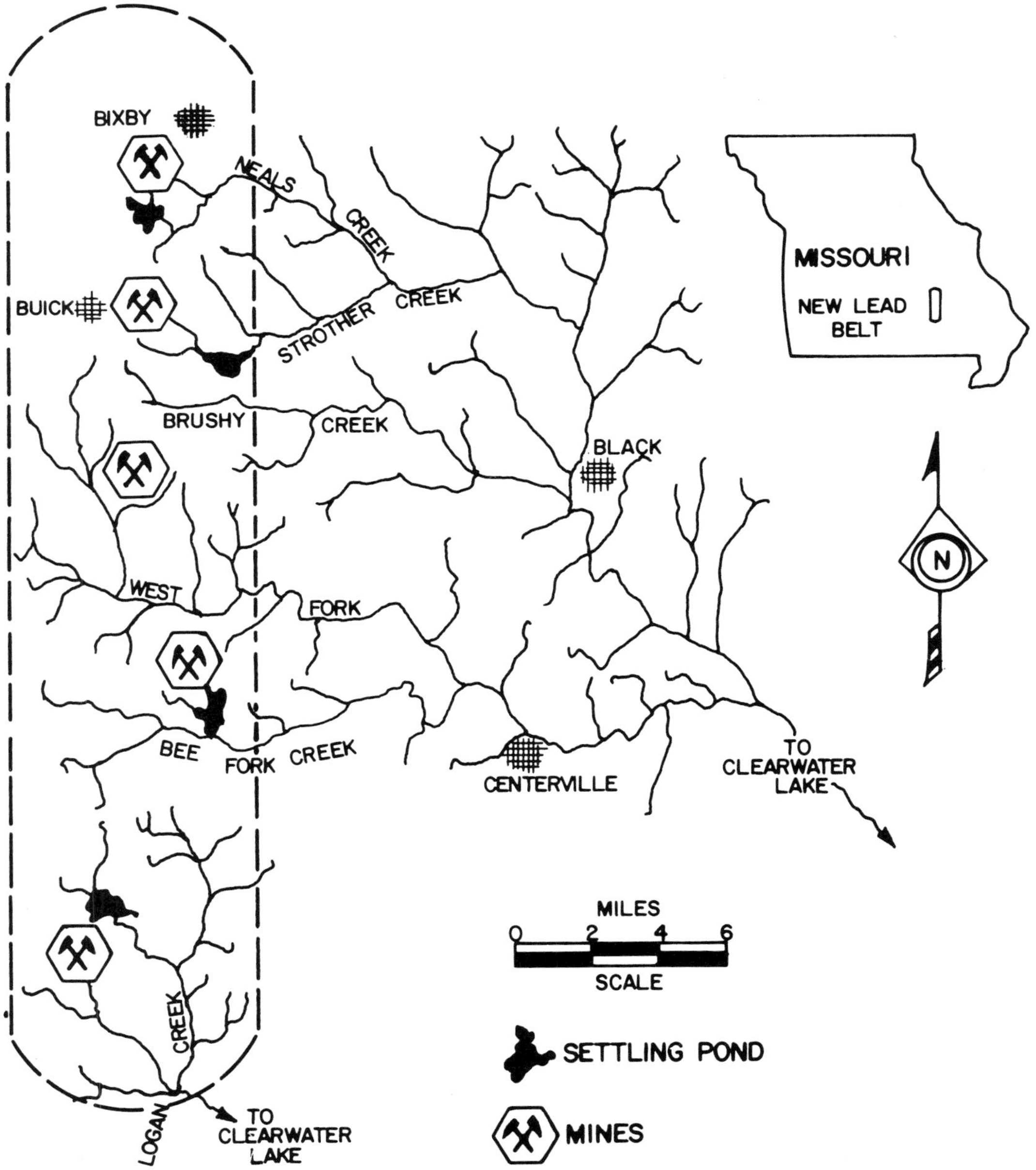

FIGURE 7-23.—Drainage pattern of the New Lead Belt of Missouri to Clearwater Lake.

Sources of Pollutants

The primary sources of heavy metals are mining and milling operations, concentrate storage, haulage by open-truck or railroad gondola car, and the smelting operations. Generally, however, wastewaters from mine and/or mill operations are the most important to the aquatic systems, although they do not normally enter the stream systems until they have been treated in at least a single tailings pond. The liquid effluent from these ponds is not chemically too different from the receiving streams, but it does contain relatively large amounts of suspended solids, which are generally associated with heavy metals. Effluent from the mining operations often contains approximately two to three times more dissolved solids than either the original mine or background stream water. [38]

Organic reagents are also released by the milling operations and may become major environmental problems. Many of these materials are known to be toxic in aquatic systems, but the levels at which toxicity occurs, particularly in the presence of heavy metals, are not known. [39] A summary of factors contributing to environmental changes in water quality are listed as follows:

A. Mine water
   1. Natural nutrient load of subterranean water
   2. Fuel spills
   3. Oil spills
   4. Hydraulic fluid spills
   5. Small mineral particles in mine effluent
   6. Blasting agents—spills and partially oxidized compounds are nutrients
   7. Highly variable mineral content of ore

B. Mill operation
   1. Chemical spills
   2. Variable mineral content of ore may cause:
      a. Excessive use of reagents and loss of toxic chemicals to effluent
      b. Low recovery of heavy metals during pulse of very rich ore
   3. Chemical reagents not adsorbed to concentrate are released in effluent
   4. Improperly placed piles of concentrate allow dispersal of heavy metals

C. Tailings ponds
   1. Improper design or placement of ponds, insufficient size or number of ponds
   2. Insufficient retention time
   3. Release of toxic milling reagents to streams
   4. Release of organic and inorganic nutrients to streams
   5. Release of finely ground rock and mineral particles to streams

Research Procedures

Heavy metal concentrations in the water (both dissolved and suspended states), stream sediments, and the aquatic biota were determined by atomic absorption methods as described in a previous chapter. Routine analyses were made by the Environmental Trace Substances Research Center at the University of Missouri. Some analyses, largely concerned with stream biota, were performed at Rolla using standard techniques coupled with occasional blind cross-checks between the two analytical laboratories.

Dissolved and suspended matter were separated by using a 0.45 μm Millipore filter; all water that passed the filter was assumed to contain essentially dissolved material. All physical and chemical parameters of water quality were determined in accordance with procedures outlined in standard methods of the American Public Health Association, [40] except for phosphorus determinations. Both total and orthophosphate were measured with procedures developed by Jankovic et al. [41] Diversity Index values, cited as the biological indicators of pollution, were measured by Ryck and Whitely [42] of the Missouri Department of Conservation by using Wilhm's expression: [43]

$$\text{Diversity Index} = \frac{\text{(No. of types of Organisms)-1}}{\text{Log}_e \text{ (total no. of organisms)}}$$

There are numerous shortcomings in the use of indices because most are distorted in some way and generally concentrate on one or more organisms. Wilhm's index is mathematically distorted to favor certain fly larvae that are carried to the species level. The index does not necessarily indicate a decline in the number of life forms, but rather a change in the basic character of the original taxonomic groupings. As such it is valuable only as a guide to indicate that changes have occurred. [44, 45]

The establishment of basic rainfall and runoff characteristics was in accordance with current hydrological practices. Two Leupold and Stevens servomanometer bubble gages were used to measure streamflow. Water samples were collected at 15 min. intervals during runoff periods. Both filtered and unfiltered aliquots were analyzed for lead, copper, zinc, and cadmium. Other water quality characteristics measured included pH, hardness, alkalinity, chemical oxygen demand, and many others.

Analytical procedures for algal analyses were detailed in a previous report. [37] Briefly, algae samples were dried at 100°C and ground. Approximately 1 g of the dried material was digested in a 100 ml Kjeldahl flask with 15 ml concentrated HNO_3 and 3 ml concentrated $HCLO_4$ until the white fumes of the perchloric acid were evident, followed by an additional 5 min. at the lowest setting of a Kjeldahl heating unit. The digested sample was picked up in 1 percent HNO_3, and the washings were added to the supernatant. The liquid samples were brought to 50 ml with 1 percent HNO_3 and analyzed for heavy metals by atomic absorption spectroscopy. Standards and blanks were prepared with 1 percent HNO_3, and the results were expressed in terms of micrograms of heavy metal per gram of dried algae.

Impact on Water Quality and Aquatic Biota

Water Quality: Most of the lead from mining and milling operations enters the streams as galena (PbS). In addition, small amounts of lead oxide (PbO), lead carbonate ($PbCO_3$), and lead sulfate ($PbSO_4$) may also be found. Little is known, however, about the chemical alterations and eventual associations of lead as it reacts with organic substances present in and around aquatic vegetation. [46] Before mining began, background concentrations of lead and copper in streams of the New Lead Belt were reported to be in a 4 to 6 μg/l range; water pH ranged normally between 7 and 8. [46]

Physical and Chemical Impact

A summary of the physical and chemical parameters of water quality at 15 selected principal sampling stations is presented in table 7-28. The most significant differences in observed values of chemical and physical characteristics of the control streams and those receiving industrial effluents are:

(1) The turbidity and suspended solids levels found in samples taken at or near industrial outfalls are far higher than those found in the receiving stream; as a general rule, the heavy metals are associated with these solids.

(2) Mine-mill effluents generally have a markedly higher level of total dissolved solids relative to the control and receiving streams.[38]

(3) Periodic elevated levels of phosphate and nitrogen are found in stream segments with occasional algal blooms or aquatic vegetation.

(4) Most of the available phosphorus and nitrogen are commonly tied up either by aquatic biota or dissolved organic compounds, many of which are probably metabolic products of the stream biota. Organic nitrogen (Kjeldahl analysis) consistently accounts for over 99 percent of the total nitrogen present.

(5) Under normal stream flow conditions, most streams of the New Lead Belt are remarkable for their high degree of clarity, purity, and aesthetic qualities.

During normal flow, both dissolved and suspended lead contents of the water were well below the 0.05 mg/l level set for drinking water by the U.S. Public Health Service and the 0.1 mg/l effluent standard set by the Missouri Clean Water Commission for this type of discharge. [44] Since streams in this region characteristically have a relatively high carbonate content and a slightly basic pH [47], any dissolved heavy metals are probably precipitated rapidly as insoluble carbonates. Since most of the particles are nearly colloidal in size, they are likely to be transported and deposited in a reservoir or any other aquatic body that provides long-term sediment conditions. It is significant that virtually all of the high measurements of heavy metals occurred during periods of storm runoff and are associated with the suspended solids fraction. This phenomenon has not been noted in most conventional monitoring pro-

TABLE 7-28.—*Summary of physical and chemical parameters of water quality*

Parameter	Stations														
	No. 1	No. 2	No. 3	No. 4	No. 5	No. 6	No. 7	No. 8	No. 9	No. 10	No. 11	No. 12	No. 13	No. 14	No. 15
pH	8.10	7.45	7.74	7.73	7.73	7.66	7.58	8.2	8.1	8.0	7.6	7.6	8.0	7.4	7.6
Temperature (°C)	0-25	0-25	0-25°C	0-25	0-25	0-25	0-25	0-25	0-25	0-25	0-25	0-25	0-25	0-25	0-25
Turbidity (JTU)	3.7	2.9	1.2	1.0	7.7	1.3	0.76	1.0	1.0	1.0	8.0	6.5	1.2	3.4	2.0
Dissolved oxygen (DO mg/l)	5.4	4.6	4.7	5.5	5.5	5.6	5.3	5.6	5.2	5.5	5.9	5.9	5.4	5.3	5.9
Alkalinity (mg/l)	180	200	198	152	150	150	150	144	156	135	152	144	145	30	80
Hardness (mg/l) calcium	150	140	210	100	95	100	100	75	80	75	110	110	110	15	50
Total	300	280	360	250	235	200	210	155	160	135	260	260	260	30	110
Chloride (mg/l)	0	0	30	20	0	0	0	0	0	0	0	0	30	0	0
Chemical oxygen demand (COD) (mgO_2 l)	30	20	50	40	75	95	35	15	20	30	31	20	50	40	20
Phosphorus (mg/l) ortho	0.1	0.1	0.1	0.1	0.1	0.1	0.1	0.1	0.1	0.1	0.1	0.1	0.1	0.1	0.1
Total	0.3	0.3	0.4	0.3	0.2	0.1	0.1	0.2	0.1	0.1	0.3	0.2	0.1	0.1	0.1
Nitrite (mg/l)	0.1	0.1	0.2	0.1	0.2	0.1	0.1	0.2	0.1	0.1	0.2	0.1	0.1	0.1	0.1
Nitrogen ammonia (mg/l)	0	0	0	0	0	0	0	0	0	0	0	0	0	0	0
Total organic	26.4	8.4	30.2	25.7	16.4	18.0	12.0	15.6	10.0	22.4	16.4	17.6	35.3	25.7	30.2
Specific conductance (μmho/cm)	360	200	260	450	100	160	340	300	380	280	750	750	740	120	340

grams since most state and federal agencies regularly use filtered samples for analyses of heavy metals.

One important finding is that lead has rarely been a problem in the effluents or in streams; however, zinc has frequently exceeded the 0.1 m/gl limits required by the state in both filtered and unfiltered samples. [37, 44-45] The highest concentrations of heavy metals have consistently occurred at stations 16 through 20, which are essentially effluent discharges from the smelter treatment facility. These are also the only stations where cadmium is regularly found, and heavy-metal concentrations in unfiltered samples at these sites consistently exceed the state's allowable limit. These effluents drain into a basin that has been studied under storm runoff conditions and will be discussed in another section.

Biological Impact: The effluent from mills and excess mining water continually pumped from the Bonneterre formation have exerted a definite impact on the productivity of small receiving streams. The amount of water that must be brought to the surface varies considerably from one mine to another, ranging from almost nothing to approximately 4,000 gal./min.

In those operations with an excess of underground water, algal blooms have occurred in some of the tailings ponds used for treatment and for several miles below the discharge point in receiving streams. These algal growths and associated rhizophytes (rooted vegetation) vary in species composition from one stream to another; in some cases, they completely cover stream bottoms. Excessive algal blooms, particularly in the warm months, have caused an unsightly appearance and altered the usual aquatic communities.

The predominant vegetation associated with streams that have algal growth problems were: *Cladophora, Oscillatoria, Mougeotia, Zygnema, Spirogyra cymbella, Achnanthes, Navicula,* and other stalked and nonstalked varities of diatoms. The typical benthic and periphytic communities display an unusually large array of diatoms, both in variety and density. These aquatic forms very efficiently accumulate nutrients in streams where their concentrations are quite low.

The dense, tangled, gelatinous communities of algae and associated microconsumers act as excellent filters, in addition to their capacity for nutrient removal and retention. The finely ground particles of rock flour or minerals that may escape the flotation and tailings reservoirs are efficiently trapped and removed by the benthic flora of the receiving streams, [48, 49] as illustrated by figure 7-24. The residence period for these accumulated materials is only temporary, since the mats are detached as they decompose or break loose during periods of storm runoff and are carried by moving water into other ecosystems.

Elevated manganese concentrations have been observed in many of the industrial effluents sampled. Patrick *et al* [50] and Gerloff and Skoog [51] have reported that manganese concentrations, similar to those observed on Bee Fork and other streams in

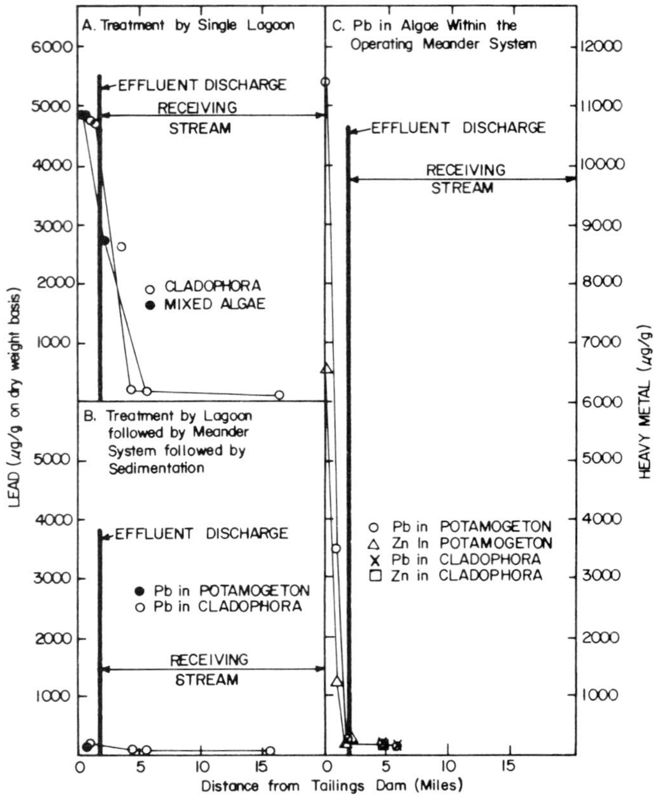

FIGURE 7-24.—Algal lead content—before, after, and within the algal meander treatment system.

the New Lead Belt, may interact with temperature to selectively favor the growth of some kinds of algae. In particular, high manganese concentrations and cold water temperatures encourage the development of diatom communities, while the opposite conditions lead to establishment of green or blue-green algae. Thus some of the differences between the predominant algal forms in the various receiving streams are partly due to differences in manganese concentrations.

A potential environmental problem has developed in the transport of heavy metals when excessive vegetational growths have occurred in the receiving streams. These dense gelatinous mats of algae and their associated aquatic populations have coated some stream beds, causing aesthetic problems, blocking photosynthetic energy input, and eliminating normal stream populations.

The quality of two typical streams as measured by diversity index values are shown in figure 7-25. [42, 52] Generally, neither a control nor a receiving stream had a diversity index of less than 4 prior to the initiation of mining operations. [42] In contrast, these index numbers decreased to levels below 4 in streams receiving wastewaters after mining was started. The initial drop, as shown in figure 7-23, occurred just before the mines started production and was probably related to premining construction activities. Since then, both the total number and type of benthic organisms (as described in Wilhm's Index) in the receiving streams have declined. With

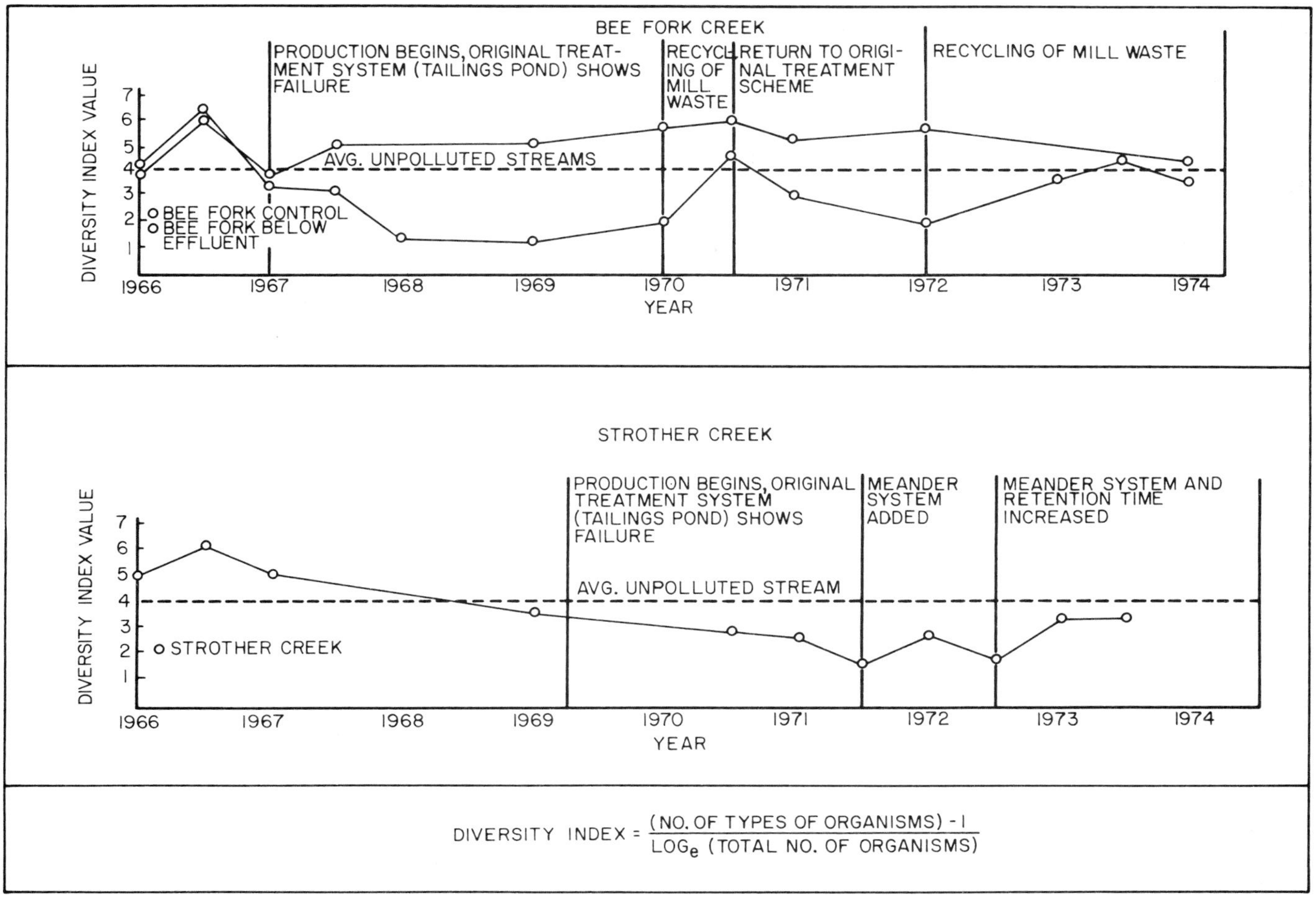

FIGURE 7-25.—Diversity index adapted from Ryck and Whitely [42] by Jennett and Wixson.

installation and operation of treatment facilities, stream water quality as measured by benthic diversity indices has improved to premining or background levels. It is important to recognize that such selected biological indices are not indicators of water pollution. They only indicate that the types and numbers of organisms used to compute the index have changed; as such, they can function as a sensitive tool for detecting possible environmental trouble areas.

Lead Transport by Runoff

The airborne transport of lead from tailings, concentrate piles, transport vehicles, and the smelter to the soil has been previously described. Research further suggested that lead is held rather tenaciously in the organic horizons and the uppermost portion of the mineral soil profile. Observations in the Old Lead Belt of Missouri indicated that lead levels in the 0 to 6 in. soil were nearly the same as those in the New Lead Belt.[37] This poses a question as to whether soil is the ultimate sink for lead or whether the isolated high values of lead in unfiltered water samples during periods of high storm runoff might be a significant transport phenomenon.

Soil column studies by Linneman [53] indicated that there was very little leaching of heavy metals at present levels of contamination, and that clay and organic soil fractions have a strong affinity for lead, effectively limiting its downward movement. Automated runoff water quality stations were also constructed on two watersheds to check this general transport hypothesis. The control watershed (approximately 5 mi^2 in area) was located 6 mi from the smelter and was assumed to be beyond inputs from industrial sources. The other watershed drained a 1½ mi^2 smelter area where the primary heavy metal input was deposited on the soil. Water samples were analyzed for lead, zinc, copper, cadmium, and 11 other physical and chemical parameters of water quality. [37, 54]

Figures 7-26 and 7-27 show rainfall intensity, runoff, and lead values for filtered and unfiltered samples during four selected storms, including large and small precipitation events, on the smelter watershed. The total lead content in the smelter watershed runoff is 30 to 50 times that of normal flow and several thousand times larger than the control; [37] the differences in basin size are insufficient to account for this. Similar patterns exist for zinc and cadmium. The coincidence of highest lead concentrations and peak runoff suggests that large amounts of material are moved during heavy storms.

Rainfall was collected in wide-mouth polyethylene

jars near the AMAX smelter during storms; its pH averaged around 6. During periods of peak runoff, the stream pH normally varied slightly. Thus it is not surprising to find that increases in dissolved metals were very small. According to Foil, [55] up to 5 to 6 mg/l of the total 25 to 30 mg/l lead occurring during peak runoff periods associated with large storms are in a dissolved form. The interesting point is the abrupt major increase in the metal content of suspended solids.

Anomalous cadmium values were also found in the runoff from the smelter watershed. This was unexpected, since it rarely occurs in natural waters of the region. Small amounts of cadmium were present in the lead ore, but that occurring in the smelter solid wastes, lagoon effluents, and leaf litter surrounding the smelter appears to be abnormally high. Similar relationships exist for zinc and copper in unfiltered runoff, indicating that runoff is a major transport mechanism for all heavy metals.

Lead in Stream Sediments

The relatively large amounts of suspended heavy metals found in streams during storm runoff pose a question as to whether they come from material washed in from the soil or from a resuspension of sediments on the stream bottom.

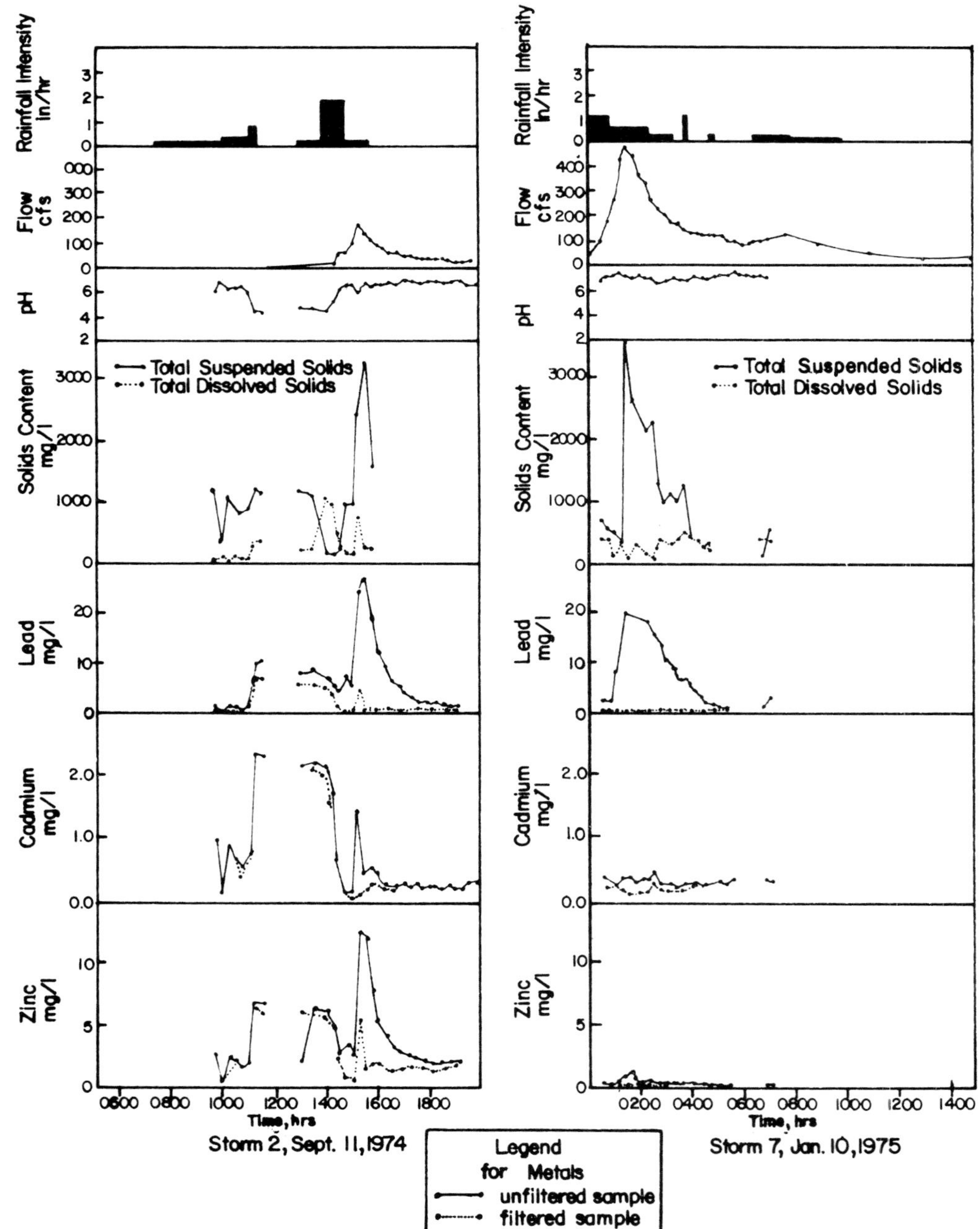

FIGURE 7-26.—Metal content of smelter runoff.

Investigations have shown that stream sediments in the New Lead Belt contain relatively low levels of lead and other heavy metals, and these concentrations are only found near effluent sources. Most of the lead generally occurs in the range of 200 mesh or finer. This is not unusual since 90 percent of the ore is ground finer than 200 mesh. In addition, the minerals containing heavy metals are generally more friable and may be crushed even finer. Also, there is virtually no clay in the stream sediments (based either on particle size or geochemical characteristics, even though soils of the region characteristically contain from 9 to 15 percent clay). Clay materials are obviously being carried in suspension by the stream water. Thus heavy metals associated with suspended solids during runoff appear to be washing directly from the soil to the streams, and thence move out of the ecosystem. This transport mechanism has been reported by studies earlier in Missouri, [49] in Illinois, [56] and elsewhere. [57, 58]

In one particular case, extraordinarily high lead levels (111,000 μg/g) were found in sediments near the ASARCO smelter at Glover, MO, just outside of the New Lead Belt. (table 7-29) This material reached the stream following attempts to reduce plant dust in compliance with requirements of the Occupational Safety and Health Act. The attempts involved installation of a sprinkler system that peri-

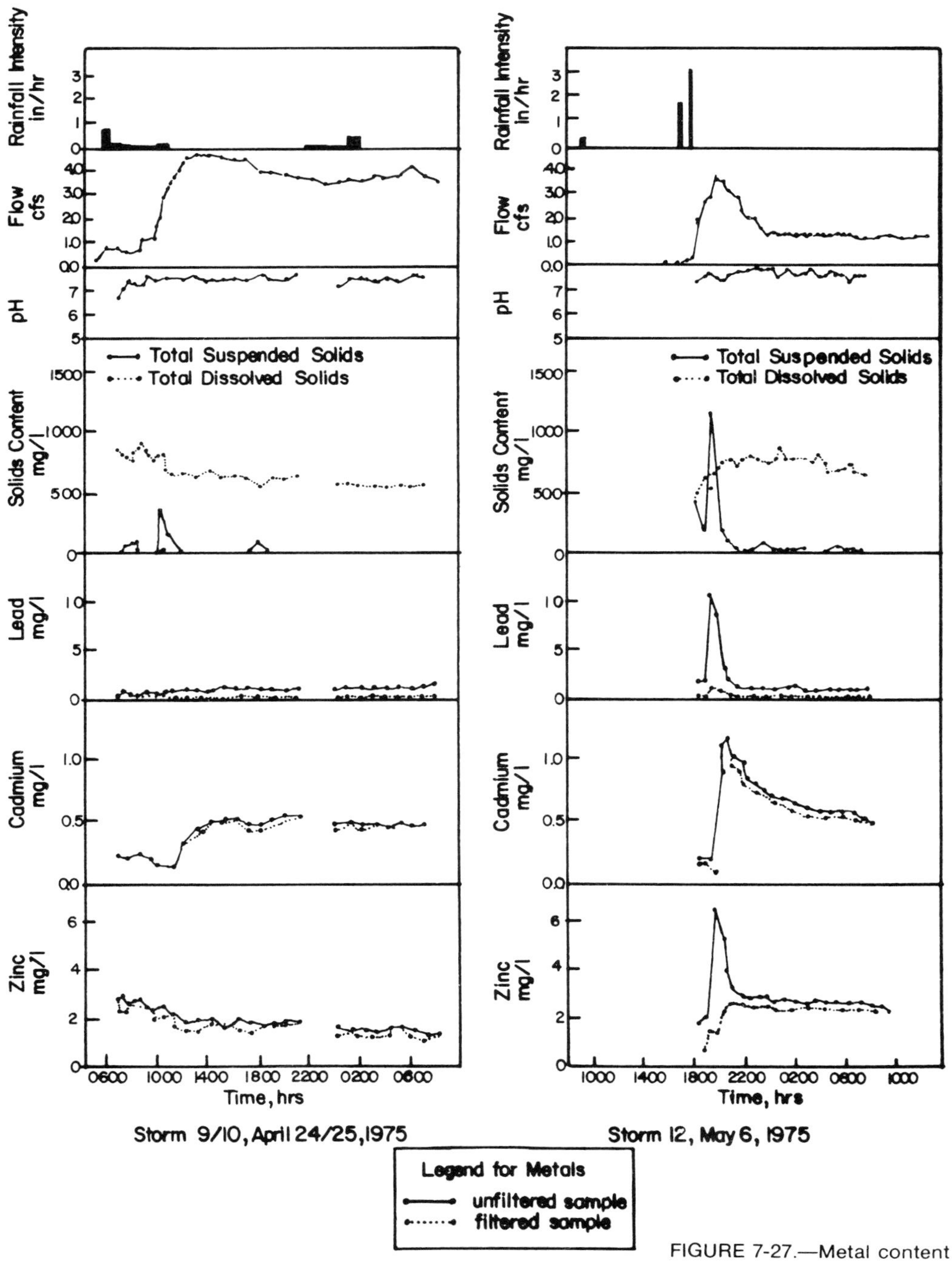

FIGURE 7-27.—Metal content of smelter runoff.

TABLE 7-29.—*Release of ASARCO smelter concentrate into Big Creek*

Sediments measured after accidental release of concentrate

Sample location	Element μg/g (average) Pb	Zn	Cu	Cd
N. Ditch on Co. Prop.	80,000	14,250	3,350	3,300
S. Ditch on Co. Prop.	116,000	95,000	2,200	3,900
1 mi upstream of effluent	10	32	10	0.5
250 yards downstream of effluent	1,720	650	64	290

Two weeks after first notice of release

Sample location	Pb	Zn	Element μg/g (average) Cu	Cd	Mn
N. Ditch on Co. Prop.	83,667	10,467	2,546	51,833	753
N. Ditch on Co. Prop.	75,167	6,533	1,740	31,000	626
S. Ditch on Co. Prop.	10,496	37,733	721	205	2,007
Stream just above effluent	180	168	20	11	941
Water (unfiltered unless marked μg/g) (average)					
N. Ditch on Co. Prop.	0.14	0.41	<0.01	0.054	0.096
N. Ditch on Co. Prop.-Filtered	0.073	0.32	<0.01	0.046	0.086
S. Ditch on Co. Prop.	2	61	0.027	25	1.27
S. Ditch on Co. Prop.-Filtered	1	56	0.015	22	1.15
Above eff. smelt Glover	<0.005	<0.01	<0.01	0.0014	0.03
Above eff. smelt Glover	0.007	0.017	<0.01	0.0024	<0.01
Below smelt eff. Glover	0.03	<0.01	<0.01	0.0076	0.048
Below smelt eff. Glover	0.014	<0.01	<0.01	0.0062	0.03

odically washed the plant down to control dust. Water from the sprinklers drained into the ditches on the north and south sides of the plant; during periods of storm runoff, material from the ditches was unintentionally washed into the streams and deposited in the sediments. This smelter does dust-smelting and processes concentrates from mines elsewhere in the world as well as from the New Lead Belt; this accounts for the very high cadmium values found along with those for lead, zinc, and copper. Although control sedimentation ponds have now been constructed to treat the plant runoff, this event does emphasize that care must be exercised to contain heavy metals and prevent their reaching streams.

Lead in Lake Sediments

Most mill and mine waste waters in the New Lead Belt ultimately flow into one of two major receiving streams, Logan Creek and Black River. Logan Creek receives wastes and mine water from only one mine, while the Black River receives the majority of wastes and mine water from the five mining operations. The velocity patterns of these streams are such that fine particles do not settle; instead, they are carried along by streamflow until quiescent conditions are found.

Both Black River and Logan Creek flow into Clearwater Lake, where retention times are favorable for sedimentation. The lake is located at Piedmont, MO, approximately 30 mi from the last of the five mine-mill discharges on the Black River and approximately 25 mi from the single mine discharge on Logan Creek. (fig. 7-28) In addition to the two arms of Clearwater Lake that receive mine-mill wastewaters, a third arm receives flow from Webb Creek. This stream drains the same forest area, but it is not exposed to any type of mining waste products.

Sediment samples were taken from various parts of Clearwater Lake using conventional dredging methods and by scuba divers utilizing hand collection devices. These, together with selected samples of aquatic biota, were analyzed to evaluate the extent of trace metal accumulation after approximately 8 years of mining operations.

Lead concentrations in the lake sediments range from less than 3 μg/g at points of stream entry to more than 60 μg/g in deeper waters near the dam. Zinc and copper concentrations follow a similar trend, ranging from 10 to 84 μg/g and 5 to 30 μg/g, respectively. Cadmium reaches detectable levels (0.3 to 0.5 μg/g) only in samples collected near the dam.

In general, lakes with long narrow arms like Clearwater, are subject to hydraulic scour and washout during storms. Under these conditions, sediments are swept up and carried toward the discharge tube located in the base of the dam. When the tube gate is opened to permit discharge from heavy inflows, the sediments and associated heavy metals are apt to be flushed from the lake, thus moving out of the ecosystem. The relatively low concentrations of heavy metals in sediments near the dam strongly suggest that this flushing action has been an active process, and it will probably continue to be so in the future. Clearwater Lake appears to be a relatively short-time sink for heavy metals. The length of residency undoubtedly depends upon the magnitude of stormflows in and out of the impoundment.

Total body lead content of bluegills, bass, goggle-eyes, catfish, and minnows collected in the lake range from undetectable levels to 14 μg/g. The mucous membranes of skin and gills show a particular affinity for heavy metals. Lead concentrations in skin and scales range from undetectable levels in catfish to 21 μg/g in bluegill. Gills from a variety of fish demonstrate lead content of 5 to 18 μg/g. Catfish gills generally contain less lead than those of bluegill. Freshwater mussels have total body lead concentrations of 25 to 30 μg/g, most of which was in the shells. Soft internal tissues (muscle, heart, kidneys, gastrointestinal organs, and reproductive organs) do not have detectable quantities of lead. These data allow some interesting comparisons with similar data collected from areas of much higher heavy-metal concentrations in the mining district. [59]

There is little evidence that surface mineralization in the mining area has elevated levels of the lead, zinc, and copper. Data from the Webb Creek arm are very consistent, ranging from 15 to 22 mg/l for lead and zinc; these are the normal background levels expected for New Lead Belt control streams. Logan Creek is only slightly higher, ranging from 12 to 40 mg/l lead. Values from the Black River arm are far more erratic, ranging from 10 to 64 mg/l for lead and zinc, with most minimum amounts above 15 mg/l. This arm is more subject to washout than the others, since it drain the largest area. Based on research evaluations of water, sediments, and selected biota, it appears that under present conditions no acute toxicity problems exist in Clearwater Lake. Limited data on similar streams in Wales [60] and Missouri [61] indicate that lead may not be as great an environmental hazard as zinc.

AQUATIC BIOCONCENTRATION OF METALS

The release of finely ground rock particles and some vagrant mineral particles from the tailings ponds is greatest during storm periods, but these materials accumulate in the benthos to a greater extent under less turbulent conditions. The magnitude of silt accumulation varies and may be complemented by additional precipitation or coprecipitation of dissolved minerals as industrial effluents mix with natural surface waters. The accumulation of silt has on occasion seriously altered normal stream communities. The turbulence of storm water flows frequently resuspends and redistributes stream sediments. Studies by Jennett and Hardie [44] have assessed the heavy metals content and distribution patterns in the sediments of receiving streams and lakes, as well as in the tailings treatment system.

The capacity of local producer organisms, living or dead, to trap and concentrate vagrant heavy metals has been observed in both terrestrial and aquatic environments. [62-65] In the aquatic environments of the New Lead Belt, quantities of lead associated with aquatic vegetation have exceeded 74,000 μg/g (dry weight) at the point of mine water discharge, and more than 8,000 μg/g at the discharge of some tailings ponds. In both aquatic and terrestrial situations, the degree of heavy metal contamination of vegetation decreases with the distance from the source. Little is known about the eventual fate of such heavy metals over relatively long periods of time.

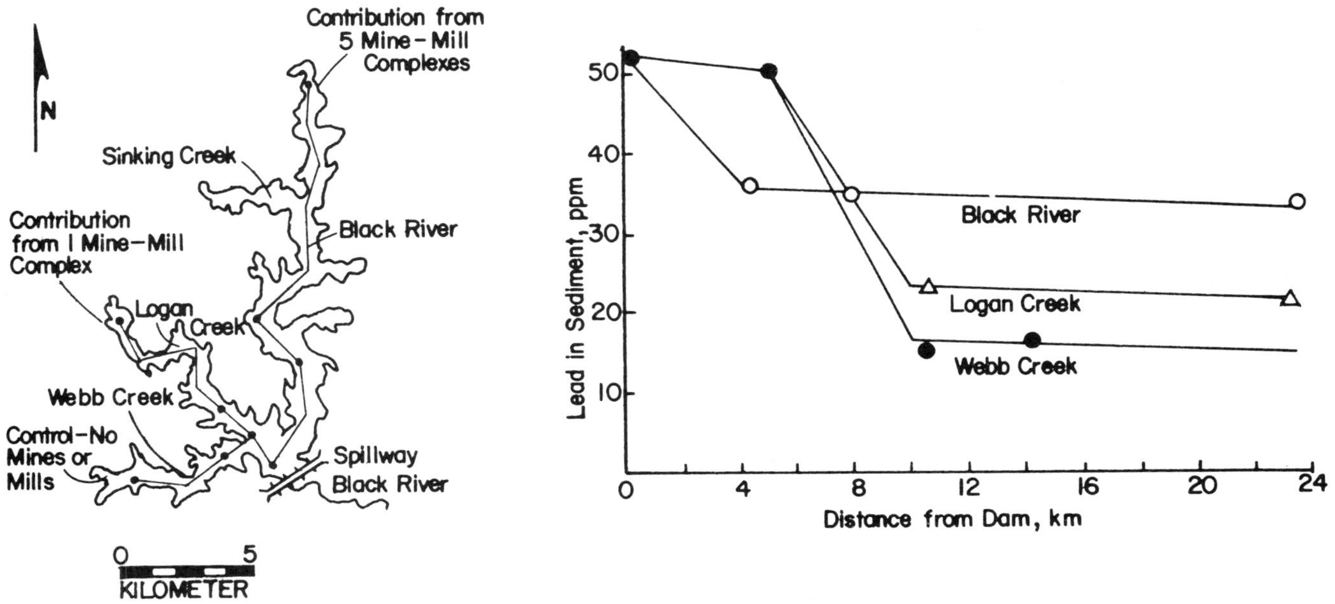

FIGURE 7-28.—Lead concentration in Clearwater Lake sediments.

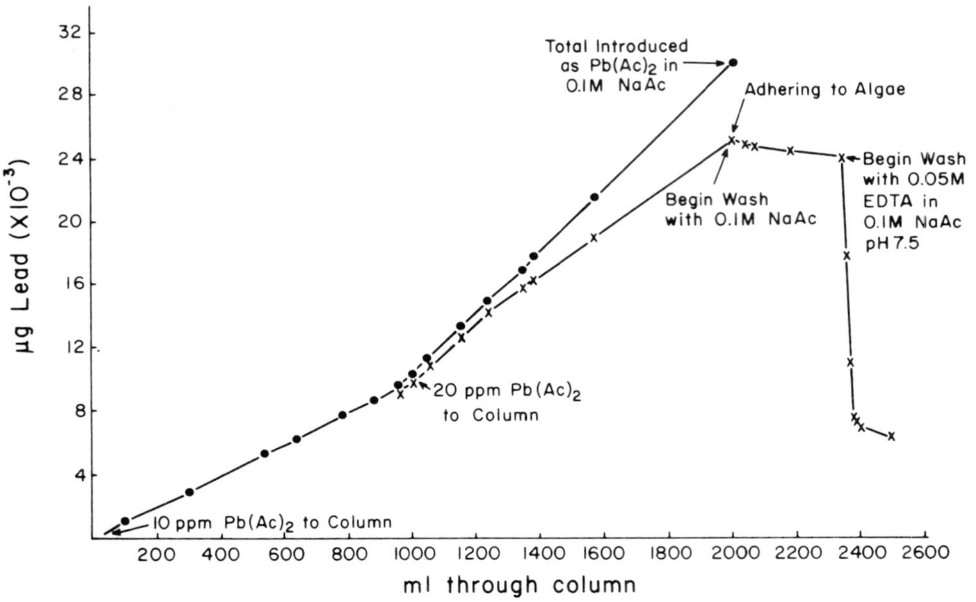

FIGURE 7-29.—Cation exchange using algae (40 g wet weight *Spirogyra*), column pretreated with 400 ml 0.1 *M* NaAc.

Heavy metals are most commonly added to streams as sulfides, carbonates, or phosphates — all of which are quite insoluble in the hard, slightly alkaline streamwater of the region. The abundant anionic sites known to be present in cell walls and surrounding matrices provide ample opportunity for ion exchange. Electron microscopic studies by Malone et al. [65] suggest that when hydroponically grown corn roots are treated with dissolved or chelated lead, insoluble lead particles appear in isolated membrane-lined vesicles within the root cells. Few vesicles containing lead are found in stems or leaves. More recent autoradiographic studies by Rule et al. [66] indicate that some of the lead applied to leaves of radishes and lettuce may enter the vascular system and be transported to other plant tissues. The lead that remains associated with aquatic vegetation in the New Lead Belt streams appears to be bound primarily to the cell membranes of algae, based on electron microscopy studies by Wixson and Gale. [67]

Extensive washing of aquatic plant material in distilled, tap, or stream water failed to remove the bound lead. However, washing with EDTA (ethylene diaminetetraacetate) at pH 7 to 7.5 in concentrations as low as 0.01 M effectively removes most of the associated lead without cell rupture. [68] Laboratory studies using filamentous algae or leaf litter packed in columns determined the capacity of plant matter to bind, by cation exchange, soluble lead salts

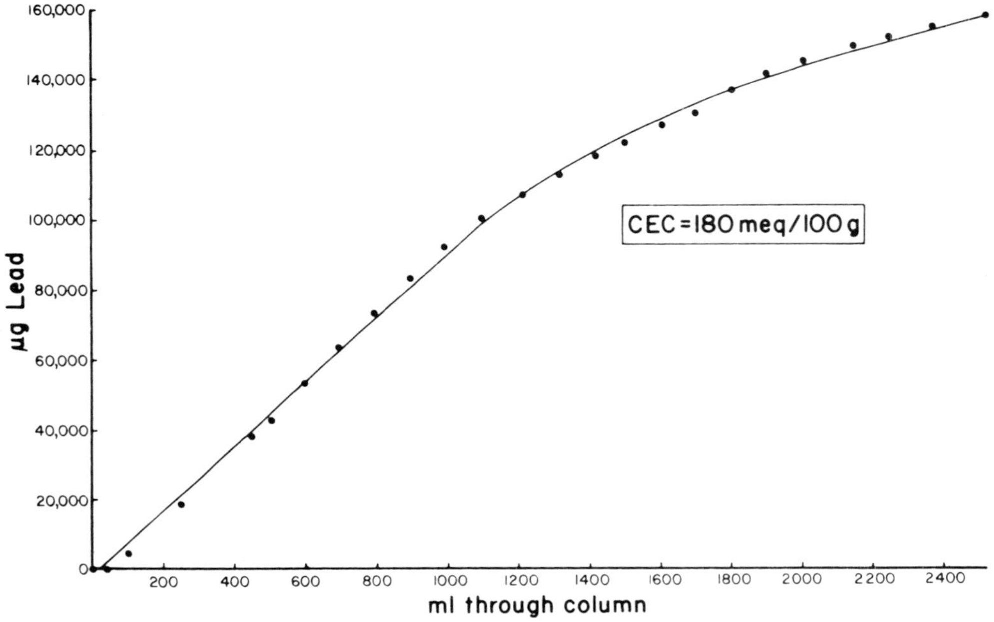

FIGURE 7-30.—Uptake of lead by leaf litter used in column studies.

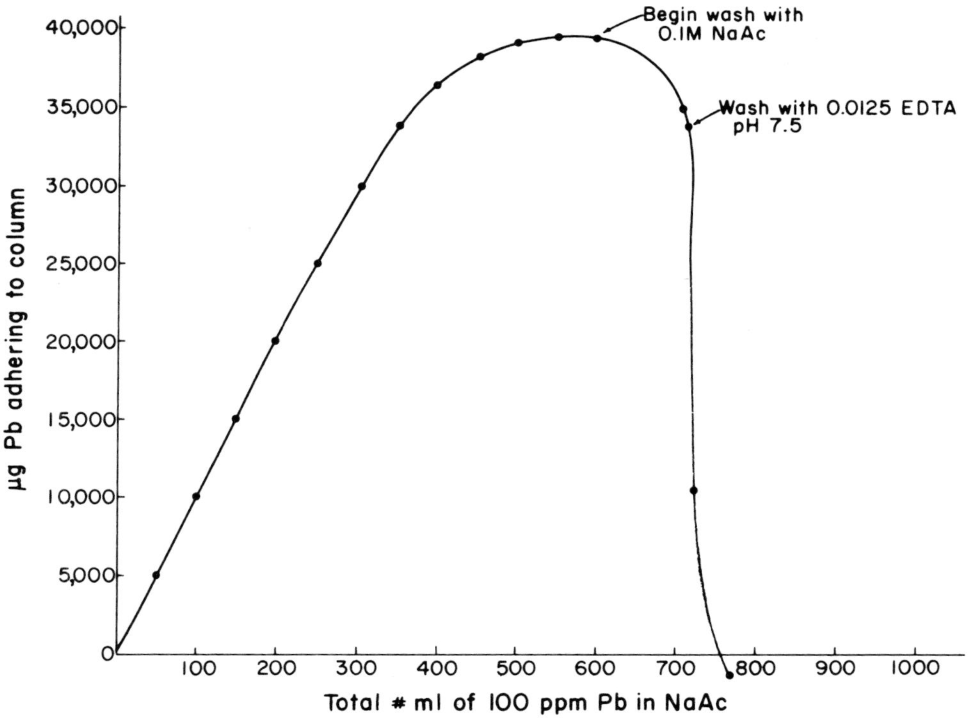

FIGURE 7-31.—Uptake and release of lead from mixed culture of algae.

in the presence of abundant monovalent cations. (fig. 7-29 to 7-32) Cation exchange capacities, based on the column studies, approached 65 meq/100 g for filamentous algae (over-dry weight basis) and 180 meq/100 g for dried forest litter material. The concentrations of lead reported in forest litter [62] and those found in aquatic vegetation often do not approximate the total cation exchange capacity of the vegetation involved. Control columns using glass wool with overlaying lead sulfide (concentrate) indicate vast differences in the leaching rates of chelated lead from insoluble particulate material (PbS) and the removal of lead bound to the surface of plant material by washing with EDTA.

Effects of Mine-Mill Complexes on Stream Biota

The amount and variety of producer organisms vary remarkably from one stream to another. In some locations, seasonal algal blooms have reached problem proportions, especially below the confluence of mine-mill wastewater and the natural receiving stream. In some cases, unsightly conditions have been created for several miles downstream due to unusual and dense algal bloom communities that

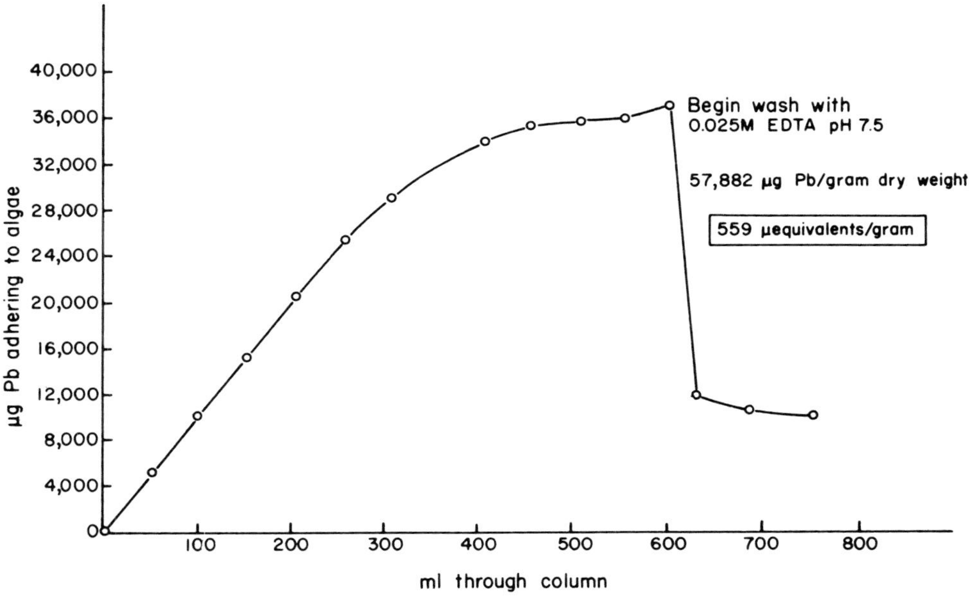

FIGURE 7-32.—Uptake and release of lead from autoclaved culture after washing with EDTA.

become coated with associate gangue. In some instances, this increased productivity has encouraged the production of game fish. Studies on the environmental impact of seepage from a century-old tailings pile in the Old Lead Belt of Missouri indicate that sediment and biota contain elevated amounts of zinc due to stream contamination. [69]

Figure 7-33 shows the schematic of the tailings ponds and meander system installed below the AMAX Buick mine and mill that produce lead and zinc concentrates. The principal milling reagents employed are shown in table 7-30. This table illustrates differences between mills employing a copper circuit in the flotation process. Since completion of the meander system and final settling pond, this operation has successfully eliminated most nuisance algal blooms in the receiving stream and has markedly improved retention of particulate material and heavy metals within the system. [54]

Growth of aquatic vegetation within the meander system and final settling pond effectively traps and binds most of the vagrant mineral particles released from the tailings lagoon, and the overall retention time in the system permits effective degradation or dilution of any remaining milling reagents. Aquatic vegetation at the head of the meander system normally displays lead contents of several thousand parts per million; these values drop to a few hundred within the fnal settling pond.

Normal consumer organisms—including aquatic insects, snails, tadpoles, crayfish, bluegills, bass, carp, and catfish— abound in the lagoons and meanders. Total body analyses for heavy metals in the larger consumers reflect the high content of metals in consumed vegetation and mucous coverings of insect larvae that comprise a major part of the diet of higher trophic levels. Considerable quantities of lead and zinc are found in the gills, mucous membranes, and bones of fish. (tables 7-31 and 7-32) The fleshy portions of fish likely to be consumed by humans are relatively free of heavy-metal contami-

TABLE 7-30.—*Principal milling reagents employed in New Lead Belt flotation processes*

Reagent	AMAX-Buick mill (no copper circuit)	Fletcher mine-mill (with copper circuit)
Frother	Methyl isobutyl, carbinol propylene, and glycol-methyl ethers	C_6-C_9 aliphatic alcohols
Promoters	Sodium diethyl-dithiophosphate and sodium isopropyl-ethylthionocarbamate	Sodium isopropyl-ethylthiono-carbamate
Collectors	Sodium ethyl xanthate	Sodium isopropyl xanthate
Zinc depressant	$ZnSO_4$ + cyanide	$ZnSO_4$ + cyanide
Zinc activator	$CuSO_4$	$CuSO_4$
Iron depression and pH control	Lime	Caustic soda
Lead depression and copper flotation	--	SO_2 + starch

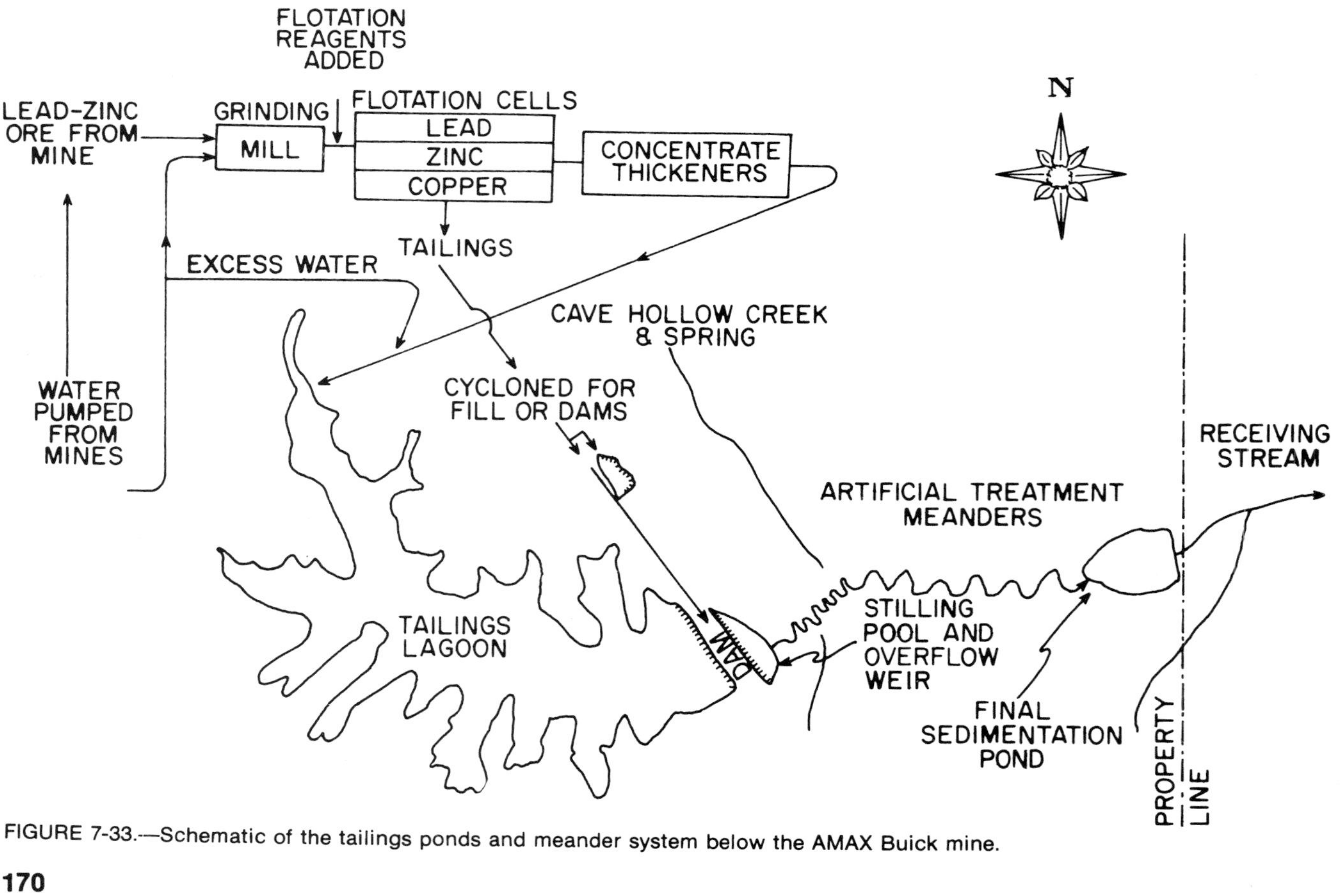

FIGURE 7-33.—Schematic of the tailings ponds and meander system below the AMAX Buick mine.

TABLE 7-31.—*Heavy metals in fish, mill A*

Location/fish part	Pb (μg/g)	Zn (μg/g)	Cu (μg/g)
	Tadpoles from meanders		
100 yards down meanders			
Eviscerated body	213	256	20
Intestine and contents	4,419	2,926	234
At Cave Hollow Junction			
Liver and heart	22	67	91
Eviscerated body	93	240	15
Intestine and contents	7,329	4,696	260
Total body	4,139	2,808	169
	Bluegills from meanders		
Muscle	1	45	11
Bone	104	152	21
Skin and scales	132	229	21
Intestine and contents	390	381	53
Gills	161	109	19
Total body	84	157	24
	Bass from meanders		
Muscle	<1	30	13
Bone	142	89	22
Skin and scales	194	176	27
Intestine and contents	259	279	159
Gills	160	105	24
Total body	7	117	28
	Bluegills (final tailings pond-spring)		
Muscle, bone, skin, and scales	20	82	7
Intestine and contents	626	437	37
Washed intestine	9	77	13
Liver	<1	83	12
Ovaries	<1	98	8
Gills	52	93	7
Total body	50	100	10
	Bluegills (final tailings pond-summer)		
Muscle	<1	42	20
Bone	15	77	14
Skin and scales	22	121	20
Intestine and contents	30	61	16
Gills	38	82	16
Total body	32	98	13

nation, since most of the heavy metals actually are contained in fecal matter within the intestional tract. Heavy-metal contents of the intestines and gills vary seasonally, reflecting climatic conditions and differences in species diet. There is no evidence of significant food chain concentration of heavy metals in the higher trophic levels of aquatic communities.

A schematic for the mine and mill tailings and recycling system at the St. Joe Minerals Corp. Fletcher Mine and Mill is shown in figure 7-34. This operation produces lead, zinc, and copper concentrates and has been considerably more confined by the space available for tailings treatment. The principal milling reagents employed are included in table 7-32. Due to space restrictions and efforts to improve conditions in the receiving stream, this operation has separated the mine and mill wastewaters and is recycling the latter. However, total recycling

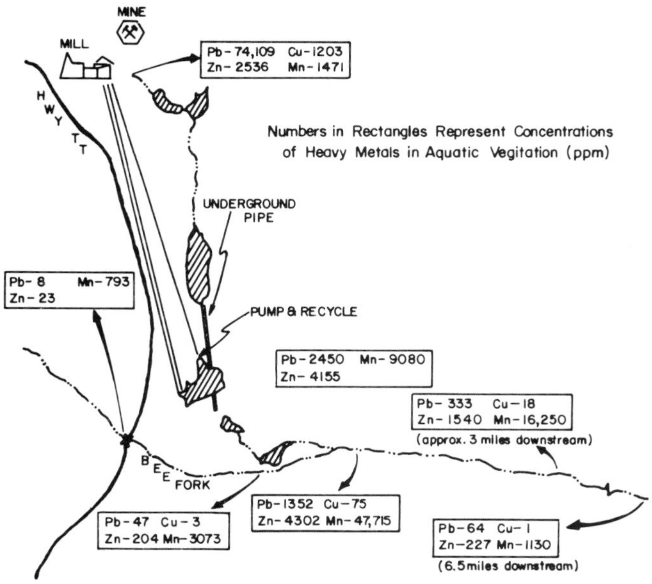

FIGURE 7-34.—Schematic of New Lead Belt mine-mill recycle treatment system.

of mill wastes has not been achieved due to some leakage from the milling wastewater retention lagoon.

Growth of aquatic vegetation in the short discharge stream below the final (mill) tailings pond has been limited to seasonal blooms of blue-green algae, diatomaceous mats, and a few pond weeds. Development of aquatic vegetation in the stream receiving the Fletcher operation discharge has usually been depressed in the early spring, as compared with other control streams or the same stream above the point of effluent discharge. Normal aquatic communities have been replaced by dense diatomaceous mats during the summer months, especially under low-flow conditions. Relatively high manganese content, along with other residual milling reagents in the effluent, has been cited as one possible contributing factor to the unusually dense diatom populations and exclusion of the normal aquatic growths.

Dominant aquatic consumers in the New Lead Belt area include minnows, sunfish, and bass. Other consumer organisms common to these Ozark streams—such as snails, crayfish, tadpoles, and various aquatic insects—are extremely rare or missing. Since heavy-metal retention within most tailings sys-

TABLE 7-32.—*Heavy metals in fish, mill B*

Fish part	Pb (μg/g)	Zn (μg/g)	Cu (μg/g)
	Stilling pool: bluegills		
Muscle and bone	17	91	2
Gills	28	90	2
Intestine and contents	22	151	10
	Final effluent: bluegills		
Muscle and bone	<1	90	3
Gills	<1	80	1
Intestine and contents	<1	81	6
Total body	6	165	3

tems is good, the possibility must be considered that residual milling reagents in the effluents may contribute to the sparcity of many consumer organisms.

Toxicity of Milling Reagents

The toxicity of the various reagents utilized in the ore milling process is neither fully understood nor appreciated. Persistent identifiable odors and the appearance of froth in the final industrial effluents of some operations provide clues to the possible reasons for serious changes in aquatic communities downstream from the point of discharge. Unfortunately, few data are available on the toxicity of milling reagents to aquatic organisms, and even less is known about the actual concentrations of residual reagents in industrial effluents. The various 96 h median toxic limits, as determined by Hawley [39], the U.S. Department of Interior Fish Pesticides Laboratory, and Gale [68] are summarized in table 7-33. These data indicate the range of concentrations in which a 50 percent mortality of the test organisms occurred. More precise determinations are difficult, since considerable variations may be observed from one group of organisms to another, depending on age, nutritional state, season, sex, water hardness, etc. These data are additionally prejudiced by the obvious shortcomings of short-term static toxicity studies. They offer little information regarding toxicity under conditions of continuous flow, long-term effects, nor embryological or teratogenic effects. Despite these deficiencies, the results provide an approximate comparison of relative toxicity.

Many of the reagents tested have absorption maxima in the ultraviolet range, but the extinction coefficients are often too small to permit detection at the very low concentrations known to be toxic to aquatic organisms. [70] The lack of adequate analytical techniques has been a serious factor in determining concentrations of milling reagents at various points in the mill circuit, in the tailings ponds, or final tailings effluents. Some of the reagents remain bound to the mineral particles and thus never enter the tailings. Other may not bind to ore particles, but remain in the final tailings. In mills that use a copper circuit, the addition of SO_2 to separate the lead and copper minerals may, under certain conditions, also cause the release of considerable quantities of xanthate collectors; these may be detected in the thickener overflows. Degradation rates vary considerably [71] and may be affected by pH, temperature, or redox potential.

Addison et al. [72], working in New Brunswick with mill effluents from sulfide ore, identified xanthate residuals of 0.2 to 1.2 mg/l; dithiophosphate residuals from 0.3 to 2.7 mg/l; and isopropyl ethylthionocarbamate residuals of 1.8 mg/l. Limited analyses of effluents from a New Lead Belt mill process found xanthate concentrations of 0.3 mg/l in thickener overflows, but undetectable levels (less than 0.1 mg/l) in the final effluent. Long-chain alcohols have been determined to exceed 200 mg/l in the final effluent on one occasion, and at another time were undetectable (less than 10 mg/l).

More detailed data are needed relative to the final effluent concentrations of these toxic substances. Circumstances that would permit toxic concentrations of milling reagents to enter receiving streams should be eliminated. In those situations where problems exist and mill effluents are suspect, every precaution should be taken to ensure adequate treatment of tailings and effective isolation of effluents known to contain toxic levels of reagents. Where possible, reagents should be selected on the basis of favorable degradability properties or reduced toxicity.

TABLE 7-33.—*Toxicity of various milling reagents, 96-h median toxic limits*[1]

Reagent	Fathead minnow	Emerald shiner	Water fleas	Bluegill	Catfish	Snails	Tadpoles	Crayfish	Golden shiners	Algal photosynthesis
Methyl isobutyl-carbinol	100-1,000	–	–	–	–	–	–	600-1,000	–	–
Propylene glycol methyl ethers	>1,000	–	–	>1,000	–	100-500	–	>100	–	–
Long-chain aliphatic alcohols	100-1,000	–	–	50-100	100-500	100-1,000	10-100	10-100	<23	100-200
Potassium or sodium amyl xanthate	1.8 -18	10-100	0.1-1.0	100-200	–	1.0-10	–	100-200	–	100-200
Sodium isopropyl xanthate	0.18-1.8	0.01-0.1	0.1-1.0	0.01-0.1	>10	10-100	10-100	1-10	–	10-100
Sodium ethyl xanthate	0.18-1.8	0.01-0.1	0.1-1.0	10-100	–	100-1,000	–	100-500	–	100-200
Isopropyl ethyl thionocarbamate	10-100	–	–	10-50	10-100	10-100	–	100-200	–	100-200
Sodium diethyl-dithiophosphate	–	–	0.1-1.0	660-1,000	–	>1,000	–	–	–	–
Minerec B (dixanthogen)	–	–	–	1.0-10	–	100-200	–	–	–	–
Aero-130 (thiocarbanalide)	–	–	–	>1,000	–	100-1,000	–	–	–	–

[1] Data on fathead minnows, emerald shiners, and water fleas from Hawley [68]; on catfish from U.S. Department of Interior Fish Pesticide Laboratory (Columbia, MO); remainder by Gale [39].

Summary

Under present conditions, lead and other dissolved heavy metals do not present a significant biological hazard to aquatic organisms of the New Lead Belt. (Zinc may be an exception and is still under study.) This fortunate circumstance is the result of the alkaline pH and the elevated hardness characteristic of the surface and subterranean waters of the region, combined with pollution control procedures initiated by industry.

Aquatic vegetation within the various mine-mill treatment systems and in the receiving streams is efficient in trapping and binding heavy metals that may be present in the particulate material entering the water. Most of the heavy metals accumulated by algae or other aquatic vegetation appear to be bound to the surface of the plants and may be removed by washing with solutions such as EDTA. Aquatic vegetation is only a temporary host for the metals, since they are carried downstream during storm runoff periods.

Suspended solids in streams are a major transport mechanism for heavy metals. These heavy metals originate from smelters, mine-mill tailings ponds, and tailings dams, or directly off contaminated soil. Most of this transport occurs during periods of storm runoff, when the soil and leaf litter are washed into the stream, and the treating ponds are subjected to turbulence. Since streams in the region contain very little fine sediment and clay, most of the heavy metals are transferred out of the New Lead Belt ecosystem.

Clearwater Lake is the only long-term impoundment in which heavy metals can accumulate to any extent. Research indicates that heavy-metal levels are increasing, but do not present a problem under existing conditions. Since this lake is subject to flooding that results in resuspension and flushing out of sediments, it should not be considered as a permanent sink for heavy metals.

Analyses for heavy metals within the tissues of normal consumer organisms have not indicated any significant biomagnification of lead, zinc, or copper. Total body content of these heavy metals is considerably more elevated in consumer organisms near tailings dams or within the treatment systems than for those in control streams. Research indicates that present metal levels do not represent a serious hazard to streams in the New Lead Belt. Game fish are common within the tailings ponds and meanders of some mine-mill operations where the aquatic vegetation is heavily laden with lead and zinc.

The ore concentration mill, with its attendant process reagents and finely ground tailings, represents the greatest concern for the aquatic ecosystem. Control must be maintained through adequate tailings systems for the neutralization of toxic chemicals and the removal of fine particulate matter. The elimination of toxic chemicals and effluent particulate matter from the final tailings pond effluents would ensure minimum adverse effects from the industrial operations on the receiving stream. In addition, it would control the release of heavy metals (as particulates), which is not a major problem under present conditions. Research should continue for the development of better analytical methods to detect organic milling reagents found to be toxic to stream organisms.

REMOTE SENSING AND DATA PROCESSING

The Missouri lead-belt study provides an excellent opportunity for the utilization of remote sensing techniques; and because of the large size of the sampling program, new techniques for data processing and presentation were developed. The results of these investigation are briefly described below.

Remote Sensing

The primary objectives of the remote sensing efforts were to develop effective techniques for remotely sensing heavy-metal concentrations in soil and vegetation and to develop efficient image processing techniques for detecting the effects of trace metals in the environment. This is a mature technology, [73] and undoubtedly success could be achieved using the highly sophisticated multichannel data collection and optimal data processing techniques currently available. The project objective, however, was to determine the degree of success that could be achieved by using inexpensive 4-channel multispectral photography, a technique easily within the reach of local industries and agencies.

In order to produce preliminary results in a reasonable period of time, the initial study area was limited to an area 10 km by 10 km, centered on the AMAX lead smelter near Boss, MO. Ground truth consisted of approximately 250 soil samples and 110 vegetation samples obtained in this area. These samples were interpolated using a linear interpolation algorithm [74] to form spatial distribution for the soil and vegetation. The spatial distribution, with the resulting contour maps for copper, lead, zinc, and sulfur concentrations in soil, are shown in figure 7-35. Vegetation samples yielded similar distribution. The techniques for developing these distributions will be briefly discussed.

The study area was overflown twice during the spring (May 17) and fall (Oct. 5) of 1973. A Hasselblad 70 mm, 4 camera cluster, operated by the University of Kansas Center for Research, Inc., was used to obtain imagery in the blue, green, red, and near-infrared bands. The camera array and aircraft mounting are illustrated in figure 7-36. Film specifications are given in table 7-34. Each mission consisted of one flight at an altitude of 11,000 ft using a wide-angle 40 mm lens and one flight at 6,000 ft using a 100 mm lens. High altitude coverage using a wide-angle 40 mm lens was desirable to prepare an image mosaic covering the area of interest, since with fewer frames to splice, higher geometric fidelity is obtained. In return for this coverage, resolution is sacrificed. The low-altitude coverage obtained high-resolution imagery for identification and mapping purposes.

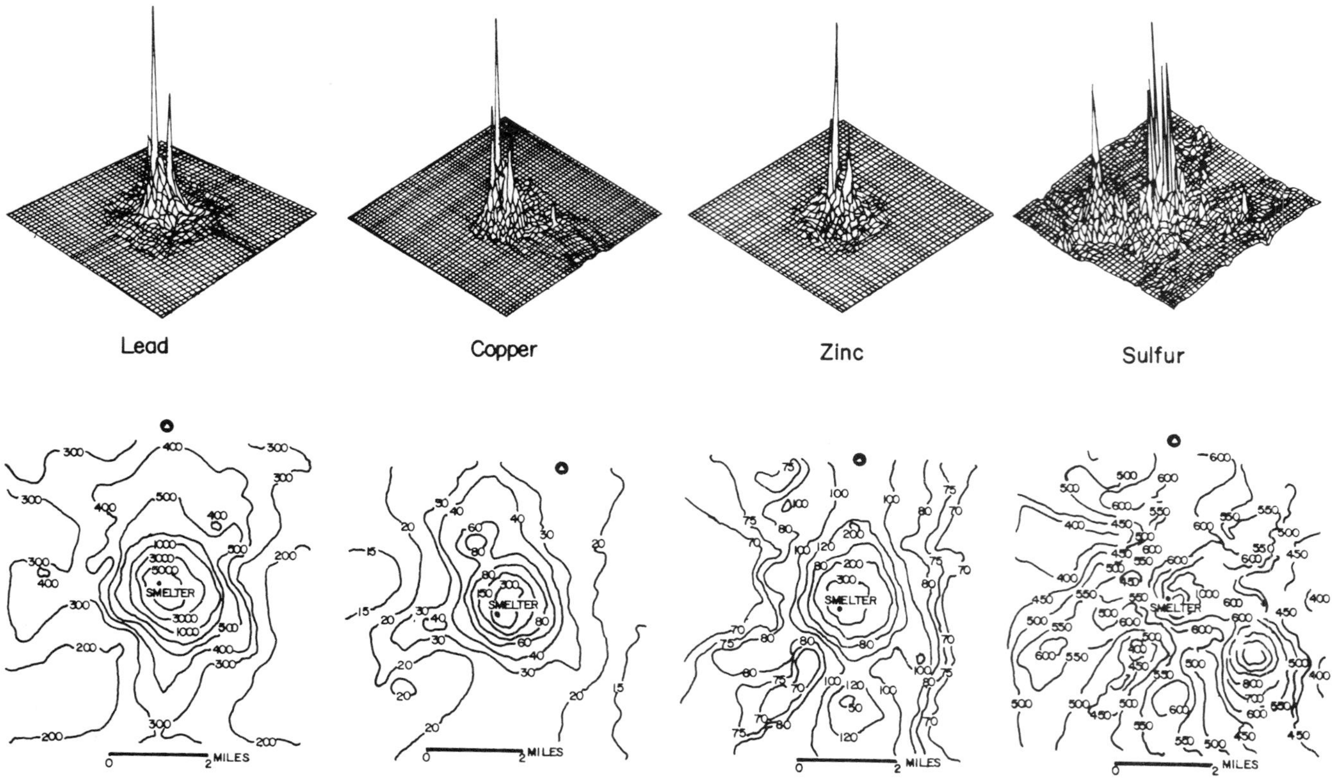

FIGURE 7-35.—Spatial distributions and contour maps for soil data.

TABLE 7-34.—*Hasselblad camera film specifications*

Image band (nm)	Filter (Kodak)	Film (Kodak)
Blue (400-500)	47 B	B & W 2402 (Plus-X Aero)
Green (500-600)	58	B & W 2402
Red (600-700)	25	B & W 2402
Infrared (700-900)	89 B	IR B & W 2402 (Infrared Aero)

Examination of the imagery indicated that the red band was the most suitable in terms of image contrast and reflectance variations across an individual frame due to incidence angle differences. Atmospheric scattering effects limit the contrast quality of the blue band. The same effect is also responsible for the low contrast observed near the edges of the

FIGURE 7-36.—Aircraft and camera mount.

(a) Side view

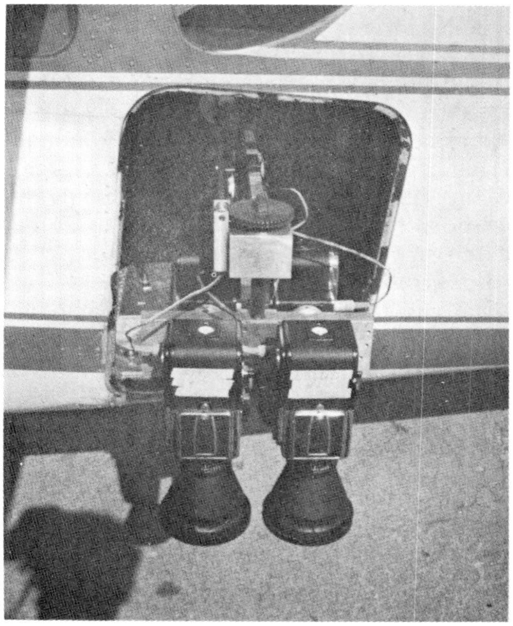

(b) Closeup of camera mount.

green band frames due to increased atmospheric mass at these angles. In the case of the infrared band, atmospheric scattering is negligible. The target reflectance, however, appears to be sensitive to incidence angle, thereby causing difficulty in making a mosaic of adjacent frames due to differences in density levels. Although the red band mosaic possesses the best qualities, all four bands were used in the color-combination and correlation studies.

Two techniques were used for data analysis and interpretation. First, color-combined mosaics were generated and compared visually with the ground-truth contour maps. Then, correlation coefficients were generated between each set of ground-truth data and each color band.

The color combiner consists of three slide projectors positioned in such a way that their beams, upon reflection by a mirror, are superimposed on a ground glass screen. Each of the beams is passed through one of three color filters: blue, green, or red. Furthermore, the intensity of each light beam can be controlled. Hence, by inserting three slides of a given scene corresponding to the same or different spectral bands, the color combiner will produce a color image of the three original black-and-white images.

About 40 different false-color images were produced and analyzed. Although some of these new false-color images greatly enhanced the discrimination between different terrain features, visual inspection revealed no apparent correlation between patterns of the false color images and the metal concentration maps. This technique was consequently used on the imagery from the first flight only. However, color combining was quite successful to discriminate between different features in the tailings pond. The resulting false color displays allowed the field engineers to efficiently select pertinent sampling sites. Mapping of surface features and area parameters for water, wooded areas, fields, and cultural features was also successful.

The correlation studies were performed using the IDECS (Image Discrimination, Enhancement, and Color Combination System) at the University of Kansas. First, the image mosaic from each mission was quantized and compressed into a 64 by 23 array. (The aerial photography covered the entire north-south extent of the 10 km by 10 km metal-concentration area, but only about one-third of the east-west width.) The 64 by 23 image array mapped onto rows 1 through 64 and columns 26 through 48 of the metal concentration array. With the two arrays quantized and registered, they were then cross-correlated.

Upon examining the preliminary result of cross-correlating the complete arrays, no discernible pattern was apparent. All factors that could have possibly led to this result were then eliminated. These include extrapolation of ground data to form the 10 km by 10 km metal concentration maps and quantization of the ground data and the imagery into discrete values. Out of the 64 by 23 image cells, only those with corresponding ground data points were next considered. Scattergrams of these new paired sets were then plotted and correlation coefficients calculated using actual concentration and image density values (no quantization). Similar scattergrams were also produced between the ground data and the density ratio of each of the four bands. None of these efforts produced any conclusive results.

The correlation coefficients between the soil and air-craft photographic data are given in table 7-35. All correlation coefficients are small, being less than 0.2 in absolute value. There are, however, several interesting observations that can be made. It appears that in late spring, blue band imagery and soil data correlate best. In early fall, infrared and green correlate best with the soil data. [75]

In an effort to achieve higher correlation between photographic and ground-truth data, the ratios of two color bands were taken. All ratio tests resulted in low values of the correlation coefficient.

Vegetation data were then correlated with the imagery using only field samples and image density values (no array extrapolation and no quantization). These results are somewhat more encouraging. (table 7-36) They indicate that additional studies are warranted. A perturbation analysis of the sampling site locations resulted in rather large changes in the correlation. Thus it was felt that the precision used in locating the sampling sites was not satisfactory for this use, and this segment of the study was discontinued.

Data Processing

One of the first problems encountered in the Missouri research project, which collected such a large quantity of data, was to establish a display format so that large-scale features would be evident upon quick examination. This problem was solved by the use of spatial data displays as illustrated in figure 7-32. Such data displays may be quickly constructed if

TABLE 7-35.—*Correlation coefficients-single color band (single color band - soil ground truth)*

Ground truth	Color band	ρ May 17	ρ October 5
Copper	Infrared	0.080	−0.170
Lead	Infrared	0.037	−0.139
Zinc	Infrared	0.055	−0.149
Sulfur	Infrared	0.076	−0.136
Copper	Red	0.074	−0.061
Lead	Red	0.060	−0.074
Zinc	Red	0.067	−0.069
Sulfur	Red	0.033	−0.003
Copper	Green	0.070	−0.174
Lead	Green	0.065	−0.163
Zinc	Green	0.073	−0.181
Sulfur	Green	0.058	−0.080
Copper	Blue	0.154	−0.091
Lead	Blue	0.125	−0.054
Zinc	Blue	0.151	−0.068
Sulfur	Blue	0.128	−0.093

TABLE 7-36.—*Correlation coefficients-single color band (single color band-vegetation ground truth)*

Ground truth	Color band	ρ May 17	ρ October 5
Post oak (lead)	Infrared	0.303	−0.080
Post oak (sulfur)	Infrared	−0.109	0.045
Pine needles (lead)	Infrared	0.330	0.096
Pine needles (sulfur)	Infrared	0.163	−0.017
Post oak (lead)	Red	−0.009	0.368
Post oak (sulfur)	Red	0.138	0.139
Pine needles (lead)	Red	−0.359	0.501
Pine needles (sulfur)	Red	0.224	0.342
Post oak (lead)	Green	0.126	0.165
Post oak (sulfur)	Green	0.091	0.090
Pine needles (lead)	Green	−0.068	0.333
Pine needles (sulfur)	Green	0.052	0.214
Post oak (lead)	Blue	0.177	0.055
Post oak (sulfur)	Blue	0.228	0.169
Pine needles (lead)	Blue	0.083	0.231
Pine needles (sulfur)	Blue	0.183	0.059

adequate computer systems are available. The spatial displays illustrated were generated with a CALCOMP platter with an IBM 3601/65 computer. In addition to the program for generating the spatial displays, techniques have also been developed for generating contour maps directly from field data. [76]

The routines for generating spatial displays and contour maps involve the process of interpolating field data to generate artificial sample values; through the use of these interpolating routines, sample values can be estimated. Statistical tests may be used to estimate the accuracy of these artificial samples, and if this accuracy is sufficiently high, field samples can be eliminated or collected at a point that will yield more meaningful information. The use of these techniques has been documented, and the results show that a potential exists for significantly reducing the cost of a field sampling program by reducing the required number of field samples for area parameterization. [77]

CONCLUSIONS

Comprehensive studies in the New Lead Belt of southeast Missouri have (1) investigated the movement of lead and other heavy metals from mining, milling, and smelting operations through, and out of, the surrounding forest ecosystem; and (2) determined the impact of these heavy metals on environmental quality. Two principal source-to-sink routes have been identified from these operations.

Smelting operations are a source of concentrates and stack emissions containing heavy metals. Fugitive dusts, which are moved by wind and deposited on the surrounding vegetation and soil, are the major source of contamination within 3 mi of the smelter. Beyond 3 mi, stack emissions are the predominant source as airborne particles that are discharged and gradually settle to the earth, carrying along associated heavy metals.

Mining and milling operations are another major source of heavy metals. Finely ground materials from the tailings dams or concentrate piles are windblown into the forest or wash directly into receiving streams.

The transport of finely ground ores and concentrates from mines and mills to the smelter provides another source of contamination within the ecosystem. When materials are transported in open trucks and railroad cars, considerable amounts are blown away and deposited on adjacent vegetation and soil. Although concentrations of lead can be quite high adjacent to the roadways, levels approach background within a few hundred feet of the haulage roads.

Lead and other heavy metals that are deposited on the forest floor tend to concentrate in the organic layers, with very little moving into the underlying mineral soil horizons. Geochemical analyses of leaf litter appear to be a much more sensitive means of evaluating the extent of pollution than are air quality measurements.

Although large quantities of lead have accumulated in the organic soil horizons near smelters, there appears to be little danger of movement downward. Soil, particularly the clay and organic fractions, have a strong affinity for lead and effectively prevent its downward movement.

The accumulation of lead by vegetation, especially vegetables and forage crops consumed by humans and animals, is a considerable cause for concern. The greatest accumulations are near smelters or in areas where crops are grown on soils near old mines or on soils that have been amended with dolomitic limestone mine wastes. The deaths of several horses grazing near a smelter have been attributed to lead poisoning.

Storm runoff is the major transport mechanism in the movement of lead. Evidence indicates that heavy metals do not concentrate in stream sediments, but relatively large amounts are carried by suspended solids associated with stormflows and temporarily deposited downstream in Clearwater Lake. Heavy metals are accumulating in the lake sediments, but research indicates that most of the metals are carried out of the lake during storms and do not pose a problem under existing conditions.

The transport of lead by stormflows may well limit its accumulation in the ecosystem. Present evidence suggests that lead levels in the New Lead Belt, which is only 6 to 8 years old, are approaching those found in the Old Lead Belt, where operations have existed for about 100 years. Thus it is likely that accumulations may stabilize within a few years.

Data from soil and vegetation studies show that smelting operations produce greater heavy-metal anomalies in terrestrial systems than do mining and milling operations. Within aquatic systems, however, the mine-mill complexes are the major sources of contamination. Aquatic problems in the New Lead Belt have been ameliorated by fortuitous stream characteristics. The alkaline reaction of natural waters retards the solubilization of the predominantly lead sulfide ores. In time, problems could arise from the more soluble forms of lead in smelter emissions.

It is entirely possible that zinc may cause more problems than lead in aquatic environments. Also, the organic reagents used in milling operations, some of which are quite toxic, may be environmentally damaging. In addition, some of these reagents have produced excessive algal blooms and growth of diatomaceous mats. Although unsightly, these aquatic growths have trapped large quantities of lead and other heavy metals. This characteristic has been utilized in the development of meander systems designed to enhance the quality of water released from tailings ponds.

Recommendations for pollution control and research application have been utilized where possible by industries and agencies for various aspects of air, soil, vegetation, and surface water in the New Lead Belt of southeastern Missouri. [78]

A technological and cooperative approach must be continued between industry, regulatory agencies, the engineering and scientific community, and the general public if mineral resources are to be produced without a detrimental environmental impact.

REFERENCES

1. H.M. Wharton, J.A. Martin, A.W. Rueff, C.E. Robertson, J.S. Wells and E.B. Kisvatsanyi. Missouri Minerals—Resources, Production, and Forecast. Missouri Geological Survey and Water Resources, Rolla, MO, No. 1, 303, (December 1969).
2. Missouri Mineral News, Missouri Geological Survey and Water Resources, Rolla, MO, *14(2):* 22 (1974).
3. The World's Number One Lead Producer. Missouri Mineral News, Missouri Geological Survey and Water Resources, Rolla, MO. *11(3):* (1971).
4. Lead Industry in July, 1972. Mineral Industry Surveys, U.S. Department of the Interior, Bureau of Mines, Washington D.C. (December 1972).
5. Missouri Mineral News, Missouri Geological Survey and Water Resources, Rolla, MO. *14(3):* 39 (1974).
6. B.G. Wixson. In Developments in Water Quality Research. H.I. Shuval, ed. Ann Arbor Science Publishers, Ann Arbor, MI. 199 (1970).
7. An Interdisciplinary Investigation of Environmental Pollution by Lead and Other Heavy Metals from Industrial Development in the New Lead Belt of Southeastern Missouri, Vol. I and II, B.G. Wixson and J.C. Jennett, ed. Submitted by the Interdisciplinary Lead Belt Team, University of Missouri-Rolla to NSF-RANN, (June 1, 1974).
8. J.C. Jennett, B.G. Wixson, E. Bolter, and J.O. Pierce. Environmental Problems and Solutions Associated With the Development of the World's Largest Lead Mining District. In Pollution: Engineering and Scientific Solutions, E.S. Barrekette, ed. Plenum Press, New York, NY 320 (1973).
9. B.G. Wixson. Development of a Cooperative Programme for Environmental Protection Between the Lead-Mining Industry, Government, and the University of Missouri. In Minerals and the Environment. M.J. Jones, ed. The Institution of Mining and Metallurgy, London. 3 (1975).
10. J.C. Jennett, E. Bolter, N.L. Gale, W. Tranter, and M. Hardie. The Viburnum Trend, Southeast Missouri: the Largest Lead-Mining District in the World—Environmental Effects and Controls. In Minerals and the Environment. M.J. Jones, ed. The Institution of Mining and Metallurgy, London. 13 (1975).
11. B.G. Wixson, E. Bolter, N.L. Gale, J.C. Jennett, and K. Purushothaman. The Lead Industry as a Source of Trace Metals in the Environment. Proceedings of Cycling and Control of Metals, U.S. Environmental Protection Agency, Cincinnati, OH. 11 (February 1973).
12. W. Fulkerson, Personal communication, Oak Ridge National Laboratory, Oak Ridge, TN (January 1973).
13. I. Vander Hoven. Deposition of Particles and Gases. Meteorology and Atomic Energy, TID-24190, U.S. Atomic Energy Commission. 202 (July 1968).
14. K.D. Hage. Particle Fallout and Dispersion Below 30 km in the Atmosphere. Report SC-DC-64-1463, Sandia Corp. (1964).
15. P.M. Shah. Distribution of Ambient Particulates and Particulate Trace Metals Around the AMAX Smelter. M.S. thesis, University of Missouri-Rolla, Rolla, MO (May 1974).
16. An Interdisciplinary Investigation of Environmental Pollution by Lead and Other Heavy Metals by Industrial Development in the New Lead Belt in Southeastern Missouri. Vol. I, B.G. Wixson and J.C. Jennett, ed. Progress report submitted to NSF-RANN by the University of Missouri-Rolla (1974).
17. R.R. Brooks. Geobotany and Biogeochemistry in Mineral Exploration. Harper and Row, New York, NY (1972).
18. D.L. Butherus. Heavy Metal Contamination in Soil around a Lead Smelter in Southeast Missouri. Unpublished Master's thesis, University of Missouri-Rolla, Rolla, MO (1975).
19. E. Bolter, T. Butz, and J.F. Arseneau. Mobilization of Heavy Metals by Organic Acids in the Soils of a Lead Mining District. Proceedings of 9th Annual Conference on Trace Substances in Environmental Health, University of Missouri, Columbia, MO (1975).
20. E.A. Bondietti and E. Bolter. Proceedings of 2nd Annual NSF-RANN Trace Contaminants Conference, Asilomar, CA (1974).
21. Ecology and Analysis of Trace Contaminants, Progress Report for Period October 1973 - September 1974. ORNL-NSF-EATC-11. Oak Ridge National Laboratory, Oak Ridge, TN (1974).
22. A.P. Watson. Trace Element Impact on Forest Floor Litter in the New Lead Belt Region of southeastern Missouri. Proceedings of 9th Annual Conference, on Trace Substances in Environmental Health, University of Missouri, Columbia, MO (1975).
23. D. Djuric, A. Kerin, L. Graovac-Leposavic, L. Novak, and J. Kop. Arch. Environ. Health. *23:* 275 (1971).
24. J. Schmitt, G. Brown, E. Devlin, A. Larsen, E. McCausland, and J. Seville. Arch. Environ. Health. *23:* 185 (1971).
25. G.T. Goodman and T.M. Roberts. Nature. *231:* 287 (1971).
26. H.L. Cannon and J.M. Bowles. Science. *137:* 765 (1962).
27. R.H. Daines, H. J. Motto, and D.M. Chilko. Environ. Sci. Technol. *4:* 318 (1970).qlJ.V. Lagerwerff and A.W. Specht.
28. J. V. Lagerwerff and A. W. Specht. Environ. Sci. Technol. *4:* 583 (1970)
29. A.J. MacLead, R.L. Halstead, and B.J. Finn. Can. J. Soil Sci. *49:* 327 (1969).
30. H.L. Motto, R.H. Daines, D.M. Chilko, and C.K. Motto. Environ. Sci. Technol. *4:* 231 (1970).
31. W.H. Smith. For. Sci. *17:* 195 (1971).
32. A. Haney, J. A. Carlson, and G.L. Rolfe. Trans. Ill. Acad. Sci. *67:* 323 (1974).
33. J.J. Connor, H.T. Shacklette, and J.A. Erdman. U.S.G.S. Professional Paper, 750-B, B151. (1971).
34. Bolter, E. In An Interdisciplinary Investigation of Environmental Pollution by Lead and Other Heavy Metals by Industrial Development in the New Lead Belt of Southeastern Missouri, vol. 2, B.G. Wixson and J.C. Jennett, ed. Annual report submitted to NSF-RANN by the University of Missouri-Rolla (1974).
35. D. D. Hemphill, C. J. Marienfeld, R. S. Reddy, and J. O. Pierce, Arch. Environ. Health. *28:* 190 (1974).
36. D. D. Hemphill, C. J. Marienfeld, R. S. Reddy, W. D. Heidlage, and J. O. Pierce. J. Ass. Offic. Anal. Chem. *56:* 994 (1973).
37. An Interdisciplinary Investigation of Environmental Pollution by Lead and Other Heavy Metals by Industrial Development in the Lead Belt of Southeastern Missouri, 2 vol. B. G. Wixson and J. C. Jennett, ed. Annual report submitted to NSF-RANN by the University of Missouri-Rolla. (1974).

38. A. Callier. M. S. thesis, University of Missouri-Rolla (1975).
39. J. R. Hawley. The Use, Characteristics and Toxicity of Mine-Mill Reagents in the Province of Ontario. Ministry of the Environment, W. Toronto, Ontario (1972).
40. Standard Methods for the Examination of Water and Wastewater, 13th ed. American Public Health Association, Washington, D.C. (1971).
41. S. G. Jankovic, D. T. Mitchell, and J. C. Buzzel. Water Sewage Works. *114:* 471 (1967).
42. F. E. Ryck and J. R. Whitely. Pollution Abatement in the Lead Mining District of Missouri. 29th Purdue Industrial Waste Conference (1974).
43. J. L. Wilhm. J. Water Pollut. Contr. Fed. *39:* 1673 (1967).
44. J. C. Jennett and M. G. Hardie. Proceedings of 2nd NSF-RANN Trace Contaminants Conference, Asilomar, CA (1974).
45. N. L. Gale, P. Marcellus, and G. Underwood. Proceedings of 2nd NSF-RANN Trace Contaminants Conference, Asilomar, CA (1974).
46. An Interdisciplinary Investigation of Environmental Pollution by Lead and Other Heavy Metals by Industrial Development in the New Lead Belt of Southeastern Missouri. 2 vol. Report submitted to NSF-RANN by the University of Missouri-Rolla (1972).
47. E. Bolter, J.C Jennett, and B. G. Wixson. Geochemical Impact of Lead Mining Waste Waters on Streams in Southeastern Missouri. Transactions of the 27th Purdue Industrial Waste Conference, Lafayette, IN 679 (1972).
48. N. L. Gale, M. G. Hardie, J. C. Jennett, and A. Aleti. Transport of Trace Pollutants in Lead Mining Waste-waters. Proceedings of 6th Annual Conference on Trace Substances in Environmental Health, Columbia, MO (1972).
49. J.C. Jennett, B.G. Wixson, E. Bolter and N. L. Gale. Transport Mechanisms of Lead Industry Wastes. Transactions of 28th Purdue Industrial Waste Conference, Lafayette, IN 496 (1973).
50. R. Patrick, B. Crum, and J. Coles. *Bot. 64:* 472 (1969).
51. G. C. Gerloff and F. Skoog. Ecol. *38:* 551 (1957).
52. J.C. Jennett and B. G. Wixson. Industrial Treatment of Heavy Metals to Protect Aquatic Systems in the New Lead Belt Area. Proceedings of 30th Purdue Industrial Waste Conference, Lafayette, IN (1975).
53. S. Linnemann. M. S. thesis, University of Missouri-Rolla (1974).
54. W. J. Ernst. M. S. thesis, University of Missouri-Rolla (1976).
55. J.L. Foil. M. S. thesis, University of Missouri-Rolla. (1975).
56. G.L. Rolfe and J. C. Jennett. Environmental Lead Distribution in Relation to Automobile and Mine and Smelter Sources. Transactions of Seminar on Heavy Metals in the Aquatic Environment, Vanderbilt University (1973).
57. S. J. de Groot and E. Allersma. Field Observation on the Transport of Heavy Metals in Sediments. Transactions of Seminar on Heavy Metals in the Aquatic Environment, Vanderbilt University. (1973).
58. J.C. Jennett and B.G. Wixson. New New Lead Belt: Aquatic Metal Pathways-Control. Proceedings of International Conference on Heavy Metals in the Environment, Toronto, Ontario, (October 1975, in press).
59. N. L. Gale, E. Bolter, and B. G. Wixson. Investigation of Clearwater Lake as a Potential Sink for Heavy Metals From Lead Mining in Southeast Missouri. Proceedings of 10th Annual Conference on Trace Substances in Environmental Health, Columbia, MO (1976, in press).
60. K.E. Carpenter. Ann. Appl. Biol. *12:* 1 (1963).
61. N. Gale, B.G. Wixson, M.G. Hardie, and J.C. Jennett. Bull. Aer. Water Resources Assoc. *9:* 673 (1973).
62. E. Bolter. Proceedings of 2nd NSF-RANN Trace Contaminants Conference, Asilomar, CA (1974).
63. D. D. Hempshill. Environ. Health. *28:* 190 (1974).
64. Environmental Pollution by Lead and Other Metals. Institute for Environmental Studies-University of Illinois, Urbana-Champaign. Progress report, Nov. 1, 1972 to Apr. 30, 1974.
65. C. Malone, D. E. Koeppe, and R. J. Miller. Plant Physiol. (1974).
66. J.H. Rule, D. Hemphill, and J.O. Pierce. Proceedings of 2nd NS-RANN Trace Contaminants Conference, Asilomar, CA (1974).
67. B.G. Wixson nd N.L. Gale. Some Limnological Effects of Lead and Associated Heavy Metals from Mineral Production in Southeast Missouri, U.S.A. Proceedings of International Symposium on Interaction Between Water and Living Matter, Odessa U.S.S.R. (October 1975, in press).
68. N.L. Gale, M.G. Hardie, J. Whitfield, and P. Marcellus. The Impact of Lead Mine and Mill Effluents on Aquatic Life. Proceedings of the University of Minnesota-AIME Mining Symposium, Duluth, MN (1974).
69. R.L. Kramer. M. S. thesis, University of Missouri-Rolla (1976).
70. I. Iwasaki. and S. R. B. Cooke. Mining Eng. 1267 (1957).
71. Aero Xanthate Handbook. American Cyanamid Co. Minng Chemicals Department, Wayne, NJ (1972).
72. Addison, Schnare, and Gordon. Analysis of Northeastern New Brunswick Mine Wastes for Xanthate, Isopropyl, Ethylthionocarbamate (DOW Z-200) and Dithiophosphate (Cyanamid Solium Aerofloat). Fisheries Research Board of Canada, Marine Ecology Laboratory, Bedford Institute, Dartmouth, Nova Scotia (1972). Cited in reference 3.
73. S. Wenderoth and E. Yost. Multispectral Photography for Earth Resources. Science Engineering Research Group, C. W. Post Center, Long Island University (1974).
74. W.H. Tranter and J.L. Sandvos. A Scheme for the Determination of Optimum Sampling Sites for Soil and Vegetation. Proceedings of the 2nd NSF-RANN Trace Contaminants Conference, Asilomar, CA 149 (August. 1974).
75. W. H. Tranter, J. L. Sandvos, F. T. Ulaby, and S. Perkins. Correlating Photographic Data and Heavy Metal Concentration in the Soil of the New Lead Belt of Missouri. University of Missouri-Rolla, Technical Report CSR-74-6 (September 1974).
76. W. H. Tranter and J. L. Sandvos. Surface Interpolation and Automated Contour Mapping. University of Missouri-Rolla, Technical Report CSR-74-5 (June 1974).
77. M. D. Turner. Computer-Aided Sampling and Surface Interpolation. M.S.E.E. thesis, University of Missouri-Rolla (1975).
78. B. G. Wixson. *Environ. Sci. Technol. 9:* 1128 (1975).

PART III

Effects of Lead

CHAPTER 8

EFFECTS ON MICROORGANISMS, PLANTS, AND ANIMALS

T. G. Tornabene

Department of Microbiology
Colorado State University

N. L. Gale
Department of Life Sciences
University of Missouri-Rolla

D. E. Koeppe
Department of Agronomy
University of Illinois

R. L. Zimdahl
Department of Botany and Plant Pathology
Colorado State University

R. M. Forbes
Department of Animal Science
University of Illinois

INTRODUCTION

A contaminant has been defined as anything added to the environment that causes a deviation from the *mean* chemical composition that a particular phase of the environment would have in the absence of human activity. In a more derogatory sense, a contaminant is further classified as a pollutant if it adversely affects man or something he values. It is possible that all contaminants may also be pollutants, but a prerequisite to this judgment is the definition of effects in a qualitative and quantitative sense.

Lead is a natural component of our environment, but the levels of natural contamination are unknown, and it has been difficult to determine the preindustrial level of environmental lead. The concern is over possible hazards to the aggregate of life as it exists in the world resulting from the increased and widespread dissemination of lead. Within this general context is the particular concern about the more subtle effects of chronic exposure to lead pollution. [1]

This chapter will present and evaluate knowledge on the biological effects of lead contamination, excluding references to the clinical aspects of lead toxicity in humans. It is divided into four sections: bacteria, aquatic flora in fauna, plants, and animals. The National Academy of Sciences report [2] has detailed the effects of atmospheric lead, and other reviews have discussed the effects of lead and other heavy metals on biological systems. [3-6] The literature on the toxicity of lead in mammals is largely devoted to eucaryotic systems and has been reviewed by Passow et al. [7]

EFFECTS ON BACTERIA

The adverse effects of metals on biological systems have recently been reviewed by Vallee and Ulmer [4] and do not need to be repeated here in detail. For convenience, however, some of the results will be included in this brief evaluation of effects of lead on bacteria.

In studies on the environmental impact of lead, it has been shown that for most naturally occurring bacteria lead can be readily tolerated without being lethal. [4, 8-20] The observed effects from lead-ion exposures have been multiform. There are reports that lead ions can stimulate the growth of a bacterium identified as *Micrococcus flava* Strevisan, [15, 16] producing an insoluble lead sulfide metabolite;

lead ions can also be moderately inhibitory on anaerobic activated sludge bacteria, [17] aerobic river-water bacteria, [8, 18] and marine sulfate-reducing bacteria. [20]

These reports indicate that lead has a relatively causal relationship with bacterial cells, with no specific inhibitory role. However, these results must be accepted with a degree of caution as they may be the products of various incomplete reflections that one generally obtains from studies involving diverse experimental procedures and research goals. For example, the moderately inhibitory effects of greater than 50 μg/g of lead on anaerobic activated sludge bacteria [17] may not at all be representative of the effects of lead on bacterial systems. The experimental designs were for the sole purpose of evaluating treatment methods for plant wastewater that may contain ihibitory metals; therefore, unusually short incubation periods were employed. The effects of lead were determined by the number of cells detected by standard turbidimetric and plate-count methods. The relatively fewer cells found in lead-containing samples were moderately contradictory to the respirometry data collected for the same samples and thus may just be an indication of a brief period of metal shocking of the cells.

The studies on inhibitory effects of lead on viability of marine sulfate-reducing bacteria [20] may not have been due directly to lead but to the accumulation of the H_2S. The experiments showing stimulation of growth [15, 16] were for cultures grown through complete growth cycles. The longer cultivation periods provide the likelihood of strain selection of obtaining lead-resistant cultures. On the other hand, the immobilization and nature of the lead in bacterial membranes could possibly increase or decrease the permeability properties and result in lead stimulation [15, 16] or inhibition [8, 17, 18, 20] of bacterial growth. A probable mechanism could be based on the synergistic effects that monovalent and divalent cations have on softening or stiffening of membranes. [21] The basis for the resistance of bacteria to lead is apparently genetically related. The moderately inhibitory, but not lethal, effects of lead in a few bacterial systems studied were attributed to a series of extrachromosomal resistant factors; [12, 14, 22] similar resistant factors in several bacteria have been recently proposed for mercury [22, 23] and cadmium. [23] In lieu of the obscurity of the modus operandi of lead, more work is necessary to determine its effects on naturally occurring bacteria.

Recent studies [10] on *Micrococcus luteus* Schroeter, Cohn cells, cultivated in a medium containing lead salts, exhibited a sequence of changes in the quantity of total cellular lipids, with essentially no changes from normal cellular yields. The lipid composition of cells cultivated five to seven times was reduced by as much as 50 percent (phase II). Cells cultivated more than seven times in lead-containing media had progressively greater quantities of lipid (phase III), approaching that found in control cells. These cells with reestablished lipid contents showed no further effects from more prolonged exposure to lead salts. Chromatographic studies of total lipids of cells of each lipid phase revealed relatively complete lipid compositions. These results indicated that lead is affecting a common biochemical parameter that is involved early in the biosynthesis of lipids. Only moderate changes were observed in the relative intensities of individual components in both the nonpolar and polar lipids in each lipid phase. The most notable changes were the decrease in aliphatic hydrocarbons, with concommitant increases in the diglycerides and components tentatively identified as a family of ketones. More work is necessary to establish the specific site(s) of interaction(s).

Analyses of lead interactions with cellular subfractions of *Micrococcus luteus* and *Azotobacter* sp. grown in media containing substantial quantities of lead revealed that virtually all of the lead taken up by the cells was associated with the cellular envelopes. Close to 90 percent of the bound lead was associated with the membrane fraction of *M. luteus.* [9] Little or no lead was found in the cytoplasmic fraction. A major conclusion of the study was that the microbial systems investigated are highly capable of abstracting substantial quantities of inorganic lead and that the lead does not appear to interfere with the normal functions of the cells. An extension of this work with studies on the interaction of lead with specific cellular components has indicated that individual membrane lipids do not provide specific stable binding sites for lead, but that natural membrane lipid mixtures may provide an environment suitable for nucleation of lead. [19] The results demonstrate a causal relationship between lead and lipids, but the impetus for lead immobilization is in the bacterial membrane. Few additional data exist on lead-immobilizing sites in bacteria. Studies on lead interactions with cellular components have been largely restricted to eucaryotic systems, proteinaceous materials, [24-27] and, to a much lesser degree, with diketones and alcohols [28] and phospholipids. [19, 29] However, it is generally expected that the basis for lead's general uptake and interaction with cells can be more readily explained by the fact that lead has a strong affinity for any chemical group capable of associating with hydrogen ions.

The enormous literature on the metabolic toxicity of lead is restricted primarily to eucaryotic systems. Much of this has been reviewed by Passow et al. [7] The little knowledge that does exist on the metabolic effects of lead on bacterial systems was derived as a consequence of studies directed towards other goals. As a result, the establishment of possible biochemical sites of lead action in bacteria remains somewhat vague. It is fairly safe to assume, however, that lead introduced into an enzymatic system *in vitro* will generally result in some interference, even though it is generally less inhibitory than many other metals. [7, 11-14, 30] The knowledge of specific biochemical lead-action sites in bacteria is generally limited to the reports that lead ions interfere with incorporation of ^{14}C-leucine into *Escherichia coli t*-RNA, by an unknown mechanism [31] and to the biochemical sites of lead action in heme synthesis, [4, 31-33] i.e., lipoamide dehydrogenase, certain ATPases; δ-aminolevulinic acid dehydrase; coproporphyrinogenoxidase; and ferrochelatase.

A major concern of the public has been the possibility of bacterial transformation of lead in the environment. The occurrence of biological conversions of lead has not yet been established; however, evidence for possible biomethylation of lead [13] is just now beginning to be reported, showing promise that identification of biotransformation mechanisms of lead is probable in the very near future.

The adverse effects of lead on bacterial cells are limited to a surprisingly small number of experiments on whole cells in the presence of lead; to a study on membrane uptake of lead; and to a few reports on the specific biochemical sites of lead action. Thus the effects of lead on bacterial cells and bacterial populations are largely unexplored, and their manifestations of toxicity are virtually unknown.

EFFECTS ON AQUATIC FLORA AND FAUNA

While there is a dearth of literature regarding the effects of lead on lower trophic levels of aquatic communities, the reports available suggest that dissolved lead salts are generally toxic in aquatic environments. The degree of toxicity, as well as the degree of solubility, are greatly dependent upon the attendant chemical and physical conditions. Moreover, the literature indicates considerable variability in the tolerance of individual aquatic organisms to the presence of lead.

Few studies have been reported on the effects of lead on photosynthetic organisms found in natural waters. Whitton [34] recently reviewed the literature dealing with the toxicity of heavy metals to freshwater algae. This review provides some useful information about field observations made on a number of rivers in Great Britain near mining and smelting operations, along with related laboratory studies aimed at determining toxic levels of lead, zinc, and copper. The observed alterations in normal flora, and especially fauna, in those British streams were attributed to the combined effects of silt accumulation and the presence of zinc, copper, and lead salts. Because of the reduced solubility of lead salts, especially in hard waters, greater concern was apparent for zinc and copper salts, which are not only more soluble under similar conditions, but also may display some synergistic effects.

Whitton [35] studied the toxicity of zinc, copper, and lead salts to *Chlorophyta* from natural waters under one set of standard test conditions. The different species of algae were found to respond with varying degrees of resistance to a wide range of heavy-metal concentrations. It is pertinent to note that all the species tested by Whitton were considerably less sensitive to lead than to zinc or copper.

The toxic effects of lead on the representative *Chlorophyta* studied by Whitton were observed with between 3 and 60 mg/l of lead, added as the chloride. Among the genera tested, *Cladophora* was the most sensitive to lead, by nearly an order of magnitude, and may be a possible indicator organism for heavy metals in stream waters. *Microspora*, isolated from streams known to receive effluents from lead mines and mills, was found to be quite resistant to lead toxicity, while others—including *Stigeoclonium, Ulothrix, Spirogyra,* and *Mougeotia*—were of intermediate resistance. Whitton reported that resistance to metals is an individual characteristic, and resistance to one metal did not necessarily imply resistance to others.

Observations in the New Lead Belt of Missouri by Gale et al. [36] have shown many similarities with those reported for the British streams receiving effluents from lead mining activities. Both situations have shown that relatively high concentrations of lead and zinc in the stream bottom sediments do not have much effect on algal growth in relatively hard natural waters. Under these conditions, the dissolved lead salts are in very low concentrations and well below the limits of tolerance of most algae, including the sensitive *Cladophora*. Extensive blooms of *Cladophora* have been observed in one stream where the trapped and bound lead associated with the filaments surpassed 5,000 μg/g. [36] Approximately 90 percent of the lead associated with these algal filaments remained bound to the algae through repeated washings in streamwater, tapwater, or distilled water, but was easily removed by one or two washings with 0.1*M* EDTA (ethylenediaminetetraacetic acid) at pH 7.0. These results indicate that the lad, though chelated or bound to the cell envelopes, apparently had no marked physiological effect under the existing natural conditions.

It is generally recognized that the lead mining industry has contributed to significant changes in flora and fauna of receiving streams, but, judging from low concentrations of dissolved lead and recent field observations, it is very doubtful that lead alone is responsible for all observed changes. It is much more probable that other substances attending lead mining and milling operations (zinc, copper, manganese, toxic milling reagents, silt, and general nutrient conditions), [37] operating independently or synergistically, should be considered suspect in those streams demonstrating serious alteration.

Because of the extremely low solubility products of some lead salts involving anions of metabolic importance (sulfates, phosphates, and carbontes), it is possible to speculate that much of the inhibition attributed to lead may be due to the simple removal of essential anions requisite to proper metabolism. Christiansen et al. [38] and Gale et al. [36] have shown that the addition of phosphate markedly increases the amount of CO_2 fixed by diatom communities suspended in distilled water or in natural streamwater where phosphate is the principal limiting factor.

Whitfield et al. [39] subsequently found that light-dependent fixation of $^{14}CO_2$ in communities of diatoms isolated from the streams of the New Lead Belt of southeast Missouri is significantly inhibited by concentrations of lead acetate or nitrate in excess of 5 mg/l lead. This effect was noted under conditions of limited nutrients and probably applies to normal stream conditions in that geographical area. In other experiments using commercial algal broth and apparently nonlimited nutrients, inhibitory effects were not seen up to approximately 20 mg/l lead.

For most common stream consumer organisms, except fish, information in the lterature is extremely scarce. The effects of lead on bacteria discussed earlier might also apply to microorganisms, with the understanding that pH, hardness, dissolved gases, and general nutrient conditions of natural waters can modify solubility and toxicity of lead salts.

The toxic concentrations of lead reported for bacteria, flagellates, and infusoria associated with sewage treatment are 0.5 to 1 mg/l, although some results indicate that levels as low as 0.1 mg/l may be inhibitory to bacteria that decompose organic matter. [40-43]

Protozoans are adversely affected by the presence of lead, as shown by Wang. [44] He found that approximately $10^{-4} M$ lead acetate was lethal for *Paramecia aurelia.* Gray and Ventilla [45] showed that low concentrations of lead lowered the growth rates of the bactivorous sediment-inhabiting ciliate, *Cristigera.*

The Environmental Protection Agency [46] cites the work of Shaw and Grushkin [47] showing lead toxicity to *Daphnia magna* (waterflea) and to *Bufo valliceps* (tadpoles) at 10 mg/l. However, Ellis [48] found that 1.6 mg/l of lead nitrate was deleterious to the growth of tadpoles and that 3.3 mg/l was lethal. An earlier study by Anderson [49] using Lake Erie water as the diluent indicated that 0.01 to 1 mg/l lead chloride can be toxic to *Daphnia magna.*

Whitley [50] studied the TL_m (median tolerance limit) levels for tubificid worms at pH values ranging from 5.8 to 9.7. Within this pH range, lead nitrate was found to be toxic at concentrations of 28 to 49 mg/l. The toxicity was more pronounced at pH extremes of 6.5 and 8.5 than at 7.5.

Brown and Ahsanullah [51] found that among the various heavy metals, lead was the least toxic to worms (*Ophryotrocha*) and brine shrimp (*Artemia*). The worms were not visibly affected by lead concentrations up to 10 mg/l while some inhibition of the growth rate of brine shrimp was observed with 5 mg/l lead.

Reviews on the toxicity of heavy metals to fish have been made by Doudoroff and Katz, [52] McKee and Wolf, [53] and the Environmental Protection Agency. [46] Data included in these reviews indicate a great variation among genera in their resistance to lead salts, and also from one set of chemical and physical conditions to another. It is very difficult to extrapolate some of the laboratory studies into the real world of natural waters and the attendant conditions of hardness, pH, oxygen concentration, etc.

A few studies pertaining to the toxicity of lead are discussed below to give some idea of the species of dish studied and the range of heavy metals toxicity.

The toxic effects of 0.1 to 0.4 mg/l lead to minnows, sticklebacks, and trout have been noted in distilled and soft water. [54-56] Minnows *(Leuciscus)* were killed within 48 h by 0.4 mg/l lead at pH 6.4 to 6.6 in distilled water and riverwater, while goldfish survived indefinitely in 1.0 but not in 10 mg/l lead solutions. Other minnows *(Phoxinus)* and sticklebacks were not visibly affected after 3 weeks exposure to 0.7 mg/l lead in harder water with a calcium content of approximately 51 mg/l. Sticklebacks were killed in soft tap water containing 0.1 to 0.2 mg/l lead. Exploratory tests by Tarzwell and Henderson [57] have shown that toxicity of lead chloride for fathead minnows in soft and hard waters was 2.4 and >75 mg/l lead, respectively, as the 96 h TL_m.

Ellis [58] found that some goldfish were killed within 4 days by 100 mg/l lead nitrate (62.6 mg/l lead) prepared in hard water at pH 6.8. A 10 mg/l lead nitrate solution was not fatal to goldfish at pH 7.4, but 1,000 mg/l was rapidly fatal at pH 6.4. It is obvious that some of these higher concentrations of lead salts could not possibly be achieved at neutral or alkaline conditions in hard water.

A concentration of 3 mg/l lead nitrate in freshwater killed common American Killifish *(Fundulus heteroclitus)* in 12 h, [59] and 10 mg/l killed trout in 2 h; [60] Ebeling [61, reported however, that 1 mg/l lead in tap water did not kill trout in 48 h. Dilling et al. [62] reported that 17 mg/l lead was the minimum lethal concentration for goldfish, whereas catfish [63] could withstand 50 mg/l lead acetate in tap water, although prolonged exposure was detrimental. Ellis [58] suggested that lead sulfide, which is only slightly soluble in water, may have a cumulative effect on goldfish kept in aquaria with the bottoms covered with galena. Exposed fish died within 61 days, while all controls survived. On the other hand, Carpenter [55] failed to observe any injurious effect of galena on fish in experiments that lasted more than 6 weeks. Other work by Crandall and Goodnight [64] suggests that chronic lead toxicity is a definite possibility in fish (e.g. guppies), even with sublethal concentrations in hard water. This suggests that it will be very difficult to establish "safe" concentrations of lead and other heavy metals for aquatic animals.

Preliminary data reported by Birge [65] indicate that concentrations of lead chloride as low as 10 μg/l produce significant lethality and morphological deformities in the developing embryos of trout. It is probable that the tolerance limits of embryonic stages to lead and other heavy metals are narrower than those of adult organisms. More information is obviously needed regarding teratogenic effects of lead, as well as long-term or chronic toxicity studies to determine effects of long exposure of individual organisms and populations to trace quantities of heavy metals.

Wallen et al. [66] tested the toxicity of insoluble lead oxide toward the mosquito fish (*Gambusia affinis*) and found a 96-h TL_m in excess of 56,000 mg/l. McKee and Wolf, [53] therefore, state that insoluble lead is not toxic to fish.

Carprenter [67] reported a general paucity of fauna, especially fish in those rivers of Wales known to be polluted by lead-mining and lead-washing activities. She correlated this with the presence of lead, zinc, and copper salts. Gale et al. [36] confirmed this in some of the streams affected by mining and milling operations in Missouri's New Lead Belt, but they stated that toxic organic milling reagents may also be involved.

It is of interest to compare the lead content of fish from different localities, but toxic levels cannot be determined from the available data. Pakkala et al. [68] analyzed 419 fish of various species sampled in 1969 from various New York State waters. Most of the samples showed lead contents of 0.3 to 1.5 mg/l, but a few had levels up to 3 mg/l. They reported no correlation between lead concentration and size, species, or sex of the fish; and lead did not accumulate in lake trout of known age up to 12 years. Fulkerson and Goeller [69] found concentrations of lead in freshwater fishes ranging from 0.5 to 2.0 μg/l, and marine fishes were reported to contain approximately 0.5 μg/l. The muscle of some species of fish from Lake Michigan ranged from 0.005 to 0.560 μg/l, with an average of 0.165 for 46 specimens. [70] In the New Lead Belt, Gale et al. [71] found total body lead values for fish ranging from <0.1 (undetectable amounts) to 192 μg/l (dry weight). In general, fish of the New Lead Belt reach much higher lead values than those reported from other natural waters. Some of this difference can possibly be attributed to the intestinal content of fish in the New Lead Belt, especially those herbivorous or grazing fish known to feed on algae that may retain lead up to several thousand parts per million. Preliminary results confirm that most of the body burden of lead is in the gills and gastrointestinal tract of typical specimens.

For comparative purposes, the review of Doudoroff and Katz [52] points out that zinc concentrations of 0.3 to 0.4 mg/l can be decidedly toxic to fish. Copper is toxic to many species of fish in concentrations of 0.01 to 0.02 mg/l; 0.25 to 1.0 mg/l copper has also been reported to be a lethal or maximum safe level for some species of trout, perch, and several other fish. Generally, it has been found that all metal salts are likely to be less harmful to fish in hard, alkaline, well-oxygenated waters.

Heavy metals probably exert their toxic effects on fish by reacting with the mucous on the surface of the gills, causing precipitation or coagulation and thus interfering with the normal exchange of gases. Fish may become acclimated to the presence of heavy metals, although this has not been documented. Because of the variable response to heavy metals and varying physical conditions that may strongly influence toxicity, Doudoroff and Katz [52] strongly recommend that bioassays be used to determine the proper levels of dilution, extent of treatment, etc., in those industrial sites where potential hazards exist.

Aronson [72] states that evidence is lacking to show that lead is present in natural waters of the United States at levels likely to be toxic, or to constitute a health problem for fish. He points out that much of the man-dispersed lead is eventually washed into natural waters and is probably precipitated by carbonates, hydroxides, and organic ligands in the water; there is no evidence that lead thus precipitated to the bottom of natural waterways is harmful to fish.

EFFECTS ON PLANTS

Before considering the effects of lead on plants, it is necessary to reiterate that under certain conditions plants do take up lead from soils, and that this lead is translocated to various tissues. (See ch. 5.) A portion of the literature on this subject is included in the National Academy of Sciences report [2] and will not be repeated here. Other sections of this book discuss the mechanism of lead accumulation at the cellular level in roots, and the chelation of lead and its effect on uptake and distribution within the plant. (See ch. 5).

While there is no evidence that lead is essential for the growth of any plant species, conflicting reports of growth stimulations and reductions due to lead are plentiful. Growth stimulations, mostly at low lead concentrations, have been reported for various plants by Keaton, [73] Weiler, [74] Berry, [75] Scharrer and Schropp, [76] Stoklaso, [77] Stutzer, [78] and Voelcker. [79] It should be noted that these studies were conducted before optimal conditions for plant growth were well defined, and the stimulation may have been due to a lead-associated effect, for example to the nitrate of lead, and not to the lead itself. Growth reductions under varied conditions have also been reported by many authors. [75, 80-84]

Other reports describe the reduction of vital plant processes such as photosynthesis, [85-88] mitosis, [89] or water absorption. [85] Lead may also contribute to copper deficiency. [90] Most of the experiments reporting lead toxicity have been conducted with plants grown in artificial nutrient cultures. However, Weiler [74] reported lead-induced plant toxicity on acid moor soils. Hooper [84] found that *Phaseolus* sp. (French bean) was damaged by 30 μg/g of lead as the sulfate alone or in combination with other nutrient salts in solution culture. No increase in toxicity from 0 to 30 μg/g was detected. Wilkens [81] reported that 3 μg/g was toxic to *Festuca ovina* (fescue). From all reports, the concept has emerged that the effect of lead, whether stimulatory or inhibitory, is dependent upon a number of possible environmental variables. These include associated anions ana cations, within the plant and in the growth media, and the physical and chemical characteristics of the soil itself. Different soils or species of plants may be very important in the overall amount of lead that becomes associated with the plants. Wallace and Romney [91] reported that soil pH, soil temperature, calcium availability, heavy metal availability, phosphorus supply, and soluble silicon may be important in the uptake of any heavy metal. Warren et al. [92] explained variations in plant uptake of lead to be due to pH, soil organic matter content, soil type, climate, topography, pollution, and geologic background of the soil. Lead toxicity to corn (*Zea mays* L.), beans, lettuce (*Lactuca* spp.), and radishes (*Raphanus* spp.) was greater when grown under slightly acid soil conditions than when grown on calcareous soils. MacLean et al. [93] found that the concentration of lead in oats (*Avena sativa* L.) and alfalfa (*Medicago sativa* L.) increased with decreased pH and organic matter, and that the addition of phosphate reduced the uptake of lead. Liming of soils has generally decreased lead uptake [94, 95] while phosphate-deficient corn accumulated several times as much lead as normal plants. [83]

Experiments have been conducted in which corn was grown in native Illinois soils selected to give a range of cation exchange capacities (CEC), phosphorus levels, and pH [96] Lead uptake and its subsequent effect (J.E. Miller, unpublished results) on 4 week old corn shoots grown in these soils with lead amendments were dependent on the level of lead in the soils relative to their capacity to sorb lead. Lead uptake decreased with an increase in soil pH, cation exchange capacity, and available phosphorus.

Lead affects plant processes only at relatively high tissue or growth culture solution concentrations, but the condition of the plant at the time of lead impact may also be of importance. Young tobacco (*Nicotiana tabacum* L.) seedlings accumulated lead more rapidly than older plants, and broad beans absorbed lead more rapidly from $10^{-4}M$ $PbCl_2$ solutions when the cotyledons were removed. [80] Similarly, Hunter [97] found that the lead content of bracken fern (*Pteridium aquilinum* L.) fronds decreased with age. Mitchell and Reith [98], reported however, a large increase in the lead content of pasture herbage in the fall. This same phenomenon was found in senescent leaves of some deciduous trees [99] and in wild oats (*Avena fatua* L.). [100]

Because of the interaction of lead with many environmental factors, it is extremely difficult to correlate its concentration with its effects. Baumhardt and Welch [101] reported that lead concentrations as low as 10 μg/g reduced corn root growth by 30 percent in nutrient culture medium, but no effect on plant growth was found when 270 μg/g of ammonium-acetate-extractable lead was present in typically nutrient-rich soils of central Illinois. Corn grown in sand with added nutrients accumulated more lead when the plants were phosphate deficient and was adversely affected at concentrations as low as 400 μg/g; however, lead concentrations as high as 2,000 μg/g in phosphate-sufficient plants had little effect. [83]

Rolfe [102] reported that lead amendments to three native Illinois soils resulted in lead uptake by 2 year old tree seedlings. In all instances, high phosphorus additions to the soils affected significant decreases in the uptake of lead. When plants are grown under natural conditions with adequate fertilization, scattered reports suggest that in all but sandy soils, soil lead concentrations of at least 1,000 μg/g must be present before effects on growth processes are observed. At high lead concentrations, reductions of photosynthesis, [85, 87, 88, 103] transpiration, [87, 103] root-tip motosis, [89] water absorption, [85] and induced apparent deficiencies of other nutrients [83, 104] have been reported. At tissue concentrations of 193 μg/g, the photosynthesis of detached sunflower (*Helianthus annus* L.) leaves was inhibited by 50 percent. [88]

Lead particulates are deposited naturally on plant leaf surfaces, with pubescent leaves accumulating seven times more than those with smooth surfaces. [105] Carlson and coworkers (unpublished data) at Illinois found that pubescent soybean leaves accumulated as much as 1,100 μg/g (aerosol dry weight) after several days of aerosol fumigation. This surface lead was not translocated and did not affect the rate of leaf photosynthesis, either immediately after or 2 weeks following fumigation.

There have been a limited number of reports of subcellular effects of lead on plant tissues. These indicate that lead does bind with membranes, and that this binding is likely associated with the detrimental effects observed. [106, 107] Studies of binding lead to membranes of isolated mitochondria indicate a very high passive affinity of lead for these membranes, with the number of sites being greater than for calcium, but generally comparable to other trace elements such as manganese, nickel, cadmium, or zinc. [106]

Some of the enzymatic reactions associated with two isolated cellular organelles, mitochondria and chloroplasts, have been most extensively studied. In both instances, lead has been found to affect the flow of electrons in electron transfer reactions [86, 108-110] and, as such, to have profound detrimental effects on the entire processes of respiration and photosynthesis. In work with isolated corn mitochondria, concentrations as low as 1 μg/g were found to reduce the oxidation of succinate. [108] Additional effects on general membrane integrity have been noted in work with isolated mitochondria, which influence several energy conversion processes associated with membranes. [106] Reports by Goyer and coworkers [111-114] and Scott, et al. [115] detailing the effects of lead on animal mitochondria, suggest a strong similarity in the effects of lead on animal and plant mitochondria.

Considerable information is available on the detrimental effects of lead on subcellular and enzyme systems isolated from animals. While the amount of corresponding work on plants is meager in comparison, that which has been done clearly indicates similar effects of lead at this level of experimentation and suggests that the potential for lead poisoning is present in plants.

The fact that lead poisoning has not been observed in plants growing under natural conditions can be explained in several possible ways. The first, and most obvious, reason is that environmental concentrations of lead are presently not high enough to cause any effect. The second possible explanation is that there are lead-binding systems in soils and plants, thus preventing lead from reaching enzymatic sites where it would affect and likely be detrimental to growth and yield.

For most land areas, the first explanation would certainly hold. Lead concentrations here are scarcely above baseline levels established before the environmental insult of lead from leaded gasolines. However, in more limited localities such as urban areas, strips of rural land immediately adjacent to highways, and areas around lead mines and smelters, the second explanation (that lead is presently bound in an inactive or unavailable mode) is more likely to hold. This second situation is potentially dangerous since environmental conditions (weather and soil type) vary greatly from one locality to another. Research is beginning to show that modifications in these conditions can greatly affect the amount of lead bound by soils and plants in an "inactive" form. The concentration of lead in a soil is, in itself, of little value in predicting its influence on plants and associated components of ecosystems. Soil type, amount of precipitation, light intensity, and temperature are all factors to be considered in this prediction. Research has now shown that 500 μg/g lead under one set of environmental conditions will have no effect on plant growth, while under other conditions plants may be severely affected.

If predictive models are to be at all useful, future research must resolve the role of varying environmental conditions and their effect on lead "inactivation" within the ecosystem.

EFFECTS ON ANIMALS

Within animals, the effects of lead are primarily manifested in the nervous and hematopoietic systems and in kidney tissue. These effects have been studied, primarily in rats, at the enzyme, subcellular, cellular, and tissue morphology levels, as well as at the systemic physiological and biochemical function levels. Some tissues other than those cited above have been studied in lead toxicity investigations. Some of the concepts derived from the many studies available in the literature are presented here.

Hematopoietic System

Although impaired functional capacity of the hematopoietic system, as evidenced by depressed levels of circulating hemoglobin, is a late development of the lead toxic syndrome, it is also true that this system can adapt to the effects of lead. However, it has received a great deal of attention. The most cogent reasons for this are the early appearance in blood and urine of changes in levels of metabolic constituents of the heme synthetic pathway, ready access of these fluids for examination, and good correlation between lead dose and response. [116-122]

The principal steps in the heme synthetic pathway are well known, and evidence from rat studies has accumulated to identify major sites at which lead interferes with this process. Lead toxicity increases δ-aminolevulinic acid (ALA) excretion in the urine and decreases δ-aminolevulinic acid dehydrase (ALAD) activity in blood and a number of other tissues. ALAD catalyzes the formation of the pyrrole, porphobilinogen, from two molecules of ALA. It is a sulfhydryl-containing enzyme strongly inhibited *in vitro* and *in vivo* by lead. [118] Hernberg and Nikkanen, [118] on incomplete evidence, surmise that lead acts, not at the active site, but by allosteric inhibition of the enzyme. The inhibition occurs quickly *in vivo*, as noted "during the very first days" of industrial lead exposure in humans, [118] within 1 day following intravenous administration of as little as 5 μg of lead per kg body weight in rabbits, [117] and within 12 h after feeding a single dose of 100 μg Pb in a nutritionally complete diet to 100 g rats. [122]

The inhibition of ALAD in mature red cells is probably of little direct physiological significance. It seems likely that ALAD present in circulating blood represents residual nonfunctional activity in the mature red cells. It is also likely that blood ALAD inhibition reflects inhibition in other tissues where heme synthesis occurs continuously for the production of heme-containing enzymes such as cytochromes. [121-123] A good correlation between lead dosage and inhibition of ALAD in blood, kidney, and liver was found by Cerklewski and Forbes [123] in young rats fed purified diets containing 0, 50, 500, or 1,000 μg/g lead (as the acetate) for 3 weeks. At the 500 μg/g dose level, the percent ALAD inhibition was 81, 53, and 30 for blood, kidney, and liver, respectively. In spite of these inhibitions in ALAD activity, the heme content of the blood was not affected by 500 μg/g of lead, thus indicating a considerable reserve capacity of ALAD activity above that necessary to synthesize heme for incorporation into hemoglobin.

One consequence of inhibition of ALAD is the increase of its substrate, ALA, in the urine. This occurs rapidly, [122] appearing within 12 h after a single oral dose of lead. In the event of continued exposure to lead, ALA excretion [123] remains elevated, yet blood hemoglobin may remain within normal limits. It thus appears that δ-aminolevulinic acid synthetase (ALAS) activity is not inhibited by lead to the same extent as ALAD. [122] Two other possible loci for lead interference with heme synthesis are coprogenase and heme synthetase, which incorporates iron into the protoporphyrin nucleus. [124]

The significance of these studies in terms of the practical effects of lead toxicity on heme synthesis is not clear. Maxfield et al. [125] reported that lead-poisoned dogs responded in normal fashion to the hematological stress of acute hemorrhage. In this experiment, dogs were given diets containing (1) no added lead, (2) lead as the acetate at levels of 100 μg/g for 46 weeks, or (3) 500 μg/g for 30 weeks followed by 1,000 μg/g for 16 weeks. At this time, blood lead in group 3 was 10 fold above the controls, and blood ALAD activity was nearly 0, indicating a significant lead body burden. There were no differences due to lead treatment for 60 weeks in the recovery rate of hemoglobin, red cell, or hematocrit numbers, and these were all within previously experienced normal rates.

The persistence of apparently normal rates of heme synthesis in these dogs is in agreement with a series of long-term experiments with rats, dogs, and monkeys. [126-129] In these experiments, as in the preceding one, lead acetate was added up to a level of 1,000 μg/g to a "stock" diet, otherwise undescribed, but probably containing calcium in excess of the needs of the animals. In rat and dog experiments employing purified diets containing 300 μg/g lead acetate with borderline or deficient calcium, [127-131] a significant anemia appeared within a few weeks.

In spite of the sensitivity of ALAD to lead, there is evidence that neither "normal" ALAD levels nor decreases in ALAD activity in response to lead are correlated with ultimate lead toxicity. [132]

Other Enzymes

A recent review by Vallee and Ulmer [4] has concisely summarized the present knowledge of biochemical consequences of lead exposure in both *in vivo* and *in vitro* systems. While *in vitro* tests show that lead has a great affinity for and inhibition of sulfhydryl enzymes, in most instances the concentrations of lead necessary to demonstrate inhibition are well above those obtainable *in vivo.* Notable exceptions to this are ALAD and lipoamide dehydrogenase. The latter has been studied *in vitro* by Ulmer and Vallee, [31] but not *in vivo.* If lipoamide dehydrogenase activity were to be depressed by lead, it might conceivably alter the availability of succinyl CoA for heme synthesis; however, it might also have profound influences on total oxidative processes of the cells, since it is involved in both pyruvate and α-ketoglutarate metabolism.

The effects of lead on extra hematopoietic heme enzymes have also been noted. Cytochromes aa_3 and *b* contents of rat kidney mitochondria were reduced in young rats fed 1 percent lead acetate for 10 weeks. [133] Mitochondrial protein decrease and moderate anemia were noted, and it is possible that the effects were produced either by a defect in heme or protein synthesis.

A stimulation of DNA synthesis in rat kidney tubular cells has been reported by Choie and Richter. [134] H^3-thymidine incorporation into tubular cells was measured following intraperitoneal injection of lead acetate at three, 48-h intervals. Each injection was followed by a brief wave of increased incorporation to 20 to 40 times normal. The significance of this finding is unknown; it appears not to be a response to tubular damage, although there must be increased cell turnover. Chronic toxicity does not lead to markedly enlarged kidneys, although it is accompanied by increased DNA synthesis and mitotic indices.

A number of reports have indicated not only a disturbance of niacin metabolism by lead toxicity, but also a diminution of lead toxicity symptoms through administration of niacin above normal requirements in rabbits and humans. (cited in reference [3]) A more recent publication by Pecora et al. [135] has clearly demonstrated that in rabbits the parenteral administration of nicotinic acid at the rate of 30 mg/day greatly limited the effect of 200 mg of lead acetate given daily by mouth. This is in contrast to results obtained with rats by Kao and Forbes [129] in which a 10 fold increase in niacin intake did not affect the progress of lead toxicity. Another example of species difference in controlling lead toxicity is the lower rise in xanthurenic acid and coproporphyrin excretion and free erythrocyte protoporphyrin following a tryptophan blood test in lead-toxic rabbits treated with large doses of pyriodoxine. [132] Kao and Forbes [129] also demonstrated the failure of pyridoxine to influence *in vitro* heme synthesis in lead-toxic rats. Pyridoxine is known to catalyze the formation of ALA and also to convert tryptophan to niacin.

Neural Tissue

Defects in function of the nervous system are serious consequences of acute and severe chronic exposure to lead. It is apparent that both brain and peripheral neural tissues are affected, based on morphological and biochemical studies, as well as on clinical observation of disturbed neural functions in humans. [136] The literature in this area was concisely reviewed by Tepper and Pfitzer, [5] unfortunately without citation of references.

Animal studies in this complex field have yielded diverse results that which have not yet been synthesized into a coherent picture, perhaps because of variations in mode and degree of exposure to lead, [136, 137] differences in age and species of animal used, and in criteria by which neural dysfunction was judged. Acute effects are well established, but the significance of low-level effects are not.

In a study of the effects of lead on learning and memory in rats, Brown, et al. [138] reported that animals 8 to 35 days old were given intraperitoneal injections of lead acetate (100 mg/kg) for 3 to 4 days. Thirty percent of the animals failed to survive, and all the survivors had locomotor problems. However, there was no difference between the lead-toxic and control animals in their ability to learn or remember a water-filled swimming T-maze. In contrast, a report by Kringman et al. [139] stated that lead toxicity in young rats permanently inhibited their ability to perform in an avoidance task, and reduce brain growth and neural lipid content. Other recent studies by Krall et al. [140] with young rats have shown that rats nursing lead-fed dams (4 percent lead acetate) may show tremors and a deformed cerebellum. A severe depression in mitochondrial phosphorylation that could be reversed by magnesium was also seen. Silbergeld [141] conducted an *in vitro* test of the effect of lead on the response of nerve and muscle to stimulation; he concluded that the effects of lead in increasing the latent period between stimulation and contraction are mediated at the presynaptic level. A 25 μM concentration of lead was required in the fluid bathing the nerve-muscle preparation to demonstrate a lead inhibition of response. This is much higher than physiologically attainable levels.

A recent study by Stowe et al. [131] has revealed a significant accumulation of lead in the brain of lead-toxic dogs fed a marginal level (0.3 percent) of calcium. Although the lead exposure (100 μg/g in the diet) of these animals was below that which would produce notable toxic symptoms in animals fed a more liberal supply of calcium, the animals showed a severe depression in feed intake and body weight gain over the 12 week experiment. Neural dysfunction was not mentioned among the many observations made on these animals. All parts of the brain except the cerebellum contained significantly elevated lead concentrations. The authors suggested that increased brain lead might displace calcium in neural membranes and produce hyperirritability. No direct evidence on this point is available, but in another study by Huffman and Weber [142] an increase in brain calcium was noted in acute lead-toxic rats.

Kidneys

The unique effects of lead on kidney morphology and function have been recognized in both clinical and experimental situations for many years. [2] The major histologic manifestations consist of early development of intranuclear inclusion bodies and swelling of mitrochondria in the cells of the proximal portion of the renal tubule. [111, 143] The nuclear inclusion bodies isolated from rat kidneys have been shown to contain protein and lead at concentrations up to 7 percent of the weight of the protein. [144] The origin of the protein matrix to which lead is bound in the inclusions is as yet unknown, although Choie and Richter [145] suggested that the protein is synthesized within the tubular cells.

Functionally, renal glycosuria, generalized hyperaminoaciduria, and hypophosphatemia are well recognized consequences of severe lead toxicity. These point to a defect in tubular reabsorption processes. A correlation between aminoaciduria and mitochondrial swelling has been documented by Goyer *et al.* [146] The kidney mitochondria isolated from lead-toxic rats were found to have a decreased rate of respiration and partially uncoupled oxidative phosphorylation, thus providing evidence of a metabolic alteration in energy metabolism consonant with reduced functional capacity of the tubules.

Reproduction

Reports of lead effects on the reproductive system have appeared over many years, but this subject has not received a great deal of orderly attention. Scattered references to decreased fertility, increased abortions, and teratogenicity appear in the earlier literature of both clinical and experimental lead toxicity. Several recent reports dealing with reduced fertility and with teratogenicity are worthy of mention.

A carefully planned study by Stowe and Goyer [147] has revealed effects of lead toxicity on both maternal and paternal reproductive efficiency. Part of a group of female Sprague-Dawley rats were fed 1 percent lead acetate added to a commercial rat ration prior to breeding with normal males. The control group was not fed any lead. The resultant pups were raised on the same rations as their dams, at breeding age, they were mated so that the maternal, paternal, and combined effects of lead could be ascertained. Significant decreases in litter size, birth weight, and survival were observed in matings involving one lead-toxic parent. These effects were greater when the dam, rather than the male, was lead-toxic, and greatest when both parents were lead-toxic.

Reduced reproductive performance by hens fed 1 percent lead acetate in a normal ration has been reported by Stowe et al. [148] The lead-treated birds laid fewer eggs than the controls, and 26 percent of the eggs had soft and malformed shells. Normally formed eggs laid by the lead-treated and control birds were of equal size, fertility, and hatchability. Eggs of the lead-fed birds contained 5 μg of lead, most of it in the yolk.

Further implications of the effects of lead on male and female reproductive ability have been reported by Hilderbrand et al. [149] in an experiment with much lower lead exposure. Mature male and female rats were given 0, 5, or 100 μg lead acetate for 30 days in addition to *ad libitum* amounts of food and water. The authors stated that the described lead treatment increased prostate weight (and hyperplasia of prostate) significantly to as much as twice the control values. Lead-exposed males refused to copulate with estrus females, although each male was given five opportunities. Among the lead-exposed male rats, 20 percent had testicle weights 70 percent below control values. These animals had histologically observable damage of seminiferous tubules, and spermatogenesis was halted. Female rats exposed to 5 μg of lead acetate experienced irregular estrus, and those given 100 μg exhibited persistent vaginal estrus and reduction in formation of corpora lutea. No mortality or overt signs of toxicity were seen. In both sexes, urinary excretion of ascorbic acid was reduced by both lead exposures, and the reduction was dose-related. This was interpreted by the authors as evidence for reduced hepatic detoxification mechanisms via decreased microsomal enzyme activity, a view supported by observed prolonged sleeping times of the lead-dosed rats following pentobarbital treatment. These intriguing observations would be more meaningful had further information on the nature of the basal diet been provided. The refusal of males to mate is in contrast to the experience reported by Stowe and Goyer [147] as a result of feeding 1 percent lead acetate. Whether this variation is a result of differences in the basal diet, strain of rats used, or length of exposure to lead cannot be ascertained from the available data.

The degree of placental transport of lead and its effects on embryo development have not been extensively investigated. In one study with rats and mice, [150] placental lead transport was very limited. Single intravenous doses of lead given to pregnant animals reduced fetal size, increased fetal resorptions, and produced hydronephrosis and nonossified cervical centri. Intraperitoneal or oral lead dosage did not produce teratogenic effects, presumably because of low absorption into the blood. More recently, the same investigators [151] have shown that organoiead compounds such as tetraethyl, tetramethyl, and trimethyl lead chloride are essentially nonteratogenic in rats when administered orally or intravenously during pregnancy. Only doses that produced severe maternal toxicity had marked effects on fetal development. It does appear that the placental barrier to lead is quite effective, and that lead effects on the total reproductive process are more likely to arise from effects on sperm and ova during their development than on the developing fetus. Evidence from Muro and Goyer [152] that lead may produce chromosome damage in developing cells lends credence to this concept.

Endotoxin Hypersensitivity

If bacterial endotoxins cross the mucosal barrier of the intestine and enter the tissues, they are thought to normally be inactivated by macrophages of the reticuloendothelial system. Increased sensitivity to bacterial endotoxins as a result of lead toxicity has been observed in rats, mice, and chicks, but only from experiments using intravenous injection.

Studies by Filkins and Buchanan [153] and Trejo et al. [154] have shown extensive functional and ultrastructural changes in the reticuloendothelial system, liver, and spleen of rats given single intravenous doses of 5 mg of lead acetate. Lead reduced intravascular and hepatic phagocytosis, as well as the ability of liver and spleen homogenates to detoxify endotoxin. The morphological changes were thought to be late manifestations of earlier biochemical impairment induced by lead and responsible for hypersensitivity to endotoxin.

A paper by Hoffman et al. [155] has investigated effects of lead acetate on phagocytic and endotoxin detoxifying activities of the reticuloendothelial system and the effect on sensitivity of rats to several parameters of endotoxin susceptibility. The authors concluded that the interaction of lead and endotoxin does not involve detoxification directly, but involves other metabolic responses of the tissues to the endotoxin.

SUMMARY

The adverse effects of lead on bacterial cells are largely unexplored, and the manifestations are virtually unknown. Although there have been few quantitative studies on the effects of lead in aquatic communities, it appears less toxic than either copper or zinc; there is considerable variability among genera and species of algae and fish. Toxicity may be related to reactions between lead and certain essential elements. Reaction with the mucous of fish gills, with resulting precipitation and coagulation, has been suggested as a mechanism of toxicity.

The uptake of lead by plants is a proven fact; but effects on plant growth appear minimal with soils concentrations less than 1,000 μg/g, especially in soils with a high cation-binding capacity. However, profound sub-cellular effects have been observed at low concentrations. Apparently buffering mechanisms exist in whole plants that inactivate lead before it reaches critical sites such as mitochondria or chloroplasts.

A large number of abnormal effects have been observed in laboratory animals when lead are ingested, and none of these are beneficial. The decrease of ALAD, the enzyme that regulates hemoglobin production, is the most significant parameter in lead intoxication of rats. The practical significance of this effect is unclear since animals can withstand long periods of very low ALAD without becoming anemic; anemia is a late manifestation of lead toxicity. Other observed effects of lead toxicity on a variety of laboratory animals include defects in nervous system development; in the structure and function of kidneys; and in decreased reproductive capacity and teratogenicity. However, most of these effects are related to amounts of lead attainable only under experimental conditions.

Lead poisoning of intact organisms is a rare occurrence under most existing natural conditions, yet organisms are frequently exposed to environmental concentrations high enough to produce a toxic effect, based on results of laboratory experiments. The questions thus arise: Why is the environmental effect of lead almost negligible? And will it remain so if environmental lead concentrations continue to increase in the future as they have over the last half century?

To answer these questions, several properties of lead must be considered that have been reported from laboratory and field experiments. Environmentally, the most important of these is that lead, compared with other metals, has a strong affinity for organic and inorganic surfaces. It is readily precipitated by many common anions, and desorption and replacement by other cations are extremely slow. Most environmental lead is in the soil and relatively nonmobile to other components of the ecosystem. Under certain environmental conditions it can, and does, move. When it moves into plants or animals, lead is almost always nontoxic because of mechanisms that bind it in an inactive form before it reaches enzyme sites where it can be detrimental.

When considering the future environmental impacts of lead, attention must be given to factors responsible for its inactivation. The organic and inorganic fractions of soil vary greatly throughout the world, and even from farm to farm in an agricultural ecosystem. As these fractions vary, so do their capacities to bind or immobilize lead. Organisms also vary in this capacity.

The ability to predict future effects of lead is further complicated by a lack of knowledge as to how environmental conditions (e.g., weather) affect the capacity of soil and organisms to withstand future lead insult. An organism under stress is weakened, and controlled research reports indicate that in most instances it cannot withstand other stresses imposed upon it as well as a normal (nonstressed) organism. The question arises, can a plant suffering from a nutrient deficiency "inactivate" lead as well as a normal (nondeficient) plant? Little research has been devoted to such basic questions. The relevance becomes apparent since, in nature, most organisms are placed under certain types of environmental stress at one or more times in their life cycle.

Thus the potential is now present for the effects of lead on organisms to become manifest. However, there is no current evidence of widespread effects. Clearly, caution must be used in extrapolating present experience to all future situations. The effect of lead on any organism will be determined by the combination of environmental factors to which an organism is exposed. Until we know more about how these are interrelated and affect living organisms, we should not be too complacent about the apparent lack of visible effects of lead on plants and animals.

REFERENCES

1. Cleaning Our Environment—The Chemical Basis for Action. American Chemical Society, Washington, D.C. (1969).
2. Lead; Airborne Lead in Perspective. Committee on Biological Effects of Atmospheric Pollutants, Division of Medical Sciences, National Research Council, National Academy of Sciences. (1972).
3. A. deBruin. Arch. Environ. Health. *23:* 349 (1971).
4. B. L. Vallee and D. D. Ulmer. Ann. Rev. Biochem. *91:* 127 (1972).
5. L. B. Tepper and E. A. Pfitzer. Clinical and Biochemical Approaches to the Study of Lead at Low Levels. Report of a symposium. National Technical Information Service PB196-767, (1970).
6. R. A. Goyer and B. C. Rhyne. Int. Rev. Exp. Pathol. *12:* 1 (1973).
7. H. Passow, A. Rothstein, and T. W. Clarkson. Pharmacol. Rev. *13:* 185 (1961).
8. J. Beryer, M. J. Johnson, and W. H. Peterson. J. Biol. Chem. *124:* 395 (1938).
9. T. G. Tornabene and H. W. Edwards. Science. *176:* 1334 (1972).
10. S. L. Peterson, L. G. Bennett and T. G. Tornabene. Appl. Microbiol. *29:* 10 (1975).
11. L. E. Den Dooren De Jong. Antonie van Leeuwenhoek. *37:* 119 (1971).
12. R. P. Novick and C. Roth. J. Bacteriol. *95:* 1446 (1968).
13. P. T. S. Wong, Y. K. Chau, and P. L. Luxon. Nature. 253: 263 (1975).
14. G. Peyru, L. F. Wexler, and R. P. Novick. J. Bacteriol. *98:* 215 (1969).
15. J. P. Devigne. Cr. Acad. Sci. D. *267:* 935 (1968).
16. J. P. Devigne. Arch. Inst. Pasteur. Tunis. *45:* 341 (1968).
17. W. R. Hass and S. Miller. Effects of Various Metals on Aerobic Microorganisms. Final Report No. IITRI - C8 213-2. ITT Research Institute, Chicago, IL (1971).
18. W. D. Deason. Ph.D. dissertation, Colorado State University, Fort Collins, CO (1970).
19. T. G. Tornabene and S. L. Peterson. Appl. Microbiol. *29:* 20 (1975).
20. Y. Hata. Norinsho Sulsan Koshusko Kenkyu Hokoku. *9:* 363 (1960).
21. J. F. Manery. Fed. Proc. *25:* 1804 (1966).
22. A. O. Summers and E. Lewis. J. Bacteriol. *113:* 1070 (1973).
23. I. Kondo, T. Ishikawa, and H. Nakanara. J. Bacteriol. *117:* 1 (1971).
24. R. A. Goyer and B. C. Rhyne. In International Review of Experimental Pathology, Vol. 12, G. W. Richter and E. Epstein, ed. Academic Press, New York, NY (1973).
25. F. S. Hsu, L. Krook, T. H. Shirley, and J. R. Duncan. Science. *181:* 447 (1973).
26. H. Skaar, O. E. Ophus, and B. M. Gullvag. Nature. *24:* 215 (1973).
27. J. F. Moore and R. A. Goyer. Environ. Health Perspec. *7:* 121 (1974).
28. G. P. Caonni, A. Liberti, and R. Palombari. J. Chromatogr. *20:* 278 (1965).
29. J. T. Hoogeveen. In Effects of Metals on Cells, Subcellular Elements and Macromolecules, J. Maniloff, R. Coleman, and M. W. Miller, ed. C. C. Thomas, Springfield, IL 207 (1970).
30. Effects of Metals on Cells, Subcellular Elements and Macromolecules, J. Maniloff, J. R. Coleman, and N. W. Miller, ed. C. C. Thomas, Springfield, IL (1970).
31. D. D. Ulmer and B. L. Vallee. Trace Substances in Environmental Health-II. Proceedings of Annual Conference on Trace Substances in Environmental Health, 2nd ed., D. D. Hemphill, ed. University of Missouri, Columbia, MO (1969).
32. G. KiKuchi, A. Kumar, R. Talmidge, and D. Sherman. J. Biol. Chem. *233:* 1214 (1958).
33. J. Lascelles. Tetraphyrrole Biosynthesis and Its Regulation. Benjamin, New York, NY (1964).
34. B. A. Whitton. Phykos. *9:* 116 (1970).
35. B. A. Whitton. Arch Mikrobiol. *72:* 353 (1970).
36. N. L. Gale, M. G. Hardie, J. C. Jennett, and A. Aleti. Proceedings of 6th Annual Conference on Trace Substances in Environmental Health, University of Missouri, Columbia, MO (1972).
37. N. L. Gale, P. Marcellus, and G. Underwood. Life, Liberty, and the Pursuit of Lead: the Impact of Lead Mining and Milling Activities on Aquatic Organisms. Paper presented at 2nd Annual NSF-RANN Trace Contaminants Conference, Asilomar, CA (August 1974).
38. S. Christiansen, Y. Manuel and N. L. Gale. University of Missouri-Rolla. Unpublished data (1973).
39. J. Whitfield, P. Marcellus, and N. L. Gale. University of Missouri-Rolla. Unpublished data (1973).
40. N. M. Kalabina, K. A. M. Viss, A. S. Razumov, and T. I. Rogovskaja. Gigiene (U.S.S.R.) No. 9, 10 (1944). Water Pollut. Abstr. *21* (February 1948).
41. F. Meinck, H. Stooff, and H. Kohlschutter. Industrie Abwasser, 2nd ed. Gustav Fischer Verlag, Stuttgart. 536 (1956).
42. M. M. Kalabina. Water Sewage Works. *93:* 30 (1946).
43. A. F. Zaitseva. Hyg. Sanit. (U.S.S.R.), No. 3, 7. Chem. Abstr. *47:* 10,778 (1953). Water Pollut. Abstr. *27:* 925 (1954).
44. H. Wang. Proc. Soc. Exp. Biol. Med. *101:* 682 (1959).
45. J. S. Gray and R. J. Ventilla. Mar. Pollut. Bull. *2:* 39 (1971).
46. Water Quality Criteria Data Book, Vol. 3: Effects of Chemicals on Aquatic Life. Prepared by Batelle's Columbus Laboratories for U.S. Environmental Protection Agency. (1971).
47. W. H. R. Shaw and B. Grushkin. Arch. Biochem. Biophys. *67:* 447 (1967).
48. M. M. Ellis, U. S. Dept. of Commerce, Bur. of Fisheries Bull., 22 (1937).
49. B. G. Anderson. Trans. Amer. Fish. Soc. *78:* 96 (1948).
50. L. S. Whitley, Hydrobiologia. *32:* 193 (1968).
51. B. Brown and M. Ahsanulla. Mar. Pollut. Bull. *2:* 182 (1971).
52. P. Doudoroff and M. Katz. J. Water Pollut. Contr. Fed. *25:* 802 (1953).
53. J. E. McKee and H. W. Wolf. Water Quality Criteria, 2nd ed. State Water Resources Control Board of California (1971).
54. J. R. E. Jones. J. Exp. Biol. *15:* 394 (1938).
55. K. E. Carpenter. Ann. Appl. Biol. *12:* 1 (1925).
56. K. E. Carpenter. Brit. J. Exp. Biol. *4:* 378 (1927).
57. C. M. Tarzwell and C. Henderson. Ind. Wastes. *5:* 12 (1960).
58. M. M. Ellis. Bulletin No. 22, U. S. Bureau of Fisheries. *48:* 365 (1937).
59. A. Thomas. Trans. Amer. Fish. Soc. *44:* 120 (1965).
60. W. Rushton. Salmon and Trout Magazine. *37:* 203 (1922).
61. G. Ebeling. Zeits. Fischerei. *26:* 49 (1928).

62. W. J. Dilling, C. W. Henley, and W. C. Smith. Ann. Appl. Biol. *13:* 168 (1926).
63. A. B. Dawson. Biol. Bull. *68:* 335 (1935).
64. C. A. Crandall and C. J. Goodnight. Limnol. Oceanogr. *7:* 233 (1962).
65. W. J. Birge. Lethal and Teratogenic Effects of Metallic Pollutants on Vertebrate Embryos. Paper presented at 2nd Annual NSF-RANN Trace Contaminants Conference, Asilomar, CA (August 1974).
66. I. E. Wallen, W. C. Greer, and R. Lasater. Sewage Ind. Wastes. *29:* 695 (1957).
67. K. E. Carpenter. Ann. Appl. Biol. *11:* 1 (1924).
68. I. S. Pakkala, M. N. White, G. E. Burdick, E. J. Harris, and D. J. Lisk. Pest. Monit. J. *5:* 348 (1972).
69. W. Fulkerson and H. E. Goeller, ed. Report No. NSF-EP-21, Oak Ridge National Laboratory, Oak Ridge, TN (1973).
70. Lake Michigan Open Water and Lake Biological Report. U. S. Environmental Protection Agency, Springfield, IL (1970).
71. N. L. Gale, B. G. Wixson, M. G. Hardie, and J. C. Jennett. Amer. Water Resources Ass. Bull. *9:* 673 (1973).
72. A. L. Aronson. J. Wash. Acad. Sci. *61:* 124 (1971).
73. C. M. Keaton. Soil Sci. *43:* 401 (1937).
74. A. Weiler. Mitt. Forstwirtsch N. Forstwiss. (Hanover). *9:* 175 (1938).
75. R. A. Berry. J. Agric. Sci. *14:* 59 (1923).
76. K. Scharrer and W. Schropp. Stschr. Pflanzenernahr, Dungung, v. Bodenk. *43:* 34 (1936).
77. J. Stoklaso. Compt. Rend. Acad. Sci. *156:* 153 (1913).
78. A. Stutzer. J. Landu. *64:* 1 (1916).
79. J. A. Voelcker. J. Royal Agric. Soc. England. *75:* 306 (1914).
80. S. Prat. Amer. J. Bot. *14:* 633 (1927).
81. D. A. Wilkens. Nature. *180:* 37 (1957).
82. W. F. Childers. Proc. Amer. Soc. Hort. Sci. *38:* 157 (1941).
83. R. J. Miller and D. E. Koeppe. Proceedings of Annual Conference on Trace Substances in Environmental Health, 4th ed., D. D. Hemphill, ed. University of Missouri, Columbia, MO (1970).
84. M. C. Hooper. Ann. Appl. Biol. *24:* 690 (1937).
85. T. H. Keller and R. Zuber. Forstwissenschaftliches Centralblatt. *51:* 368 (1964).
86. University of Illinois. An Interdisciplinary Study of Environmental Pollution by Lead and Other Metals. NSF Grant GI-31605, progress report (1972).
87. F. A. Bazzaz, G. L. Rolfe and P. Windle. J. Environ. Qual. *3:* 156 (1974).
88. F. A. Bazzaz, R. Carlson, and G. L. Rolfe. Environ. Pollut. *7:* 241 (1974).
89. F. S. Hammet. Protoplasma. *4:* 183 (1928).
90. T. J. Ganje and A. L. Page. Calif. Agric. *26:* 7 (1972).
91. A. Wallace and E. M. Romney. Agron. Abstr. 130 (1970).
92. H. V. Warren, R. E. Delavault, K. Fletcher, and E. Wilks. Proceedings of Annual Conference on Trace Substances in Environmental Health, 4th ed., D. D. Hemphill, ed. University of Missouri, Columbia, MO (1970).
93. F. J. MacLean, R. L. Halstead, and B. J. Finn. Can. J. Soil Sci. *49:* 327 (1969).
94. W. J. Cox and D. W. Rains. J. Environ. Qual. *1:* 167 (1972).
95. M. K. John and C. Van Laerhoven. J. Environ. Qual. *1:* 169 (1972).
96. J. E. Miller, D. E. Koeppe, and J. J. Hassett. Commun. Soil Sci. Plant Anal. (1975 in press).
97. J. G. Hunter. J. Sci. Food Agric. *4:* 11 (1953).
98. R. L. Mitchell and J. W. S. Reith. J. Sci. Food Agric. *17:* 437 (1966).
99. M. M. Guha and R. L. Mitchell. Plant and Soil. *24:* 90 (1966).
100. D. W. Rains. Nature. *233:* 210 (1971).
101. G. R. Baumhardt and L. F. Welch. J. Environ. Qual. *1:* 92 (1972).
102. G. L. Rolfe. J. Environ. Qual. *2:* 153 (1973).
103. F. A. Bazzaz, G. L. Rolfe, and P. Windle. J. Environ. Qual. *3:* 156 (1974).
104. F. A. Gilbert. Advan. Agron. *4:* 147 (1952).
105. J. B. Wedding, R. W. Carlson, J. J. Stukel, and F. A. Bazzaz. Environ. Sci. Technol. *9:* 151 (1975).
106. J. Bittell, R. J. Miller, and D. E. Koeppe. Plant Physiol. *30:* 187 (1974).
107. H. J. M. Bowen. Trace Elements in Biochemistry. Academic Press, New York, NY (1966).
108. D. E. Koeppe and R. J. Miller. Science *167:* 1376 (1970).
109. C. D. Miles, J. R. Brandle, D. J. Daniel, O. Chu-Der, P. O. Schnare, and D. J. Uhlick. Plant Physiol. *49:* 820 (1972).
110. M. B. Bazzaz and Gouindjee. Environ. Lett. (1974, in press).
111. R. A. Goyer. Lab. Invest. *19:* 71 (1968).
112. R. A. Goyer, A. Krall, and J. P. Kimball. Lab. Invest. *19:* 78 (1968).
113. R. A. Goyer and A. Krall. J. Cell Biol. *41:* 393 (1969).
114. R. A. Goyer, D. L. Leonard, J. F. Moore, B. Rhyne, and M. R. Kringman. Arch. Environ. Health. *20:* 705 (1970).
115. K. M. Scott, K. M. Hwang, M. Jurkowitz, and G. P. Brierly. Arch Biochem. Biophys. *147:* 557 (1971).
116. R. L. Zielhuis. Arch. Environ. Health. *23:* 299 (1971).
117. D. Prpic-Majic, P. K. Mueller, T. Beritic, and R. Stanley. International Symposium on Environmental Health Aspects of Lead, Amsterdam (1972).
118. S. Hernberg and J. Nikkanen. Pracov. Lek. *24:* 77 (1972).
119. B. Haeger-Aronsen, M. Abdulla, and B. I. Fristedt. Arch. Environ Health. *23:* 440 (1971).
120. K. I. Campbell, W. M. Busey, N. K. Weaver, J. A. Taylor, and A. A. Krum. J. Amer. Vet. Med. Ass. *159:* 1523 (1971).
121. J. A. Millar, V. Battistini, R. L. C. Cumming, F. Carswell, and A. Goldberg. Lancet *3:* 695 (1970).
122. R. L. C. Kao and R. M. Forbes. Proc. Soc. Exp. Biol. Med. (1973).
123. F. L. Cerklewski and R. M. Forbes. University of Illinois. Unpublished data (1972).
124. M. Kreimer-Birnbaum and M. Grinstein. Biochim. Biophys. Acta. *111:* 110 (1965).
125. M. E. Maxfield, G. J. Stopps, J. R. Barnes, R. D'Snee, and A. Azwar. Amer. Ind. Hyg. Ass. J. *33:* 326 (1972).
126. D. C. Jessup and L. D. Shott. Lead Acetate, 22-Month Chronic Toxicity Study—Rats. National Technical Information Service Report. Pb201-135. (1969)
127. D. C. Jessup and L. D. Shott. Lead Acetate, 22-Month Dietary Administration—Dogs. National Technical Information Service Report. Pb 201-136. (1969).
128. D. C. Jessup, L. D. Shott, and W. M. Busey. Lead Acetate, 22-Month Chronic Toxicity Study—Monkeys. National Technical Information Service Report. Pb 201-137, (1970).
129. R. L. C. Kao and R. M. Forbes. Arch. Environ. Health. (1973, in press).
130. K. M. Six and R. H. Goyer. J. Lab Clin. Med. *76:* 933 (1970).
131. H. D. Stowe, R. A. Goyer, M. R. Kringman, M. Wilson, and M. Cates. Arch. Pathol. *95:* 106 (1973).
132. B. Garber and E. Wei. Bull. Environ. Contam. Toxicol. *9:* 80 (1973).
133. B. C. Rhyne and R. A. Goyer. Exp. Mol. Pathol. *14:* 386 (1971).
134. D. D. Choie and G. W. Richter. Proc. Soc. Exp. Biol. Med. *142:* 446 (1973).
135. L. Pecora, A. Silvestroni, and A. Brancaccio. Pan. Med. *8:* 284 (1966).
136. D. Bryce-Smith. Chem. Brit. *8:* 240 (1972).
137. P. K. Thomas. Proc. Royal Soc. Med. *64:* 295 (1971).
138. S. Brown, N. Dragann and W. H. Vogel. Arch. Environ. Health. *22:* 370 (1971).
139. M. R. Kringman, S. A. Butts, E. L. Hogan, and P. G. Shinkman. Fed. Proc. *31:*665, Abstr. No. 2536 (1972).
140. A. R. Krall, C. Pesavento, S. J. Harmon, and R. M. Packer. Fed. Proc. *31:* 665, Abstr. No. 2537 (1972).
141. E. K. Silbergeld. Fed. Proc. *32:* 262, Abstr. No. 275 (1973).
142. N. E. Huffman and L. J. Weber. Fed. Proc. *32:* 262, Abstr. No. 276 (1973).
143. R. A. Goyer, A. Krall, and J. P. Kimball. Lab. Invest. *19:* 78 (1968).
144. R. A. Goyer, P. May, M. M. Cates, and M. R. Kringman. Lab. Invest. *22:* 245 (1970).
145. D. D. Choie and G. W. Richter. Science. *177:* 1194 (1972).
146. R. A. Goyer, D. L. Leonard, J. F. Moore, B. Rhyne, and M. R. Kringman. Arch. Environ. Health. *20:* 705 (1970).
147. H. D. Stowe and R. A. Goyer. Fertility and Sterility. *22:* 755 (1971).

148. H. D. Stowe, R. A. Goyer, and M. M. Cates. Fed. Proc. *31:* 734, Abstr. No. 2916 (1972).
149. D. Hilderbrand, M. Olds, R. Der, and M. S. Fahin. Conference on Trace Substances in Environmental Health, 4th ed., D. D. Hemphill, ed. University of Missouri, Columbia, MO (1972).
150. R. M. McClain and B. A. Becker. Fed. Proc. *29:* 347 (1970).
151. R. M. McClain and B. A. Becker. Toxicol. Appl. Pharmacol. *21:* 265 (1972).
152. L. A. Muro and R. A. Goyer. Arch. Pathol. *87:* 660 (1969).
153. J. P. Filkins and B. J. Buchanan. Proc. Soc. Exp. Biol. Med. *142:* 471 (1973).
154. R. A. Trejo, N. R. DiLuzio, L. D. Loose, and E. Hoffman. Exp. Mol. Pathol. *17:* 145 (1972).
155. E. O. Hoffman, R. A. Trejo, N. R. DiLuzio, and J. Lamberty. Exp. Mol. Pathol. *17:* 159 (1972).

CHAPTER 9

HUMAN HEALTH IMPLICATIONS

Paul B. Hammond
Department of Environmental Health
University of Cincinnati

The research activities reported in this book characterize the movement of lead from auto exhaust and from a lead mining and smelting operation through the geomass and biomass. Possible adverse consequences to many elements in the biomass also are examined. The study of man is not included. That being the case, the human health implications are rather limited.

The human health effects of lead have been investigated intensively for the past 50 years. The payoff in terms of human health has been dramatic. Through application of knowledge acquired over that period of time, the incidence of clinical lead poisoning in the populations at risk has been drastically reduced. The beneficiaries have been mainly workers in the lead-using industries and, more recently, infants and young children. But undue lead exposure of these two groups persists, and a commensurately high level of research activity continues.

For workers in the lead-using industries, the sources of lead are well known and too numerous to be documented here in any great detail. Suffice it to say that the major exposure problems occur in lead smelting and refining and in the manufacture of electric storage batteries. For these and other industrial operations, the level of lead exposure even today far exceeds the level for the general population. The problem of industrial lead exposure has recently been documented in great detail by a WHO Task Force [1] and in more abbreviated form by Hernberg. [2] Almost invariably, the transfer of lead from the environment to the subject is by inhalation of dusts and fumes.

Through application of engineering control measures, institution of better housekeeping measures, and the use of respirators in selected circumstances, the level of human exposure in the lead-using industries has been reduced considerably in recent years. The health effects have consequently become progressively less frequent and serious. But application of newer, more sophisticated diagnostic methods continues to reveal previously unknown effects occurring at relatively low levels of exposure. Thus industrially exposed workers in Finland have been shown to exhibit a slowing of peripheral nerve conduction. [3, 4] Although no functional impairment (e.g., paralysis) has been associated with this effect, this clearly is an undesirable situation.

Even more recently, examinations of workers in a variety of occupations have uncovered evidence of impaired renal function at levels of lead exposure previously considered safe. [5, 6]

The situation with regard to lead exposure in infants and young children is similar. Recent studies indicate that harmful effects may occur at exposure levels previously considered to be safe. Many uncertainties surround the problem of lead exposure in children. The population at risk is predominantly children 1 to 5 years of age. Cases of clinical poisoning occur mainly in the inner cities of the East and Midwest, among residents of old, poorly maintained housing. The source of poisoning usually is chipped or cracked paint and plaster of high lead content that youngsters eat in toxic quantities. Although toxic effects involve a variety of organs and systems, the major concern has been with regard to effects on the central nervous system. The neurological effects may be serious or even fatal. Worse, severe, permanent or long-lasting sequelae may occur, such as convulsive seizures, idiocy, and hydrocephalus. More commonly, the sequelae are of a more subtle nature, such as reduced learning ability, sensorimotor disturbances, and reduced attention spa. These and other subtle sequelae to pediatric lead poisoning were observed even in the absence of prior encephalopathy as early as 1943. [7] There is some recent evidence that subtle neurological damage occurs at levels of lead exposure that do not cause other obvious clinical manifestations of poisoning. [8, 9] Thus as with workers in industry, earlier concepts as to what constitutes a safe level of lead exposure are undergoing a process of downward revision.

In a similar vein, certain other biological effects of

lead exposure have recently been noted to occur in the range of human lead exposure generally considered safe. These are biochemical disturbances of the hemesynthetic mechanism that are seen in the blood. Two occur at levels of lead exposure below those that cause any observable decrement of hemoglobin. They are inhibition of aminolevulinic acid dehydrase (ALAD) [10, 11] and increase in protoporphyrin. [12] The significance of these findings for human health remains to be established. Although the hemoglobin concentration in blood is not impaired in the lower range of exposure at which these effects are seen, subtle effects on human health that have not yet been perceived may be occurring.

The degree of human exposure to lead at which one or another affect is seen is not readily established. The sources of lead encountered by an individual are variable and fluctuate considerably in intensity from hour to hour and from day to day. The concentration of lead in the blood (Pb*B*) is used most commonly as an index of total exposure. Long experience has shown that the correlation is fairly good between Pb*B* and the intensity of several biochemical effects, e.g. ALAD inhibition and the concentration of protoporphyrin in the blood. Thus Pb*B* is considered to reflect the magnitude of the current and recently acquired biologically active fraction of the body burden of lead. The relationship between Pb*B* and the more significant human health effects of lead is not at all clear, however. The degree of one or another impairment of body function may be the result of an earlier exposure not well reflected in Pb*B* at the time impairment is measured. Or it may be that the impairment is more a function of duration of exposure than of intensity of exposure. Thus a profile of total exposure over time may be important, as by repeated determination of Pb*B*.

In spite of some obvious limitations, Pb*B* assumes a very important role in the identification of individuals and populations unduly exposed to lead. Thus lead-screening programs have been instituted in many large cities whereby Pb*B* determinations in children have served to identify hazardous home environments before the level of accumulated exposure of the child caused clinical lead poisoning.

Pb*B* determinations have also played a key role in establishing the relative importance of various environmental sources of lead. A good example is provided by a recent study of the effect of ambient air lead concentration on Pb*B*. [13] Air lead exposure of individuals was monitored continuously for 2 to 4 weeks. An estimate was made of the contribution of air lead to Pb*B* over the range encountered by adults in the general population. (fig. 9-1) In numerous other cases in studies of the significance of air as a source of human lead exposure, Pb*B* has been correlated with air lead or with proximity to freeways. [14-16] To a lesser extent, the contributions of food and beverages to Pb*B* also have been examined. [1]

A major consideration in any study of environmental problems is the reliability of the analytical methods. For the determination of lead in blood, interlaboratory comparison programs have revealed that less than half of the laboratories doing blood lead analyses perform satisfactorily. [17-19] The numerous contributions of the Colorado State University group (detailed in ch. 2) to the resolution of problems of lead analyses are noteworthy. This is particularly true in regard to analyses by atomic absorption spectrometry, the method that in all likelihood will continue to dominate studies of human health effects for some years in the future.

As indicated earlier, the sources that may give rise to excessive lead exposure among occupationally exposed adults are both well known and controllable. Such is not the case in regard to sources of lead in infants and young children. The U. S. children in inner cities who have undue lead exposure probably number in the hundreds of thousands. Blanksma et al. [20] found that the percentage of Chicago slum children having Pb*B*'s in excess of 49 μg per 100 g was 8 percent and 4 percent in 1967 and 1968, respectively. Fine et al. [21] observed that 18.6 percent of Illinois children had Pb*B*'s greater than 39 μg per 100 g and that 3.1 percent had Pb*B*'s greater than 59 μg per 100 g. An even higher incidence of excessive lead exposure was reported by Pueschel et al. [22] Most studies of lead exposure in children are not representative of the general population. They usually focus on old inner city neighborhoods where past experience has shown that the incidence of excessive exposure is greatest. Nevertheless, the number of children with excessive exposure is still quite high.

Lead-base paint has long been recognized to be the major environmental source of excessive exposure in young children. [23, 24] Accessible indoor and outdoor painted woodwork is usually suspect. But the concentration of lead in the soil immediately adjacent to houses may be extremely high due to weathering of painted surfaces, [25, 26] and eating soil is not uncommon among young children. [22] It is, of course, possible that the source of excessive lead in soil is often auto exhaust rather than paint. The potential hazard to children of lead in city street dust and soil from auto exhaust has been cited. [27] But one study indicated that auto exhaust was not the source of lead in any of eight cases of excessive lead absorption due to oral intake. [28] It is not possible to generalize from negative findings in such a small series, however.

Tracing the sources of lead in infants and young children is extremely difficult. It is generally assumed that the route of entry is by licking, chewing, and swallowing foreign objects, a habit commonly referred to as "pica." That being the case, it is to be expected that the ingestion of undue amounts of lead would be reflected in elevated fecal lead excretion. In the study of Ter Haar and Aranow, [28] fecal lead excretion was elevated in the cases of poisoning. They reasoned that if the lead originated from atmospheric fallout of auto exhaust, the fecal excretion of some marker associated with fallout should also be elevated. The marker used was a radioactive decay product of radon gas. Radon emanates from the ground into the atmosphere at a steady rate and decays to ^{210}Pb, which enriches fall-out dust and thereby serves as a marker of atmospheric particulates. Since ^{210}Pb was no greater in the feces with high lead concentrations than in the normal feces, it was concluded that the lead did not originate from auto exhaust. This ingenious approach to the source-tracing problem needs to be expanded to include methodology for determining the contributions of one or more the numerous specific sources that may exist in a child's total environment. One fruitful avenue of marker investigation is elemental analysis of lead-containing particles. Each lead source has unique characteristics. Some of these may retain their integrity during passage through the gastrointestinal tract. The investigations conducted by the Illinois and Colorado groups may prove to be applicable in tracing lead from human feces back to the sources as they are actually consumed by children. The preconcentration techniques developed by Skogerboe and associates at Colorado State and the work done at Illinois by Natusch and associates concerning the elemental composition of particles may prove useful in tracing sources of lead back from man to his environment. (See ch. 1.)

Although the major hazardous sources of lead for children seem to be substances foreign to the normal diet, it must be borne in mind that the normal regimen of food and beverages is of itself a substantial source of lead. In an adult population whose air lead exposure is negligible (<0.1 μg/m³). Pb*B* was found to be 12.3 μg per 100 g in men and 7.9 μg per 100 g in women, presumably as a result of dietary intake only. [29] Similarly, in children living in rural Ireland, Pb*B*'s were below 13 μg per 100 g, but with 55 percent of the cases were greater than 10 μg per 100 g. [30] Estimates of dietary lead intake in adults and children are quite variable, probably reflecting differences in sampling techniques as well as real differences related to dietary habits. [1]

It is interesting to note that this significant source of lead has never received the same degree of scrutiny as air lead. Certainly the contribution of soil and atmospheric deposition of lead to plant foods varies only slightly within the range of concentrations attained in agricultural areas. (See chs. 4 and 5.) But this kind of assessment ignores the contribution of food processing. The differences in lead concentration in "natural" milk versus processed milk are noteworthy. Human breast milk contains <5 to 12 μg per l [31, 32] and milk taken directly from cows contains 9 μg per l. [33] By contrast, whole bulk market milk contains 20 to 40 μg per l. [34, 35] More reduction of human lead exposure may be possible by improved food technology than by removing lead

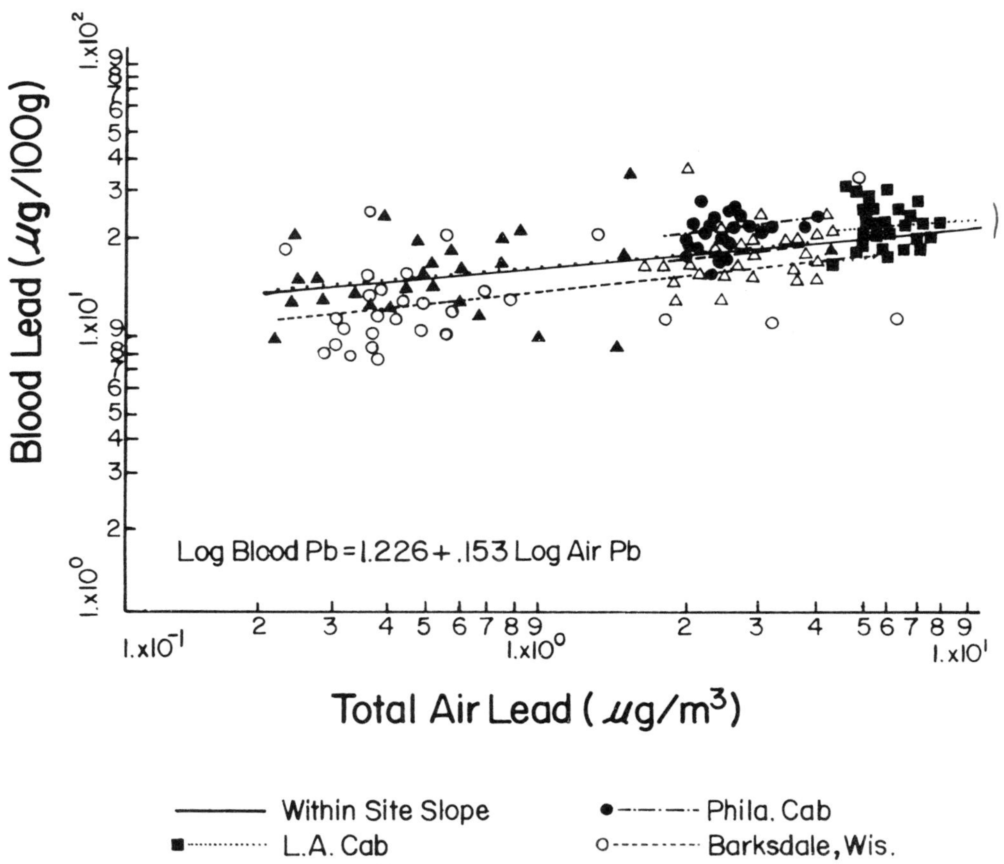

FIGURE 9-1.—Blood Lead versus Total Air Lead.

from gasoline. But the possibilities would only become known as a result of tracing changes in the lead content in food from the fields to the dinner table. This apparently has not been done to any significant extent.

While the environmental studies reported in other chapters have seemingly limited application to the solution of immediate problems of human lead exposure, it must be borne in mind that human welfare is not solely a matter of human health in the physical sense. The integrity of the total environment in which we live is of equal importance. Further, the wealth of expertise developed by the RANN participants will surely find further application, some of which may well be concerned with consideration of the direct effects of lead on man.

REFERENCES

1. Environmental Health Criteria for Lead. Report of a WHO Task Group. (To be published in 1976).
2. S. Hernberg. Work, Environ. Health. *10:* 53 (1973).
3. A. M. Seppalainen and S. Hernberg. Brit. J. Ind. Med. *29:* 443 (1972).
4. A. M. Seppalainen, S. Tola, S. Hernberg, and B. Kock. Arch. Environ. Health. *30:* 180 (1975).
5. L. F. Vitale, M. M. Joselow, R. P. Wedeen, and M. Pawlow. J. Occup. Med. *17:* 157 (1973).
6. R. P. Wedeen, J. K. Maesaka, M. M. Lyons, B. Weiner, and G. A. Lipat. Amer. J. Med. (1975 in press).
7. R. K. Byers and E. E. Lord. Amer. J. Dis. Child. *66:* 471 (1943).
8. R. E. Albert, R. E. Shore, A. J. Sayers, C. Strehlow, T. J. Kneip, B. S. Pasternack, A. J. Friedhoff, F. Covan and J. A. Cimino. Environmental Health Perspectives, Experimental Issue No. 7. p. 27. (May 1974).
9. B. de la Burdé and McL. S. Choate. J. Pediat. *87:* 638 (1975).
10. S. Hernberg, J. Nikkanen, G. Mellin, and H. Lilius. Arch. Environ. Health. *21:* 140 (1970).
11. J. L. Granick, S. Sassa, S. Granick, R. D. Lavere, and A. Kappas. Biochem. Med. *8:* 149 (1973).
12. H. A. Roels, J. P. Buchet, and R. R. Lauwerys. Int. Arch. Arbeitsmed. *34:* 97 (1975).
13. A. Azar, R. D. Snee, and K. Habibi. Environmental Health Aspects of Lead. Proceedings of a symposium held in Amsterdam, Holland, Oct. 2-6, 1972. Commission of the European Communities, Luxembourg. 581. (1973).
14. F. Coulston, L. Goldberg, T. A. Griffin, and J. C. Russell. The Effect of Continuous Exposure to Airborne Lead, II: Exposure of Men to Particulate Lead at a Level of 10.9 g/m^3. Final report to the U.S. Environmental Protection Agency (1972).
15. H. V. Thomas, B. K. Milmore, G. A. Heidbreder, and B. A. Kogan. Arch. Environ. Health. *15:* 695 (1967).
16. R. J. Caprio, H. L. Margulis, and M. M. Joselow. Arch. Environ. Health. *28:* 195 (1974).
17. J. F. Keppler, M. M. Maxfield, W. D. Moss, C. Tietjen, and A. L. Linch. Amer. Ind. Hyg. Ass. J. *31:* 412 (1970).
18. A. Berlin, P. del Castilho, and J. Smeets. Environmental Health Aspects of Lead. Proceedings of a Symposium held in Amsterdam, Holland, Oct. 2-6, 1972. Commission of the European Communities, Luxembourg. 1033 (1973).
19. A. Berlin, K. H. Schaller, and J. Smeets. International symposium, CEC-EPA-WHO Paris (June 1974).
20. L. A. Blanksma, H. K. Sachs, E. F. Murray, and M. J. O'Connell. Pediatrics. *44:* 661 (1969).
21. P. R. Fine, C. W. Thomas, R. H. Suhs, R. E. Cohnberg, and B. A. Flashner. J. Amer. Med. Ass. *221:* 1475 (1972).
22. S. M. Pueschel, L. Kopito, and H. Schwachman. J. Amer. Med. Ass. *222:* 462 (1972).
23. J. J. Chisolm, Jr., and H. E. Harrison. Pediatrics. *18:* 943 (1956).
24. H. Sachs. Environmental Health Perspectives, Experimental Issue No. 7. 41. (May 1974).
25. F. S. Fairey and J. W. Gray. J. S. C. Med. Ass. *66:* 79 (1970).
26. J. R. Bertinuson and C. S. Clark. Interface. *2:* 6 (1973).
27. Lead: Airborne Lead in Perspective. National Academy of Sciences. Washington, D. C. A report of the Committee on Biological Effects of Atmospheric Pollution, Division of Medical Sciences, National Research Council. (1972).
28. G. Ter Haar and R. Aranow. Environmental Health Perspectives. Experimental Issue No. 7. 83. (May 1974).
29. C. H. Nordman. Doctoral thesis, University of Helsinki (May 1975).
30. H. Grimes, M. H. P. Sayers, A. Berlin, P. Recht, and J. Smeets. Commission of the European Community, Luxembourg. Report V/F/1491/75 (April 1975).
31. S. H. Lamm and J. F. Rosen. Pediatrics. *53:* 137 (1974).
32. Y. K. Murthy and U.S. Rhea. J. Dairy Sci. *54:* 1001 (1971).
33. P. B. Hammond and A. L. Aronson. Ann. N.Y. Acad. Sci. *111:* 595(1964).
34. D. C. Mitchell and K. M. Aldous. Environmental Health Perspectives, Experimental Issue No. 7. 59 (May 1974).
35. R. A. Kehoe. J. Royal Inst. Pub. Health Hyg. *24:* 1 (1961).

CHAPTER 10

INADVERTENT WEATHER MODIFICATION

Myron L. Corrin
Department of Atmospheric Sciences
Colorado State University

INTRODUCTION

This discussion will deal with the possible effects of lead-containing aerosols on precipitation modification and will be directed toward the possible behavior of such materials as condensation nucleants and ice nucleants. In the atmosphere, the transformation of water vapor to liquid water in cloud droplets occurs through a condensation process requiring the presence of condensation nuclei. The transformation from vapor to ice or the freezing of liquid requires again the presence of ice nucleants. These latter may be natural (as clays) or artificial (such as silver iodide used in "cloud seeding").

Precipitation may be formed in warm clouds (temperatures greater than 0° C) by a coalescence process; to obtain efficiency, a portion of the cloud droplets must be larger than 20 μm. If the cloud is formed in the presence of a large excess of condensation nuclei, the available water vapor is distributed over a large number of drops, and the droplet size is correspondingly small. In order to obtain the larger drops required to initiate and hence to further the coalescence process, some large and hygroscopic particles must be present. These will form droplets that grow at the expense of the smaller drops due to the reduction in vapor pressure produced over saturated solutions. On the basis of ideal solution behavior (a fairly good approximation in dilution solution), the reduction of vapor pressure by solution of lead-containing species would be less than 0.1 percent for all lead-containing species with the exception of lead nitrate; the solubilities are generally far too small to permit a major effect. Lead nitrate is not emitted and, as far as we are aware, does not exist in the atmosphere in measurable quantity. Thus lead compounds emitted or formed in the atmosphere would not be expected to alter the stability of warm clouds and, hence, affect percipitation from this source.

Interest in possible inadvertent weather modification effects due to atmospheric lead arises from the suggestions of Schaefer [1] and Morgan and Allee [2] that lead iodide, a known ice nucleant, may be formed in the atmosphere by a reaction involving iodine, which is present in trace amounts in the atmosphere. In the natural atmosphere, enough natural ice nucleants are normally present to convert the water droplets to ice in a supercooled liquid cloud at temperatures lower than −25° C. At warmer temperatures, artificial ice nuclei (such as silver or lead iodides) may be used as effective ice nucleants and thus permit the transformation from vapor or liquid to ice. Two possibilities exist if such artificial nucleants are used. First, given the proper concentration of ice nucleants, ice crystals are formed that grow at the expense of the water droplets. These crystals reach sizes that permit fallout. If the temperature is sufficiently low below cloud base, snow is obtained; if the temperature is high, the ice crystals melt and rain occurs. Thus precipitation is enhanced. The second possibility is that, if the combined number of natural and artificial ice nucleants per unit volume is too high, the supply of water vapor and liquid water is insufficient to permit the growth of the ice crystal to sizes that permit fallout; or if fallout occurs, the sizes are too small to resist sublimation below cloud base. Thus little precipitation reaches the ground, and the addition of ice nucleant suppresses precipitation.

LEAD-CONTAINING AEROSOLS AS ICE NUCLEANTS

Grant, et al. [3] conducted a set of preliminary experiments at Colorado State University (CSU) to examine automotive exhaust as a possible source of ice nucleants. Two automobile engines were employed; one was operated for several weeks on lead-free gasoline, while the other was fueled with "regular" gasoline containing a lead antiknock additive. The lead-free gasoline contained less than 1 μg/g lead. Exhaust samples were diluted in the CSU vertical wind tunnel and injected into the CSU isothermal cloud chamber. Samples from both vehicles indicated little ice nucleation activity (essentially within the noise level of the chamber). The exhaust samples were then treated with iodine vapor prior to injection; the pressure of iodine vapor was the saturation vapor pressure over the solid (0.9 torr). Both samples indicated an increase in concentration of ice nucleants over the noniodine-treated material by about three orders of magnitude. Within the limits of experimental error, there was no difference between the exhausts from the two vehicles. These experiments are open to question, since there was no assurance that the lead was purged from the one engine by several weeks' operation with nonleaded fuel.

In a second, and more comprehensive, investigation, [4] two new 5-horsepower engines were used. These were thoroughly cleaned, and one was reserved for use with nonleaded fuels. The nonlead gasoline selected had a lead content of less than 0.5 μg/g. The two motors were connected to a common load and operated during tests under the same conditions.

At nucleation temperatures ranging from $-12°$ to $-20°$ C, the number of active ice nucleants produced after iodine treatment was always greater in the exhaust from the use of nonleaded fuel. This in effect confirmed the previous preliminary study. A plot of the observations is given in figure 10-1. It was also noted that (1) filtered exhaust does not react with iodine vapor to produce active nucleants; (2) if iso-octane is used as fuel, no nucleants are produced upon treatment with iodine vapor; (3) a PbBrCl aerosol does not act as a nucleant or produce a nucleant upon treatment with iodine vapor; (4) if lead tetraethyl and no scavengers are added to lead-free gasoline, copious quantitites of ice nucleants are produced upon treatment with iodine vapor; and (5) a synthetic lead iodide aerosol is an effective ice nucleant.

Thus it may be concluded that (1) the species in exhaust active toward reaction with iodine to form ice nucleants are particulate; (2) the particulate formed in the combustion of iso-octane is not active, suggesting that the active species are particulates formed from the combustion of aromatic constituents in the fuel (nonleaded gasoline is higher in aromatic content than leaded fuel); (3) PbBrCl does not react with iodine vapor to form lead iodide; (4) metallic lead or lead oxide may react with iodine vapor to produce lead iodide; and (5) lead iodide is an effective ice nucleant.

These results are supported by some thermodynamic calculations. Consider, for example, the reaction between $PbBr_2$ and I_2

$$PbBr_2(s) + I_2(g) = PbI_2(s) + Br_2(g)$$

The standard free energy change for this reaction is +16.8 kcal, and the equilibrium constant is 4.5×10^{-13}. Thus a ratio of gaseous iodine to bromine of 2×10^{12} would be required to form PbI_2. Such ratios are not encountered in the atmosphere. Similar effects are calculated for $PbCl_2$, PbBrCl, and $PbSO_4$; these substances are stable with respect to PbI_2 formation. Lead monoxide gives a standard free energy change of -2.2 kcal for the reaction

$$2PbO(s) + 2I_2(g) = 2PbI_2(s) + O_2(g)$$

The equilibrium constant at 298 K is 42.5. But at oxygen partial pressures in the atmosphere of 152 torr, a partial pressure of iodine vapor of 2 torr would be required to form PbI_2. Such partial pressures do not occur in the real atmosphere, nor were they simulated in the experiments described above. Lead dioxide does not react with iodine vapor; the equilibrium constant is 3×10^{-5}. One reaction does occur:

$$Pb(s) + I_2(g) = PbI_2(s)$$

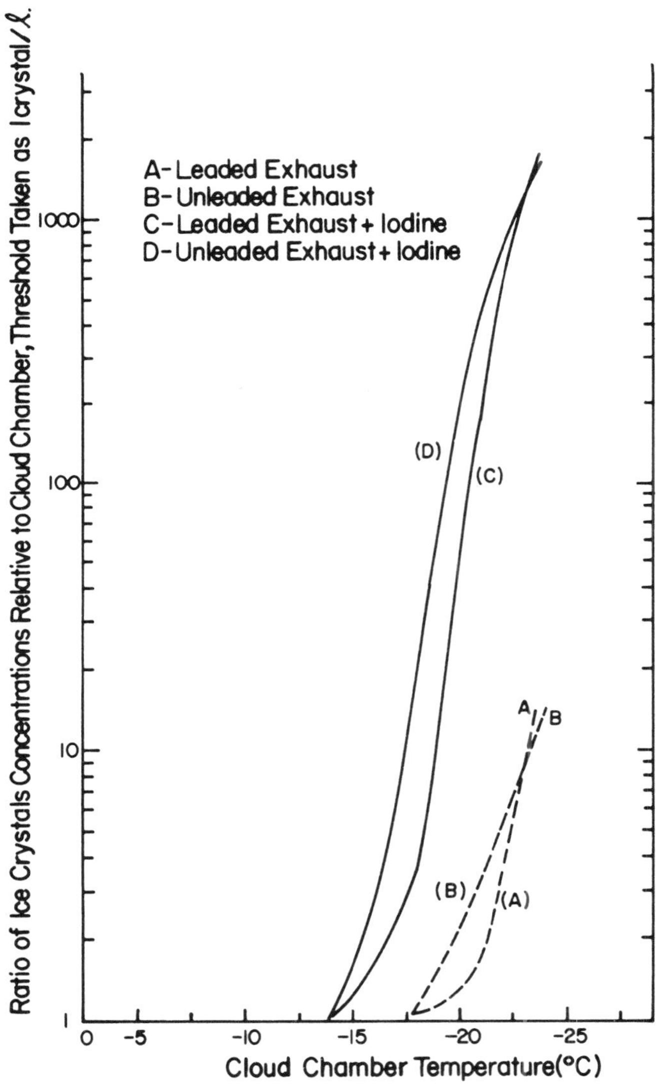

FIGURE 10-1.—Ratio of concentration of ice crystals formed from raw and iodine-treated auto exhaust produced from leaded and lead-free gasoline.

The equilibrium constant is 7×10^{33}, and the required iodine pressure is 2×10^{-37} torr. This pressure was exceeded in the laboratory simulation and may be exceeded in the atmosphere. It is concluded that metallic lead was the active species formed in the combustion of fuel to which lead tetraethyl was added in the absence of scavenger.

These observations cited above thus support the conclusion that (1) the combustion of both leaded and nonleaded fuel forms a particulate species of possible high aromatic content that react with iodine vapor to form an effecive ice nucleant; and (2) the lead-containing aerosols in the exhaust from an internal combustion engine fueled with a lead additive mix containing scavenger are not active ice nucleants, nor do they react with iodine vapor to form such nucleants.

REFERENCES

1. V.J. Schaefer. Science. *154:* 1555 (1966).
2. G.M. Morgan, Jr., and P.A. Allee. J. Appl. Meteorol. *7:* 241 (1968).
3. L.O. Grant, M.L. Corrin, and R. Rinehart. Unpublished observations.
4. TO. Grant and M.L. Corrin. J. Weath. Mod. *5:* 238 (1973).

PART IV

Control Strategies

INTRODUCTION TO PART IV

Harry W. Edwards
Department of Mechanical Engineering
Colorado State University

These chapters focus upon methods for controlling contamination of the environment by lead. Measure involving both regulation of lead usage and control of lead release are considered in the context of reducing lead inputs to the environment. Sources considered are the lead mining, milling, and smelting industries, and combustion of leaded gasoline. The nature and dispersal of lead released from these two types of sources are quite different; the controls are therefore considered separately. In the case of industrial lead emissions, the controls are identified for specific plant operations. Application of modern control technology and good housekeeping practices can be very effective in minimizing environmental contamination by these lead industries. In considering the role of regulatory agencies relative to controlling industrial lead emissions, advanced planning and industry-agency cooperation are desirable in developing stable and effective policies that clearly specify industry responsibilities.

In the case of automotive lead emissions, the issue of continued widespread use of lead alkyls is not readily separated from other health effects, environmental quality, and energy conservation aspects of automobile and automotive pollution. In terms of the development of a long-range policy regarding automotive lead, economic considerations, in addition to the technological factors, must enter strongly. A number of alternative controls are therefore examined. These include: reducing the lead content of gasoline, use of particulate traps to reduce lead release, and effective urban planning. Other factors that influence lead distribution in an urban environment are also subject to some degree of control. They include: traffic routing, parking, building and greenbelt geometry and placement, and mass transport systems. Physical modeling is particularly useful for predicting the effects of new urban structures and traffic routes on dispersion of lead and other airborne pollutants. No single control strategy seems ideal for all situations, and many problems will undoubtedly best be solved by a combination of approaches involving the use regulation, release control, and effective urban planning to control lead exposure and accumulation, especially in urban areas.

CHAPTER 11

Control of Industrial Emissions of Lead to the Environment

J. Charles Jennett Bobby G. Wixson, and Ivon H. Lowsley*
Department of Civil Engineering
University of Missouri-Rolla

INTRODUCTION

As lead ore is mined, milled, smelted, refined, and manufactured into a wide variety of useful products, large quantities of wastes are produced. The control of wastes from the lead industries, and particularly the abatement of those from the primary production processes, are discussed herein, and the New Lead Belt in southeast Missouri is used as a frequent example. This recent development is presently the largest lead mining region in the world. It represents the newest and most modern industrial development of its kind. The amount of lead-containing waste from the operation is large. This waste tends to be highly concentrated, difficult to control, and quite obvious because of its great volume.

The U.S. Environmental Protection Agency (EPA) is presently working on new requirements and standards for the lead mining industry. Industry effluent characteristics and treatment patterns are being investigated, and standards related mostly to the control of liquid effluents are scheduled to be established in 1976. Both the lead and zinc mining industries will be grouped into a single category; mines will be classified as lead, zinc, lead-zinc, or ther, depending on whether lead or zinc is recovered.

The various levels of treatment technology that are to be achieved are questionable. Level 1 technology, which must be achieved by all plants in each industry not later than July 1, 1977, will be based upon the average of the best existing performance by plants of various sizes, ages, and unit processes within each industrial category. This means that the standards will be based on the best current technology within exemplary operations of the mining industry in total, and not confined to the lead/zinc industry alone.

*Presently Chairman, Department of Civil Engineering, Syracuse University.

PROBLEM AREAS OF LEAD EMISSION CONTROL

Solid Waste

Lead-containing solids are initially generated during underground mining operations where large quantities of mineral-bearing rock are removed, initially crushed underground, hauled to the surface, and processed to a size fraction finer than 200 or 300 mesh. The finely ground material is passed through the flotation operation and processed into concentrates containing between 70 and 85 percent lead. This concentrate is then transported to primary smelters for conversion into 99.99 percent lead metal. The disposal of waste material from the grinding and milling process, known as gangue or tailings, is one of the most critical problems faced by the mining industry. [1] The amount of rock waste varies with the percentage of recoverable ore, but ranges from 6.5 to 50 times the final tonnage of metal produced. The current practice is to dispose of this material in nearby valleys or on land surrounding the operation. The tailings are also used to build dams for the tailings ponds; these operate as large sedimentation basins and as biological treatment devices for the liquid wastes from the mining and milling operations.

The finely ground tails contain high concentrations of lead (depending on mill efficiency, concentrations may vary from 500 to 20,000 $\mu g/g$) and may become an environmental hazard if washed directly into streams or blown into forests surrounding the operation. In some locations, the windblown tailings have been deposited and concentrated in the top 1 in layer of the forest floor to levels as high as 30,000 μg/g as lead mainly in the less-than-80-mesh fraction. [1]

A similar control problem is related to the lead, zinc, or copper concentrates that are often stored in open piles at the mill and smelters prior to processing. As with tailings, the finely ground concentrates are subject to water or wind action. Environmental problems may be more critical with the concentrates since they contain from 70 to 80 percent lead.

Another problem presently exists in the transportation of lead, zinc, or copper concentrates from the mill to the smelter in open railroad cars or trucks. Unless sufficiently moist, the finely ground material blows onto roadsides and results in high concentrations of soil lead along haulage routes.

The disposal of sludges containing dangerous quantities of heavy metals from some waste neutralization processes is a major problem, and studies are needed to evaluate possible detrimental environmental aspects.

Airborne Emissions

Control problems faced by both the mining and smelting sources are similar, even though primary or secondary smelting operations are involved. Most primary lead smelters in the United States essentially use the processing steps of sintering, reduction in a blast furnace, and refining, although there are differences in equipment and operational details.

The emissions from most domestic lead smelting operations are likely to be a combination of particulate matter and gases. Sulfur dioxide is generated during the sintering process and discharged directly to the atmosphere by approximately half of the six domestic lead smelters operating in the United States at the present time. Other smelters convert 50 to 70 percent of the SO_2 to sulfuric acid and discharge the gases remaining to the atmosphere. Carbon monoxide and carbon dioxide are also discharged from the blast furnaces.

Dust generated in the smelting processes is collected by baghouse filters, electrostatic precipitators, venturi scrubbers, and various other wet scrubbing mechanisms that are 95 to 99 percent effective. The remaining 1 to 5 percent dust load is discharged through the plant stack as PbS and PbO. Fugitive dust from concentrate stock piles and dust-handling operations (again either PbS or PbO) may become airborne by high winds or through vehicular traffic.

Initially, a mixed charge consisting of lead concentrate and fluxing materials is sintered to remove the sulfur and produce a cohesive, sporous mass. During this process, off-gases containing 4 to 6 percent SO_2, tail gases, dust, and fumes are emitted. The sinter machine tail gases may or may not be recirculated. The porous and consolidated sinter is passed through a crusher-sizer and a material handler, where additional dust is generated. The sinter is then mixed with coke and reduced to metallic lead in the blast furnace. The flue gases emitted by the blast furnace contain particulates, fumes, and variable amounts of SO_2. [2] The metallic lead from the blast furnace is then refined in large kettles and dross reverberatory furnaces. A very small amount of SO_2 and metallic fumes are emitted from these refining operations. [3] A schematic of this process is presented in figure 11-1.

Principal emissions of concern from the smelter operations are: stack emissions, fugitive dusts, and SO_2.

Stack Emissions: The stack normally discharges off-gases containing particulate lead from the blast furnace and tail gases from the sintering operation, if they are not recirculated within the machine. These effluent gases may be passed through a cyclone, baghouse, or an electrostatic precipitator before being emitted into the atmosphere.

Fugitive Dusts: Fugitive dusts are one of the major sources of lead pollution close to the smelter, [4] a point long recognized by industry. [5] Significant and often neglected sources of fugitive dusts include material-loading and unloading areas, concentrate and slag piles in the plant area, and open railroad cars and trucks transporting materials to and from the plant. The amount of dust produced also depends upon meteorological and operational factors.

Sulfur Dioxide Emissions: In order to prevent air pollution problems and meet the stringent emission and ambient air quality standards, the strong off-gases and weak tail gases from the sintering machine are often fed into a sulfuric acid plant; 60 to 70 percent of the SO_2 from the blast furnace is also discharged through the stack after passing through some type of particulate control equipment.

Liquid Waste Emissions

Mining: Water quality problems can be generated by mining operations, milling operations, or combined mine-mill complexes. Situations concerned with mines only are rare. If a mine is dry, it has no water quality problems. However, most lead mining operations in Missouri are wet, and the underground water must be brought to the surface. In many cases, mine water cannot be recycled and must be treated prior to discharge into area streams.

The amount of lead and other heavy metals dissolved in the water produced underground is primarily a function of the composition of the minerals associated with the ore body, the pH of the mine water, and its alkalinity. The more acidic the water, the more likely it is to contain large quantities of lead and other associated metals. Alkalinity affects the buffering capacity of water, and, therefore, variances in pH. The suite of associated minerals is not usually a major factor, since most of the lead in the United States is mined as galena (PbS). However, the mineral suite becomes important if pyrites are present and produce acid mine drainage due to oxidation. In the New Lead Belt of southeast Missouri, the pH of both the surface and underground water varies frum about 7.9 to 8.1; the water, contains only small quantities of dissolved lead and zinc. [6] However, the secondary problem of suspended

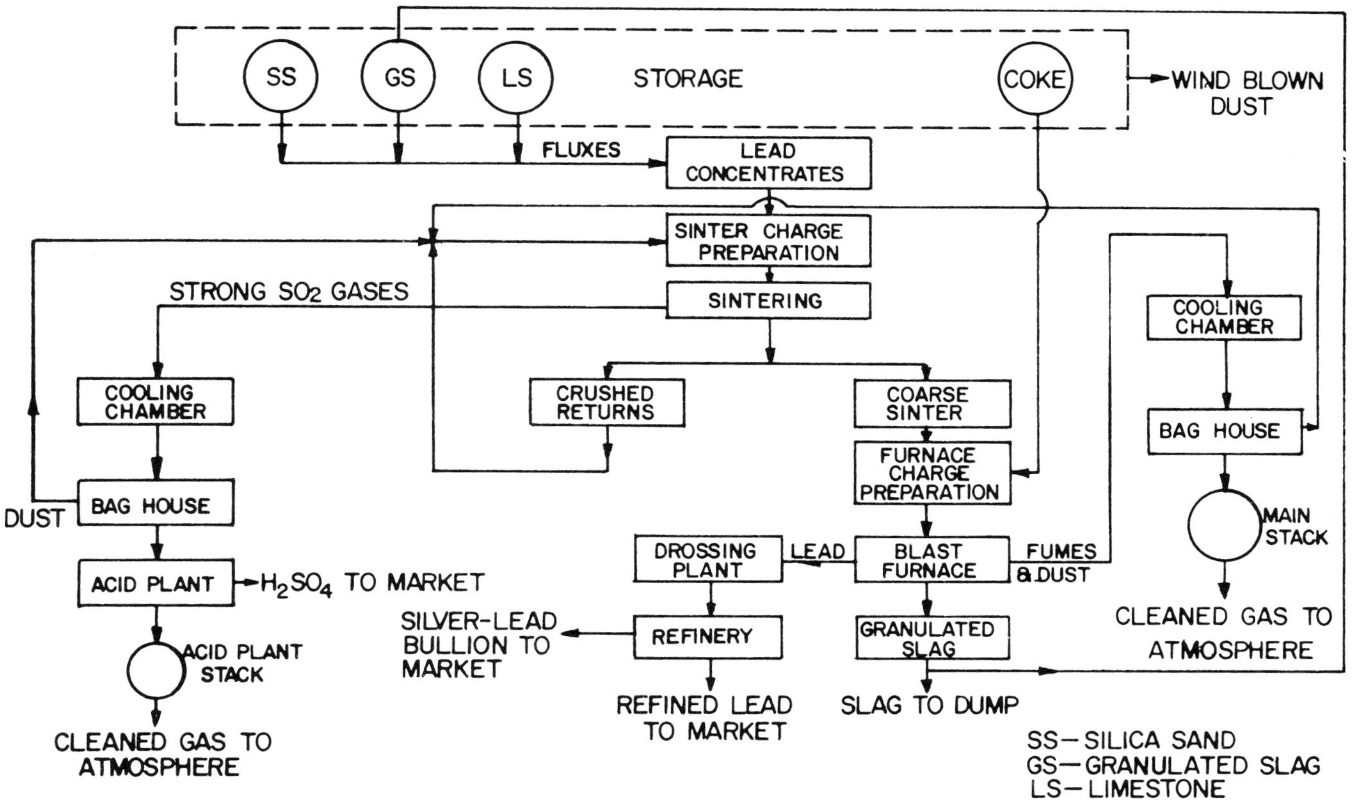

FIGURE 11-1.—General flow sheet of lead smelting.

heavy metals may be a major factor in the release of lead from mine water. [7] These materials should, however, precipitate with adequate time in sedimentation ponds.

In cases where acid mine drainage problems exist, substantial amounts of dissolved heavy metals, particularly lead, zinc, copper, and cadmium, may be present. In these cases, wastes must be subjected to far more rigorous treatment than that required in the New Lead Belt.

Milling: Effluents from lead-zinc milling operations are a second, but generally larger, source of lead in the aquatic environment. [7-9] In the milling process, chemical agents are added to the water containing ore particles, and air is bubbled through the mixture to extract the various minerals through a flotation process. [10] The flotation agents usually consist of chemical collectors, frothers, depressants, and activants. [7, 9, 10] Lead, zinc, and sometimes copper minerals from the flotation process are pumped into thickeners; there the concentrate settles, leaving the water, excessive flotation reagents, and colloidal or supracolloidal minerals to be discharged as effluent for treatment in sedimentation lagoons. This procedure is illustrated in figure 11-2.

Most undesirable water problems are due to a lack of coordination between the mine and mill operations. For example, if high-grade ore is sent to the

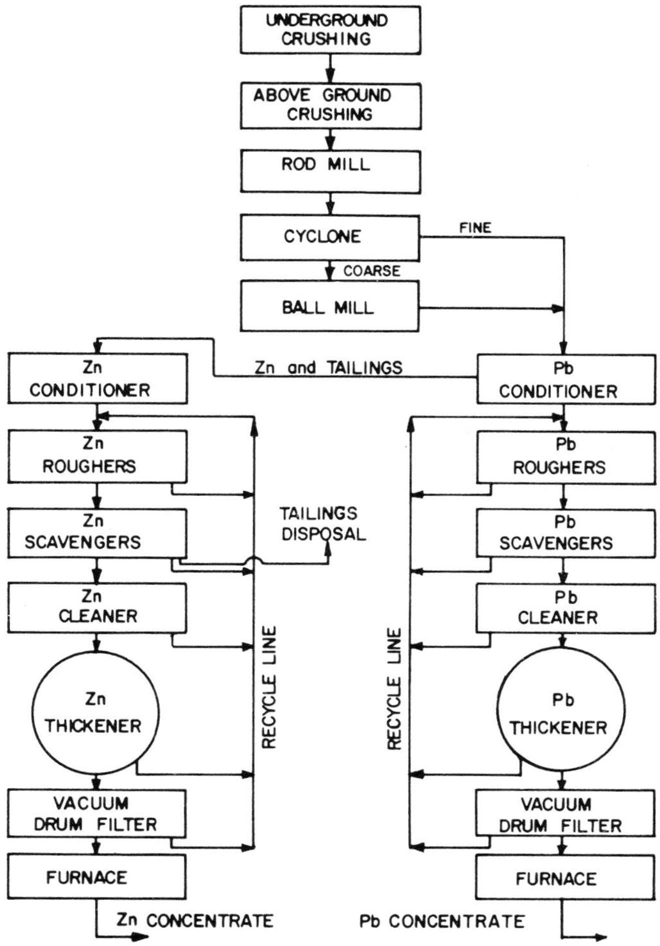

FIGURE 11-2.—Typical flow process used for milling and concentrating lead and zinc in southeast Missouri (After Ernst). [13]

mill, and insufficient flotation reagents are used, excessive amounts of dissolved, suspended, heavy metals are released to the treatment ponds. If a low-grade ore is sent to the mill, excessive organics and flotation reagents may be used and subsequently released to the environment. Another major problem within the milling process is the possibility of an accidental chemical reagent spill that may be acidic in nature.

The treatment of milling and mine-mill wastewater generally occurs in tailings ponds. If the water is not allowed to short-circuit, thus releasing partially treated wastes, [6, 11] these ponds are relatively effective sedimentation devices as long as they are operated with sufficient retention time, are not subjected to acid spills, and are properly designed and constructed.

The tailings ponds are frequently considered by industry to be effective biological treatment devices that decompose residual organics from the milling process. Unfortunately, they are not effective for a number of reasons that have been discussed by Jennett and Wixson [9] and Gale et al. [12]

Storm Runoff: Storm runoff contributes to lead pollution from milling activities in several ways. Due to dissolved carbon dioxide, the pH of receiving streams tends to drop during periods of excessive runoff. The organic acids that leach from the soils enter the stream systems and may increase the solubility of metals in the water and sediments. Research in the New Lead Belt has indicated that streams draining the smelter areas have normal background lead levels of 10 to 30 μg/l. These streams may decrease by 1 to 2 pH units during periods of stormwater runoff, and the dissolved heavy metals may temporarily increase to as high as 4 mg/l. [7] During heavy rainstorm, colloidal and suspended solids are picked up and transported downstream from the tailings dam and concentrate storage piles surrounding the mills and smelters. Storms also create turbulence in the tailings ponds and reduce the effective retention time by disturbing the solids, reducing settling conditions, and allowing the solids to be carried through the system into the receiving stream. [13]

PRESENT CONTROL TECHNOLOGY

SOLID WASTES

The practice of transporting lead, zinc, and copper concentrates in open railroad cars or trucks should be discontinued. The cars could be covered with a suitable material or sprayed with some type of protective layer such as latex, which could be easily removed during smelting. A modest cost would be associated with the operation. Industry has argued that this control is trivial compared with other problems, and that the concentrate is relatively insoluble lead sulfide. However, for the relatively small expenditure of time and effort required, it would seem wise to control this particular source of pollution.

The concentrate storage piles at the mills and smelters should be covered or contained in some manner to prevent windblow. [14] An alternative procedure would be to continuously keep the piles damp. However, any drainage from these piles would have to be retained and treated by sedimentation before being released. The AMAX Lead Co. of Missouri has designed and installed a sprinkler system to reduce blowing dust within the smelter area. The site has been contoured so that most dust can be washed from the grass and plant roads into a central drain for treatment. This procedure represented an expensive addition to the plant; initial installation would have been more economical.

Windblown materials from the tailings dams contribute the largest single volume of waste rock dust from the mining operations and require some type of control treatment. One possible method being tested by several companies is to cover the tailings with a layer of soil and establish a cover of vegetation. Treating the tailings dams with a relatively inexpensive chemical such as water glass (sodium silicate) to render them less suceptible to the action of both water and wind may also be worth further investigation.

Airborne Emissions.

Control of Particulate Emissions and Process Dusts: The most commonly used techniques to reduce dust and fumes from the sinter machine gas stream involve settling in large flues, or the use of electrostatic precipitators or baghouses. Collection efficiencies of 96 percent for precipitators and 99.5 percent for baghouses have been obtained. [15] The dusts generated in the sinter crushing and sizing operations are removed by impingement-type wet scrubbers or fabric filter bags.

Blast furnace and dross reverberatory furnace off-gases, as well as sinter machine tail gases, are cleaned in baghouses using wool or fiber glass bags or by processing them through electrostatic precipitators. Where a sulfuric acid plant is used to recover SO_2, the particles from the strong flue gases are removed by the acid plant baghouse.

Examples of processes and control methods used for lead smelter operations in Germany are summarized in table 11-1.

Control of Gaseous Emissions: Sulfur dioxide is the major pollutant in the gaseous emissions from the

TABLE 11-1.–*Processes and control methods for lead smelter operations in Germany*

Process	Control devices	Efficiency (percent)	
		Primary	Secondary
Sinter machine	Centrifugal EP[1], BF[2]	80-90	95-99
Blast furnace	Centrifugal EP	80-90	95-99
Reverberatory furnace	Waste heat EP, boilers, tubular BF coolers	70-80	95-100

[1] Electrostatic precipitator.
[2] Bag filter.

sintering machine, blast, and dross reverberatory furnace. Most SO_2 produced in concentrations of 4 to 6 percent can be converted to sulfuric acid in an acid plant. The remaining gases may then be discharged to the atmosphere with sufficient dilution to meet most federal or state emission standards. Another SO_2 control method for flue gases is dispersal through a tall stack (600 ft) coupled with a remote weather telemetry system that allows operations to be terminated when inversions are expected. This method requires a properly designed stack in order to comply with ambient air quality standards.

The SO_2 in the weak tail gases from the sinter machine is controlled either by stack discharge or by recirculating and concentrating the SO_2 within the machine. The SO_2 contained in the blast furnace off-gases is discharged through the stack without reduction. With the present technology, the weak SO_2 gases cannot be efficiently recovered by an acid plant or by other types of treatment (such as limestone scrubbers) that are expensive and inefficient.

Processes and Control Methods: The processes and control methods for lead smelter operations practiced in the United States are presented in table 11-2. Investigations at the U.S. Bureau of Mines' Metallurgical Research Center in Salt Lake City have been successful in converting most of the SO_2 to elemental sulfur. [16, 17] This so-called "citrate process" removes 90 to 99 percent of the SO_2 from industrial gases and consists of the following steps: washing the flue gas to remove particulates and SO_3, adsorption of the SO_2 in citric acid or other carboxylate solutions, reaction of the H_2S loaded solution in a closed system to precipitate the absorbed SO_2 as elemental sulfur, and separation of the sulfur from the regenerated solution by oil flotation and melting. [18, 19]

TABLE 11-2.—*Processes and control methods for lead smelter operations in the United States*

Process	Control device	Pollutant controlled
Sintering machine		
Strong gases	Single-adsorption sulfuric acid plant, tall stack with meteorological control, acid plant baghouse	SO_2, particulates
Weak gases	Baghouse, electrostatic precipitator	Particulates
Sinter crushing and handling	Impingement-type scrubber, bag filter	Particulates
Blast furnace	Fume catcher, baghouse, electrostatic precipitator, Venturi	Particulates
Dross reverberatory furnace	Baghouse, electrostatic precipitator, Venturi	Particulates

Control by Proper Designing: Designers of future lead smelting and refining plants should consider the location of the material-receiving area, the isolation and confinement of storage piles, the paving of areas around the smelter, the transfer of materials on enclosed belt conveyors, and process modifications. Most industrial planners are aware of these problems and are taking steps to ensure that future smelters will utilize all possible improvement.[5]

In selecting control equipment, plant managers should apply the best available, practical, and economical control technology compatible with the emission and ambient air quality standards set by appropriate regulatory agencies.

The present use of primary and secondary equipment to remove particulates will continue until better devices are developed. The primary equipment removes larger particulates and usually involves centrifugal methods. Submicrometer materials are removed in the secondary process either by a fabric filter or an electrostatic precipitator.

The control of SO_2 to meet both emission and ambient air quality standards remains one of the lead smelting and refining industry's major problems. Although several emission control methods are available, no one method has been generally suited to all conditions or uniformly economical in operation, maintenance, and efficiency of control.

Water Quality Control

Many different systems have been used throughout the world to treat wastewaters from lead mining and milling. Effluents from most mines are treated by retaining the combined mine and mill wastewater in large, deep ponds constructed by building tailings dams across nearby valleys.

Often considered a simple treatment method, tailings ponds are actually deceptively complex, since a number of simultaneous functions must be performed. These functions include: removal of tailings solids by sedimentation, acid neutralization, formation of heavy metal precipitates, sedimentation of metal precipitates, long-term retention of settled tailings and precipitates, stabilization of oxidizable constitutents, balancing of effluent quality and quantity fluctuations, and balancing stormwater storage and flow. [20]

Lagooning operations vary in shape, depth, number of lagoons, and retention time. This type of treatment has not been effective in treating residual organic milling reagents due in part to (1) dilution of the organic-rich mill reagents by the mine water to levels that prevent effective biological treatment in lagoons; (2) lagoons too deep for effective light penetration; (3) inadequate retention times produced by lagoon shape, depth, temperature inversions, and other factors that short-circuit the treatment; and (4) premature filling of the lagoon. [1, 7, 11] Research by Bolter et al. [21] indicates that a retention time of at least 9 days is required for the effective removal of excess concentrations of lead, copper, and zinc. Hawley [10] suggests that months rather than days are required for effective treatment.

The lagoons should be divided into at least two (preferably three) cells to facilitate better operation, and they should be shallow enough to promote bio-

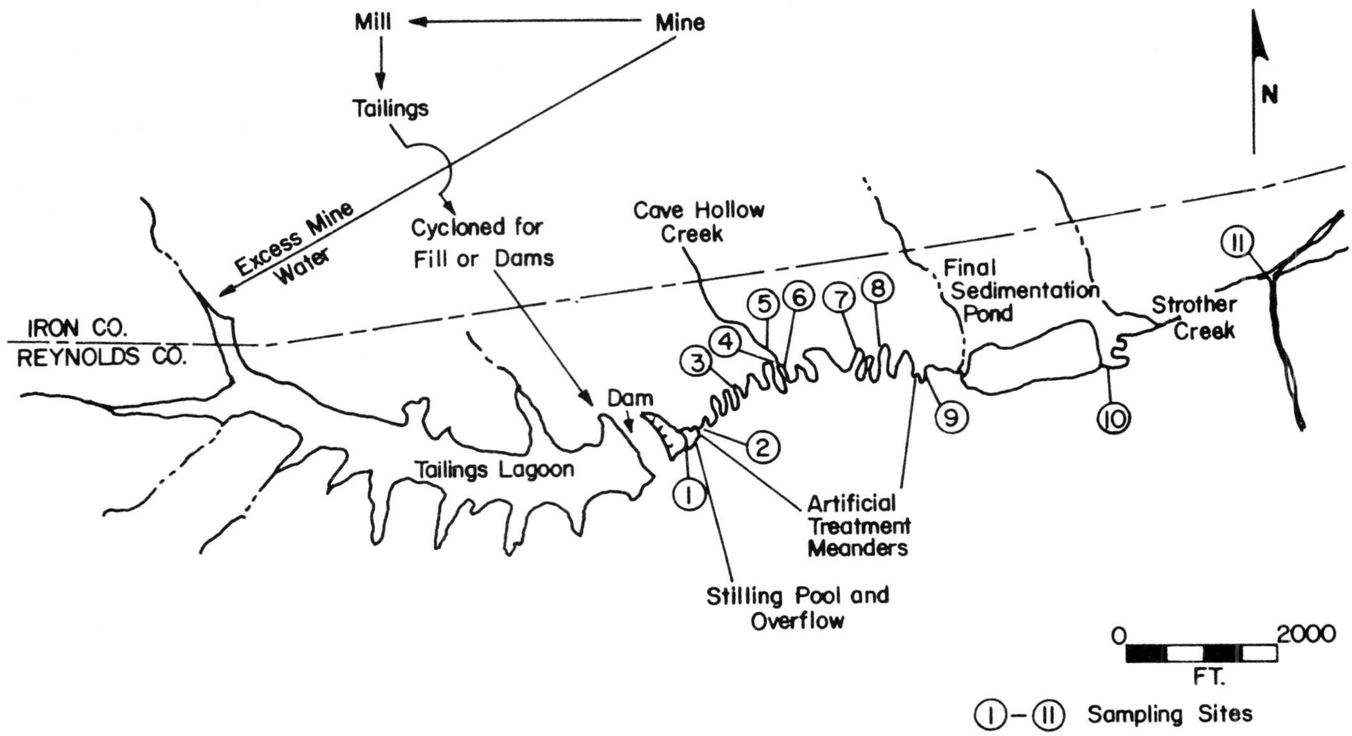

FIGURE 11-3.—Diagram of experimental meander system used to iologically treat mining wastes in southeast Missouri. [13]

logical reduction of oils and other organic materials. [1, 7] Lagoons remain popular primarily because they produce effluents that meet most state guidelines with a minimum of maintenance and operator supervision.

Additional waste treatment may be accomplished by following the lagoon system with an extensive series of shallow meanders, about 1 m depth. Such a system was installed at the AMAX Buick Mine below the final tailings dam. Algae growing in the shallow meanders successfully trapped the very fine particles of metals and rock flour prior to discharge into area streams. It was necessary to terminate the meanders with a well-baffled final sedimentation basin to contain the floating algal mats, thus increasing the sedimentation time of the system. (fig. 11-3) This meander system has successfully lowered the frequency of algal blooms, as well as the heavy metal content of the algae and plants normally found in the receiving stream. [22, 23] (fig. 11-4) There was also an overall reduction in the effluent-heavy metals ratio, defined by the State of Missouri as the measured concentration of an element divided by the state's allowable effluent concentration for that element. [24] The measurement is generally based on filtered samples. The elements and their allowable concentrations in Missouri are:

Iron	7.0 mg/g	Nickel	0.80 mg/g
Manganese	1.0	Copper	0.05
Barium	1.0	Zinc	0.15
Chromium	1.0	Lead	0.10
Cadmium	0.01		

The heavy metals concentrations at various points in the AMAX Buick Mine sedimentation and biological meander treatment system are shown in figure 11-5. Below the treatment system, Strother Creek rarely exceeds the allowable concentrations for heavy metals except during storms when algae in the meanders are detached and the final sedimentation pond is less efficient due to temporary turbulence. The effectiveness of this particular meander system has been further described by Ryck and Whiteley. [23]

Operational problems may limit the efficiency of the meander system at other sites, and such a system may not be applicable for some mining operations. These systems require large amounts of otherwise usable level land directly below the tailings ponds. Artificial meanders are also subject to washout during intense storms that may remove the algal mats. During the time required for algae regrowth, the capacity to trap fine particles is limited. At present, there are no existing design criteria for meander systems, and their establishment is essentially an experimental procedure. Based on research findings by Ernst [13] and others, [24, 25, 38] it appears that the meander treatment system does offer considerable promise in treating mine-mill wastes; however, additional studies are needed to optimize the design or to allow transfer of this procedure to different locations.

Patterson and Minear [18] reviewed the physicochemical processes used to treat wastes containing heavy metals and summarized the theory, practice, and costs involved in such treatments. Data published by Adams et al. [26] strongly indicate that biological systems, particularly activated sludge treatment techniques, can tolerate large quantities of heavy metals and may be effective in their removal.

Research strongly suggests that the ideal procedure for treating mine or mill effluents involves the use of recycled water in all operations. (fig. 11-6) Recycling offers a number of advantages: a zero-effluent discharge eliminates problems in receiving streams and the need for permits and impact statements; in-house spills of acids, milling reagents, oil, etc., can be contained and treated; and the water quality of the receiving streams is not dependent on the operational efficiency of the treatment system. However, the desirability and economics of recycling continue to be highly controversial topics in the lead mining industry.

One problem is that the return of mine water underground may actually take the form of deep-well injection and may not be desirable as a recycling method or recharge. Many industrial operators have been concerned about using a closed loop or recycling in the milling operation because of in-house operational problems. However, many mines are

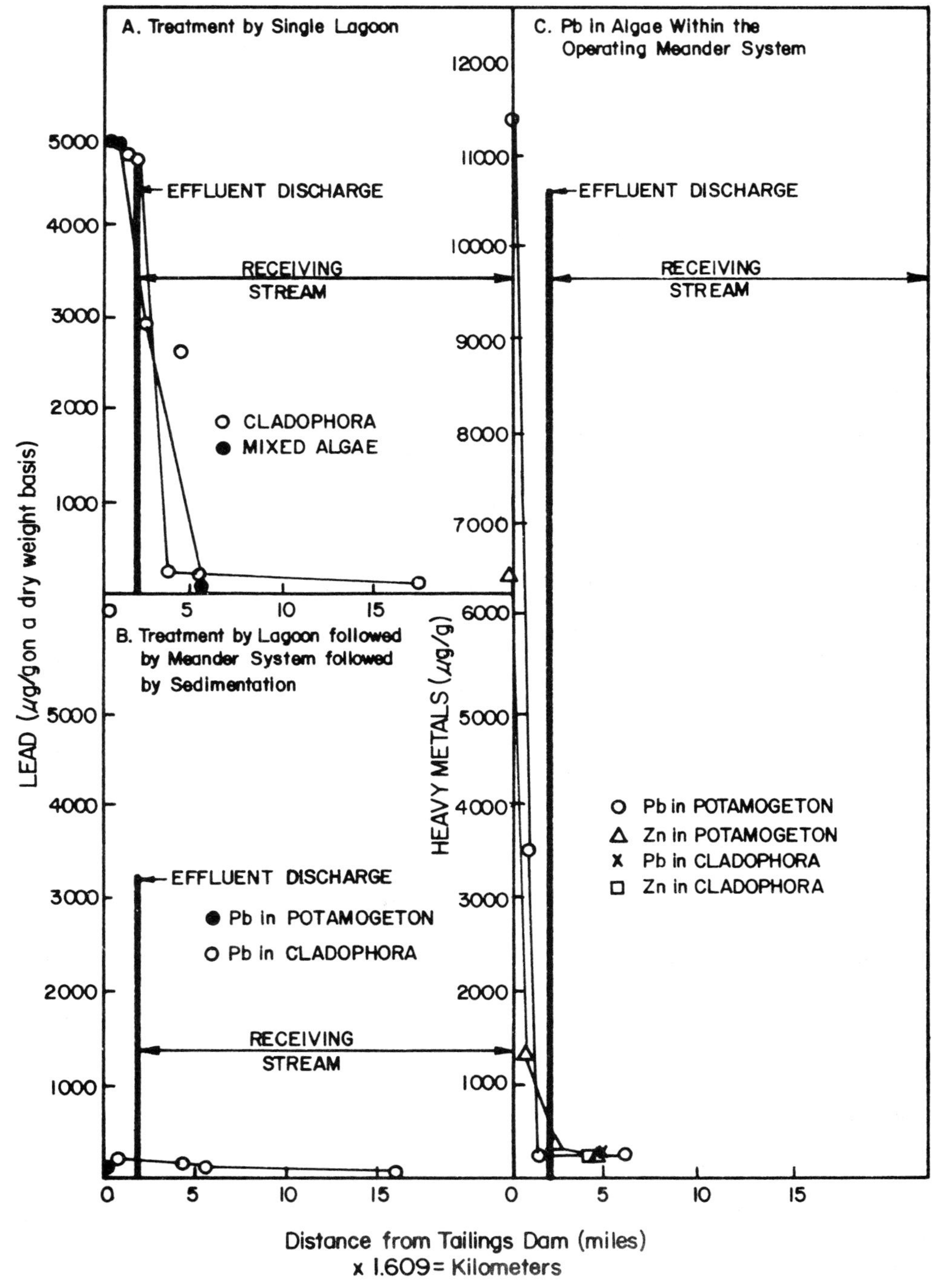

FIGURE 11-4.—Effects of meander treatment on algal lead content before, after, and within the system

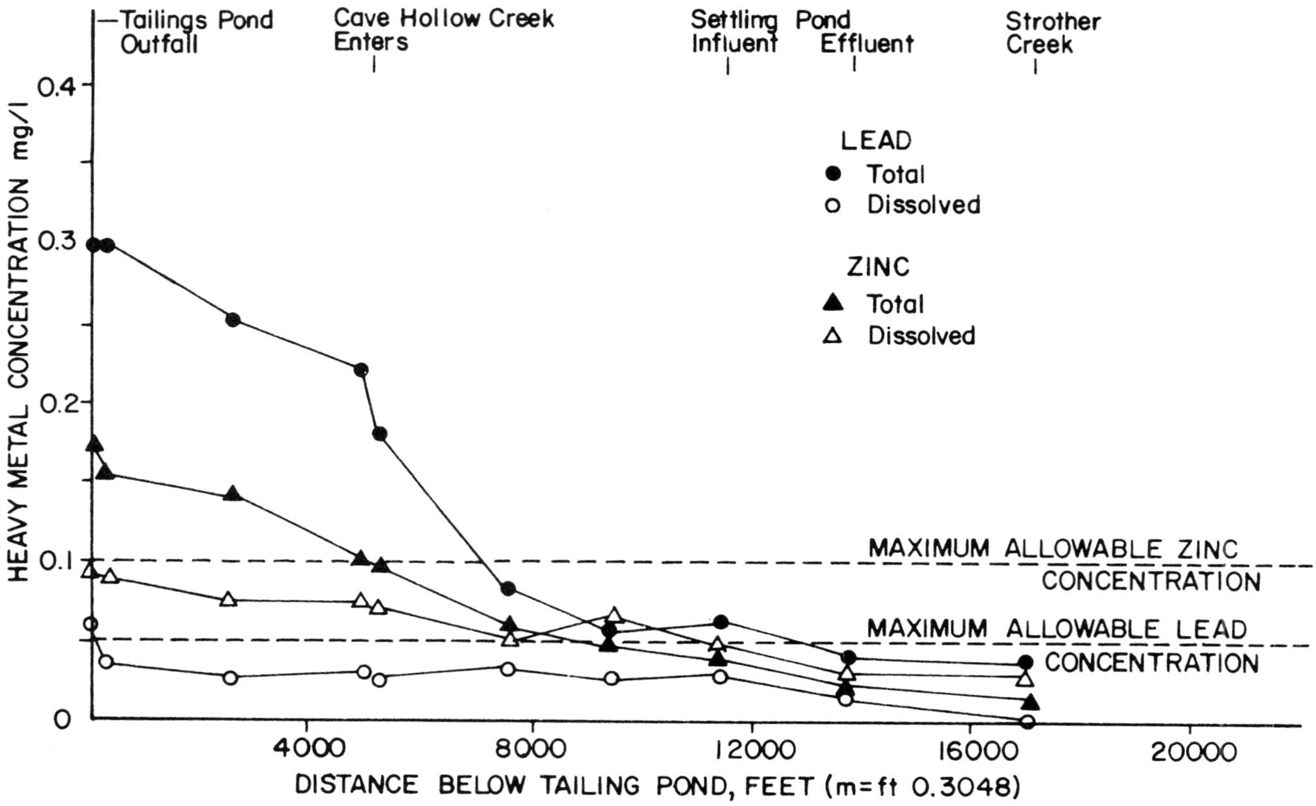

FIGURE 11-5.—Concentrations of lead and zinc at various locations within the AMAX Buick Mine meander treatment system (after Enrst). [13]

either dry or produce less water than is required for the milling process, so that recycled water must be used.

Two mining companies in the New Lead Belt are presently using recycled water in the lead-zinc milling process. St. Joe Minerals recycles water at the Viburnum, Fletcher, and Brushy Creek operations as an environmental protection measure. The company recycles tailings pond water into a separate reservoir that is later used in the milling complex. The water is pumped from the surface of the tailing ponds through portable barges that can be moved to appropriate locations. Recycling rates range from 25 to almost 100 percent, depending upon the mill make-up demand. Cominco American, Inc., recycles water at its Magmont operation because of a scarcity of water from the mine. The small amount of water lost from this operation is excess rainfall that may overflow the recycling ponds. Biological studies by the Missouri Department of Conservation have shown that the quality of receiving stream water has not been damaged by occasional storm-related overflow from the Magmont Mine. Recycling has also restored water quality to approximately pre-mining conditions at the St. Joe operation. [23]

Some of the operational problems associated with recycling mill water have been discussed by Sharp and Clifford. [30] They observed that recycling at some operations did not cause any unsolvable metallurgical problems and that an economic benefit may be gained by decreasing the amounts of flotation reagents required.

A summary of water recycling in Canadian mills by Pickett and Joe [31] indicates that recycling is feasible in the gravity and magnetic circuits, as well as in the simple flotation circuits used by most Canadian mills. Problems have arisen in the Pb-Zn-Cu selective flotation circuits with regard to controlling chemical dosages in the recycled water. Water recycling in the milling process may be limited by the buildup of complex chemical compounds that decrease metal removal, thereby causing operating problems. In Canada, [31] the best overall solution involved the use of a cascade pond system. There the reclaimed water from thickeners, filters, and tailings ponds was matched with the water quality requirements for each part of the flotation circuit to give the most effective combination in the process.

If mine water cannot be recycled, it must be treated and handled separately from the milling system to prevent dilution of the organic milling reagents and thus to facilitate their breakdown by biological actions. [27] Control of pH or the addition of precipitants may be required at some operations, depending on the nature of the ore deposit. The New Brunswick lead mining area in Canada, for example, has a major problem with acid drainage because of pyrites in the ore. Rainwater passing through the piles of waste rock on the surface creates acidic leachates that must be contained and neutralized for pollution control. Hawley [28] has provided an excellent review of acid mine drainage, particularly as related to heavy metals mining.

Effluents from the mining or milling operations should be analyzed for both dissolved and suspended heavy metals. This has not been practiced, primarily because data from unfiltered samples are highly variable. However, research in the New Lead Belt indicates that suspended lead constitutes the largest portion transported through the ecosystem. [7, 14] Rolfe and Jennett [29] reported similar findings for lead transport in the sediments of streams draining areas in central Illinois, where the major source of lead is automobile exhaust fumes. If lead is carried into a stream system where the natural pH is low, or if an acid spill occurs, the suspended lead could be solubilized and become a potential environmental hazard. For this reason, a record should be kept of all suspended and dissolved heavy metals released by a given system.

It is important to note that virtually all information on heavy metals in streams of the United States and other countries is based on dissolved forms. Research dindings at the University of Missouri and the University of Illinois indicate that this practice ignores a major transport mechanism for heavy metals. [29] The significance of this mode of heavy-metal transport remains unclear. It is instructive to recall that mercury in streams was long considered to be inert while, in fact, it was converted into a highly toxic species by microbial populations in the sediments.

Two recent reviews are particularly useful in relation to the mine water problem. [28, 32] Hawley [28] states that, in general, all waste treatment systems currently used by the mining industry discharge effluents that contain small but still undesirable quantities of metals and nonmetals. Further, most, if not all of the systems experience substantial increases in the total dissolved solids levels in the effluent from treatment operations, particularly if neutralization is required.

Hawley also feels that environmental regulatory agencies will soon require essentially no discharge of specific substances, particularly metals from mining areas where the substances in question are a problem. He lists the following advanced waste treatment techniques that could accomplish no-discharge goals in treating acid mine drainage and the associated metals: distillation, evaporation, freezing, reverse osmosis, solvent extraction, electrodialysis, absorption, ion exchange, and biological systems. Hawley suggests that only ion exchange, electrodialysis, and reverse osmosis have received significant attention and hence offer any immediate application to mine and mill wastewater treatment. EPA [32] lists the same three procedures, but also includes data on multistage flash evaporation (a distillation process) and freezing.

Control of Metal-Bearing Waste Waters from Nonmining Sources

Patterson and Minear [33] reviewed the relatively small amount of literature concerned with the treatment of lead-bearing wastewaters and pointed out that lead is usually precipitated by adding lime or dolomite; the highly insoluble hydroxides and carbonates thus formed can be removed by sedimentation or filtration. They also listed a number of processes that can be use to recover and recycle lead, rather than specifically treat the effluent material. Watson [34] has reviewed the problems of control-

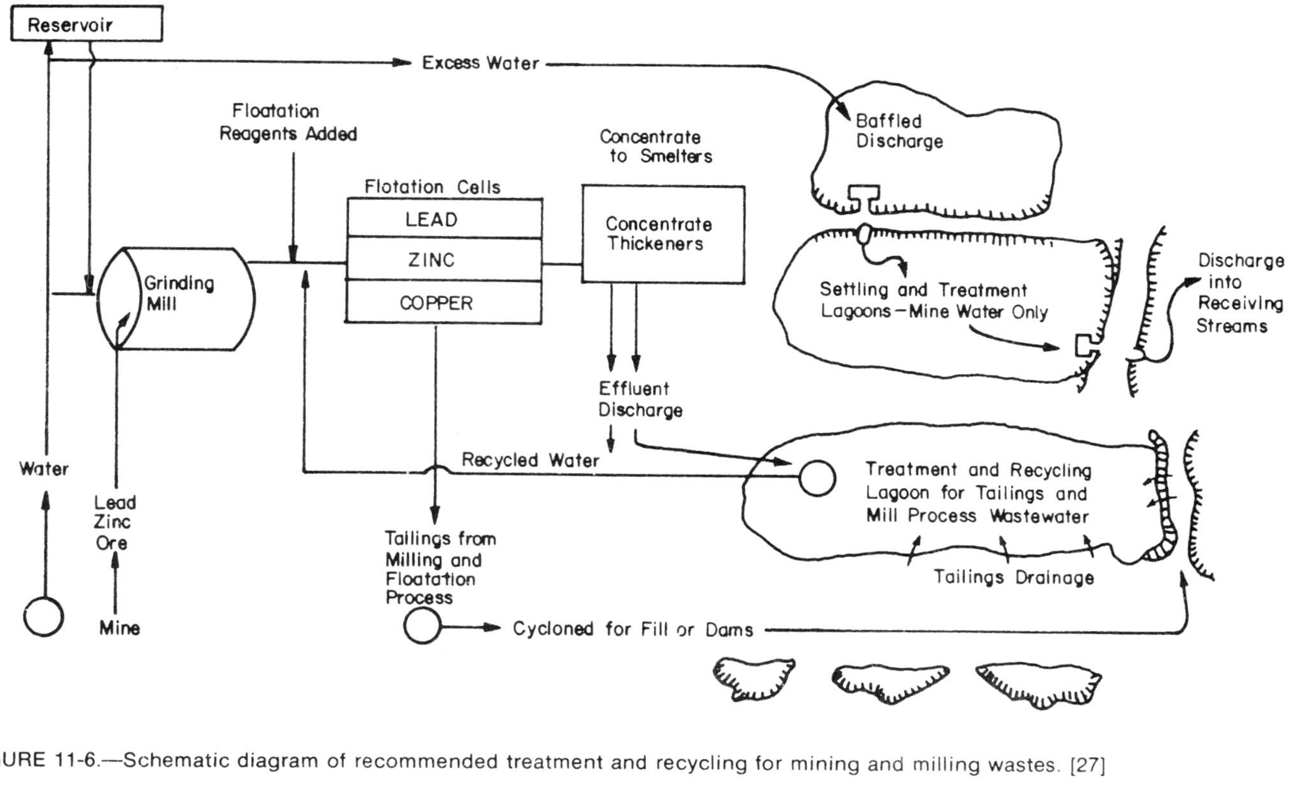

FIGURE 11-6.—Schematic diagram of recommended treatment and recycling for mining and milling wastes. [27]

ling pollution due to metal-finishing wastes. He described the nature, characteristics, and impacts of metal-finishing wastewaters on sewer systems, treatment plants, streams, and lakes. Also discussed were methods for treating these wastewaters and in-plant control procedures. While the report did not specifically involve lead, it covered a large number of heavy metals, and the information may be applied to the treatment of lead-bearing wastes.

Ching et al [35] have reported on the physical-chemical interactions in activated sludge. At low concentrations, metals are complexed by organic compounds; at high concentrations, precipitation of the metal and entrapment by the floc may occur. They suggest that the high-molecular-weight polymers of the flocculating material provide many functional groups that act as a binding site; and that lead is adsorbed and removed in preference to copper, cadmium, or nickel. However, the susceptibility of these sludges to anaerobic digestion was not discussed.

The data are also limited regarding the fate of heavy metals in an anaerobic environment. Since soil and stream sediments are frequently anaerobic or microaerophilic, more work is required in this area to determine the effects of heavy metals on the microbes, with emphasis on how their metabolic products might alter the mobility of heavy-metal ions.

The effective removal of heavy metals by soils has been shown repeatedly, [36] although the mechanism of this removal is not fully understood. Precipitation by basic anions or chelation to phosphate groups has been proposed but not yet proven. Phosphate precipitation of lead has been proposed by several researchers. [37, 38] Addition of phosphate to soil does not increase its binding capacity for lead. [39]

Cooperative and continuing research efforts between the Metropolitan Sanitary District of Chicago and the University of Illinois have focused on the feasibility of disposal on agricultural land of sludges containing large quantities of heavy metals. Although few data have been published, it appears that the metals are concentrating and accumulating in the top layers of the soil.

Other Control Techniques.

A number of promising methods for treating heavy-metal-retaining wastes are described in the literature. Many of these processes have been used successfully to treat acid mine drainage and heavy-metal-containing wastes from mining, milling, and industrial operations; none, however, has been investigated sufficiently to be considered a standard method. Processes that may have applicability for treatment of wastes containing both dissolved heavy metals and refractory organics are summarized in table 11-3. The latter category is included because such compounds are used by most lead industries, and there is a growing awareness that organics and organometallics may be as toxic as their associated metals or even more so. Some of the processes that may eventually apply to effluents from lead mining and processing are briefly discussed below.

TABLE 11-3.–*Unit operations and processes applicable to controlling lead in industrial wastewaters (Adapted from Heckworth [40])*

Process	Dissolved metals	Dissolved refractory organics
Air stripping		X
Chemical oxidation		X
Chemical precipitation	X	X
Distillation	X	X
Electrodialysis	X	P[1]
Electrolytication		X
Extraction	X	X
Fluidized bed incineration		X
Foam separation	P	X
Freezing	X	X
Gas hydration	X	X
Carbon adsorption		X
Ion exchange	X	
Ozonation		X
Reverse osmosis	X	X

[1] P - Possible under certain conditions.

Ion Exchange: In summarizing ion exchange methods, Hawley [28] points out that the recovery of valuable metals from solutions and the treatment of wastewater are likely to continue as a major application of the method. However, recovery of metal ions is usually from acidic solutions as in hydrometallurgy and treatment of pickling wastes, dilute leachates, and acid coal mine drainage. The treatment of tailings pond effluents from base metal mines represents a different situation. The effluents are generally alkaline in reaction, and the value of the metals to be removed is negligible due to their extremely low concentration.

Hawley also stresses that strongly acidic cation exchangers (such as sulfonated styrene copolymers) provide rather high exchange capacities that are not influenced significantly by the solution pH. He states, however, that the organic compounds used as flotation agents in the mine-mill circuit may influence performance during the lifetime of the resin.

Research at the pilot plant stage has demonstrated that both ferrous and ferric iron can be removed from acid mine drainage (AMD). [41, 42] Holmes and Schmidt [43] found that ion exchange does not lower the concentration of manganese in typical AMD, but that Mn can be precipitated if the effluent is raised to pH 9.9 by the addition of lime. Although ion exchange can be used to treat metal-containing wastewater, its direct application to the removal of lead and other heavy metals has not been established.

Freezing and Distillation: Both freezing and distillation are future treatment processes, but freezing should be more economical since it requires about 1/6 as much energy to operate. [44]

Based on the batch studies conducted, it appeared that various metals and acid could be reduced approximately 85 to 90 percent by partial freezing. [45]

The Office of Saline Water has investigated the applicability of salt water conversion processes to the treatment of AMD; its multistage flash distillation process and the design of plants have been studied by other investigators. [46] Schroeder and Marchello [47] calculated a cost of $1.07 to treat 1,000 gallons of AMD by multistage flash distillation. Applied Science Laboratories, Inc., has conducted a number of experiments on AMD in Pennsylvania. EPA has summarized the costs for a 5 million gal./day flash evaporation plant and states that freezing should be feasible in AMD treatment, although the process has not been fully tested. [43]

Electrodialysis: Electrodialysis drives cations and anions through differential membranes to differently charged poles; operational costs are almost directly proportional to the amount of salts removed. If the entire waste stream is treated by electrodialysis, the resultant effluent is essentially pure water. [47]

EPA studies [32] have shown that the membranes quickly fouled in AMD treatment due to the presence of ferric ions. However, further tests on mine drainage previously treated by coagulation and filtration to remove the ions were successful.

Reverse Osmosis: Hawley [28] believes that reverse osmosis has considerable potential in treating mine wastes whether they are acidic or not. An EPA publication [48] provides a flow diagram of a reverse osmosis system that may be used alone to treat AMD or in combination with chemical neutralization, coagulation, and/or disinfection to produce potable water.

In a Minnesota study, [48] five different membranes rejected more than 90 percent of nickel, zinc, iron, and copper salts. Most of these membranes were also effective in treating chromic acid rinse waters. They were, however, sensitive to hydrolyzation at pH below 2.5. In Canada, [49] nickel recovery of 99 percent has been achieved with a cellulose acetate membrane operating at 450 psi. The reverse osmosis process has been successful in removing silver from the washout affixing baths, and in the case reported it seems to pay for itself. [50] Iwai et al. [51] reported that reverse osmosis can reduce the concentrations of zinc and mercury in a secondary effluent from 5.14 mg/1 to 0.22 mg/1. Witmer [52] discussed the application of reverse osmosis to a number of products and suggested that future applications of the process should be expected to treat wastewater effluents and in metal finishing.

In summary, reverse osmosis represents a waste treatment process that is rapidly becoming more widely accepted for treating toxic wastes. The only two limiting factors are the costs of energy and the development of an improved membrane with a higher flux rate. Current advances in technology strongly indicate that further breakthroughs for the reverse osmosis process are quite possible in the near future.

Carbon Adsorption: Carbon adsorption is not designed to remove heavy metals from wastes, but rather to remove extremely small quantities of organics from water. The most common use of the method is to remove taste and odor producing trace organics from wastes and contaminated drinking water supplies.

Deep-Well Disposal: Deep-well injection of wastes may be defined as "injection of liquid wastes, or gases dissolved in liquid, into the subsurface strata containing noncommercial brines." [53] This procedure is not commonly used for handling industrial wastes, possibly because it has frequently been referred to as "an environmental time bomb." [54] Deep-well injection is not a treatment and, in many cases, not even permanent disposal. It is more a matter of waste storage for months, years, or an indefinite period of time.

E.I. DuPont de Nemours & Co. [55] has treated acid wastes by injecting them into porous formations with a neutralizing capability. Also, Midwest Steel [56] has disposed of about 20 percent of its total waste pickling acid, oil emulsions, chromic acid, and orthosilicate wastes into an underground system. Nemerow [57] lists a number of other similar acidic wastes that have been disposed of underground.

The primary problem related to injection of mining wastes into the ground is that they must be free of both suspended solids and heavy metals. However, both could be removed by treatment prior to injection. Based on available information, it appears that the use of deep-well injection by the mining industry is questionable, since it is not a useful form of waste treatment.

Control by Advanced Planning.

Environmental quality is a relatively new parameter for both industry and the governmental agencies that were created to develop and enforce regulatory measures. The complexity of environmental problems was not clearly understood by either government or industry, and conflicts have arisen over the development of regulations, especially with regard to their rigidity, practicality, and compliance schedules. Now that the reality of environmental problems is recognized, it is essential that industry and government develop advanced planning procedures that will minimize the efforts and expenditures required to meet environmental standards.

Research is an essential part of advanced planning. Without an adequate data base, realistic standards cannot be developed; neither can industry respond effectively with necessary modifications in plant designs and procedures. The research needs are of such magnitude that they cannot be met by any one sector acting alone. Instead, joint efforts by government, industry, academic institutions, and private research groups are essential. In addition, there must be close cooperation and liaison to prevent repetitive or overlapping efforts.

There is little doubt that advanced planning based on sound research will foster cooperative, rather than antagonistic, attitudes between industry and government. This has been demonstrated in the New Lead Belt where the NSF-RANN research program has served as a catalyst in developing cooperative interactions between industry and regulatory agen-

cies. [58] This cooperation has eliminated redundant research, produced data underlying stringent and reliable industrial controls, and provided insights into future problems. Comparable opportunities exist with other industries throughout the country.

REFERENCES

1. J. C. Jennett, B. G. Wixson, E. Bolter, and J. O. Pierce. Society of Engineering Science First International Meetıng, Tel Aviv, Israel, June 12, 1972. In Pollution: Engineering and Scientific Solutions, E. Barrakette, ed. Plenum Press 320. (1973).
2. K. J. Semrau. J. Air Pollut. Contr. Ass. *21:* 185 (1971).
3. A. E. Vandergrift. et al. Particulate Pollutant System Study—Vol. III. National Technical Information Service, APTD-0745, Springfield, VA (1973).
4. An Interdisciplinary Investigation of Environmental Pollution by Lead and Other Heavy Metals from Industrial Development in the New Lead Belt of Southeastern Missouri, Vol. I and II, submitted by the Interdisciplinary Lead Belt Team, University of Missouri-Rolla, to NSF-RANN. B. G. Wixson and J. C. Jennett, eds. (June 1974).
5. M. N. Anderson and B. G. Wixson. Proceedings of 7th Annual Conference on Trace Substances in Environmental Health, D. D. Hemphill, ed. Columbia, MO 205 (1973).
6. E. Bolter and N. H. Tibbs. Water Geochemistry of Mining and Milling Retention Ponds in the "New Lead Belt" of S. E. Missouri. Completion Report to OWRR, Missouri Water Resources Research Center, Columbia, Mo. Project No. A-032-Mo. (1971).
7. J. C. Jennett, B. G. Wixson, E. Bolter, and N. L. Gale. Proceedings of 28th Purdue Industrial Waste Conference. 496 (1973).
8. N. L. Gale, M. G. Hardie, J. C. Jennett, and A. Aleti. Proceedings of 6th Annual Conference on Trace Substances in Environmental Health, D. D. Hemphill, ed. Columbia, MO 95 (1972).
9. J. C. Jennett and B. G. Wixson. J. Water Pollut. Contr. Fed. *44:* 2103 (1972).
10. J. R. Hawley. The Use, Characteristics and Toxicity of Mine-Mill Reagents in the Province of Ontario. Ministry of the Environment, Toronto, Ontario (1972).
11. B. G. Wixson and E. Bolter. Proceedings of 5th Annual Conference on Trace Substances in Environmental Health, D. D. Hemphill, ed. University of Missouri-Columbia 143. (1971).
12. N. L. Gale, B. G. Wixson, M. G. Hardie, and J. C. Jennett. Photosynthetic Organisms and the Nutritional Impact of Mine and Mill Effluents in the New Lead Belt of Southeastern Missouri. Paper presented AIME Annual Meeting, Chicago, IL Published by the Society of Mining Engineers of AIME (March 1973).
13. W. J. Ernst, Jr. M. S. thesis, University of Missouri-Rolla, MO (1976).
14. K. Purushothaman, J. Huang, J. C. Jennett, and B. G. Wixson. Proceedings of 23rd Annual Conference on Sanitary Engineering, University of Kansas. 113 (1973).
15. A. C. Stern. Air Pollution Vol. III. Academic Press, New York, NY (1968).
16. W. A. McKinney, W. I. Nissen, D. A. Elkins, and J. B. Rosenbaum. Proceedings of Flue Gas Desulfurization Symposium, sponsored by the U. S. Environmental Protection Agency, Atlanta, GA (Nov. 4-7, 1974).
17. L. Korosy, H. L. Gewanter, F. S. Chalmors, and S. Vasan. Proceedings of Symposium on Sulfur Removal and Recovery from Industrial Sources, Division of Industrial and Engineering Chemistry, 167th American Chemical Society National Meeting, Los Angeles, CA (1974).
18. J. W. Patterson and R. Minear. Removal of Heavy Metals by Activated Sludge. Proceedings of Conference on Trace Metals in the Aquatic Environment, Vanderbilt University, Nashville, TN (1973).
19. J. B. Rosenbaum, W. A. McKinney, H. R. Beard, Crocker, Laird, and W. I. Nissen. The Citrate Acid Process for SO_2 Recovery. U. S. Department of the Interior, U. S. Bureau of Mines, Washington, D.C. (1973).
20. A. U. Bell. The Tailings Pond as a Waste Treatment System. Can. Mining Met. Bull. April 1974.
21. E. Bolter, J. C. Jennett, and B. G. Wixson. Proceedings of 27th Purdue Industrial Waste Conference. 679 (1972).
22. J. C. Jennett and M. G. Hardie. Proceedings of 2nd Annual NSF-RANN Trace Contaminants Conference, Asilomar, CA 172 (1975).
23. F. R. Ryck and J. R. Whitely. Pollution State of the Lead Mining District of Missouri. Proceedings of 29th Purdue Industrial Waste Conference (1974).
24. Effluent Guidelines. Missouri Clean Water Commission, Jefferson City, MO (1971).
25. B. G. Wixson and N. L. Gale. Some Limnological Effects of Lead and Associated Heavy Metals from Mineral Production in Southeast Missouri, U.S.A. Proceedings of Symposium on Interaction Between Water and Living Matter, Odessa, U.S.S.R. (October 1975).
26. C. Adams, W. W. Eckenfelder, and B. L. Goodman. The Effect and Removal of Heavy Metals in Biological Treatment. Conference on Heavy Metals in the Aquatic Environment, Vanderbilt University, Nashville, TN (1973).
27. J. C. Jennett and B. G. Wixson. Control of Lead Mining and Milling Wastes. Transactions of the University of Missouri-Rolla's 100th Centennial Symposium entitled Technology for the Future to Control Industrial and Urban Wastes (1971).
28. J. R. Hawley. The Problem of Acid Mine Drainage in the Province of Ontario. Ministry of the Environment, Toronto, Ontario (1972).
29. G. Rolfe and J. C. Jennett. Environmental Lead Distribution in Relation to Automobile, and Mine and Smelter Sources." Proceedings of Conference on Trace Metals in the Aquatic Environment, Vanderbilt University. (1973).
30. F. H. Sharp and K. L. Clifford. American Institute of Mining Engineers Meeting, Chicago, IL (1973).
31. D. E. Pickett and E. G. Joe. Water Recycling Experience in Canadian Mills. Society of Mining Engineers of AIME (1973).
32. Processes, Procedures, and Methods to Control Pollution From Mining Activities. U. S. Environmental Protection Agency, EPA-43019-73-01 (1973).
33. J. W. Patterson and R. A. Minear. Wastewater Treatment Technology. National Technical Information Service, PB-204 521 (1971).
34. M. R. Watson. Pollution Control in Metal Finishing. Noyes Data Corp. Park Ridge, NJ (1973).
35. M. H. Ching and J. W. Patterson. Heavy Metal Uptake by Activated Sludge. 46th Annual Water Pollution Control Federation Conference (1973).
36. G. R. Wetnik and J. E. Etzel. J. Water Pollut. Contr. Fed. *44:* 1561 (1972).
37. J. C. Jennett, E. Bolter, N. L. Gale, W. Tranter, and M. G. Hardie. The Largest Lead Mining District in the World—Environmental Effects and Controls. Proceedings of Institute of Mining and Metallurgy International Seminar on Minerals and the Environment, London, England. 13 (1975).
38. J. O. Nriagu. Inorg. Chem. *11:* 2499 (1972).
39. S. Linnemann. M. S. thesis, University of Missouri-Rolla, MO (1974).
40. C. W. Heckwerth. Water Wastes Eng. A-6 (January 1971).
41. F. Pollio and R. Kunin. Water Wastes Eng. A-1 (August 1967).
42. C. J. Sterner and H. A. Conahan. Proceedings of 23rd Purdue Industrial Waste Conference, Lafayette, IN p. 101. (1968).
43. J. Holmes and K. Schmidt. 4th Symposium on Coal Mine Drainage Research. Mellon Institute, Pittsburgh, PA (1972).
44. R. Smith. Cost of Conventional and Advanced Treatment of Wastewaters. U. S. Department of the Interior, Water Research Laboratory, Advanced Waste Treatment Branch, Cincinnati, OH (1968).
45. Applied Science Laboratories, Inc. Purification of Mine Water by Freezing. FWPCA Water Research Series. Government Printing Office, Washington, D.C. (1971).
46. A Study of Large Size Saline Water Conversion Plants. Office of Saline Water, Research and Development Report No. 72. Washington, D.C. (1973).
47. W. C. Schroeder and J. M. Marchello. Study and Analysis of

the Application of Saline Water Conversion Process to Acid Mine Waters, Office of Saline Water, Research and Development Progress Report No. 199. Washington, D.C. (1966).
48. Ultrathin Membranes for Treating Metal Finishing Effluents by Reverse Osmosis. Report prepared by Minnesota Pollution Control Agency for U.S. Environmental Protection Agency, Minneapolis, MN (1971).
49. New Systems Control Planting Wastes. Can. Chem. Processing. *56:*29 (1972).
50. B. Leightwell. Chem. Process Eng. *52:* 79 (1971).
51. Iwai et al. Kyoto Daigaku Kogaku Kenkyusho Iho Jap. *40:* 57 (1971).
52. F. E. Witmer. Environ. Sci. Technol. *7:* 314 (1973).
53. J. C. Longfield. J. Water Pollut. Contr. Fed. *45:* 404 (1973).
54. E. S. Shannon. J. Water Pollut. Contr. Fed. *40:* 2059 (1968).
55. Deep Well Disposal Increases. Civ. Eng.-ASCE. *43:* 97 (1973).
56. C. D. Hartmann. J. Water Pollut. Contr. Fed. *40:* 95 (1968).
57. L. Nemerow. Liquid Waste of Industry: Theories and Practices and Treatment. Addison-Wesley, Reading, MA (1971).
58. B. G. Wixson. Development of a Cooperative Programme for Environmental Protection Between the Lead Mining Industry, Government and the University of Missouri, Proceedings, Minerals and the Environment, Institute of Mining and Metallurgy, London, England. 3 (July 1974).

CHAPTER 12

AUTOMOTIVE LEAD

Harry W. Edwards
Department of Mechanical Engineering
Colorado State University

INTRODUCTION

The continued widespread use of lead alkyl compounds as gasoline antiknock additives is presently under heavy attack on a number of fronts. Three major areas of concern are: the possible threat to human health posed by automotive lead, environmental consequences of lead contamination, and interference by lead with the operation of catalytic automotive pollution control devices. This chapter is concerned with the need for low-lead gasolines and technological alternatives for reducing environmental lead inputs from automotive sources.

LEAD AND AUTOMOTIVE EMISSION CONTROL TECHNOLOGY

Federal Emission Regulations

The requirements of the Clean Air Act Amendments of 1970 [1] have undoubtedly been a major factor in arousing interest in reducing the lead content of gasoline. Although this legislation does not specifically require such reduction, many automobile manufacturers are relying upon lead-intolerant oxidation catalysts for meeting exhaust emission standards for CO and unburned hydrocarbons. The original federal emission regulations [2] applicable to new light duty vehicles are summarized in table 12-1.

TABLE 12-1.–*Federal emission standards for new light-duty vehicles*

Model Year	Maximum emission levels (g/mi)		
	Hydrocarbons	CO	NO_x
1973	3.4	39.0	3.0
1974	3.4	39.0	3.0
1975	0.41	3.4	3.1
1976	0.41	3.4	0.40

Under the provisions of the legislation, the EPA Administrator is authorized to grant delays in implementing the schedule for reducing new car emissions. At the request of the automotive industry, several such delays have been granted. In March 1975, EPA Administrator Russell E. Train [3] set interim standards for hydrocarbons at 1.5 g/mi and for carbon monoxide at 15 g/mi. He recommended retaining those standards through 1979 and then proceeding in 1980 to the current California levels of 0.9 g/mi for hydrocarbons and 9.0 g/mi for carbon monoxide. The statutory levels, 0.41 g/mi for hydrocarbons and 3.4 g/mi for carbon monoxide, would be achieved in 1982. Train also recommended that the interim nitrogen oxides standard of 2.0 g/mi set for 1977 be retained through 1982, unless research demonstrates the need for a more stringent NO_x standard.

A primary basis for the March 1975 suspension decision is concern for increased sulfate emissions from catalyst-equipped cars. The oxidation catalysts that convert CO and unburned hydrocarbons to CO_2 and water vapor also oxidize SO_2 to sulfates and sulfuric acid. While SO_2 oxidation also occurs in the atmosphere, this process is sufficiently slow so that atmospheric dispersion of automotive emissions normally prevents hazardous accumulations of sulfates from automotive sources. In a November 1973 statement [4] to the Senate Public Works Committee, Administrator Train reported that the average catalyst-equipped car could be expected to emit three to five times more sulfates than 1973 noncatalyst cars. At that time, EPA announced that, in the absence of an effective sulfate emission control, more than one model-year of a catalyst-equipped car in use at a given time could cause a sufficient rise in ambient sulfate levels to produce adverse health effects. In February 1975, EPA scientists [5] reported that increased sulfate emissions from cars equipped with oxidation catalysts would probably result in a net public health risk unless the sulfur content of gasoline or particulate emissions could be controlled by other means. The scientists concluded that after four model-years of catalyst-equipped cars, the public health risks would likely exceed benefits in all areas of the Nation unless other regulatory steps were taken. EPA is examining means for controlling sulfate emissions, including the possibility of requiring widespread availability of low-sulfur gasoline.

Because minimizing CO and hydrocarbon emissions is, to some extent, at cross purposes with simultaneously minimizing NO_x emissions from most engines, [6] the optimum strategy for meeting the statutory 1976 federal standards is not yet totally clear. While technology for controlling CO and hydrocarbon emissions is fairly well developed, that for controlling NO_x emissions to the extent required for the 1976 model-year vehicles has not been perfected. Therefore, the presently favored approach is to minimize NO_x formation by reducing combustion temperatures. Generally, the results of this approach are decreased thermal efficiency, increased fuel consumption, and increased production of CO and unburned hydrocarbons. While CO and unburned hydrocarbons thus produced can be oxidized to CO_2 by using thermal reactors or oxidation catalysts, the secondary combustion process produces no useful work. Fuel economy penalties associated with the use of catalysts to achieve statutory 1975 standards have been estimated [7] at no more than 3 or 4 percent. In fact, significant gains were reported for some catalyst-equipped 1975 models, relative to their noncatalyst 1974 counterparts. [8] However, even the catalyst-equipped 1975 models do not achieve the statutory emission levels.

The catalysts' requirement for unleaded gasoline has been estimated to increase petroleum consumption by 2 percent and gasoline production costs by less than ¼ cent per gallon. [7] A fuel economy penalty of approximately 30 percent has been estimated to achieve the statutory 1976 NO_x standard of 0.4 g/mi. [9]

The technological need for low-lead gasoline may be temporary. The key factor is the extent to which the statutory 1976 model-year NO_x standard of 0.4 g/mi may be eased. Sufficient relaxation of this standard may very well mean that catalysts will not be the best approach for meeting CO and hydrocarbon emission standards. Lead-tolerant systems offer considerable promise for meeting CO and hydrocarbon emission requirements, but generally with somewhat higher than 0.4 g/mi NO_x emissions. EPA reported that by 1985 the national ambient air quality standard for NO_2 (100 $\mu g/m^3$) would be exceeded in only three air quality control regions if the 1975 automotive NO_x emission standard of 3.1 g/mi is retained, and in only one region if the 1976 standard of 0.4 g/mi is implemented. [10] Only in Los Angeles would the ambient NO_2 standard clearly be exceeded, whether or not the NO_x emission standard of 0.4 g/mi is implemented. Previously, due to errors of a factor of two or more in measuring ambient NO_2 levels, EPA classified 47 air quality regions priority I relative to NO_2 (110 $\mu g/m^3$ or greater). Accordingly, it was reported that while the 0.4 g/mi NO_x standard might produce some improvement over the 1975 standard of 3.1 g/mi, only a very few air quality control regions at most would fail to meet the national ambient air quality standard for NO_2 in either case.

While the use of unleaded gasoline is presently foreseen as a permanent requirement for vehicles equipped with catalytic converters, it seems clear that use of lead-intolerant catalysts may not represent the optimum long-range approach for reducing emissions at least cost. Any fuel economy penalties associated with the use of catalytic converters or other emission control systems must be regarded as undesirable, especially in light of present energy shortages. Any significant increase in petroleum consumption made necessary by fuel requirements of pollution control devices must be viewed with concern. The combined effects of emission control regulations on increasing petroleum consumption could have the undesirable effect of aggravating present fuel shortage problems and accelerating depletion of domestic petroleum reserves. As new approaches for realizing emission regulations without fuel economy penalties are developed and perfected, fuel requirements may change. The use of catalytic converters may be an interim approach pending perfection and production of noncatalytic systems, some of which have already provided encouraging performance reports. [7] In view of the possible temporary need for unleaded gasolines and the increased petroleum consumption required for production of unleaded gasolines, the decision to phase out the use of lead in gasoline will undoubtedly be reviewed periodically.

Lead-Tolerant Systems

Alternative, lead-tolerant emission control technologies, possibly superior to catalytic converters, have been investigated by EPA. [7] The alternate systems given primary consideration by EPA in the context of meeting 1975-76 standards are: the rotary engine, the diesel engine, and the stratified-charge engine. The manufacturers are: Toyo Kogyo with the Mazda rotary engine and thermal reactor, Daimler-Benz with a diesel engine, and Honda with a CVCC stratified-charge engine. [11] These systems do not employ catalysts and do not require unleaded gasoline. These three manufacturers have demonstrated emission levels below the 1975 model-year requirements, and Honda and Toyo Kogyo have reported satisfactory 50,000 mi durability data. They have also demonstrated emissions below or very close to the 1976 model-year requirements. In terms of the present likelihood of meeting the 1975 and 1976 model-year emission standards, these findings place the above three manufacturers in a more favorable position than manufacturers relying on catalytic devices. No catalytic system has yet been developed that has been shown capable of meeting all of the 1976 model-year emissions regulations, including the 50,000 mi durability requirement.

EPA pointed out that while the diesel engine can achieve superior performance in terms of fuel economy and emission control, widespread use of diesel engines in light-duty vehicles could create problems in terms of odor and particulate emissions. Although 1975 emission standards have been realized with the Toyo Kogyo rotary engine, these engines operate with a fuel economy penalty of between 15 and 17 percent compared with conventional engines. The 1975 emission standards have been met with the Honda CVCC stratified-charge engine without fuel economy penalties. However, EPA reservations were the following: little is known about the Honda engine; it is not clear that the same approach will work for larger engines; and even if the Honda approach were to be applied to a majority of new vehicles produced, it would take consideraby more than 5 years to change over existing production capabilities. In light of this, EPA reported [7] that "control of emissions to anything like statutory 1975 levels will therefore almost certainly depend on the use of a catalytic converter on a large number of vehicles for a substantial period of time." This was before precipitation of the energy crisis.

Two noncatalytic systems developed by domestic manufacturers are also noteworthy, although neither has been demonstrated capable of meeting the 1976 model-year NO_x emission standard. Both involve the use of thermal reactors to oxidize hydrocarbons and CO. The first system is the Total Emissions Control System (TECS) of E.I. duPont deNemours & Co. [12,13] The system consists of three primary components: an exhaust-manifold reactor to promote oxidation of unburned hydrocarbons and CO to CO_2, an exhaust gas recirculation system to reduce NO_x levels, and a trapping system to remove particulate matter, including lead. A disadvantage of the du Pont system is that attaining low NO_x emission levels with exhaust gas recirculation results in some loss in fuel economy. The second system is that developed by the Ethyl Corp. [14,15] The Ethyl system is designed for lean operation at air/fuel ratios of between 17 and 19. Features of the Ethyl lead reactor system are special carburetion for lean operation, an insulated exhaust-manifold reactor, and a particulate trap. Some loss in fuel economy is associated with inclusion of exhaust gas recirculation for lowering NO_x emissions.

Particulate Traps

Devices have been developed and tested that prevent release of a major portion of the lead contained in gasoline. The simplest and most practical devices are particulate traps. The usual application calls for replacement of the muffler with a device that promotes particle agglomeration and collects the larger particles by an inertial or impaction technique. The Ethyl system first cools the exhaust gas to promote particle nucleation and agglomeration, and then passes the exhaust through an inertial separator for particle collection. [15,16] Test data indicate that substantial reductions in exhausted lead and other particulates can be achieved with relatively inexpensive and simple designs that can replace standard mufflers. The Octel [17] device contains alumina-coated, stainless-steel wool that reduces lead emissions up to 80 percent compared with standard cars equipped with conventional mufflers. The duPont lead trap [18] contains a bed of ¼-in. alumina pellets for particle agglomeration and cyclonic inertial separators for particle collection. Tests indicate that, compared with conventionally equipped cars, trap-equipped cars emit 84 percent less lead over 100,000 mi. The duPont tests indicate that trapping efficiency is no more than a few percent less for small particles (less than 1 m) than for larger particles. These devices are judged to be of high merit. Installation of particulate traps may offer at least a partial solution to the problem of reducing atmospheric lead levels and lead fallout without a major reduction in the lead content of gasoline.

Two final points must be made concerning lead traps. First, for vehicles requiring leaded gasoline, lead-particle trapping is not (and is not represented to be) suitable for removing sufficient lead from the exhaust for subsequent use of lead-intolerant catalysts for CO and hydrocarbon oxidation. The second point is that, in terms of widespread use of particulate traps on vehicles requiring leaded gasoline, adequate provisions must be made for collecting and disposal or recovery of lead contained in such devices. A recent study indicates that it is technically feasible to recover and recycle lead, bromine, chlorine, and stainless steel contained in these devices. [19]

Health and the Environment

While this chapter does not deal specifically with health effects of automotive lead, a brief examination of the current controversy is useful for perspective. The literature indicates that although there is no proven health threat to the general population attributable directly to present environmental levels of automotive lead, there is considerable disagreement concerning the potential medical significance of present levels. In its examination of the biological significance of airborne lead, the National Academy of Sciences' Lead Panel [20] reported that no definite health hazards can now be ascribed to atmospheric lead burdens. Potential hazards were then considered by the NAS Panel in terms of ingestion by children of airborne lead fallout and in terms of greater exposures to airborne lead in certain occupations. While inhalation of airborne lead and ingestion of dust and street dirt contaminated by automotive lead, especially by children, were not identified by the NAS group as causes of clinical lead poisoning, questions were raised concerning the long-term effects of low-level exposures.

EPA's position on health hazards of airborne lead is not identical with that of the NAS Panel. In its most recent (Nov. 28, 1973) position paper [21] on airborne lead health effects, EPA identifies automotive lead as the most important remaining controllable source of environmental lead. EPA reported that sustained airborne exposures in the range of 5.0 to 6.7 g/m^3 could elevate blood-lead levels to a region considered medically undesirable; further, EPA reported that present lead exposures through air and dusts in some urban areas leave little or no margin of safety compared with concentrations thought by some to be associated with biomedical harm. A number of studies were cited by EPA concerning the possibility that lead exposures may be linked to various subclinical effects, especially in children; no conclusions linking airborne lead to these effects could be drawn, however. The basic EPA position seems to be that since lead has no known beneficial biological function and is toxic in sufficient doses, it would be prudent to reduce airborne lead exposures. This position evolved following an international symposium to assess health hazards of lead cosponsored by EPA and the Commission of the European Communities [22] and an EPA-sponsored symposium on low-level toxicity of lead. [23]

The ultimate resolution of questions pertaining to health effects of automotive lead must rest with the medical community, and these questions are not addressed in this report. Nevertheless, it seems appropriate to observe that very little attention is given to possible health effects of lead substitutes in either the NAS lead report or in the most recent EPA lead report. In developing a policy concerning health effects of automotive lead, it would seem desirable to examine possible health effects of substitute fuels so that the various medical tradeoffs can be identified and evaluated.

Earlier chapters of this report provide an evaluation of the environmental impact of automotive lead. The general picture that emerges is that although man has substantially contaminated some portions of the environment with lead from automotive sources, no drastic environmental consequences have been identified for present levels. Data representing the findings of the three university research teams and other groups range from the effects of lead at the cellular level [24] to effects of lead on weather. [25] The NAS Lead Panel, [20] in its examination of the biological effects of lead, posed a number of key questions concerning the environmental impact of automotive lead. Data from the three universities participating in the NSF-RANN lead program should prove highly significant in answering these questions, especially in the areas of gathering and sampling aerometric data, lead chemistry in nature, analytical methodologies, and effects of lead on biota.

LEAD SUBSTITUTES

High Aromatic—Content Gasoline

Addition of lead alkyls to gasoline prevents engine knocking and permits the use of higher compression ratios. They are generally beneficial in terms of efficiency and performance. Without lead, the alternatives for conventional engines include: reduced compression ratios, substitution of other antiknock additives, and gasoline reblending. Reformulation of gasoline to produce equivalent octane ratings in the absence of lead could be accomplished in several ways, but the presently favored approach is to increase the aromatic hydrocarbon content of gasoline feedstocks. Typical percentages for the aromatic content of gasoline range from about 22 percent for leaded regular to about 45 percent for unleaded premium. [26] While increasing the olefin content and the fraction of lighter, branched-chain paraffinic hydrocarbons would also increase octane ratings. The result could be a substantial increase in photochemically reactive evaporative emissions in the absence of effective controls. Moreover, the octane improvement resulting from a fractional increase in aromatics is often much higher on a mass basis.

A number of serious technological problems have been identified with a switchover to highly aromatic gasolines. The first problem is that the present refining capabilities would require substantial modification and expansion to produce highly aromatic gasolines on a national basis. In its first evaluation of the impact of removing lead from gasoline, the U.S. Department of the Interior estimated that the switchover would require 10 additional new refineries of the average size now being constructed. [27] The second problem is that the reforming process used for increasing the aromatic content yields somewhat less gasoline per barrel of crude petroleum. The Interior Department estimated that the combined effects of diminished gasoline yield and reduced fuel economy resulting from proposed lead restrictions would increase crude petroleum requirements by 1.2 million barrels a day, about seven percent of the total daily U.S. consumption. [27]

Another potential problem that has not been fully resolved is the possibility that the widespread use of high-aromatic-content gasolines may increase ambient levels of harmful nonlead substances. The problem of increased sulfate emissions associated with oxidation catalysts is discussed in an earlier section of this chapter. Attrition of trace metals from catalytic beds has been reported. [28] Additionally, increased production of carcinogenic polynuclear hydrocarbons is linked to increased gasoline aromaticity. [29, 30] While effective oxidation catalysts would presumably control the release of polynuclear aromatic hydrocarbons as well as other hydrocarbons, there is presently no assurance that sufficiently durable catalysts are forthcoming. A relevant question seems to be what fraction of catalyst failures can be tolerated with respect to harmful nonlead emissions.

Other Additives

An organic manganese compound, 2-methylcyclopentadienyl manganese tricarbonyl, has antiknock properties and is currently under consideration as a replacement for lead alkyls. The effects of increasing ambient manganese concentrations on health and on chemical processes that occur in the atmosphere have received very little study. Of particular concern are the catalytic properties of manganese compounds, especially in terms of possible catalysis of atmospheric sulfur oxidation reactions. It is also not clear whether manganese antiknock compounds would be compatible with all automotive oxidation catalysts.

REGULATORY ACTION

In February 1972, proposed regulations concerning the use of lead and phosphorus additives in gasoline were published by EPA. [31] These regulations called for a phased reduction in the lead content of gasoline and required that one grade of lead-free and phosphorus-free gasoline be made generally available by July 1, 1974. Based on input received from the scientific and industrial communities and from the general public, EPA proposed a revised plan for regulating fuels and fuel additives. [32] The plan includes two separate pieces of regulatory action. Other countries have also taken steps concerning the use of lead additives.

Widespread Availability of Low-Lead Gasoline

The final regulations apply to U.S. retailers and provide for the general availability of unleaded gasoline (less than 0.05 g of lead/gal. and not more than 0.005 g of phosphorus/gal.) of not less than 91 research octane number after July 1, 1974. In areas above 2,000 ft in elevation, the octane number may be reduced by one for each succeeding 1,000 ft up to a maximum of three octane numbers.

Phased Reduction in Gasoline Lead Content

The proposed regulations based on health effects of airborne lead provide for reducing the lead content of all grades of leaded gasoline. These regulations apply to refiners and call for a maximum quarterly average lead content of gasoline of 2.0 g/gal. after Jan. 1, 1975; 1.7 g/gal. after Jan. 1, 1976; 1.5 g/gal. after Jan. 1, 1977; and 1.25 g/gal. after Jan. 1, 1978.[1] A period of 60 days following publication of the proposed regulations was allowed for interested persons to submit comments. This regulation, especially, drew considerable criticism. [27, 33-35] Lead additive manufacturers appealed EPA's decision to implement the phased reduction to the U.S. Court of Appeals (District of Columbia). A three-judge panel of the court decided in favor of the petitioners and set aside the phased reduction. [36] EPA has appealed the decision and was granted the opportunity to argue the case again before the full court in May 1975.[2]

International

Although limits for the lead content of gasoline have been observed in most countries for years, a number of governments have recently made proposals or adopted regulations that would have the effect of reducing the lead content still further. [37] There are some notable contrasts. In the United Kingdom, regulation is by industry-government agreement, rather than by statute, and provides for ultimately reducing the lead content to 0.45 g/l.[3] A maximum of 0.40 g/l has been proposed for members of the European Economic Community. Legislation of the Federal Republic of Germany provides that the lead content of gasolines must not exceed 0.15 g/l after Jan. 1, 1976. A similar provision has been proposed in Sweden. A maximum of 0.45 g/l has been proposed in France. There is no specified maximum in Ireland. Japan has proposed a regulation that permits only "lead-free" gasoline beginning in April 1974. In the U.S.S.R. the lead content of gasoline is restricted to not more than 0.40 g/l, and use of gasoline containing lead is prohibited in Moscow, Leningrad, and in certain capitols and resorts.

CONTROL ALTERNATIVES

The removal of lead from gasoline would, from a purely conceptual point of view, appear to the simplest approach to reducing environmental inputs of automotive lead. However, the technological and economic impacts of completely banning lead additives may not be acceptable, quite apart from unresolved questions concerning possible beneficial or deleterious health consequences. It seems clear that the development of a rational and stable policy concerning automotive lead will require input on technological, economic, environmental, and medical factors. Even though tradeoffs may resist quantification, known effects can be separated from uncertainties, and minor impacts can be distinguished from major disruptions. It must also be recognized in developing policy that the gasoline lead issue is only one aspect of the broader problem of achieving and maintaining environmental quality at acceptable economic and social costs. It is beyond the scope of this chapter to formulate policy recommendations. However, it is possible, based on the information examined, to identify a number of technically feasible alternatives for reducing environmental inputs of automotive lead.

Partial Reduction in Lead Content of Gasoline

A partial reduction in the lead content of gasoline would permit some continuing benefit from the function of lead as a petroleum conserver and stabilize or reduce lead inputs. There are indications that atmospheric lead levels are rising with time in certain urban areas, [38] and there is widespread agree-

---

[1]Final regulations would require reduction of the lead content of the total gasoline pool to 0.5 g/gal. in 1979. Fed. Reg., *38:* 33734 (1973).

[2]The full court decided in favor of EPA. Decision upheld by Supreme Court June 14, 1976. Compliance date may be moved back to 1981.

[3]The proposed U.S. limit of 1.25 g/gal. corresponds to 0.33 g/l.

ment that a long-term trend in this direction would be undesirable.

Widespread Use of Particulate Traps

Existing data indicate that a major portion of the lead in gasoline can be retained in a vehicle equipped with a particulate trap. Widespread use of particulate traps should be accompanied by implementation of procedures for trap collection and lead recovery to prevent trapped lead from reentering the environment dissipatively.

A Two-Lead Strategy

Whether by agreement or by regulation, gasoline supplied to areas where automotive lead buildup is greatest could be of lower lead content than gasoline supplied to other areas.

Sound Urban Planning

Lead exposures and buildups in urban areas are subject to some degree of control by means of sound urban planning. Traffic routing, parking, building and greenbelt location and geometry, street washing, and mass transit systems are factors that can enter in planning urban development to reduce exposure to lead and other pollutants.

No single strategy seems ideal for all situations. Because the technological need for unleaded gasoline may be temporary, a long-range policy concerning the control of environmental lead inputs from automotive sources should reflect the broad spectrum of technological, environmental, economic, and medical impacts associated with the proposed controls. Some factors remain insufficiently defined, and there is a clear need for continuing environmental monitoring. Controls that are implemented should therefore reflect both the state of knowledge of these factors and the need for flexibility in dealing with problems that may be specific to certain combinations of traffic, geometric, and meteorological conditions.

REFERENCES

1. Public Law 91-604, 91st Congress, 2nd Session (Dec. 31, 1970).
2. Fed. Reg. *37:* 24250 (Nov. 15, 1972).
3. Anon. J. Air Pollut. Contr. Ass. *25:* 540 (1975).
4. R.E. Train. Statement presented by Administrator of U.S. Environmental Protection Agency to U.S. Senate Committee on Public Works (Nov. 6, 1973).
5. J.F. Finklea, W.C. Nelson, J.B. Moran, G.G. Akland, R.I. Larsen, D.I. Hammer, and J.H. Knelson. Estimates of the Public Health Benefits and Risks Attributable to Equipping Light Duty Motor Vehicles with Oxidation Catalysts. U.S. Environmental Protection Agency, Research Triangle Park, NC (Feb. 1, 1975).
6. Aerospace Corp. An Assessment of the Effects of Lead Additive in Gasoline on Emissions Control Systems Which Might be Used to Meet the 1975-76 Motor Vehicle Emission Standards. Final Report PB-205-981, National Technical Information Service, Springfield, VA 1971.
7. Fed. Reg. *38:* 10317 (Apr. 26, 1973).
8. J.R. Pierce. Scient. Amer. *232*(1): 34 (1975).
9. National Academy of Sciences. Report by the Committee on Motor Vehicle Emissions. Washington, D.C. (Feb. 15, 1973).
10. Fed. Reg. *38:* 15174 (Jun. 8, 1973).
11. J.L. Bascunanna. Divided Combustion Chamber Gasoline Engines—A Review for Emissions and Efficiency. Paper 73-74, 66th Annual Meeting, Air Pollution Control Association, Chicago, IL (Jun. 24-28, 1973).
12. E.N. Cantwell, R.A. Hoffman, I.T. Rosenlund, and S.W. Ross. A Systems Approach to Vehicle Emission Control. SAE Paper 720510, National SAE Meeting, Detroit, MI (May 1972).
13. E. I. du Pont de Nemours & Co., Inc. "TECS - du Pont's Approach to Automotive Emissions Control." Wilmington, DE (1973).
14. W.E. Adams, H.J. Gibson, D.A. Hirschler, and J.S. Wintringham. J. Auto. Eng. *2*(10): 12 (1971).
15. D.A. Hirschler, W.E.Adams, and F.J. Marsee. Lean Mixtures, Low Emissions and Energy Conservation. Paper AM-73-15, National Petroleum Refiners Association Annual Meeting, San Antonio, TX (Apr. 1973).
16. G.L. Ter Haar, D.L. Lenane, J.N. Hu, and M. Brandt. J. Air Pollut. Contr. Ass. *22:* 39 (1972).
17. Associated Octel Co., Ltd. The Elimination of Vehicle Particulate Emissions. Serial No. OP 72/5, London, England (Dec. 1972).
18. E. I. du Pont de Nemours & Co., Inc. Design and Performance of du Pont Lead Trap Systems. PLMR-6-73. Wilmington, DE (Aug. 30, 1973).
19. E.S. Jacobs. Personal communication. (Dec. 20, 1973).
20. Lead: Airborne Lead in Perspective. National Academy of Sciences. A report of the Committee on Biological Effects of Atmospheric Pollution, Division of Medical Sciences, National Research Council. Washington, D.C. (1972).
21. U.S. Environmental Protection Agency. EPA's Position on the Health Implications of Airborne Lead. Washington, D.C. (Nov. 28, 1973).
22. Environmental Health Aspects of Lead. Proceedings of an International Symposium held in Amsterdam, Holland, Oct. 2-6, 1972. Published by the Commission of the European Communities, Luxembourg (May 1973).
23. Symposium on Low-Level Lead Toxicity. Sponsored by EPA and NIEHS, Raleigh, NC (Oct. 1-2, 1973).
24. T.G. Tornabene and H.W. Edwards. Science, *176:* 1334 (1972).
25. L.O. Grant and M.L. Corrin. J. Weath. Mod. *5:* 238 (1973).
26. B.H. Eccleston and R.W. Hurn. Comparative Emissions from Some Leaded and Prototype Lead-Free Automobile Fuels. U.S. Bureau of Mines Report 7390 (May 1970).
27. Communication to Casper Weinberger from the Office of the Secretary, U.S. Department of the Interior, Washington, D.C. (1973).
28. W.D. Balgord. Science. *180:* 1168 (1973).
29. E.S. Jacobs, P.J. Brant, C.S. Hoffman, Jr., G.H. Patterson, and R.L.Willis. Polynuclear Aromatic Hydrocarbon Emissions from Vehicles. Paper presented at the National Meeting of the American Chemical Society, Los Angeles, CA (Mar. 1971).
30. A.P. Altshuller. Effects of Reduced Use of Lead in Gasoline on Vehicle Emissions and Photochemical Reactivity, (with appendix and list of Errata). U.S. Environmental Protection Agency, Research Triangle Park, NC (February 1972).
31. Fed. Reg. *37:* 3882 (Feb. 23, 1972).
32. Fed. Reg. *38:* 1254 (Jan. 10, 1973).
33. E.E. David, Jr. Office of Science and Technology, Executive Office of the President, Washington, D.C. Communication to D.E. Crabill, Office of Management and Budget (Nov. 1, 1972).
34. International Lead Zinc Research Organization, Inc. Response to Request for Comment on Notice of Proposed Rule Making on Regulation of Fuels and Fuel Additives. New York, NY (Mar. 9, 1973).

35. Ethyl Corp. Critique of EPA's Position on the Health Effects of Airborne Lead. Ferndale, MI (Mar. 9, 1973).
36. Ethyl Corp. et al. versus EPA, U.S. Court of Appeals, District of Columbia, No. 73-2205 (Jan. 28, 1975).
37. Internatinal Lead and Zinc Study Group. Lead in Gasoline. New York, NY (Mar. 1973).
38. L.B. Tepper and L.S. Levin. A Survey of Air and Population Lead Levels in Selected American Communities. Final report, University of Cincinnati, OH (Dec. 1972).

CHAPTER 13

URBAN PLANNING

Jack E. Cermak
Department of Civil Engineering
Colorado State University

*Roger S. Thompson**
Colorado State University

INTRODUCTION

Historical Perspective

Selections of sites for early cities were based upon a few simple criteria. The chosen locations were usually near a river or lake to satisfy the need for a water supply. Desirable topographical settings were fertile valleys for farming or ocean bays to serve as seaports. As these cities grew larger, industry and the associated consumption of energy increased. It has become apparent that such settings strongly influence both the local meteorology and atmospheric transport.

The number of variables considered in developing or evaluating urban or land-use plans has been increasing rapidly over the years. One of the latest to appear is the environmental effect of new land developments and urban renewal projects. In recent years, public awareness of the hazards of air pollution has grown to the extent that federal legislation (Clean Air Act of 1970) requires the proponents of proposed construction to submit an environmental impact statement to the U.S. Environmental Protection Agency. This statement is an evaluation of the environmentally deteriorating effects of the proposed construction. The current practice is to write such a statement after developing a plan that best meets all other design criteria, rather than to consider the environmental impact as a parameter to be minimized in the design stage. State-of-the-art techniques for analysis or prediction of environmentally harmful effects are often inadequate to provide the conclusive information needed to prepare the impact statement—particularly with regard to the increase of lead and other automotive emissions in the ambient air. Considerable effort is currently being devoted to the development, improvement, and validation of these techniques.

Factors Affecting the Environmental Impact of Urban Developments

The effect of new urban developments on the dispersion and fate of pollutants (more specifically lead) released in automobile exhausts is addressed in this section. Traffic distributions (in time and space) and individual automobile emission rates define the source locations and strengths of pollutants. Hence, street location, parking area siting, and traffic routing are important factors. Local topography and meteorological conditions are factors that influence the overall mixing and ventilation of the urban area. Urban design factors such as building shapes and sizes and the relative locations or arrangements of building complexes may either inhibit or enhance the mixing and ventilation.

The effects of the foregoing factors on the concentration level and distribution of automotive lead have not been described in a general formulation. Because of the complex interactions between the atmospheric motions and urban and topographic geometry, a physical model of specific cases gives the most reliable and complete information on potential lead concentrations for a planned urban development.

Purpose and Scope

The discussion of this section on control strategy is focused on how some of the factors identified as affecting environmental impact may be evaluated in the urban design process. Emphasis is on their effects upon the concentration magnitude and distribu-

*Current address, U.S. Environmental Protection Agency, Research Triangle Park, NC.

tion of lead released into the environment by automotive traffic and industrial sources and how these effects can be predicted for different combinations of the variables. The primary predictive technique to be described is the physical model.

FACTORS AFFECTING LEAD CONCENTRATION IN THE URBAN ENVIRONMENT

Automotive Traffic

Automotive traffic patterns in urban areas determine the source or entry of automobile emissions into the atmosphere. Typical weekday patterns exhibit high levels of traffic during "the morning rush hour" (7:00 to 8:00 a.m.) caused by the commuting labor force and an afternoon peak (4:00 to 5:00 p.m.). [1] The spatial distribution of traffic throughout the street and freeway network of the area defines the location and strength of the pollutant sources.

Emission rates of automobiles are usually expressed primarily as a function of vehicle speed. The locations of parking lots and traffic signals where vehicles spend much time stopped or moving slowly with the motors idling are also areas of high emissions.

To describe clearly the source of pollutant emission into the atmosphere, traffic counts and average vehicle speeds for streets and freeways throughout the area must be obtained. Emission rates as a function of vehicle speed must be available for the mix of automobile types using the roadway.

Transport Characteristics

Once the distribution of sources is defined, the problem becomes one of determining transport characteristics for the contaminants throughout the urban environment. Meteorological considerations are fundamental to an understanding of the processes involved. Wind speed, wind direction, and the vertical temperature structure of the atmosphere are the most influential meterorological parameters. For most existing cities, information on the frequency of average wind speeds of different magnitudes for at least eight wind directions can be obtained from the National Climatic Center at Asheville, NC. Information on the frequency and type of temperature inversions can also be obtained from the same source. Low wind speed and a temperature profile with a low-level inversion produce the most severe buildup of pollutants. This combination usually occurs concurrently with the morning traffic rush hour, compounding the adverse effect during that time period.

The fallout rate of lead-containing compounds and particles determines the travel time and thus the distance traveled from the point of emission.

Urban Geometry

Urban geometry plays an important role in the dispersion of pollutants, as well as in determining the source distribution. Clearly, an arrangement of streets and freeways that minimizes travel time of automobile trips (which in itself is desirable) also decreases the total amount of contaminants emitted into the atmosphere. However, this roadway network may not be desirable in terms of the locations of areas of high emissions. For economic and aesthetic reasons, freeway routes usually bypass urban areas rather than pass through them. This is also preferable in the sense that the emission sources are farther from the urban areas of high-density pedestrian traffic. Traffic patterns are also influenced by the location of parking facilities, which, incidentally, may be areas of high emissions.

The shapes and sizes of individual buildings and combinations of buildings and street geometries strongly influence the dispersion characteristics of the area. Wind speed and direction may be modified by channeling down street canyons. Vortices are created by building corners and edges. The spacing of buildings and the relative amount of free space such as parks and parking lots affect the turbulent or mixing ability of the atmosphere. Measurements have shown that urban areas are heat islands and may be at temperatures 5° to 10° C above the surrounding areas. [2] This temperature difference creates an updraft over the urban center, which produces an inflow from the outer portions of the area toward the center. Pollutant dispersion is strongly influenced by all of these effects.

Topography

The local and surrounding topography plays an important role in determining the local meteorology and dispersion characteristics. Mountains, hills, and valleys have an ever present influence on wind speed and direction patterns. Nearby large bodies of water produce the familiar sea-breeze, land-breeze wind pattern. Since rivers follow a path through the lowest region of a valley, gravity winds tend to converge down the valley sides to the river. Cities built in large topographic basins usually build higher concentrations of pollutants than those developed on ridges or flat terrain.

PHYSICAL MODELING OF ATMOSPHERIC TRANSPORT IN URBAN AREAS

In recent years, it has become possible to determine atmospheric transport characteristics in urban areas through the use of wind tunnels designed to simulate the atmospheric boundary layer. [3] The features of these facilities and the simulation requirements and capabilities of existing facilities are presented in the following paragraphs.

Wind Tunnels for Simulation of Atmospheric Boundary Layers

Wind tunnels with a capability for simulating characteristics of the atmospheric boundary layer to a satisfactory degree of accuracy differ from conven-

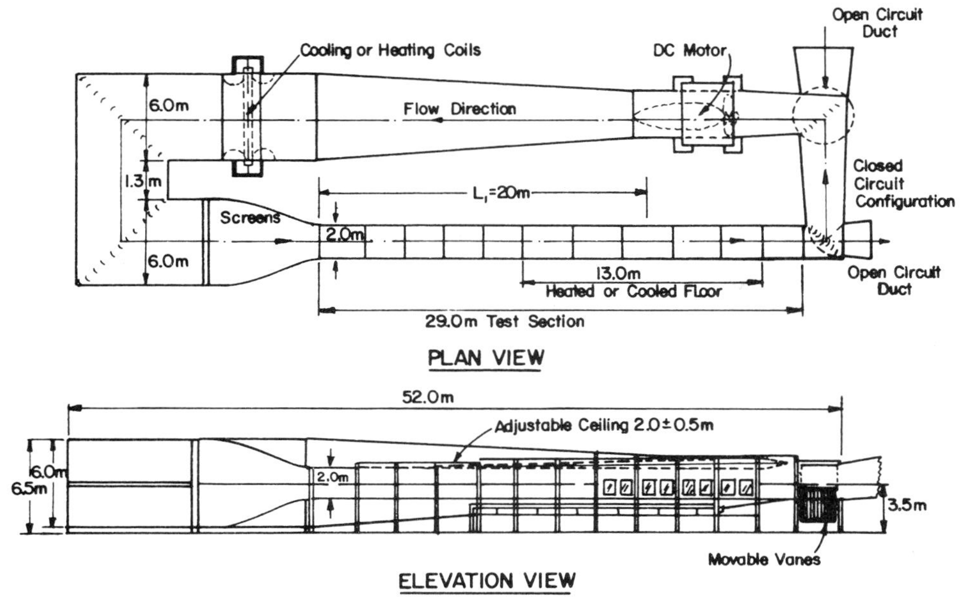

FIGURE 13-1.—Meteorological wind tunnel—Fluid Dynaimcs and Difussion Laboratory, Colorado State University.

tional aeronautical wind tunnels in several respects. [4] The meteorological wind tunnel requires a long test section to develop thick turbulent boundary layers; a cross-sectional area adjustment to accommodate flow with zero longitudinal pressure gradient; a steady controllable low-speed flow velocity down to 1 mi/h; and, if thermal stratification is to be reproduced, a cooling and heating system for developing either a stable or unstable thermal boundary layer.

Special wind tunnels designed for studies of atmospheric transport are shown in figures 13-1 and 13-2. The meteorological wind tunnel (fig. 13-1) is capable of developing boundary-layer thicknesses from 2 to 4 ft over the downwind portion of the test section, and bulk Richardson numbers from −0.5 to +0.5 over the same portion of the test section. Elevated inversions can also be developed by cooling the upstream part of the test section floor (by insertion of dry ice or additional refrigeration) and heating the downstream portion of the floor. The environmental wind tunnel (fig. 13-2) can also develop boundary-layer thicknesses up to 4 ft; however, vortex generators are required at the test section entrance. Thermal stratification corresponding only to an elevated inversion can be created by the introduction of about 700 lb of dry ice over the first 8 ft of the test-section floor.

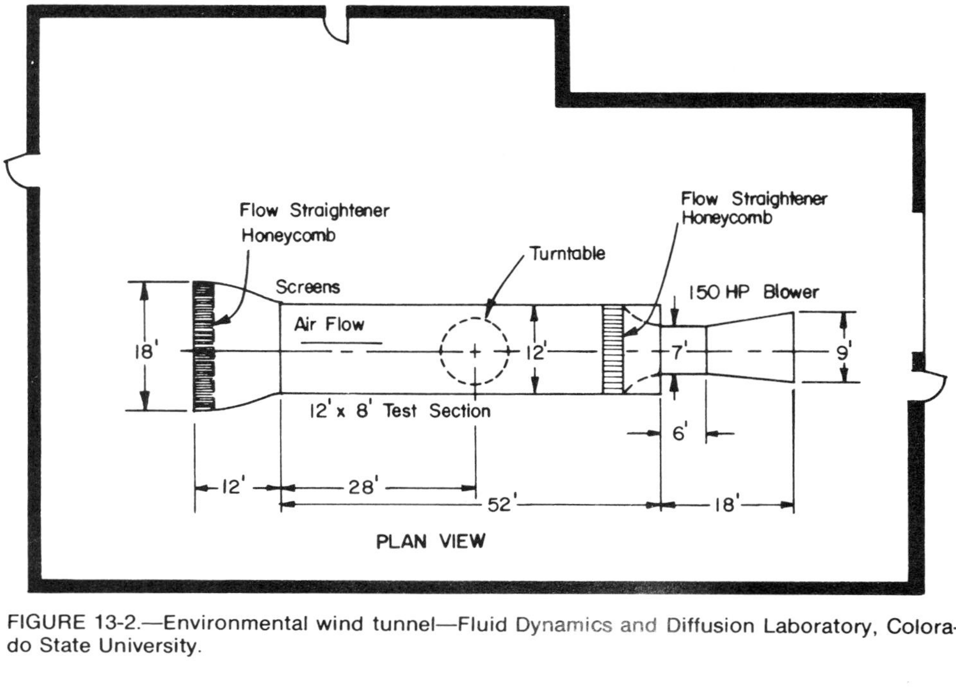

FIGURE 13-2.—Environmental wind tunnel—Fluid Dynamics and Diffusion Laboratory, Colorado State University.

Requirements and Capabilities for Physical Modeling Atmospheric Transport

A set of requirements for *exact* similarity has been established by Cermak. [4] In summary, the requirements may be stated as follows:

(1) Geometric similarity: a model of undistorted geometry will produce a flow that is compatible with the principle of mass conservation.
(2) Equal Rossby number: $Ro = U_0/(L_0 \Omega_0)$
(3) Equal gross Richardson number: $Ri = [(\Delta \bar{T})_0/T_0](L_0/U_0^2)g_0$. If the atmospheric flow is composed of two layers of different stratification, two Richardson numbers, say Ri_1 and Ri_2, are required to specify similarity requirements.
(4) Equal Reynolds number: $Re = U_0 L_0/\nu_0$
(5) Equal Prandtl number: $Pr = \nu_0/(k_0/\rho_0 C_{p0})$
(6) Equal Eckert number: $Ec = U_0^2/[C_{p0}(\Delta \bar{T})_0]$

The possibility of meeting these requirements in the laboratory and approximations introduced through partial fulfillment of the exact similarity requirements will be discussed later.

The foregoing requirements must be supplemented by the stipulation that the surface boundary conditions and the approach flow characteristics be similar for the atmosphere and its model. Similarity of initial conditions (approach flow characteristics) pertaining to the mean and turbulent velocities is often referred to as kinematic similarity.

Surface boundary condition similarity requires similarity of the following features:
(1) Surface roughness distribution.
(2) Aerodynamic effects of surface roughness elements corresponding to an "aerodynamically rough" surface.
(3) Surface temperature distribution.

Similarity of the approach flow characteristics requires similarity of the following elements:
(1) Distributions of mean and turbulent velocities.
(2) Distributions of mean and fluctuating temperatures.
(3) The longitudinal pressure gradient (should be zero).
(4) Equality of the ratio H_2/H_1 if the flow is layered.

If all the foregoing requirements were met, all scales of motion ranging from micro- to meso-scale, 10^{-3} to 10^5m, could be simulated simultaneously. The diffusion process could then be accurately reproduced over horizontal distances up to 10^5 m within the planetary boundary layer. However, all of the requirements cannot be met by existing laboratory facilities, and partial or approximate simulation must be used.

The flow characteristics and physical size of wind tunnels are such that most of the requirements for similarity with the atmosphere can only be approximated with varying degrees of accuracy. This does not eliminate the possibility of making useful studies of diffusion with small-scale models, but it limits the range of length scales and thermal conditions for which the studies are feasible. Each similarity requirement will be examined to determine the necessary approximations imposed by the physical model and the resulting limitations imposed upon air-pollution studies.

Undistorted Scaling of Geometry: Selection of equal length scales for horizontal and vertical distances results in an undistorted model. Such a model is usually used unless the scale is so small (less than 1:1,000) that the surface with scaled surface roughness elements becomes an aerodynamic smooth surface. In this case, a local distortion through increase in height of roughness elements is necessary to produce an aerodynamically rough surface. No limitations on possible air-pollution studies result from this requirement.

Equal Rossby Number: The ratio of model and atmospheric Rossby numbers is $U_{om}L_{oa}/(U_{oa}L_{om})$, since the wind tunnel (unless placed upon a rotating platform) has the same angular velocity as Earth. Accordingly, the flow effects produced by rotation of Earth is not reproduced, since the Rossby-number ratio will be in the range of 10^2 to 10^4. This limits the simulation to atmospheric flows not affected appreciably by rotation of Earth. Using a Rossby number of 10 and greater to express this condition, a broad guideline for atmospheric flow that may be simulated without substantial rotational error can be stated as follows: the time of travel for a fluid particle at a height where surface effects are small should not exceed approximately 10^3 seconds.

Equal Gross Richardson Number: Exact similarity of density stratification is essential for realistic simulation of atmospheric motion. Bulk Richardson numbers over a layer 10 cm deep can range from -0.5 to $+0.5$ in the meteorological wind tunnel to cover a wide range of atmospheric conditions. This requirement does not limit to any significant extent the atmospheric flows that can be simulated.

Equal Reynolds Number: The ratio of Reynolds number for the wind-tunnel model and the atmospheric Reynolds number is $U_{om}L_{om}\nu_{oa}/(U_{oa}L_{oa}\nu_{om})$. This ratio varies from 10^{-2} to 10^{-4} and depends primarily upon the model scale. Fortunately, flow characteristics over rough surfaces composed of blocklike buildings become independent of Reynolds number, for Reynolds numbers ordinarily attained in the wind tunnel. Therefore, the inequality of Reynolds number is not a serious error when simulating dispersion over urban areas.

Equal Prandtl Number: The requirement of equal Prandtl numbers for the two flows is met exactly when using air as the wind-tunnel fluid. Over a temperature range of $-100°$ to $100°$ C, the Prandtl number varies only from 0.75 to 0.70. Accordingly, this requirement does not restrict the flows which can be simulated.

Equal Eckert Number: The Eckert number is proportional to the Mach number squared. Therefore, for the slow flows under consideration, the equality of Eckert numbers may be relaxed without introduction of significant error.

Surface Boundary Conditions: Surface conditions are met by scaling roughness elements (buildings and trees) by the same length scale used for relating the atmospheric- and small-scale flow field, provided that the modeled surface acts as an aerodynamically rough surface. When the scale reduction is large, greater than 1:1,000, and the surface-roughness elements are small (one-story buildings, trees, etc.), the foregoing criterion cannot be met. [5] The roughness element heights in such cases must be exaggerated to regain a rough character. By adding to the scaled roughness height an amount equal to the viscous zone thickness, $\mu_v = 10\nu/\mu_*$, a surface of appropriate roughness can be obtained. Topographic relief is usually represented by undistorted geometrical scaling of the full-scale features.

Surface temperature distributions may be obtained by heating and/or cooling elements in the surface. [5] The magnitude of the surface temperature is determined by the requirement to satisfy equality of Richardson numbers for model and atmosphere.

Approach Flow Characteristics: The long test-section wind tunnel (fig. 13-2) can be used to simulate the vertical distribution of mean and fluctuating velocities and temperatures. It works very well over a wide range of scales for ground-based thermal stratifications that are either stable or unstable over the entire boundary layer. [4] When an elevated inversion accompanied by a mixing layer must be simulated, a layer of dry ice can be added on a portion of the upstream test-section floor (approximately 7 m). By varying the length of the dry-ice boundary, the downstream surface heating rate and the flow speed, the desired mixing-layer depth H_l can be obtained.

Approach-flow characteristics may be controlled to varying degrees by the introduction of devices such as jets, vortex generators, grids, roughness plates, screens, and heating or cooling elements at the test-section entrance. A review of various methods employed is given by Teunissen. [6]

In summary, wind tunnels of the type and size shown in figures 13-1 and 13-2 can simulate the transport of pollutants in urban areas with a high degree of reliability. Accordingly, quantitative information on the effects of source location, building and street geometry, and topography for a wide range of meteorological conditions is now accessible for urban planning.

PLANNING TO MINIMIZE LEAD CONCENTRATIONS

Planning must begin by recognizing the transport liabilities and assets of the natural site characteristics—local meteorological parameters and topographic features. Major sources of lead in the urban environment originate from automobile exhaust emissions and from some industrial stacks. Accordingly, planning to assure minimum human exposure to high lead concentrations in an urban environment requires locating main traffic arteries, automobile parking complexes, and industrial sites so that the prevailing wind and atmospheric transport direction are away from areas of high-density dwellers. In most urban areas, other considerations and constraints have dominated the planning process, with the result that much urban planning now centers on the modification of existing urban geometry and on traffic routing and regulation.

The scale of areas for planning varies from regional considerations, e.g. 100 km by 100 km, to local problems within areas of one block, e.g. 100 m by 100 m. Most of the discussion in the following paragraphs is confined to planning for local areas 10 km by 10 km in extent.

Street-Freeway-Parking Networks

Major traffic arteries should be located in a manner that will enable winds from the most probable directions to transport pollutants away from core areas of high-density dwellers. Slow-moving traffic through core areas should be minimized by providing parking immediately adjacent to the major traffic arteries. Plans should include provisions for the transport of people from parking areas to core areas of the urban complex. Mass transport systems such as subways, moving sidewalks (conveyor belts), electric trams, or programmable people movers can provide the necessary mobility.

In cases where a river valley dominates the topography, or substantial differences in elevation occur due to other topographic forms, the general rule of locating traffic arteries at elevations lower than the areas of concentrated pedestrian activity should be applied. Topography has significant effects upon the motion of a stably stratified atmosphere; mean motion of a stably stratified atmosphere is usually down the path of maximum grade. Therefore, maximum ventilation of such a site can be obtained by placing the traffic artery on this path of maximum grade as demonstrated by Cermak [3] (using a 1:400 scale model of downtown Denver). This is particularly important when considering air pollution produced by other automobile exhaust products such as carbon monoxide and hydrocarbons because turbulent mixing is inhibited by the stable stratification. Accordingly, large concentrations of these pollutants (including lead) can develop unless plans provide for "ventilation" channels or a reduction of the source intensity.

Heat-island effects can modify the downwind distribution of emissions released along a traffic artery. Chaudhry and Cermak [5] used a 1:4,000 scale model of Fort Wayne, IN (fig. 13-3) to study the ground-level concentration distribution downwind of a line source oriented perpendicular to the mean wind direction. Figure 13-4 shows that the surface dosages along the lateral edges of the city were much larger than if the air flow had remained two-dimensional or unaffected by the higher temperatures at the city center. Field measurements [7] confirmed findings derived from the model.

A common practice in many urban renewal projects or new buildings of major size is to construct underground parking garages beneath the development. In these cases, special attention must be given to street-level concentrations developed by exhausts

FIGURE 13-3.—Model of Fort Wayne, IN, (1:4,000 scale) in environmental wind tunnel (flow is out of the photograph).

discharged from the garage. Measurements of concentration in a small-scale physical model placed in a wind tunnel permit the only reliable means for obtaining data for planning and design. Studies of Chaudhry and Cermak [8] for a single building, and of Cermak et al. [9] for a complex of new buildings, show that elevated releases and increased exit velocity of exhaust gases are effective means for reducing local concentrations. The latter model is shown in figure 13-5. Distributions of concentration coefficients are shown in figure 13-6 for two 15 ft exhaust stacks and in figure 13-7 for the same two stacks extended to 64 ft. Concentrations resulting from these stacks serving the underground parking garage of the Yerba Buena Center in San Francisco were determined by sampling radioactive krypton, which was released from the stacks in metered quantities.

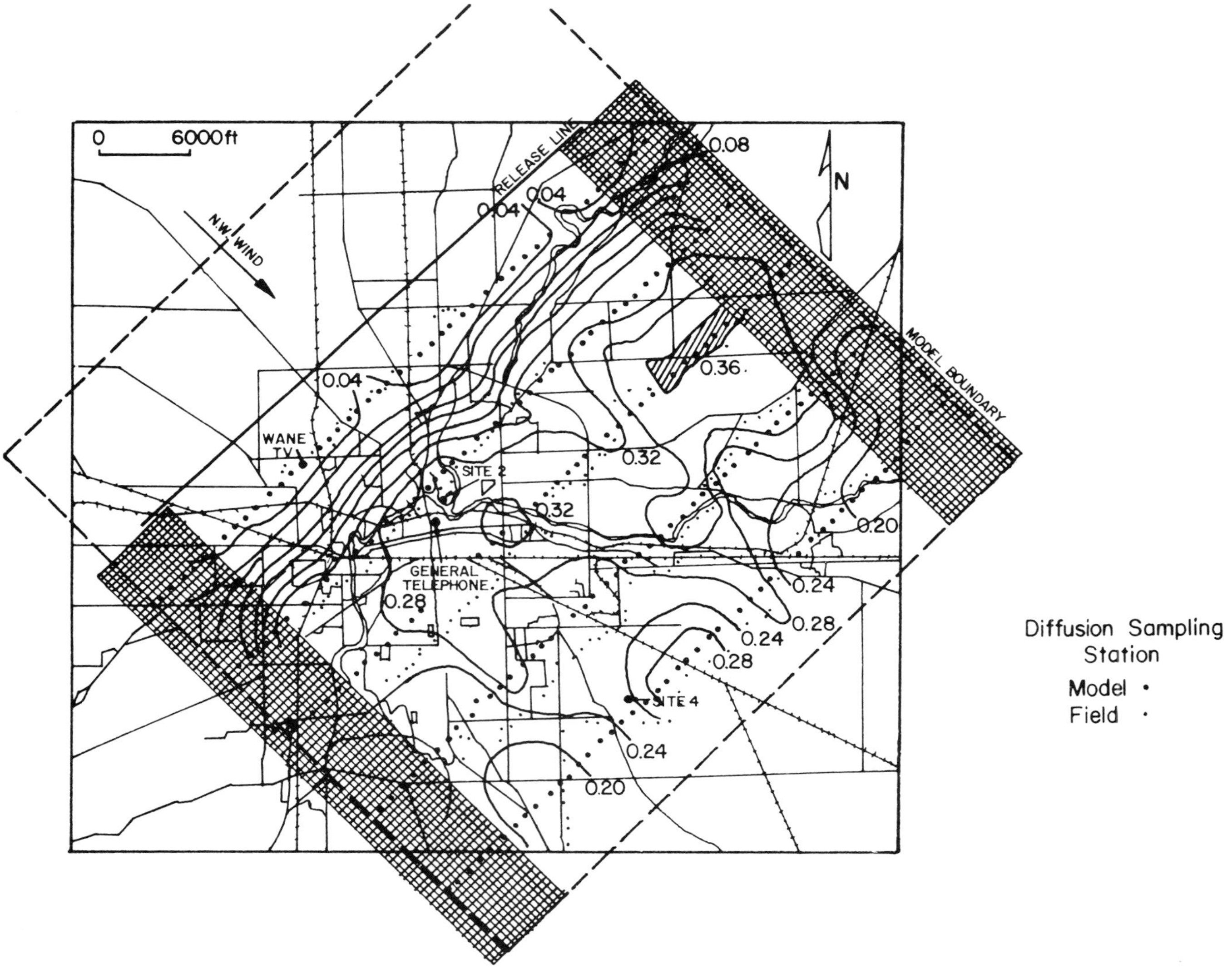

FIGURE 13-4.—Surface dosage pattern observed over the model. Numbers on the isodosage lines indicate nondimensional dosage defined in text.

Building-Complex Geometry

Building types and arrangements can influence strongly the concentrations of lead in city streets. Concentration measurements in small-scale models of a city canyon indicate that the city blocks composed of continuous lines of equal-height buildings result in maximum levels of contamination. A physical model to investigate diffusion in an urban complex. is shown schematically in figure 13-8. This study, currently in progress at Colorado State University under NSF Grant AEN74-17624 A01, was confined initially to an "idealized" city of uniform blocks. Figures 13-9 and 13-10 show the distribution of concentration coefficients for wind perpendicular and parallel to the street containing line sources of radioactive krypton. Current studies are being made to determine the effect of nonuniform distributions of buildings of different heights.

Greater atmospheric mixing rates can be obtained by leaving gaps and/or constructing buildings with varied heights in a block. A major consideration in future planning is the concept of replacing a block or several blocks of many low buildings by a single tall structure. The tall-building concept permits a substantial area to be kept open at street level. Urban geometry composed of tall buildings with large spacing maximizes atmospheric mixing and minimizes concentrations for the same traffic density. Other advantages of this type of urban development are improvements in the overall street-level environment—more sunlight, lower temperatures, and greater aesthetic appeal. However, these benefits would be obtained at the expense of pedestrians being subjected to more intense wind gusts. Strong support for the tall-building, open-space urban plan was presented by leading architects and structural engineers of the United States at a recent meeting in Mexico City. [10]

FIGURE 13-5.—Model of Yerba Buena Center, San Francisco (1:240 scale). [9]

Lead contamination may occur in city buildings if lead-contaminated air from the streets enters

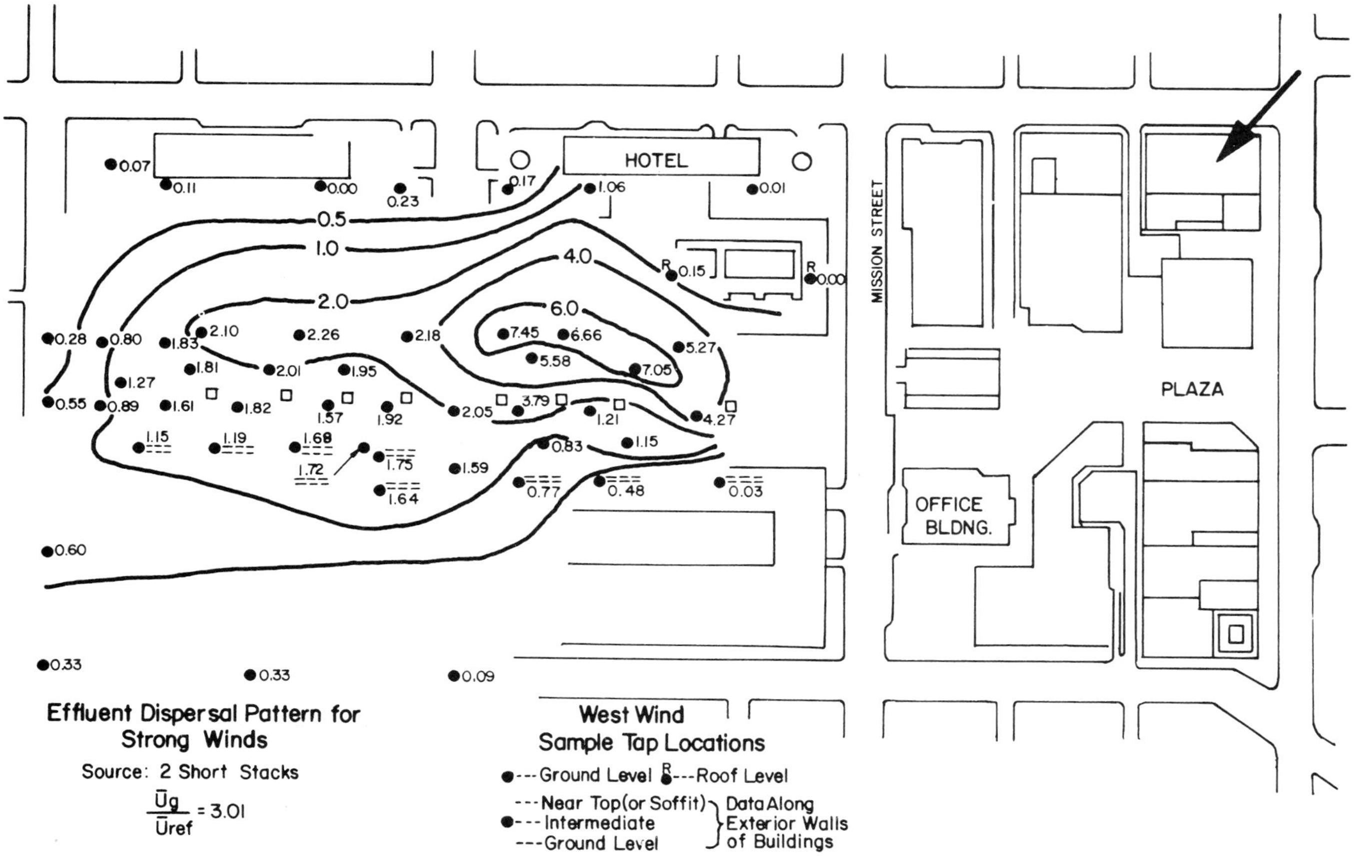

FIGURE 13-6.—Diffusion from two short stacks in a strong west wind.

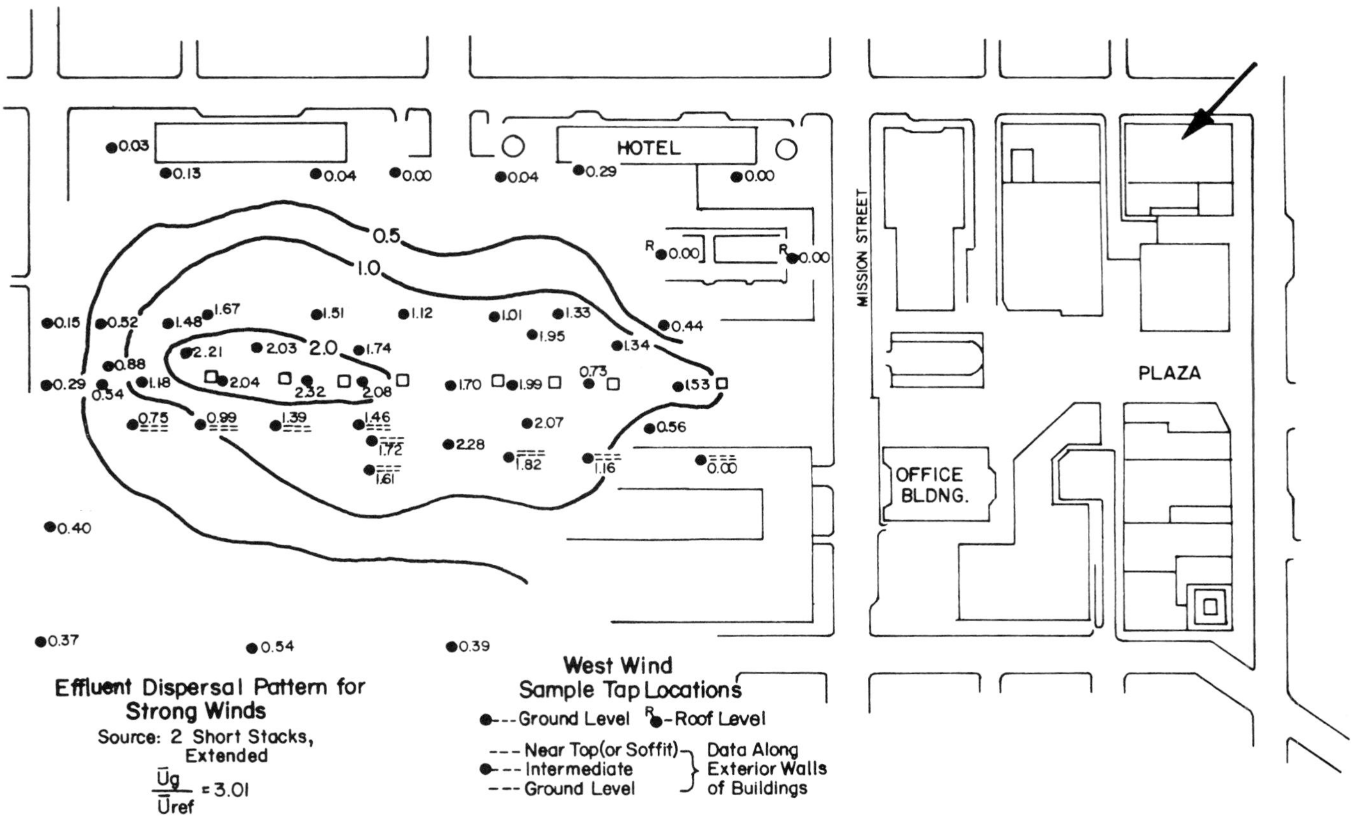

FIGURE 13-7.—Diffusion from two extended stacks in a strong west wind.

through intakes of the ventilation system. This type of pollution can be best avoided by using tracer-gas movement studies in a physical model during the planning stage. Chaudhry and Cermak [8] used this technique to determine local transport properties around the Children's Hospital, National Medical Center. Here, the dominant source of contaminants was a parking garage beneath the building with exhaust ports discharging at ground level. As a result of this investigation, air intakes were removed from one side of the building and certain windows were installed for permanent closure.

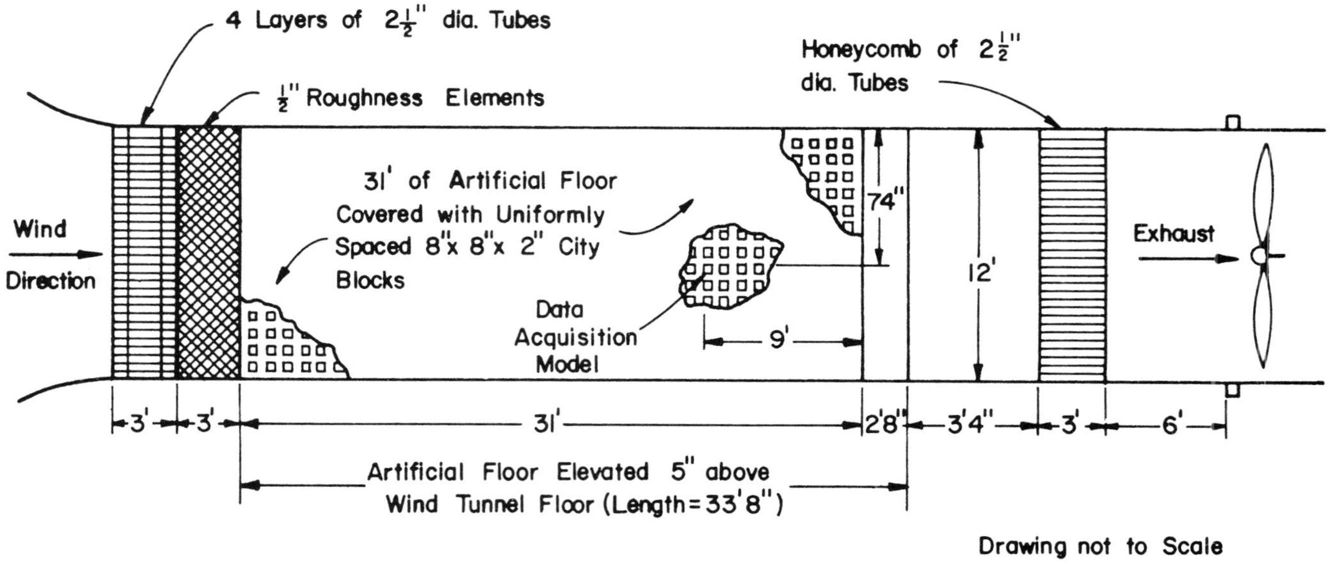

FIGURE 13-8.—Floor plan for modeling of idealized city in environmental wind tunnel.

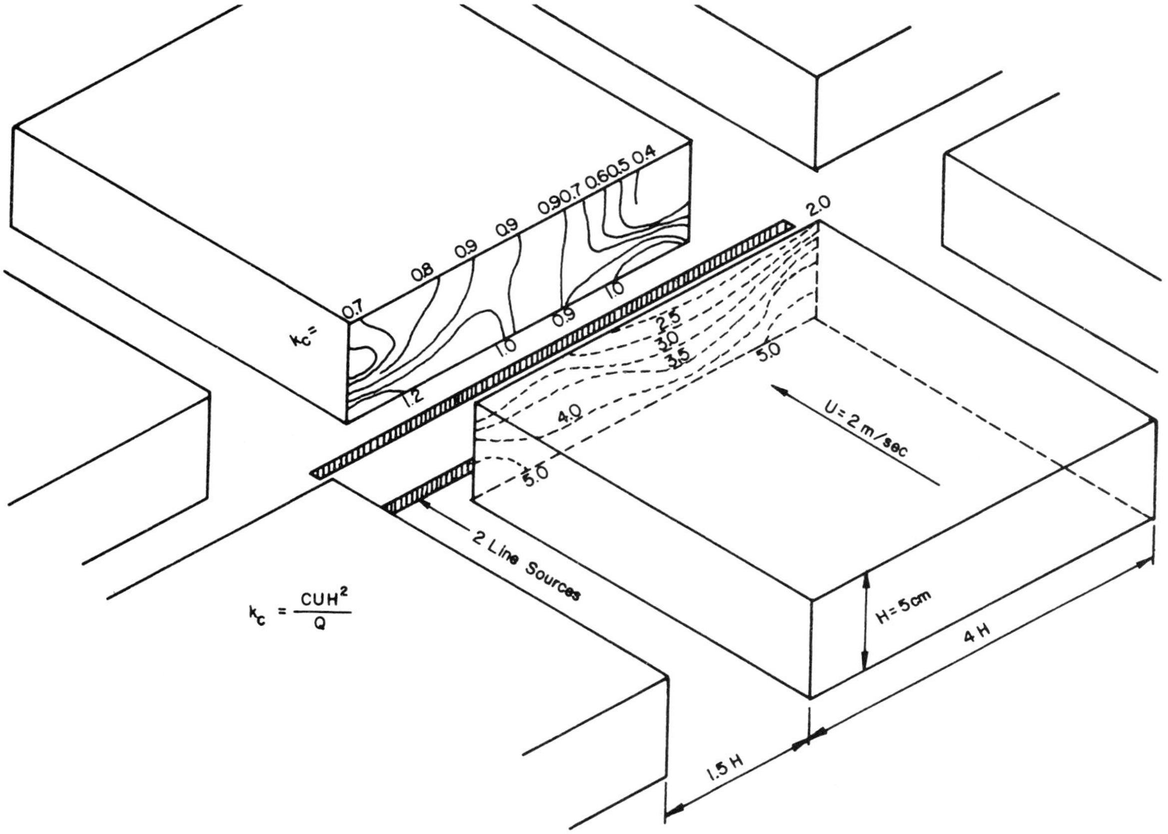

FIGURE 13-9.—Concentration coefficients measured on walls of street canyons. Wind direction is perpendicular to street. $k_c = CUH^2/Q$

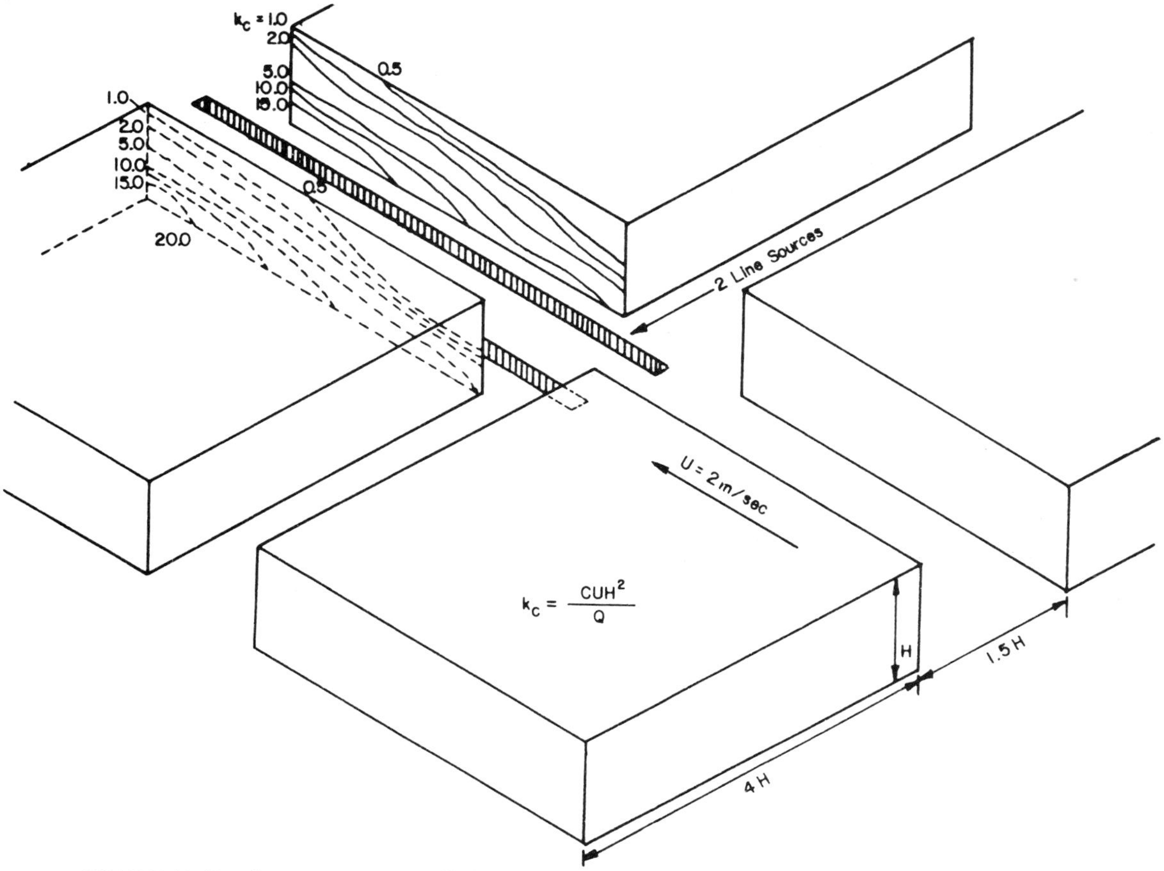

FIGURE 13-10.—Concentration coefficients measured on walls of street canyons. Wind direction is parallel to street. $k_c = CUH^2/Q$

SUMMARY

Urban planning to minimize concentrations of lead introduced into the environment by automotive traffic and industry requires coordinated use of information that describes very complex factors and interactions. The first of these is the intensity and distribution of lead sources in space and time, which are closely associated with the network of traffic arteries, streets, and parking areas. The meteorological variables profoundly affect the transport of lead from its sources through the urban complex. Finally, the urban geometry itself, and the relationship of buildings and streets to local topographic features, determine how the atmosphere distributes matter in time and space.

Physical modeling of these phenomena in meteorological wind tunnels provides the only technique by which all the primary factors and interactions affecting transport of environmental contaminants can be accounted for on the scale of urban complexes. This technique becomes more indispensable as the need for concentration prediction becomes more local in character. Using concentration data obtained from the physical model for a range of specific meteorological variables, together with probability densities for source characteristics and the meteorological variables, enables establishing for a particular urban plan the probability that the concentration will exceed a prescribed level in a specified time interval and location.

Information obtained from physical modeling accomplished up to this time and basic consideration of mass transport substantiate that use of the following general guidelines will result in reduced lead concentrations in urban centers:

(1) Maximize the distance between high-density pedestrian activity and lead sources by the development of automobile parking near traffic arteries on the periphery of urban areas; also provide public mass transit to the urban center.

(2) Maximize mass transport rates for stably stratified atmosphere by locating new urban centers on areas of maximum elevation relative to the local topography and by locating high-density traffic routes parallel to the lines of maximum elevation changes.

(3) Maximize mass transport rates by the development of urban centers in which individual tall structures are spaced to maintain 40 to 60 percent open space at ground level.

(4) Maximize the source-pedestrian distance for lead sources such as parking garages and industrial operations by the use of elevated release to the atmosphere with strong vertical exhaust velocity.

NOMENCLATURE

C = concentration, M/L^3
C_p = specific heat at constant pressure, L^2/t^2T
g = gravitational acceleration, L/t^2
h = roughness element height, L
H = height of building, L
H_1, H_2 = height of inversion, L
k = thermal conductivity of fluid, ML/t^3T
k_c = concentration coefficient, CUH^2/Q
K_M = eddy diffusivity for momentum, L^2/t
K_s = equivalent sand roughness, L
l = reference length, L
l_v = viscous zone thickness, L
q_0 = surface heat flux, M/t^3
Q = source strength, M/t
t = time, t
T = instantaneous temperature, T
T_w = mean temperature at $z = 0$, T
ΔT = $\overline{T} - \overline{T}_w$, T
$(\Delta \overline{T})_0$ = $\overline{T}_0 - \overline{T}_w$, T
u^* = shear velocity, $(\tau_0/\rho)^{1/2}$, L/t
U = reference mean velocity, L/t
z = vertical coordinate, L
z_0 = roughness length, L
δ = boundary layer thickness, L
δ_T = thermal boundary layer thickness, L
ρ = mass density, M/L^3
ν = kinematic viscosity, L^2/t
τ_0 = surface shear, ML/t^2
Ω = angular velocity, t^{-1}

Subscripts:

$(\ )_0$ = reference quantity

Superscripts:

$(\ )^*$ = nondimensional quantity
$(\overline{\ })$ = time average
$(\ )'$ = instantaneous fluctuation from time average

Dimensions:

L — length
M — mass
t — time
T — temperature

REFERENCES

1. Highway Research Board. Highway Capacity Manual. Special Report 87. Washington, D.C. (1965).
2. J.T. Peterson. The Climate of Cities: A Survey of Recent Literature. National Air Pollution Control Administration, U.S. Government Printing Office, Washington, D.C. (1969).
3. J.E. Cermak. Air Motion in and Near Cities—Determination by Laboratory Simulation. Proceedings of Western Resources Conference, Denver, CO (Jul. 8-10, 1970).
4. J.E. Cermak. AIAA J. *9:* 1746 (1971).
5. F.H. Chaudhry and J.E. Cermak. Wind-Tunnel Modelling of Flow and Diffusion Over an Urban Complex. Project THEMIS Report No. 17. Colorado State University, Fort Collins, CO (1971).
6. H.W. Teunissen. Simulation of the Planetary Boundary Layer in a Multiple-Jet Wind Tunnel. Report 182. Institute for Aerospace Studies, University of Toronto, Ontario (1972).
7. G.R. Hilst and N.E. Bowne. A Study of the Diffusion of Aerosols Released from Aerial Line Sources Upwind of an Urban Complex. Final Report, Contract No. DA 42-007-

AMC-38(R) to Dugway Proving Ground. The Travelers Research Center, Inc., Hartford, CT (1966).
8. F.H. Chaudhry and J.E. Cermak. Study of Wind Pressures and Air Quality Around Children's Hospital, National Medical Center. Technical Report, Leo A. Daly Co., Washington, D.C. (Mar. 1971).
9. J.E. Cermak, F.H. Chaudhry, A.C. Hansen, and J.A. Garrison. Wind and Air-Pollution-Control Study of Yerba Buena Center, Technical Report, McCue-Boone-Tomsick, San Francisco, CA (Jun. 1972).
10. J.E. Cermak, F.H. Chaudhry, A.C. Hansen, and J.A. Garrison. Civ. Eng. *43*(6): 98 (1973).

SUMMARY OF PART IV

H.W. Edwards

The general picture that emerges in considering control of environmental contamination by lead is that the highest lead concentrations, hence the greatest probabilities of potential environmental damage, are generally highly localized to regions near the sources. While lead can be detected in air, water, and biota at great distances, no drastic environmental consequences have thus far been established for these lower levels of lead contamination. The emphasis in control technology is therefore placed upon reducing lead release. This will reduce lead buildups and possible effects thereof in the immediate region of the source.

In considering industrial lead emissions, there is a clear need for preplanning in selecting sites for new lead mills and smelters. Standards applicable to industrial activities should be firmly based on research and should reflect the need for stable regulations that clearly specify long-term industry responsibilities. Cooperation of industry and government is desirable in developing regulations. Application of modern control technologies and good industrial housekeeping practices can be very effective in minimizing environmental contamination by lead industries. Fugitive dust is a primary source of airborne lead emissions from mines, mills, and smelters. Concentrate piles can be covered or kept damp; concentrate in transport can be covered or otherwise protected from blowing winds. Because of lead fallout from smelter stack plumes, tall stacks along are insufficient to prevent lead contamination. Air pollution control devices should be utilized and their performance monitored on a regular basis. Treatment of mill effluent is most important to maintain high water quality in receiving streams. Water quality should be measured on unfiltered as well as on filtered samples. Sediment and suspended solids transports are mechanisms of major importance.

In considering control of automotive lead emissions, the issue of continued widespread use of lead alkyls is not readily separated from other technological aspects of automotive emission control. The technological need for low-lead gasolines may be temporary and depends largely upon whether use of lead-intolerant oxidation catalysts becomes widespread. Other considerations (including economic, environmental, and health factors) enter strongly; and there are uncertainties in each of these areas, especially for the effects of lead substitutes. Because of these uncertainties, there is a clear need to continue environmental monitoring. There is, however, widespread agreement that a long-term trend of rising airborne lead concentrations would be undesirable.

A major energy concern is that the combined effects of requiring low-lead gasolines and oxidation catalysts may significantly increase petroleum consumption in the United States. Because of the lack of definition in certain of these areas, a number of technological measures are considered in the context of reducing environmental inputs and buildups of automotive lead. A partial reduction in the lead content of gasoline appears technically feasible in terms of maintaining some degree of continuing benefit from the role of lead as a petroleum conserver. A two-lead strategy, in which lower lead-content gasolines would be supplied to areas where automotive lead buildup is greatest, would also be a possibility. Tests indicate that installation of particulate traps on automobiles would retain most of the lead in gasoline within exhaust system components. Particulate traps could replace standard mufflers at an incremental cost that does not appear prohibitive. Proper attention would have to be given to disposal of discarded muffler-traps, so that accumulated lead would not reenter the environment dissipatively.

Urban lead exposures and buildups are also subject to some degree of control through sound urban planning. No single control strategy seems ideal for all situations, and controls that are implemented should reflect both the state of knowledge and the need for flexibility in dealing with problems that may be specific to certain combinations of traffic, geometric, and meteorological conditions. In a more general sense, the technological factors associated with the issue of lead in gasoline should be viewed in perspective as facets of the broader problem of achieving and maintaining environmental quality at acceptable economic and social costs.

There is widespread agreement that reducing the lead content of gasoline will result in increased gasoline prices. Producing gasoline of suitable octane rating without lead will require more expensive and more energy-consuming refining processes. High-octane fuels are essential for efficient operation of today's high-compression engines, and reducing engine octane requirements translates into reduced fuel economy and increased petroleum consumption for automotive transportation. The economic benefits of lowering the lead content of gasoline have not been assessed and should be evaluated within the larger context of the nationwide program to improve air quality. Possible benefits would include reductions in the costs of air pollution damages to health, materials, vegetation, and the number of cases of lead poisoning. A major difficulty of the present approach to meeting national air quality standards is that automobile manufacturers have become locked into a technology that is unproven and may not be cost-effective. There are also indications that widespread application of this technology may reduce operating cost differences between large and small cars. The incentive to purchase smaller cars with better fuel economy may diminish as a consequence. This strategy for controlling automotive pollutants, requiring removal of lead from gasoline, seems likely to add to current petroleum shortages.

Urban planning to minimize lead exposures and buildups requires the coordinated use of information concerning a number of complex factors. The temporal and spatial distributions of traffic and industrial sources must be evaluated. Meteorological factors and urban geometry profoundly affect ventilation conditions and, therefore, dispersal of lead and other pollutants. Physical modeling with wind tunnel facilities is currently the only technique whereby all of the primary factors affecting dispersal can be accounted for on the scale of urban complexes. Physical modeling enables detection of areas where the buildup of lead and other pollutants is likely to be high and permits an evaluation of the probable effects of new buildings and traffic routes. A number of general guidelines emerge from physical modeling carried out thus far. Lead concentrations in urban centers can be reduced by using mass transit systems to transport people to and from parking areas located on the periphery of urban areas near traffic arteries. New urban centers should be located on areas of maximum elevation relative to local topography to maximize pollutant transport under stably stratified flow conditions. Pollutant transport will also be enhanced when individual tall structures in urban areas are spaced to maintain a proper proportion of open area. In the case of industrial and parking-garage lead sources, the effective pedestrian-source distance can be increased by the use of elevated release with strong vertical exhaust velocity.

PART V

Economic Aspects of Control

CHAPTER 14

ECONOMICS OF AUTOMOTIVE LEAD CONTROL

George Provenzano
Institute for Environmental Studies
University of Illinois

INTRODUCTION

Under the provisions of the Clean Air Act of 1970, the U.S. Environmental Protection Agency (EPA) has promulgated two sets of regulations that restrict the use of lead additives in the production of gasoline. In the first set of regulations, EPA ordered major gasoline retailers to sell one grade of unleaded gasoline beginning July 1, 1974. [1][1] In the second set EPA established a 5 year, phased-reduction schedule beginning January 1, 1975, that reduces the maximum allowable lead content of leaded grades of gasoline. [2][2]

The general availability of unleaded gasoline has enabled automobile manufacturers to install lead-intolerant catalytic emissions control systems on 1975 and later model cars. Auto makers are currently using control systems incorporating oxidation catalysts to meet Federal automobile emissions standards for hydrocarbons and carbon monoxide; they may also use systems containing reduction catalysts to meet increasingly stringent nitrogen oxide standards. The effectiveness of the catalytic activity in both of these systems is severely reduced when the combustion products of lead additives are present in the engine exhaust.

EPA's purpose in phasing-down the lead content of leaded grades of gasoline is to directly reduce automotive lead emissions. The combustion of leaded gasoline is currently the largest source of lead that is emitted into the atmosphere. A large fraction of this lead quickly settles out of the air, contaminating soil and dust along streets and highways. Airborne lead can be inhaled and absorbed into the body through the lungs. Lead-contaminated dirt and dust are a potential source of ingestible lead, particularly for children. Because of the toxic nature of lead, EPA regards the present levels and pervasiveness of automotive lead emissions as a hazard to public health. [5]

The full implementation of both sets of regulations will significantly reduce the widespread use and, hence the cost and energy-saving benefits, of lead additives in the production of high octane gasolines. Because of the importance of these benefits, EPA's actions have generated many questions concerning the economic and technical prudence of supporting an automotive emissions control strategy that demands lead removal. Some of these questions point to apparent and growing conflicts between policies for protecting environmental quality and those for conserving energy.

Some of the economic-energy effects of reducing the use of lead in gasoline are examined in this chapter. In the beginning, lead reduction regulations are discussed within the overall context of EPA's program for controlling automotive emissions. Succeeding sections then examine the technical and economic impacts of lead removal on automotive fuel economy and the domestic petroleum industry.

It is important to note that many of the economic cost implications of lead removal are discussed without presenting any economic evaluation of the corresponding social benefits of reducing automotive lead emissions. Thus only one side of a cost-benefit analysis of controlling automotive lead emissions is presented. In this chapter, as in the context of a cost-benefit analysis, the term "costs" refers to the net increases in expenditures of resources that stem from the implementation of lead additive controls. The term "benefits" refers to reductions in the social costs or damages that automotive lead emissions impose on public health, crops, animals, materials, and aesthetics. In this context, it is important to note that the benefits—measured in terms of a re-

[1] All gasoline retailers who sell more than 200,000 gal. of gasoline annually or gasoline retailers located in rural areas with less than 50 persons/mi^2 who sell 150,000 gal. annually must offer for sale one grade of unleaded gasoline. [3] This includes about 160,000 service stations or approximately 45 percent of the branded retail outlets in the U.S.

[2] Because of court challenges, EPA will not begin enforcing the phased-reduction schedule for all refiners until October 1, 1979. Enforcement will begin at the 0.5 g/gal. average lead level. [4]

duction in the number of lead poisonings, in the number of individuals with elevated blood lead levels, in adverse environmental effects, or in the risks of adverse human health and environmental effects—of reducing the use of lead additives have yet to be assessed in a formal manner. [3,4]

LEAD ADDITIVE CONTROLS FROM THE PERSPECTIVE OF THE FEDERAL AUTOMOBILE EMISSIONS CONTROL PROGRAM

Both regulations for controlling the use of lead additives are part of a larger Federal program for eliminating automobile-related air pollution. Because of this fact, the underlying rationale for each restriction can be best understood if explained within the context of this larger program. The following sections discuss the different bases for the orders requiring the general distribution of unleaded gasoline and for the implementation of the phased-reduction schedule.

Engineering-Economic Basis for Unleaded Gasoline

The 1970 Amendments to the Clean Air Act, Public Law 91-604, have provided EPA with the statutory means to establish stringent automobile emissions standards. To help ensure that these standards (table 14-1) would be met, Congress also authorized Federal regulation of fuels and fuel additives. Under the Clean Air Act, EPA may control or prohibit the menufacture and sale of any fuel or fuel additive for use in motor vehicles if the emissions products of the fuel or fuel additive significantly impair the performance of any emissions control device that is in general use, or that, because of its state of development, is likely to be in general use.

EPA ordered the general availability of unleaded gasoline in order to support automobile manufacturers' decisions to use lead-intolerant catalytic emission control systems that can also meet these standards. Because EPA does not prescribe which types of systems the automobile industry must use in meeting standards, an explanation of manufacturers' reasons for installing lead-sensitive control systems is essential to this discussion of the economics of removing lead from gasoline. Several summaries of the technical status of emissions control systems for spark-ignition internal combustion engines and of the effects of lead emissions on these systems have been published by the National Academy of Sciences [10-11] and EPA. [12-16] This section will not repeat the findings of those studies. Instead, the discussion will focus on three points: (1) the economic-engineering reasons for the automobile industry's selection of lead-intolerant catalytic systems for meeting 1975-76 statutory standards; (2) the reasons for the industry's installation of these systems on 1975 model cars even after 1975-76 emissions standards had been relaxed to the point that permits the use of lead-tolerant systems; and (3) the future prospects of the industry's adoption of lead-tolerant emissions control systems.

From a purely technical standpoint, the complex and interdependent relationships between automotive engine design, fuel composition, and exhaust characteristics offer several options for controlling emissions. Some of the possibilities include:

(1) *Modifying combustion in order to reduce emissions from the bare engine:* Examples of modi-

TABLE 14-1.–*Automobile emissions standards (using the 1975 Federal test procedure)*

EMISSION STANDARDS	Emissions (g/mi) Hydrocarbons	Carbon monoxide	Nitrogen oxides
Uncontrolled (pre-1968 vehicles)	8.7	87.0	3.5
1973/74 Federal standards	3.0	28.0	3.1
1974 California standards	2.9	28.0	2.1
1975 Federal statutory standards[1,2]	0.41	3.4	3.1
1975 Federal interim standards[2,3] (extended through 1976)	1.5	15.0	3.1
1975 California standards[2,3] (extended through 1976)	0.9	9.0	2.0
1977 Federal standards (ESECA)[3] (suspended)	0.41	3.4	2.0
1977 Federal interim standards[3]	1.5	15.0	2.0
1977 California standards[3]	0.41	9.0	1.5
1976 Federal statutory standards[1,3] (deferred until 1978)	0.41	3.4	0.4

[1] In the 1970 Amendments to the Clean Air Act, Congress mandated the implementation of emissions standards for hydrocarbons (HC), carbon monoxide (CO), and nitrogen oxides (NO_x) that—beginning with the 1975 model year for HC and for CO, and the 1976 model year for NO_x—would be 90 percent less than 1970 standards. Although they have been deferred and suspended, the levels specified in these standards can only be changed by an act of Congress. Hence, they are known as statutory standards.

[2] On April 11, 1973, USEPA Administrator, William Ruckelshaus granted the automobile industry's application for a 1 year suspension of the 1975 statutory emissions standards. Because the Clean Air Act requires that interim standards be set if a suspension is granted, in his decision Ruckelshaus promulgated two sets of interim standards: one set for the state of California and one for the rest of the U.S. [7]

[3] The Energy Supply and Environmental Coordination Act of 1974, (ESECA) Public Law 93-313, maintains 1975 Federal interim and California standards through the 1976 model year and mandates an NO_x standard of 2.0 g/mi for the 1977 model year. The act also allows automobile manufacturers to request a 1 year suspension in 1977 Federal standards for HC and CO. In March 1975, USEPA Administrator Russell Train granted a suspension and retained 1975 Federal interim standards for the 1977 model year. [8] The suspension has not affected California's right to implement more stringent standards in the 1977 year. [9]

[3] Measuring the benefits of air pollution control is an extremely difficult and complex undertaking. It is complex in that an analysis of the benefits due to reductions in ambient concentrations of air pollutants requires a synthesis of information from several disciplines. Benefits measurements depend on (1) the availability of physical information defining relationships between emissions and ambient concentrations of pollutant; (2) the availability of physical, biological, and medical information defining relationships between ambient concentrations of pollutant and adverse effects; and (3) the availability of economic information concerning society's willingness to pay to avoid those adverse effects. The difficulty in making benefits measurements stems from the fact that much of the necessary information is not available.

[4] In a recent study, the Coordinating Committee on Air Quality Studies of the National Academy of Sciences - National Academy of Engineering developed an estimate of $5 billion annually as the value of the benefits of reducing automobile emissions. [6] This study did not give any consideration to the potential benefits of reducing automotive lead emissions.

fications that have been used since 1970 include lean combustion engines; electronic ignition; improved cold-start systems; better carburetion; spark retard; and exhaust gas recirculation (EGR). Of these, EGR is of particular importance because of its effectiveness in reducing the formation of NO_X. NO_X formation stems from high combustion temperatures. EGR acts to reduce peak combustion temperatures by cooling and recirculating part of the exhaust gases back through the intake manifold. Auto makers began using EGR on 1974 model cars.

(2) *Adding exhaust treatment devices to clean up engine emissions before discharge:* There are two basic classes of exhaust treatment devices, thermal reactors and catalytic converters. The thermal reactor is a high-temperature, engine exhaust manifold that completes oxidation of hydrocarbons (HC) and CO. The catalytic converter is a muffler-like device containing catalytic materials, usually noble metals (platinum and/or palladium), that chemically promote the oxidation of HC and CO and/or the reduction or NO_X.[5]

(3) *Developing alternative low-emissions designs for the conventional spark-ignition internal-combustion engine:* There are three engine systems that are basically different from the conventional spark-ignition system and are potential candidates for large-scale production within the next 5 years: the diesel engine, the Wankel rotary engine, and one or more versions of the stratified-charge engine.

(4) *Developing alternative low-emissions propulsion systems to replace the internal-combustion engine:* Gas turbine, steam, Stirling engine and electrically powered vehicles are all examples of alternative propulsion systems. Large-scale production of the first three is unlikely until well into the 1980's. Low-performance electric vehicles (urban cars) may be mass-produced during the next 5 years, but these cars cannot compete with the current combustion engine vehicles in highway driving.

Enactment of the Clean Air Amendments in December 1970 gave automobile manufacturers less than 4 years before introduction of 1975 models to develop emissions control systems that would meet 1975 standards. Because certification testing of 1975 model production prototypes necessarily had to begin several months before the start of the model year, production design details had to be frozen by early 1974 at the latest.

Faced with this time constraint, the domestic automobile industry was limited to the selection of an emissions control strategy incorporating engine modifications plus ''add-on'' exhaust treatment systems; this was the only feasible option for producing large numbers of vehicles that would meet 1975-76 statutory standards. Although the manufacturers did not rule out the possibility of developing a practical, low-emissions alternative to the conventional internal combustion engine, in 1970 they testified before Congress that the short timetable for compliance required by the Clean Air Amendments did not permit sufficient lead-time for development, tooling, and mass-production of an entirely new engine. [18]

Even without this stiff timetable, the automobile industry is facing certain obvious technical and economic constraints that have made the adoption of a modifications-plus-treatment-systems strategy a preferable approach to emissions control. By adopting this kind of strategy, the automobile industry has preserved as much as possible of the conventional spark-ignition, internal-combustion engine technology. In the process of protecting engine technology, the manufacturers also have protected their investments in existing engine production facilities; they have prolonged the use of existing tooling for as long as possible. Conversely, with the add-on strategy, the automobile industry has avoided making large expenditures for research and development and for new plants and equipment that would be needed to manufacture a replacement for the current version of the internal combustion engine.

In selecting an emissions control system for 1975-76, the automobile industry relied on a backlog of research that had been done prior to 1970 on two classes of after-treatment systems—lead-sensitive catalytic systems and lead-tolerant thermal reactors. As of 1970, these systems were the leading contenders for use on 1975 model cars. Of the two, catalytic systems exhibited some definite technical advantages over thermal reactors for use in meeting 1975-76 statutory standards.

First, in terms of control efficiency (i.e. the percentage of engine emissions removed by the control system before exhausted into the atmosphere), oxidation catalysts demonstrated greater facility than thermal reactors in achieving 1975 statutory emissions levels for HC and CO. A major difficulty with early versions of reactor systems involved achieving high enough reactor temperatures to oxidize these pollutants effectively. By using fuel-rich carburetion and air injection, reactor temperatures can be increased to the point where emissions levels approach 1975 statutory levels, but high temperatures cause more severe durability problems than in reactors that are operated under lean-fuel conditions. [10, 12] Operating under fuel-rich conditions also necessarily results in decreased fuel economy.[6]

[5]There are at least three distinct catalytic converter systems in use or under development: (1) the single-bed oxidation catalytic system; (2) the dual-bed, oxidation-reduction catalytic system; and (3) the three-way combination, single-bed system that simultaneously oxidizes and reduces emissions. Oxidation catalyst technology has been under development since the early 1960's and is much more advanced than the reduction catalyst technology. The durability and effectiveness of reduction catalysts also require more precise control of engine operating conditions than is achieved in current production vehicles. [17]

[6]Estimates of this fuel penalty range from approximately 25 to 30 percent relative to 1970 models [12].

Second, in terms of control flexibility (i.e. the ease of adapting 1975 emissions control systems for use in meeting 1976 standards), oxidation catalysts offered greater potential than thermal reactors for modification in 1976. With conventional engines, a combination of techniques will probably be needed to meet 1976 statutory standards. Oxidation catalysts or thermal reactors must be coupled with EGR and a reduction catalyst to achieve required NO_x levels.

Rather than attempt to develop two completely different systems for 1975 and 1976, domestic manufacturers emphasized catalytic systems because of the relative ease of modifying, manufacturing, and utilizing them to achieve greater NO_x control. [10] For example, the dual oxidation-reduction catalytic system, on which most development efforts have been concentrated, utilizes the oxidation catalyst that manufacturers began installing in 1975 vehicles. From a manufacturing standpoint, the problems of producing oxidizing and reducing catalysts are essentially the same, and production techniques and facilities are somewhat interchangeable between the two types of catalysts. Finally, from the standpoint of assembly, the dual system can be installed on vehicles as two separate catalytic beds, or both catalysts can be packaged in a single container.

Perhaps the overriding disadvantage of catalytic converters has been the fact that their performance is impaired when the combustion products of lead additives are present in the engine exhaust. [12, 19] Automobile manufacturers have known since the early 1960's that lead "poisons" catalysts. Lead particles and compounds adhere to the surfaces of the active ingredients, causing rapid deterioration of catalytic action.

Lead emission products also impair the performance of thermal reactors and EGR. [12] The metallic alloys and ceramic materials in thermal reactors erode and corrode more rapidly when the emissions products of lead and phosphorus are present in the exhaust, although there is some indication that leaded gasoline without phosphorus produces less corrosion. Lead particles reduce the effectiveness of EGR by forming deposits that alter the air-flow characteristics of the system.[7]

Given the economic-engineering advantages of the add-on strategy and, given the particular advantages of catalysts, automobile manufacturers have actively encouraged production and distribution of unleaded gasoline on a nationwide basis in order to facilitate the introduction of catalytic emission control systems. In early 1970, General Motors, Ford, and Chrysler announced that compression ratios on 1971 and later model cars would be lowered in preparation for meeting 1975-76 emissions standards; [20] engines with lower compression ratios would be able to run on lower octane unleaded fuels. During legislative hearings before the Senate Subcommittee on Air and Water Pollution, the major manufacturers also presented evidence describing the deleterious effects of lead on emissions control systems and requested that unleaded gasoline be made available. These actions essentially placed the burden on the Congress and the petroleum companies to ensure the availability of a fuel that was compatible with the kind of emissions control systems Detroit had in mind for 1975-76.[8]

In late 1971, EPA made a technical assessment that confirmed the automobile industry's commitment to catalysts for meeting statutory 1975-76 standards. [12] The major conclusions of this study were:

(1) All control systems that major automobile manufacturers were planning to use for meeting 1975-76 emissions standards incorporated catalytic converters.
(2) The emissions products of lead additives (even in small amounts) in gasoline greatly reduced a catalyst's ability to control emissions of carbon monoxide, unburned hydrocarbons, and nitrogen oxides.
(3) Lead traps for removal of lead particulates from exhaust gases prior to passage through a catalyst were not effective in preventing damage to the catalyst.
(4) Unleaded gasoline should, therefore, be made available in sufficient quantities to satisfy the demands of vehicles equipped with catalytic converters.

Based on these findings EPA ordered the general availability of unleaded gasoline. The Agency concluded that without regulatory action, the supply of unleaded gasoline would be uncertain in all parts of the country and insufficient to assure protection of catalytic devices.

In April 1973, EPA suspended 1975 statutory HC and CO emissions requirements for one year. In place of these, the Agency implemented two sets of less stringent interim standards, one set for California and a second set for the rest of the United States. The 1974 Energy Supply and Coordination Act (ESECA) maintained these interim standards through the 1976 model year. In March 1975, EPA further delayed the enforcement of the 1975 statutory requirements; for the 1977 model year, the Agency retained the 1975-76 interim emissions levels for HC and CO. The NO_x emissions standard for 1977 models was reduced to 2.0 g/mi as mandated by ESECA.

[7]Lead emission products also cause greater engine deterioration in the combustion chamber, decreased spark-plug lifetime, decreased exhaust system lifetime, and increased lubricating oil contamination.

[8]During 1970, the automobile industry's relatively active approach to the legislative and regulatory process represented a significant departure from its previous behavior toward government regulation of its products. According to White, [21] before 1970 the industry generally had resisted any regulatory attempts either to curb automobile air pollution or to increase passenger safety. In the face of regulatory constraint, the auto companies had often dragged their feet and had resisted the imposition of government restrictions altogether.

Although several noncatalytic approaches were used to meet 1975-76 emissions standards,[9] since 1975 domestic automobile manufacturers have used lead-intolerant oxidation catalysts as the primary means of emissions control. U.S. auto makers installed catalytic converters on approximately 80 to 85 percent of the 1975 model production and required the use of unleaded gasoline in over 95 percent of that year's production.[10] In 1976, the rate of catalyst installation increased to nearly 100 percent of domestic new car production.[11] For the 1977 model year, U.S. manufacturers installed catalysts or required the use of unleaded gasoline in all new cars. [16]

Manufacturers are using catalysts because, at interim requirements, catalytic add-on systems have some definite advantages over noncatalytic approaches to control emissions from a conventional internal combustion engines.[12] The major advantage is that the use of catalysts permits changes in engine operation that result in improvements in fuel economy. The installation of catalytic converters on 1975-76 models permitted substantial improvements in fuel economy over equivalent 1973-74 models.[13] In comparison, the use of a noncatalytic add-on approach to meet Federal interim standards resulted in further reductions in fuel economy from 1973-74 levels. (The relationship between fuel economy and emissions control is discussed more completely later in this chapter.)

The energy crisis of 1973-74 has raised improving fuel economy to a priority status that is above controlling emissions. Automobile manufacturers are now giving greater emphasis to improving fuel economy as a design objective. The order of priorities within the industry currently runs as follows: (1) improving fuel economy, (2) controlling emissions, (3) improving safety, and (4) reducing costs. [22]

Because of the importance of fuel economy, there is a high probability that virtually all domestic automobile production for the 1978, 1979, and 1980 model years will be equipped with catalysts. Better fuel economy is achievable with a catalyst system than without one, on any engine system that domestic manufacturers are likely to mass produce during the next 3 years. Consequently, domestic producers will use catalytic systems unless they are ruled out by regulatory action.

Although the widespread use of oxidation catalysts is likely to continue, industry's (and EPA's) commitment to this technology has been controversial for at least two reasons. First, it was discovered in 1973 that catalysts, while oxidizing HC and CO, have an unintended side effect; they also oxidize some of the sulfur dioxide in engine exhausts to sulfate particulates and sulfuric acid, with the result that the average catalyst car emits more sulfates than the average noncatalyst car. [23] Second, there is concern that the use of catalyst technology—especially reduction catalysts—may be an extremely costly approach to achieving further reductions in automobile emissions.

Although automobiles have not been considered a major source of sulfur dioxide, the sulfate discovery raised the question of whether emissions from 1975 and later model cars would present a health hazard by increasing sulfate levels in the vicinity of roadways. In Congressional hearings on this matter [24], EPA Administrator Russel Train testified that a deferral of 1975 interim standards based on the potential health effects of sulfate emissions would postpone further reductions in automotive emissions for an additional year. Train indicated that such a postponement would be undesirable, primarily because it would delay the achievement of primary ambient air quality standards.

Train pointed out that a deferral of 1975 emissions standards would also require a prohibition of the use of catalysts. This would be necessary because some automobile manufacturers had indicated that fuel economy improvements with catalysts made their use attractive even for 1974 emissions requirements. Accordingly, Train maintained the interim emissions standards as promulgated and did not take any action to inhibit the use of catalysts.

In March 1975, EPA granted a 1 year suspension of the 1977 federal emissions standards because the Agency decided that the potential health hazards of sulfate emissions from catalysts were great enough

---

[9] Chrysler and Ford certified conventional engines with EGR, spark retard, and air pumps for 1975-76 standards. Three foreign manufacturers are producing low-emission alternatives to the conventional internal combustion engine that have met these standards. These are the diesel engine (Daimler-Benz), the prechamber, dual-carburetor, stratified-charge engine with air injection and lean thermal reactor (Honda), and the Wankel rotary engine with a thermal reactor (Toyo Kogyo).

[10] All General Motors' domestic 1975 model cars were equipped with catalysts. All of Chrysler's and Ford's 1975 model cars offered for sale in California were equipped with catalysts. Approximately 94 to 95 percent of Chrysler's 49 state models had catalytic converters. At the beginning of the model year, 67 percent of Ford's 49 state production were equipped with catalysts, but all Ford cars—catalyst as well as noncatalyst—required unleaded gasoline for better HC control. Ford increased the percentage of its 1975 production with catalysts as the model year progressed in order to improve fuel economy.

[11] For 1976, all General Motors and Ford passenger vehicles were equipped with catalysts. Chrysler's 1976 production has remained essentially the same as 1975, with the exception that the company introduced a lean-burn engine that does not require the use of a catalyst. Catalyst-equipped cars will probably comprise 94 to 95 percent of Chrysler's 49 state production in 1976.

[12] The 1975 Federal interim standards are nearly four times less stringent than the original 1975 statutory emissions standards for HC and CO. At these higher levels, oxidation catalysts are also attractive for technical reasons. First, the certification test cycle is less demanding on catalytic performance to achieve the necessary control efficiencies. Second, thermal degradation of the catalyst is likely to be less of a problem because the system can achieve interim standards at lower temperatures than are necessary for statutory standards. [17]

[13] In meeting 1973-74 emissions standards, the automobile industry used a combination of EGR and spark retard. For each EGR flow rate, spark firing was retarded from the point at which optimum fuel economy is achieved in order to reduce hydrocarbon emissions. With 1975-76 models, manufacturers advanced spark timing with EGR, increasing fuel economy. Catalytic converters control the accompanying increases in HC emissions associated with spark advance.

to warrant substantial delays in implementing more stringent control requirements for HC and CO. With the oxidation catalysts now in use, the greater the control efficiency of the catalyst, the more sulfate is formed.

EPA regards the sulfate problem as serious and is considering several measures for controlling sulfate emissions, including the implementation of an emission standard and a regulation requiring the general availability of low-sulfur gasoline. A stringent emission standard may mean the demise of catalyst technology.

Since 1971, the Committee on Motor Vehicle Emissions (CMVE) of the National Academy of Sciences has monitored research and development of alternative systems for emissions control. CMVE's focus has been on assessing the technological feasibility, cost, and fuel use associated with meeting the statutory NO_x emissions standard of 0.4 g/mi.

Reports by the Academy [10, 11, 25] have indicated that the statutory NO_x standard can probably be achieved by 1978, as currently required, with catalyst approaches; however, a commitment to catalytic technology at this time may be very expensive in terms of resource utilization. Since no catalytic system has yet to achieve the 50,000 mi durability requirement, there is considerable chance that several catalyst changes will be required in 50,000 mi as part of the manufacturers' specified maintenance. The costs of these changes may add several hundreds of dollars to the lifetime maintenance costs of a vehicle. Catalytic systems are also likely to result in an increase of 300 to 400 dollars in the sticker price and a fuel economy penalty of up to 10 percent.

CMVE [10, 11] also investigated several alternative approaches for emission control, including the prechamber, dual-carburetor, stratified charge engine with thermal reactor; the lean-burn engine with thermal reactor; the diesel engine; and the direct-fuel-injection, stratified charge engine with oxidation catalyst. These systems are capable of meeting statutory HC and CO emissions standards, as well as standards for NO_x at least as low as 1.5 g/mi. At these emissions levels, these systems should provide better fuel economy than conventional engines equipped with dual or three-way catalysts. CMVE has indicated that these systems could be available for limited mass production in the early 1980's, but strict adherence to the statutory NO_x standard will discourage the development and use of these technologies.

The prechamber, dual-carburetor, stratified charge engine and the lean-burn engine are of interest for this discussion because these systems show promise of achieving substantial reductions in current emission levels without requiring a lead-free fuel.[14] Both engines operate on very lean air-fuel mixtures and make use of recent improvements in carburetion techniques in order to achieve smooth combustion. Engines that operate on the lean side emit relatively low levels of NO_x and CO. Emissions of HC tend to rise as the air-fuel ratio increases, due to misfires. Conventional lean-burn and stratified charge engines, therefore, require a thermal reactor or catalytic converter for HC control. Because of the lower exhaust temperatures of lean-burn designs, thermal reactor performance is limited, but a lean-burn engine with a reactor can probably meet statutory HC and CO requirements. [11] Lean-burn engines also have the advantage of achieving improvements in fuel economy over comparable 1970 vehicles.

Improvements in carburetion techniques have also been used as an alternative to catalysts in achieving lower emissions from conventional spark-ignition engines. [11] Combination air-fuel-and-feedback-control systems have been developed in which the feedback control monitors the composition of the exhaust or some other engine parameter and provides a corrective signal back to the primary air-fuel metering device. The corrective signal maintains the air-fuel ratio within the limits required for low emissions. Saab-Scania Aktiebolag has certified a lead-tolerant system of this kind to meet the 1975-76 California emissions standards. [15]

These new systems bracket a range of emissions standards in which lead-tolerant emissions control devices can be used. On the upper end of the bracket, conventional engines with improved carburetors can meet the California standards of 0.9, 9.0, and 2.0 g/mi. for HC, CO, and NO_x respectively. On the lower end, alternative low-emissions engines can meet levels of 0.41, 3.4, and 2.0 g/mi. for the three pollutants. If future standards are set within this range, the widespread adoption of one or more of these systems could ultimately eliminate the current need for unleaded gasoline.

The status of automobile emissions standards for 1978 and beyond is uncertain at this time. Since the spring of 1975, Congress has deliberated on amendments to the Clean Air Act that will extend the final compliance date for statutory emissions of HC and CO. A Senate bill would extend the 1977 standards until 1978, and statutory HC and CO limits would go into effect in 1979. An NO_x standard of 1.0 g/mi would go into effect in 1980, with the statutory level of 0.4 g/mi becoming a long-term research objective. The House version would extend 1977 standards for 2 years, and statutory standards would not go into effect until 1982. Although each bill was approved, both houses failed to agree on a final form of the amendments before the 1976 adjournment; amendments are expected to be reintroduced in the 1977 session.

Industry will not select the emissions control systems for 1978 and beyond until Congress has enacted standards for those years. From an economic standpoint, because long-term emissions standards are more important than a temporary delay or relaxation of current standards, the automobile industry has taken a wait-and-see attitude during these legislative proceedings. Manufacturers have stressed the

[14]The CMVE report did not discuss the recent development of a lead-tolerant oxidation catalyst by Chrysler and du Pont. Laboratory tests have indicated that these catalysts have lower than desirable hydrocarbon conversion efficiencies. [26]

importance of making near-term plans that are compatible with long-term plans, and they cannot do so without knowledge of what the long-term standards will be. Manufacturers do not want to risk entering into a development or production program that they believe will not be compatible with future standards or will not be viable if there is a sudden shift in regulatory policy.

If the domestic automobile industry adopts a lead-tolerant, low emissions engine technology, such as the lean-burn or stratified charge engine, manufacturers will not be able to convert a large percentage of production to this technology until the early 1980's. Before production can begin, manufacturers must complete the development of the engine and design, and install and debug production line tooling. Estimates of the length of time to complete these operations are on the order of 4 to 6 years. [11] Even after production begins, production and cost constraints may limit the rate of phase-in for a new engine.

Because of lead-time and investment considerations, automobile manufacturers may be hesitant to introduce a new engine—lead-tolerant or otherwise—without certain guarantees. For example, the introduction of a new engine may be feasible only if there is a freeze on the NO_x standard at 1.5 to 2.0 g/mi for 10 years. The development of new engine systems is considerably less advanced than catalytic systems for achieving NO_x emissions below 1.5 g/mi. This gives catalysts an advantage over alternative systems if the NO_x standard is reduced very rapidly. A long freeze on standards also has important economic ramifications in that it allows automobile manufacturers to obtain a reasonable rate of amortization for the substantial investment requirements.

In summary, domestic automobile manufacturers adopted a strategy of using add-on, lead-intolerant, oxidation catalysts to meet 1975 Federal emissions standards. Given the time and economic constraints, an add-on strategy provided the only feasible means of meeting 1975 standards on schedule, and the catalytic converter had demonstrated technical advantages over other alternatives. In spite of the fact that standards have been relaxed to the point that would permit the use of lead-tolerant systems, domestic automobile manufacturers continue to use catalysts because at present standards these engines can be adjusted so that fuel economy is improved.

In all likelihood, catalysts will be installed on a majority of cars manufactured in the 1977 through 1980 model years. If a new engine system were adopted, 4 to 6 years would be needed before a large percentage of vehicles incorporating that technology could be produced. Unless the NO_x standard remains above 1.5 g/mi for at least 10 years, it is unlikely that a lead-tolerant emissions system will be introduced.

Health Basis for Low-Lead Gasoline

Under the Clean Air Act, EPA also may control or prohibit the manufacture and sale of any fuel or fuel additive for use in motor vehicles if the emission products of such fuel or fuel additive will endanger public health. In contrast to the technical basis for ordering unleaded gasoline, the rationale for the implementation of the phased-reduction schedule was based entirely on public health considerations. On the basis of an evaluation of available scientific and medical information, EPA concluded that environmental lead exposure is a major public health problem, with present levels of lead exposure constituting a sufficient risk of adverse physiological effects for a small but significant portion of the urban adult population and up to 25 percent of the children in urban areas. [5]

The combustion of leaded gasoline has contributed the largest fraction of lead currently reaching the environment. From 1968 to 1974, over 250,000 tons of lead have been used each year to produce gasoline additives. [27][15] It has been calculated that about 70 percent of the lead in gasoline is emitted into the environment as particulate matter in automotive exhausts. Automotive lead emissions account for about 90 percent of airborne lead emissions except in lead mining and smelting areas. [28]

After leaving the exhaust pipe, the heaviest lead particles fall to the ground within a few hundred feet of the roadways, [28] whereas the lighter particles remain suspended in the air for some time. Lead fallout contributes to the concentrations in street dust and dirt, which may be resuspended in air and transported into homes. Lead in dirt and dust in streets and homes represents a potential source of lead, particularly for children.

Lead exposure results from a combination of sources, including lead in the air, food, water, leaded paint, and street dust; and it is difficult to determine which source(s) is (are) the most significant from a health standpoint. Although ingestion of lead-based paint chips is regarded as the major cause of overt lead poisoning in children, sources of lead other than paint may play an important role in childhood lead exposure, especially at subclinical levels of lead toxicity.

With respect to lead in dust and dirt, EPA concluded that these sources contribute to increased lead levels in human beings, both through inhalation of resuspended dusts and, in children, through inadvertent ingestion of dust and dirt. The Agency believes that the availability of lead from this source significantly reduces the quantity of lead required to produce clinical lead poisoning in a child who is exposed to other sources of lead. Finally, because automotive lead is a major contributor to lead in dust and dirt, and because lead in dust and dirt has been related to undue absorption of lead in children, EPA has advocated that lead in gasoline should be reduced to the degree possible. [5]

[15]U.S. consumption of lead for lead additives production reached a peak of 278,000 tons in 1970. It declined in 1971, but rose again to 278,000 tons in 1972. Since 1972, the use of lead in lead antiknocks has declined at an accelerating rate, with consumptions of 274,000, 250,000, and 208,000 tons in 1973, 1974, and 1975, respectively. The drop from 1974 to 1975 indicates the impact of the catalytic converter.

As promulgated, the phased-reduction schedule will reduce the lead content of gasoline by prohibiting refiners from exceeding a specified *average* lead content per gallon. Each refinery will be permitted to average its lead usage over its quarterly production of all grades of gasoline—regular, premium, and unleaded. The average lead content of the total pool must not exceed the total pool standard that will decline over a 5 year period. The objective of the phased-reduction schedule is to reduce the 1971 level of lead utilization in gasoline by 60 to 65 percent.

The proposed and revised regulations prescribing the lead content of gasoline are summarized in table 14-2. The 1974 date for introducing unleaded gasoline corresponds with the beginning of the new model year for 1975 automobiles. The revised phased-reduction schedule that averages over all grades of gasoline has been moderated somewhat in the early years, but extended for an additional year through 1979.

The health basis for reducing the lead content of gasoline has been challenged from the beginning. In 1973, during administrative hearings on the phased-reduction proposal, the National Petroleum Refiners' Association, [31] the International Lead Zinc Research Organization, [32] and the major producers of lead antiknocks, Ethyl Corp. [33] and E. I. du Pont de Nemours and Co., [34] all contested EPA's position that automotive lead emissions are hazardous to public health. After the phased-reduction schedule was implemented, Ethyl, du Pont, National Petroleum Refiners Association, Nalco Chemical Co., and PPG Industries, Inc., all filed separate petitions in the U.S. Court of Appeals requesting that the Court review and ultimately set aside EPA's basis for reducing lead in gasoline.

In December 1974, the U.S. Court of Appeals for the District of Columbia set aside the phased-reduction schedule on the grounds that EPA had not established that automotive lead emissions were a significant public health hazard. [35] In February 1975, EPA filed a petition requesting a rehearing of the case by the 9 justice Court of Appeals. The petition argued that the Court had erred in concluding that the Clean Air Act requires the Agency to establish, with a high degree of certainty, how much lead is taken into the body from automobile emissions and exactly what danger that lead causes. [36] In March 1975, the Court vacated its initial decision and subsequently reheard the case. [37]

In March of 1976, the Court of Appeals, in a 5 to 4 decision, reversed its earlier ruling and held that EPA had acted within its statutory authority in implementing the phased-reduction schedule. [37] Manufacturers of lead additives appealed the case to the U.S. Supreme Court; in June 1976, the Supreme Court denied the petition for *certiorari.*

In July 1976, EPA announced that enforcement of the phased-reduction schedule, which had been suspended while court action was pending, would begin at the 1.4 g/gal. level in accordance with the promulgated schedule for 1976. [38] Because of the delays and uncertainties caused by the litigation, EPA also provided the petroleum industry with an opportunity to once again comment on the impact of the phased-reduction schedule on refining operations. During the delay, refiners had not proceeded to install additional refining capacity to produce the larger quantities of high-octane components needed for unleaded and low-lead grades of gasoline. Specifically, EPA sought evidence concerning an assertion by refiners that because of rapid increases in gasoline consumption since 1975, enforcing the phased-reduction schedule would make it impossible to meet projected summertime demands in 1977 and 1978.

EPA received a large number of responses indicating that many refiners would not have sufficient octane-producing capability to meet anticipated demands at the interim lead phase-down concentrations. EPA contracted with a consulting firm to analyze these responses and compared the consultant's report with an independent assessment by the Federal Energy Administration of the impact on motor gasoline supplies of EPA's lead phase-down. The EPA consultant and FEA concluded that enforcing the phased-reduction schedule as promulgated would result in a gasoline shortage of about 6.6 percent in 1977 and 9.0 percent in 1978.

In view of this evidence, the Environmental Protection Agency has suspended enforcement of the phased-reduction schedule through 1977. In addition, because meeting the 0.8 g/gal. level for 1978 (table

TABLE 14-2.—*U.S. Environmental Protection Agency restrictions on the lead content of gasoline*

Research octane number	TEL (g/gal.) 1974	1975	1976	1977	1978	1979	1980
Unleaded 91[1]	0.05[2]	0.05	0.05	0.05	0.05	0.05	0.05
Regular 94[3]	2.00	1.70	1.50	1.25	1.25	1.25	1.25
Premium 100[3]	2.00	1.70	1.50	1.25	1.25	1.25	1.25
Average[4,5,6]	–	1.70[7]	1.40[7]	1.00[7]	0.80[7]	0.50	0.50

[1] The octane requirement for unleaded gasoline is not less than 91 research octane number.

[2] Federal regulations define "unleaded gasoline" as gasoline containing not more than 0.05 g of lead/gal. and not more than 0.005 g of phosphorus/gal. Phosphorus emissions products also cause deterioration of catalytic devices.

[3] Proposed on Feb. 23, 1972. [29]

[4] Proposed on Jan. 10, 1973. [30]

[5] Finalized on Dec. 6, 1973. [2]

[6] The allowable average lead content per gallon of gasoline is determined by dividing total lead by total gasoline production of regular, premium, and unleaded grades of gasoline.

[7] Suspended on Sept. 28, 1976. [4]

14-2) will require about the same construction lead time as the 0.5 g/gal. standard for 1979, EPA will suspend the 1978 phase-reduction level for those refiners who have made sufficient progress in procuring and installing new capacity so as to meet the 0.8 g/gal. level at the earliest practicable date. The 0.5 g/gal. standard will now go into effect on October 1, 1979. [4]

Prior to 1975, the national average lead content of regular and premium grades of gasoline had declined steadily. By the end of 1975, the average lead concentration in regular had declined by over 27 percent from its 1970 peak. The decline in the concentration of lead in premium was only slightly less (figure 14-1).

Lead concentrations declined because petroleum refiners built more octane-processing capacity in anticipation of the production of automobiles that would run on unleaded gasoline. Unleaded gasoline requires essentially the same high-octane components that are currently blended into premium gasoline. Because unleaded gasoline sales were very low before 1975, oil companies were able to blend part of this higher octane production into regular and premium, reducing the need for lead in those grades. (A complete discussion of the role of lead additives is presented later in this chapter.)

Since 1975, domestic automobile manufacturers have installed lead-intolerant catalytic converters on virtually all cars produced in the United States. Because these cars must use unleaded gasoline, sales of this fuel have been increasing very rapidly. In addition, improvements in economic conditions during 1975-76 have also generated an overall increase in the demand for gasoline.

Because of these two factors, gasoline is being produced near refinery capacity levels, reducing refinery processing flexibility and increasing the demand for lead as an octane booster. As a result, in 1976 the concentration of lead in premium and regular grades was expected to be higher than 1975 levels. Lead concentrations were also expected to increase slightly in 1977 and 1978, depending on the

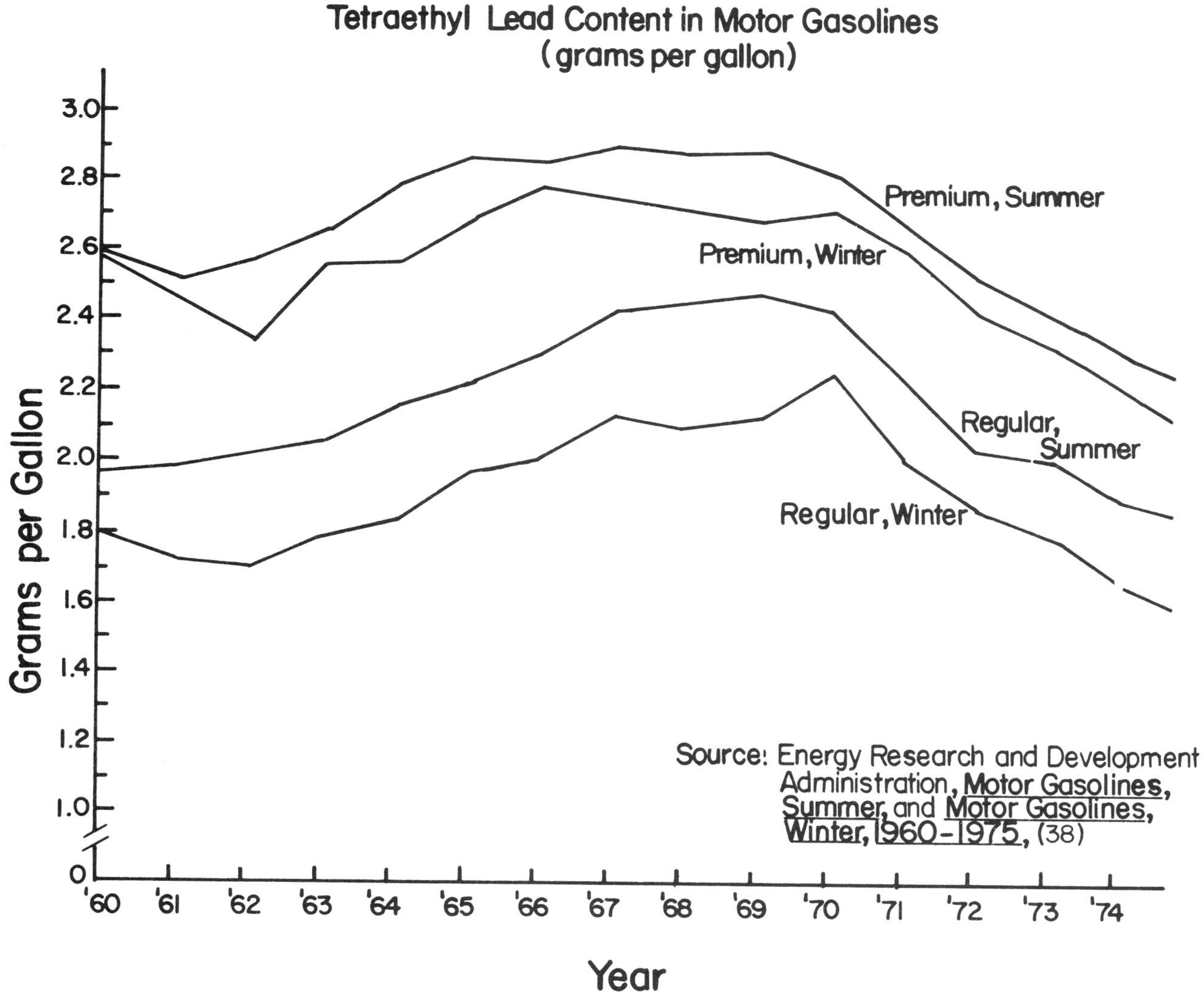

FIGURE 14-1.—Tetraethyl lead content in motor gasolines.

strength of the economy and on sales of unleaded gasoline.

The national average lead content of all grades of gasoline has, of course, declined more rapidly as sales of unleaded gasoline have increased. For the same period, the average lead content of gasoline has decreased by approximately 0.9 g/gal. or about 34 percent. Estimates of the average lead content of all gasoline for the winter and summer of 1975 were less than the requirements of the phased-reduction schedule that was not enforced during this time. (table 14-3).

As the fraction of the automobile population that is equipped with lead-intolerant catalysts increases, sales of unleaded gasoline will increase, and the average lead content of gasoline will decrease. If automakers continue to use catalysts on virtually all new cars, lead emissions will decline as old vehicles that burn leaded gasoline are retired. Finally, if the average lead content continues to decline more rapidly than the phased-reduction schedule, the EPA regulation may well be redundant for many regions of the country.

LEAD ADDITIVES, FUEL ECONOMY, AND EMISSIONS

The energy requirements for automotive transportation have grown tremendously since World War II, with the result that the family car now consumes more energy than all other modes—passenger and freight—of transportation combined. In addition, virtually all of the energy used by automobiles comes from petroleum. Because the use of lead in gasoline enabled automobile manufacturers to produce more efficient engines, removal of lead from gasoline meant reductions in automotive fuel economy. This has resulted in an adverse impact on total petroleum consumption.

Lead Additives, Fuel Economy, and High-Compression Engines

Prior to 1970, lead alkyls played an important role in improving engine fuel economy. Lead additives provided an inexpensive means of increasing gasoline octane rating, which is a measure of that fuel's ability to resist engine knock. Improvements in the knock-resistant quality of gasoline, in turn, permitted improvements in the operational efficiency of high-compression engines.

Engine knock occurs when engine temperatures become hot enough to ignite the air-fuel mixture prematurely, causing uneven combustion and inefficient engine operation in terms of fuel consumption. Increases in operating temperatures accompany increases in engine compression ratio, and because increasing the compression ratio improves efficiency, engine knock places an upper limit on efficient engine design.

Of the different engine specifications that influence fuel economy (compression ratio, displacement, and air-fuel ratio), it is well established that increasing the compression ratio, holding all other engine specifications and vehicle performance[16] constant, is one way of producing better gasoline mileage. Increases in compression ratio up to 14:1 will produce increases in an engine's ability to do work, either with greater efficiency or with greater power per gallon of gasoline consumed.

As the compression ratio increases, gains in engine efficiency or performance can be realized only if the corresponding motor fuel octane number increases. (Fig. 14-2 illustrates a relationship between octane number and the amount of work an engine can do.) Conversely, as the compression ratio decreases, engine octane requirements and engine efficiency decrease. If performance is held constant, a reduction in the compression ratio generally results in an even larger decrease in fuel economy because the most direct ways of maintaining constant performance—for example, increasing the rear axle ratio or increasing engine displacement—require additional fuel.

In the 10 years preceding 1971, compression ratios increased on all domestic model cars from a sales-weighted average of 8.8:1 in 1962 to a high of 9.3:1 in 1966 before declining slightly to 9.1:1 in 1970. [42] During this period, national average fuel economy for passenger cars declined from 14.4 mi/gal. in 1962 to 14.0 mi/gal. in 1966 and to 13.6 mi/gal. in 1970. [43] This decline of 5.5 percent in average fuel economy is rather slight, considering that the sales-weighted average inertia weight of domestic cars in-

[16]Performance refers to the acceleration capability of a car as measured by the number of seconds required to reach a particular speed.

TABLE 14-3.–*Estimated annual average tetraethyl lead content of automotive gasolines, 1968-1975*[1]

Grades	1968	1969	1970	1971 (g/gal.)	1972	1973	1974	1975 (W)[2]	1975 (S)[2]
Premium	2.80	2.79	2.76	2.63	2.48	2.38	2.26	2.10	2.23
Regular	2.29	2.31	2.35	2.12	1.96	1.92	1.80	1.58	1.85
Total[3]	2.50	2.51	2.53	2.27	2.09	2.01	1.86	1.55	1.66

[1] Grams per gallon by grade averaged by percentages of summer and winter gasoline sales by grade. See U.S. Bureau of Mines [39] and Ethyl Corp. [40]

[2] Winter and summer.

[3] Grams per gallon averaged by percentages of premium and regular to total gasoline sales. 1971 to 1974 assume 3 percent of total sales are unleaded; 1975: winter 10.6 percent unleaded, and summer 14.1 percent unleaded.

creased by 7.8 percent and that sales-weighted average engine displacement increased by 23.7 percent. [42] In addition, the number of cars equipped with energy-demanding accessories such as air conditioners, automatic transmissions, power brakes, and power steering also increased during this period. Tables 14-4 and 14-5 indicate the increases in sales-weighted average inertia weights and in the percentage of cars equipped with power accessories during the period from 1968 to 1973.

Detroit reduced compression ratios on 1971 models to accommodate low-octane, low-lead, and unleaded gasolines in preparation for the introduction of catalytic converters on 1975 and 1976 models. Further reductions in compressionratios were also made in 1972 and 1973 models. On an average sales-weighted basis, the compression ratios of domestic cars dropped from 9.1:1 to 8.65:1 in 1971, to 8.45:1 in 1972, and to 8.3:1 in 1973. [45]

These reductions in compression ratios resulted in losses in fuel economy and performance. Other changes in 1971 through 1974 models—such as the use of spark retard for better HC control and an increase in vehicle weights resulting from efforts to meet Federal safety standards—also had adverse effects on fuel consumption rates. Of these, increases in vehicle weights were the single most important factor that affected fuel economy (table 14-6).

Some studies [45-49] have attempted to isolate the relative importance of the effects of specific changes—a reduction in compression ratio or an increase in displacement for example—on fuel economy and vehicle performance. Several studies [46-48] have estimated the total combined effects of emissions controls and of changes in vehicle weight on fuel economy.

Estimates of the relative effect that a drop in compression ratio has on the fuel economy of an individual vehicle depend on vehicle weight, vehicle performance, and test conditions. Huebner and Gasser [49] reported that a 10 percent change (9.5:1 to 8.6:1) in the compression ratio of an intermediate car (3,600 lb) resulted in a loss in fuel economy of 2.6 percent (0.3 mi/gal.) to 3.0 percent (0.5 mi/gal.) and a decline in acceleration of 4 to 6 percent under urban stop-and-go driving and 70 mile-per-hour steady-state driving conditions, respectively.

These measurements correspond with other estimates of the change in fuel economy resulting from a change in compression ratio. For example, one estimate indicates that an increase in the compression ratio of one number would improve fuel economy by 3 percent and would require an octane increase of approximately 3 numbers. [50] CMVE [11] states that a reduction in compression ratio from 9.0:1 to 8.0:1 will produce a loss in fuel economy of 4 to 5 percent.

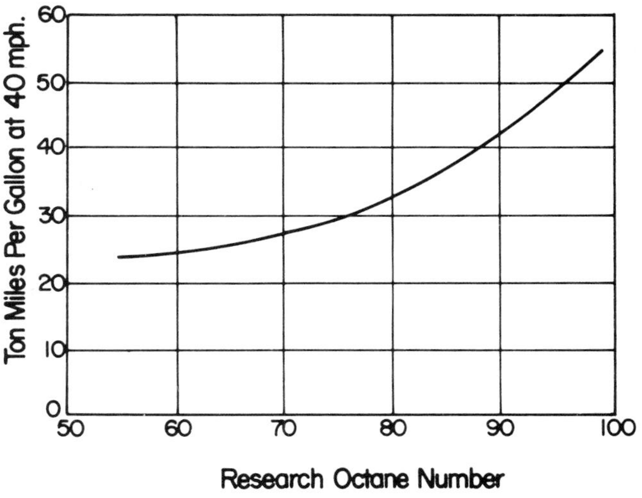

FIGURE 14-2.—Engine performance per gallon of fuel at specific octane levels [41].

TABLE 14-4.—*Characteristics of automobiles in the United States, 1968 - 1973*

Type	Sales-weighted average inertia weight by class (lb) [42]					
	Model year					
	1968	1969	1970	1971	1972	1973
Standard	4,342	4,575	4,639	4,735	4,805	4,904
Intermediate	3,737	3,748	3,985	4,043	4,049	5,285
Compact	3,274	3,202	3,225	3,331	3,383	3,406
Subcompact	2,351	2,343	2,393	2,427	2,551	2,571
Specialty	3,777	3,947	3,807	4,093	4,240	4,407
Total (average)	3,863	3,941	3,876	3,886	3,941	3,968

TABLE 14-5.—*U.S. production units with power accessories. [44]*

Accessory	Production units (percent)					
	Model year					
	1968	1969	1970	1971	1972	1973
Air conditioning	44.3	55.3	60.1	63.4	68.5	72.6
V-8 engine	86.4	88.8	88.4	80.2	79.2	83.5
Automatic transmission	89.0	90.2	91.3	91.9	90.5	93.4

TABLE 14-6.–*Inertia weight and national average fuel economy [48]*

Year	Sales-weighted annual average inertia weight (lb)	EPA City/Highway (mi/gal.)	U.S. Department of Transportation (mi/gal.)
1967	–	15.95	13.93
1968	3,863	15.79	13.79
1969	3,941	15.59	13.63
1970	3,876	15.43	13.57
1971	3,886	15.31	13.57
1972	3,941	15.15	13.49
1973	3,968	14.95	13.10
1974	4,015	14.74	–
1975	4,087	14.87	–
1976	4,032	15.07	–

Because the relative effect of a decline in compression ratio depends on vehicle size, the aggregate effect on fuel consumption depends on the sales mix of new car purchases. The sales mix has changed from one model year to another, with increasing penetration of smaller cars in the new car market in recent years. In making year-to-year comparisons, changes in aggregate fuel economy conceal the behavior of individual automobiles if the mix is allowed to vary. On the other hand, it is difficult to select an individual model that is representative of a typical automobile over a multiyear period because of abrupt changes in engine displacements, discontinuation of certain models, and differing effects of emissions control systems.

To circumvent these difficulties, La Pointe [45] held sales weights at 1967 levels in estimating changes in aggregate fuel consumption. Using test track measurements of the fuel economy of new vehicles, La Pointe estimated a total decline of 22.2 percent in fuel economy from 1967 to 1973. Of this amount, 13.2 percent (60 percent of the total decline resulted from emission controls.

La Pointe [45] estimated that the 1971 reduction in compression ratios resulted in a 3 percent decline in fuel economy over 1970 cars. He indicated that the overall 1971-73 sales-weighted reduction in compression ratios from 9.27:1 to 8.3:1 would theoretically cause a cumulative fuel penalty of 6 to 8 percent (25 to 35 percent of the total decline), relative to 1967 vehicles at constant performance. To the extent that the relative share of small cars in the automobile population has increased and to the extent that performance may have deteriorated in 1971 through 1974 models, La Pointe's estimates overstate the impact of the reductions in compression ratios.

To summarize, in preparation for the use of lead-sensitive catalytic converters, automobile manufacturers reduced engine compression ratios to accommodate lower-octane unleaded gasolines. The net result of this change was a decline in vehicle fuel economy and performance of approximately 3.0 to 4.0 percent, relative to 1970 models.

With the installation of catalytic converters, automobile manufacturers have reclaimed losses due to spark retard and have restored fuel economy to near-1970 levels. But in view of the enormous amounts of petroleum energy consumed by automobiles, even a small improvement in fuel economy will have a major impact on the demand for crude oil, domestic and imported. Therefore, the question remains as to whether or not it is feasible for the automobile industry to develop a lead-tolerant emissions control system that will meet Federal standards. If a system of this kind were feasible, it would enable manufacturers to make additional gains in fuel economy by restoring compression ratios to pre-1971 levels. There is also a related question concerning the importance of the phased-reduction schedule and whether the severity of the schedule would discourage development of an emissions control system that takes advantage of leaded gasoline. These questions are discussed below.

Lead-Tolerant Emissions Control Systems, Fuel Economy, and Emissions

The technical relationships between emissions control systems, control efficiency, and fuel economy are too extensive to discuss here. Of concern are the relationships between fuel economy and lead-tolerant emissions control systems and the possibility of increasing fuel economy by increasing the compression ratio.

The tradeoffs between engine emissions, fuel consumption, and compression ratio are also complex. While it is well established that increasing compression ratio increases NO_x formation, recent research also indicates that increases in compression ratio produce greater HC emissions. Based on laboratory studies using a single cylinder engine, Morgan and Hetrick [51] reported that fuel consumption decreased with increases in compression ratio at any given NO_x emissions level, and minimum fuel consumption was obtained at the highest compression ratio studied when HC emissions were not a constraint. However, if both HC and NO_x emissions were constrained at low levels through the use of EGR and spark retard, minimum fuel consumption was obtained at the lowest compression ratio studied. In addition, minimum fuel consumption at controlled HC and NO_x levels was obtained at the leanest air fuel mixture studied. On the basis of their research, Morgan and Hetrick concluded that the pre-1971 means of improving fuel economy by increasing compression ratio may not be feasible when both HC and NO_x must be controlled to meet emissions standards.

Emission control systems that have met or show promise of meeting different sets of emissions standards are listed in table 14-7. The probable fuel economy of each system relative to 1970 models is also listed. The lead-tolerant systems, which were discussed in the first section of this chapter, include conventional engines equipped with advanced carburetors and improved exhaust manifolds (III-b); lean-burn engines with improved carburetors and/or thermal reactors (V-f and V-g); and prechamber, dual-carburetor, stratified-charge engines (V-c).

Conventional engines with improved carburetors can meet current California emissions standards without using catalysts. At this level of control, the fuel economy of these lead-tolerant systems is about the same as with catalyst systems. [15] Any attempt to improve fuel economy through increases in compression ratio will result in increased NO_x and HC emissions. What improvements (if any) in fuel economy can be made, therefore, depend on the size of the improvement due to an increase in compression ratio relative to the reduction in fuel economy caused by incremental EGR and spark retard needed to control additional NO_x and HC.

TABLE 14-7.–*Emissions standards, fuel economy, and emissions control systems*

Standards HC/CO/NO_x (g/mi)	Fuel economy relative to 1970 model year		Emissions control system
1973-74 Federal 3.0/28.0/3.2	–6 to –15 percent for compacts to luxury cars, respectively	I.	EGR plus spark retard
1975-76 Federal 1.5/15/3.1	Approximately the same as 1970 vehicles of the same weight	II.	(a) Oxidation catalysts, EGR, electronic ignition, quick warm-up techniques.
	–2 to –20 percent for small to large cars, respectively		(b) Air pumps, spark retard, EGR
1975-76 California 0.9/9.0/2.0	–5 percent	III.	(a) Oxidation catalysts, EGR, electronic ignition, quick warm-up techniques.
	About the same		(b) Advanced carburetors plus improved intake and exhaust manifolds
1977 Federal (hypothetical) 0.9/9.0/1.0	–20 percent	IV.	(a) Oxidation catalysts, EGR, air pumps
	–5 percent		(b) Advanced carburetors, EGR improved manifold systems
	At least as good and possibly better		(c) Three-way catalyst or dual-catalyst systems
	About the same		(d) Pre-chamber, dual carburetor, stratified charge (CVCC) engines with EGR
	10 to 20 percent improvement, and from 5 to 10 percent improvement in small and large cars, respectively		(e) Direct-fuel-injection stratified charge (CCS) engines with EGR and oxidation catalysts
	10 to 20 percent improvement, and from 5 to 10 percent improvement in small and large cars, respectively		(f) Diesel engines with EGR
1977 Federal (suspended) 0.41/3.4/2.0	At least as good and possibly better	V.	(a) Oxidation catalysts, EGR, air pumps, cold start procedures
	At least as good and possibly better		(b) Three-way catalyst or dual-catalyst systems
	About the same		(c) Prechamber stratified-charge engines
	15 to 20 percent improvement		(d) Direct-fuel-injection stratified-charge engines with oxidation catalysts
	25 percent improvement		(e) Diesel engines
	About the same		(f) Lean-burn engines with improved carburetors
	Some improvement		(g) Lean-burn or prechamber stratified charge engines with thermal reactors
1978 Federal (statutory) 0.41/3.4/0.4	–2 to –3 percent	VI.	(a) Three-way catalysts plus EGR
	–5 to –10 percent		(b) Dual-catalyst systems
	–15 to –25 percent for small to large cars, respectively		(c) Prechamber, dual-carburetor, stratified charge engines plus EGR
	Approximately the same		(d) Direct-fuel-injection stratified charge engines

As CO and HC emissions levels are reduced below 1975-76 California standards, control of these pollutants becomes more difficult with both catalytic and noncatalytic systems. In addition, at any emissions level below the California standards, it is unlikely that lead-tolerant systems will exhibit superior fuel economy over lead-intolerant systems. [11, 15]

Using a noncatalytic approach to meet lower standards—for example, 0.41, 3.4, and 2.0 g/mi of HC, CO, and NO_x, respectively—involves increasing spark retard for additional HC control and using leaner carburetion, shorter choke times, and increased air injection rates for better CO control. These are the elements of a lean-burn design. The additional spark retard will probably offset any improvements in fuel economy due to extra enleanment, and this approach is expected to have a fuel penalty comparable to that incurred by vehicles meeting 1975-76 California standards. [15]

Because of the difficulty in meeting more stringent HC emissions levels from conventional engines that use catalytic or noncatalytic systems, it is unlikely that fuel economy can be increased more than marginally by increasing compression ratio. Improved catalysts or thermal reactors may provide additional control of HC emissions, but increases in NO_x emissions will place an upper limit on raising the compression ratio. Consequently, the octane requirements of conventional engines with emissions levels below 1975-76 California standards are likely to remain well below pre-1971 requirements.

Finally, stratified-charge engines may not have octane requirements that are comparable to present conventional engines. The Honda engine, as tuned to meet 1975-76 California standards, appears to have a research octane number (RON) requirement of 77 to 85, which is easily satisfied with 91-RON unleaded gasoline. These engines are lead-tolerant, but because of the low-octane requirements, the economic benefit of lead additives is considerably less than with high compression engines. Furthermore, as with conventional engines, the maximum compression ratio appears to be limited by HC emissions control, rather than by fuel octane rating. [11]

In conclusion, it is unlikely that automobile manufacturers will be able to restore compression ratios to pre-1971 levels with emissions standards at or below 1975-76 California requirements. Hence, any future fuel economy gains from increasing compression ratios will only be a fraction of the losses attributed to 1971-73 compression ratio reductions. As a result, future demands for high-octane gasoline and the economic and energy-saving benefits of lead additives will be reduced significantly from pre-1971 levels.

ECONOMIC EFFECTS OF REDUCING THE USE OF LEAD ADDITIVES

The domestic petroleum industry has used lead additives since the 1920's as a low-cost means of increasing motor-gasoline octane ratings. The implementation of regulations that limit the use of these additives has required adjustments in a number of sectors, including procurement, refining, distribution, and marketing. These adjustments have affected the economics of petroleum refining, causing increases in the unit costs of producing both unleaded and conventional grades of gasoline.

Role of Lead Additives in Manufacturing Gasoline

Leaded gasoline is a mixture of several types of hydrocarbon molecules, lead additives, and several other compounds.[17] The hydrocarbons broadly consist of saturated open-chain paraffins, cyclic paraffins, olefins, and aromatics that differ widely in specific gravity, volatility, and octane number.[18] The final composition of the hydrocarbon mixture in gasoline depends on the crude oil starting material, the available types of processing equipment, and the relative costs and economic values of producing and utilizing different hydrocarbon fractions in gasoline, fuel oil, and all other petroleum products. Octane rating is the single most important determinant of the economic value of using the different hydrocarbon streams in blending gasoline.

To increase octane rating, small amounts of a mixture of tetraethyl and tetramethyl lead are added to clear (unleaded) gasolines after refining and blending. The octane ratings for clear gasolines are usually 6 to 8 octane numbers below commercial octane requirements of 94 to 95 RON for regular and 98 to 100 RON for premium. In 1970, for example, typical clear octane ratings for regular and premium were 86 to 88 RON and 93 to 95 RON, respectively. Clear ratings were increased to commercial ratings by adding an average of 2.1 to 2.4 g of lead/gal. for regular and 2.6 to 2.8 g/gal. for premium. The actual amount of lead that is added to gasoline depends on the quality of the crude oil, season of the year (fig. 14-1), and region of the country. The above concentrations represent the range in the national average concentrations of lead in gasoline from 1965 to 1973. [39]

Figure 14-3 illustrates an important economic-engineering relationship between the concentration of tetraethyl lead and gasoline octane rating. The initial addition of lead produces a substantial increase in

---

[17]Other compounds that are added to gasoline include detergents, which prevent the buildup of materials on carburetor surfaces; corrosion inhibitors, which control rusting in pipelines; antioxidants, which inhibit the ability of the olefinic portions of gasoline to oxidize and polymerize into a product referred to as gum; and deposit modifiers, which prevent preignition by modifying engine wall deposits.

[18]There are several scales for rating the octane number of a particular gasoline blend. Research octane number (RON) refers to a specific and widely used measure of octane rating that is determined in a single-cylinder engine under laboratory conditions. The RON rating scale is calibrated with iso-octane as 100 and *n*-heptane as zero. Motor octane number (MON) is a measure of octane rating that is also determined in a single cylinder engine under slightly more severe conditions than the research octane test. The MON rating is several numbers lower than the RON rating. Road octane number is determined in actual road tests. Generally, these ratings lie between the research and motor numbers.

octane; 1 g/gal. may increase octane by 5 to 6 numbers. Successive increments of lead result in smaller increases in octane. In the economists' jargon, there are diminishing marginal returns from adding more and more lead to gasoline.

Nature of Economic Effects of Reducing Use of Lead Additives

Lead additives act as substitutes for the more expensive, higher octane blending components. Because many of these are synthesized from lower octane components, the use of lead enables refiners to produce more gasoline per barrel of crude oil. The synthesis of higher octane components also requires additional processing equipment, processing energy, and butane,[19] which are not needed if lead is used. These factors indicate the economic and energy-saving benefits of adding lead to gasoline.

Large, flexible, and efficient refineries have accommodated reductions in the use of lead antiknocks by adjusting the level and combination of processing techniques used to produce gasoline.[20] Specifically, refineries have increased the octane rating of the clear pool or the total stock of raw unleaded gasoline. To do so, refiners have produced larger volumes of higher-octane blending components, particularly aromatics and branched-chain paraffins. In very general terms, this change has been accomplished by increasing the amount of reforming and alkylation processing.

Catalytic reforming of naphtha is the primary means of synthesizing the high-octane aromatics that make up 25 to 30 percent of gasoline production. Reforming involves reshaping, under heat and pressure, heavy, low-octane hydrocarbons into higher-octane benzene ring structures. If reforming severity is increased to obtain a higher-octane stream of aromatics, the yield of reformate is reduced, and the production of less valuable LPG and fuel gas is increased. Using this process to increase clear-pool octane results in an overall loss of product volume, or a crude penalty for a given volume of crude throughput.

Alkylation is a process that links low-octane, gaseous short-chain hydrocarbons into high-octane, branched-chain blending components that make up approximately 10 percent of gasoline production.

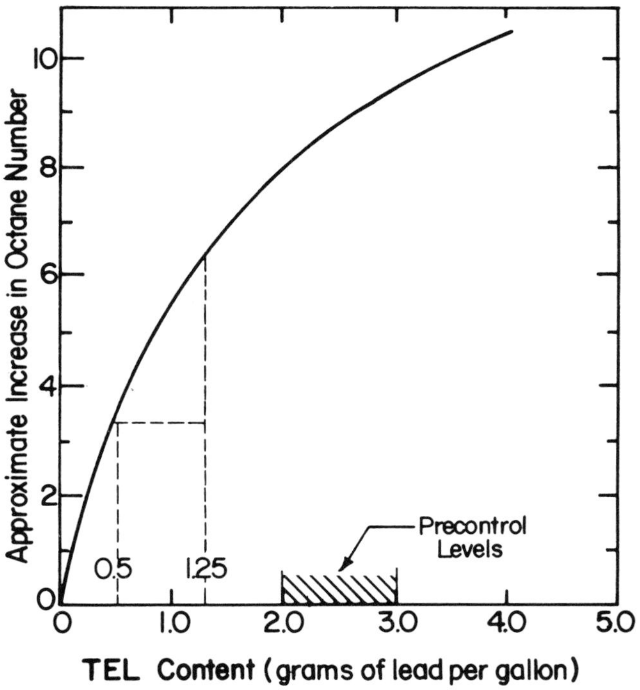

FIGURE 14-3.—Approximate relationship between tetraethyl lead (TEL) content and increase in octane number.

Increasing production from this process also involves a crude oil penalty. To provide more feedstock for alkylation, catalytic crackers must produce more gases at the expense of gasoline blending components. Because alkylation does not convert all of the gaseous feedstocks into high-octane blending components, total gasoline output is decreased for a given volume of crude input.

Hence, additional volumes of crude petroleum must be refined to meet product octane requirements for the same volume of gasoline output, if the use of lead is restricted to concentrations below the economic optimum. More crude must be refined to produce an unleaded grade of 91 RON, as well as regular and premium grades with reduced lead contents.[21]

Larger volumes of unleaded and low-lead gasolines also must be refined as the number of vehicles with catalytic converters and less efficient, low compression engines increases. As the percentage of the national automobile fleet with these characteristics increases, a larger volume of gasoline will be needed in order to travel the same number of vehicle miles.

To the extent that existing capacity is inadequate to produce larger quantities of unleaded and low-lead gasolines, new processing capacity must be added. Capacity can be increased in several ways. Additions can be made to selected processes at existing refineries; for example, alkylation and reform-

[19]Butane is a high-octane blending component that is also used to increase gasoline vapor pressure. Higher vapor pressures are needed to improve the ease of starting, particularly in winter. Increased use of butane in winter gasolines reduces the need for lead (fig. 14-1).

[20]The first stage of the refining sequence for gasoline involves processing crude oil in a distillation unit for separation into product streams by boiling range. The next stage involves further separation, cracking, synthesis, and blending of distillate streams into final products. Major processes used to increase gasoline volume and octane rating include catalytic cracking, reforming, and alkylation. Catalytic cracking, which is not discussed in the text, is used to convert heavier distillates into lighter, gasoline-boiling-range materials for blending and for feedstocks for reforming and alkylation processing. Cracked materials make up over one-third of the total average refinery production.

[21]Several studies have indicated that any additional demands for crude oil because of lead reductions or otherwise, represent increases in already growing oil import requirements. [52, 53] Domestic oil production is declining, and new supplies must come from abroad. While this aspect may be important politically, it has no bearing on an assessment of the economic effects of reducing the use of lead in gasoline.

ing capacity can be increased. Total refinery throughput capacity can be increased at existing refineries. Finally, completely new "grass roots" refineries can be constructed. The actual way in which industry makes additions to capacity has an important effect on the costs of production.

The amount of additional refining capacity required for lead removal is dependent on a number of variables that govern the demand and supply of gasoline. In terms of demand, the need for additional refining capacity depends on how rapidly gasoline demand increases relative to an "expected normal rate" of growth in consumption. This increased demand, in turn, depends on the number, fuel economy, and utilization of post-1970 vehicles.

In terms of supply variables, requirements for additional capacity depend on the speed and timing of the phased-reduction schedule and improvements in processing technology. A feasibility study [54] of several alternative phased-reduction schedules determined that capital investment requirements for the most rapid lead-removal schedules were substantially greater than requirements for the slowest schedules. To the extent that new developments in processing technology enable refiners to convert a larger volume of each barrel of crude into high-octane blending components, less capacity will be required to handle the same volume of throughput.

The costs of new capacity include the costs of the equipment as well as its construction. The refinery process industry is a highly specialized segment of the construction industry and cannot expand rapidly enough to accommodate a sudden increase in added demand for new plants. This study by Bonner and Moore [54] indicated that rapid lead removal would create a sharp business cycle, with a peak-year boom and attendent rapid decline in employment in the process-construction industry. Consideration of these limitations was an important factor in defining the severity of the original phased-reduction schedule.

Financing the construction of additional refining capacity represents the most significant impact of a lead-reduction program on the petroleum industry. The petroleum industry is very capital intensive; petroleum companies invest billions of dollars annually in new production, refining, transportation, and marketing facilities. Capital expenditures for additions to refining capacity that are over and above expenditures planned for normal expansion and replacement can only be financed through new investment funds or a reallocation of funds from other projects.

Removal of lead from gasoline results in some reductions in refining costs, and credits must be given for these reductions when estimating the *net* change in production costs. The major cost-saving feature of producing unleaded and low-lead gasolines stems, of course, from not having to purchase and add lead antiknocks. For 91 RON unleaded gasoline, the cost savings from not adding lead practically offsets the additional raw-materials and processing costs. Because catalytic reforming is also a source of aromatics for petrochemical feedstocks, some refiners may increase production of these materials as a direct consequence of having to increase production of unleaded gasoline. A cost savings occurs to the extent that unleaded gasoline production also results in increased production and sale of the economically more valuable petrochemical feedstocks.

In addition to the refining sector, that part of the petroleum industry that is concerned with distribution and sale of gasoline has been affected by decisions to remove lead from gasoline. New distribution terminal facilities such as separate tanks, pumps, and piping have been installed for handling unleaded gasoline. The requirements for additional equipment would have been much greater if unleaded fuels required a completely segregated mode of distribution; however, facilities for handling leaded gasolines can also be utilized for unleaded gasoline.

In marketing unleaded gasoline, petroleum companies have adopted a three-grade marketing strategy with low-octane, unleaded gasoline as the third grade. From the refining standpoint, this approach for removing lead is considerably less expensive than producing an unleaded regular or premium grade of gasoline. From the marketing standpoint, however, service stations have had to install new tanks and dispensing pumps in order to sell the new unleaded grade.

The petroleum companies and ultimately the consumer must pay the increase in cost for producing and distributing unleaded and low-lead gasolines. The extra cost stems largely from amortized investments for additional processing and marketing facilities and, to a lesser extent, from additional processing and product losses. Estimates of the increase in costs of removing lead from gasoline depend on many assumptions concerning, among other things, the expected normal growth in gasoline consumption, automobile fuel economy, and the severity of the phased-reduction schedule. These are discussed in the following section.

Estimates of Costs of Reducing the Use of Lead in Gasoline

EPA and the petroleum industry have made a number of estimates of the costs of reducing the use of lead in gasoline. Some of these estimates were developed through the use of rather sophisticated models. In particular, several estimates were made by simulating U.S. refining operations with one or more linear programming models of petroleum refineries. [52-55] Linear programming models were used because of their ability to determine the least-cost levels of processing and expansion that would be needed to produce specified quantities of unleaded and low-lead gasolines. The approach taken to estimate the economic effects of lead restrictions involved first making forecasts of available refining capacity and of the demands for motor gasoline and all other major petroleum products for a base case or reference scenario. Forecasts for the reference scenario were made by assuming an "expected normal rate" of growth in the demand for gasoline that would occur if there were no restrictions on the use of lead additives. Similar forecasts of gasoline demand were also made for scenarios that assumed various levels of automobile emissions control and accompanying changes in fuel economy and in the use of lead additives.

Demand forecasts for the reference case and emissions control scenarios were then entered into linear programming models; for both situations, the optimal response in terms of crude oil and new capacity requirements was determined. For the emissions control scenarios, optimal requirements were determined subject to restrictions on the use of lead additives.

A comparison of each lead-removal scenario with the reference case provided an estimate of the increase in resource expenditures caused by restrictions on the use of tetraethyl lead. The estimates of additional resource requirements that were made in this way were very sensitive to the size and characteristics of the initial demand forecasts.

The Bonner and Moore study, [54] made for EPA, examined the feasibility of 13 alternative lead-removal schedules. These schedules were grouped into two classes, one related to a two-grade marketing strategy, the other to a three-grade strategy. In all cases, a lead-free grade was required by 1974 to meet the needs of 1975 model cars equipped with catalysts. The feasibility analysis examined approximate capital costs, pool octane numbers, and the year-to-year increase in gasoline volume.

The Bonner and Moore study estimated that production of 93 RON unleaded gasoline and of low-lead gasolines in quantities sufficient to comply with EPA's original phased-reduction schedule would necessitate increasing the national clear-pool octane rating to 91.5 RON by 1980 from 88.2 RON in 1972. The study estimated that this increase would require $2.7 billion in additional refining investments by 1980, but that the cost savings from using less lead would more than offset additional processing costs. Added distribution and marketing costs were estimated at $2.9 billion. This amount varied with the marketing strategy, but not with the severity of the phased-reduction schedule. On an average cost basis, these increases would result in an added cost of 0.25 cents/gal., or about a 1.4 percent increase in the average price of gasoline delivered to service stations.[22] All costs were estimated in terms of 1970 prices.

Using back-of-the envelope calculations, Johnson and Kittrell [22] estimated a retail price for unleaded gasoline of 2.0 cents/gal. greater than leaded regular. They estimated that increased processing, distribution, and marketing costs associated with supplying a new grade of gasoline would add about 0.5 cents/gal. in terms of constant 1974 dollars. Based on historical analysis of the pump price and sales volume differentials between leaded regular and leaded premium, Johnson and Kittrell added 1.5 cents/gal. for increased retail marketing distribution costs in order to arrive at the 2.0 cent figure.

Although the Committee on Motor Vehicle Emissions [11] interpreted the extra 1.5 cents as a reflection of increased service station marketing costs required by the transition to unleaded gasoline, it is apparent that this amount represents an effort on the part of service stations to maintain a portion of the larger profit margin that they have enjoyed from selling premium gasoline.[23] A substantial portion of increased marketing costs for unleaded gasoline is non-recurring capital expenditures for additional storage tanks and pumps. Approximately 65 percent of the total industry effort to convert to a three-grade marketing system was committed by 1971 [54]; consequently, a substantial portion of these increased capital expenditures have already been depreciated.

The Bonner and Moore report [54] and three other similar studies are summarized in table 14-8. The differences in the projected economic impacts for each study are primarily a function of the different scenarios examined. If the octane level of unleaded gasoline remains at 91 RON, it is unlikely that the increases in crude oil consumption projected by the Bonner-Moore, [54] Turner-Mason-Solomon, [53] and Ethyl-Corp. [52] studies will materialize by 1980. There are several reasons for this, the most important of which is evidence indicating that the rate of growth in gasoline demand is slowing down and probably will remain below the growth rates assumed for the lead-removal scenarios analyzed in the studies (table 14-9).

From 1970 to 1973, demand grew at an average rate of 5.4 percent per year. Demand reached an average of 6.675 million barrels per day in 1973, before declining during the recession year of 1974. With economic recovery, demand has now surpassed 1973 levels, but the post-recession rate of growth in gasoline consumption has been lower than pre-1973 levels. In 1975, demand averaged 6.675 million barrels per day, or an increase of 2.1 percent over 1974. In 1976, overall demand has been running at a record pace, reaching 6.844 million barrels per day during the first half of the year. Demand was projected to average 7.020 million barrels per day by the year's end, up 5.2 percent from 1975. [59]

Rapid growth in gasoline consumption in 1976 has accompanied economic recovery; it reflects the sud-

---

[22]In interpreting the results of their study to the petroleum industry, Bonner and Moore [56] indicated that lead removal would result in an increase in the pump price of gasoline of 1.5 to 2.0 cents/gal. Added manufacturing and distribution costs would account for 0.2 to 0.9 cents/gal.; lost revenues from declining premium sales would make up the remainder. It should be noted that this latter item does not represent any increase in the resources required to produce unleaded or low-lead gasolines.

[23]Before 1974, the price differential between premium and regular gasoline at the refineries and terminals was approximately 2 cents/gal. [57] On the other hand, the price differential between these two grades at service stations was usually about 4 cents. Because motor fuel taxes and transportation costs from terminals to service stations were the same for both grades, the extra 2 cents/gal. reflected the higher profit margin for premium gasoline.

TABLE 14-8.—*Summary of economic impact studies of removing lead from gasoline*

Study	Economic impacts by 1980	Assumptions
Bonner and Moore [54] (schedule 0)	3.0 percent increase in crude oil $2.7 billion in added refining investment costs $3.0 billion in added distribution investment costs	1975-76 statutory standards; phased reduction schedule (schedule 0); catalytic converters on all 1975 and later model cars; fuel penalties: 5 percent on 1971-74 model cars and 12 percent on 1975 and later models; 9.2 percent increase in total gasoline consumption by 1980; 93 RON unleaded gasoline; 68.8 percent unleaded gasoline in 1980.
Turner, Mason, and Soloman [53]	6.9 percent increase in crude oil $3.6 billion in added refining investment costs $0.8 billion increase in added marketing investment costs	1975-76 statutory standards; phased reduction schedule (schedule 0); catalytic converters on all 1975 and later model cars; fuel penalties: 6 to 7 percent on 1971-74 model cars and 17 percent on 1975 and later models; 12.8 percent increase in gasoline consumption by 1980; 91 RON unleaded gasoline; 65.9 percent unleaded gasoline in 1980.
Arthur D. Little, Inc. [55]	$2.2 billion increase in added refining investment costs	91 RON unleaded grade; 100 percent unleaded gasoline in 1980.
Ethyl Corp. [52]	5.8 to 6.1 percent increase in crude oil	93 RON or 91 RON unleaded gasoline; fuel penalties: 7 percent for 93 RON and 9 percent for 91 RON; 100 percent unleaded gasoline in 1980.

den surge that has occurred in the economy as it has begun to move upward from a recession low. As recovery continues, the rate of economic growth will begin to decline. A corresponding decline in the growth of gasoline consumption can also be expected because changes in the level of gasoline demand and economic activity take place somewhat concurrently.

Unless there is a sustained shift back to large automobiles, the rate of growth in gasoline consumption will probably not average more than 3.0 percent for the rest of the decade. This lower rate of growth is due to the impact of automobiles that are getting more miles per gallon and to the effect of the Arab oil embargo.

Some additional reasons why the crude oil projections in the lead-removal studies are not likely to materialize include the following: first, EPA's phased-reduction schedule will not go into effect until October 1979, after a delay of nearly 5 years. During that time, the proportion of the automobile population of post-1971 automobiles that require lower-octane gasoline will have increased, causing a gradual reduction in the average leaded octane level of gasoline. (fig. 14-1) Consequently, the belated implementation of the phased-reduction schedule will result in smaller crude penalities and require smaller additions of octane-producing facilities than initially anticipated.

Second, refiners are using substitute additives for lead alkyls in both leaded and unleaded gasolines. Of the organometallics, methylcyclopentadienyl manganese tricarbonyl (MMT) has demonstrated advantages similar to tetraethyl lead in boosting octane, but the cost of manganese is much higher than lead when compared on a weight-to-weight basis. [60] With restrictions on the use of lead, the costs of MMT now compare favorably with the costs of increased processing for the purpose of raising octane. Furthermore, a manganese concentration of 0.125g/gal. does not have any adverse effects on catalytic converters according to Ethyl Corp. [61]

TABLE 14-9.—*Comparison of actual and projected annual rates of growth in the demand for motor gasoline in the United States*

Year	USBM [58] (actual)	USEPA [54]	Turner-Mason-Soloman [53]	Ethyl Corp. [52]
1971	3.97	–	4.6	–
1972	6.30	3.9	4.1	–
1973	4.39	3.8	3.9	–
1974	–2.05	3.5	3.7	–
1975	2.10	4.5	4.1	5.5
1976	5.20[1]	4.3	5.0	–
1971-76	3.14[1]	4.0	–	–
1971-80	–	3.8	3.7	5.5

[1] Projected.

Third, the application of new refining technology may permit the conversion of a larger percentage of each barrel of crude oil into gasoline. Hydrotreating technology, which is now commercially available, provides a means for producing a high-quality feedstock from the heavy residual fractions. These fractions make up the portion of a barrel of crude oil that is currently processed into heavy oil, petroleum coke, and asphalt. This feedstock can be further processed in an appropriately designed catalytic cracking unit into gasoline and distillate. [62]

It is difficult to assess whether or not the increased investment costs for lead removal have created a financial burden for oil companies. First of all, refinery construction expenditures historically have exhibited a marked cyclical pattern, and it is difficult to determine what represents an average annual expenditure. Second, the relative impacts of increased investment costs have fallen unequally on individual companies. Increased expenditures for refining capacity have varied from company to company, depending on the capacity and type of existing processing equipment.

Looking at the largest companies in the industry, it does not appear, however, that lead removal has produced any major shifts in capital expenditure patterns. From 1965 to 1969, large oil companies devoted an average of 10.3 percent of their total annual capital expenditures in the United States to refineries. For the 1970 to 1974 period, when companies began installing equipment to produce unleaded gasoline, the share devoted to refineries increased slightly to an average of 11.7 percent of total annual capital expenditures. [63] The slight difference between the two 5-year periods probably reflects the fact that oil companies plan new projects 3 to 5 years in advance of construction.

In conclusion, the incremental costs of producing and distributing 91 RON gasoline to service stations are about 0.5 cents/gal. This increase results from the need for additional refining capacity, processing, and crude oil. The crude penalty for producing unleaded gasoline is probably less than 1 percent/gal. An additional increase of 1.5 cents/gal. for service station marketing expenses is responsible for the difference of 2.0 cents/gal. between the retail prices of unleaded and leaded regular. The requirement to produce unleaded gasoline has not resulted in a major shift in refinery construction expenditures for the largest segment of the petroleum industry.

REFERENCES

1. Fed. Reg. *38*: 1254 (Jan. 10, 1973).
2. Fed. Reg. *38*: 33734 (Dec. 6, 1973).
3. Fed. Reg. *39:* 43281 (Dec. 12, 1974).
4. Fed. Reg. *41:* 42675 (Sept. 28, 1976).
5. U. S. Environmental Protection Agency. EPA's Position on the Health Implications of Airborne Lead, Washington, D.C. (Nov. 28, 1973).
6. National Academy of Sciences. Air Quality and Automobile Emission Control, Vol. 4—The Costs and Benefits of Automobile Emission Control. Prepared for the Committee on Public Works, U.S. Senate. U. S. Government Printing Office, Washington, D.C. (1974).
7. W. Ruckelshaus. Decision of the Administrator on Removal from the U.S. Court of Appeals for the District of Columbia (Apr. 11, 1973).
8. Fed. Reg. *40:* 11900 (Mar. 5, 1975).
9. Fed. Reg. *40:* 24350 (Jun. 5, 1975).
10. National Academy of Sciences. Report by the Committee on Motor Vehicle Emissions. Washington, D.C. (Feb. 1973).
11. National Academy of Sciences. Report by the Committee on Motor Vehicle Emissions. Washington, D.C. (Nov. 1974).
12. Aerospace Corp. Final Report: An Assessment of the Effects of Lead Additives in Gasoline on Emission Control Systems Which Might Be Used to Meet the 1975-76 Motor Vehicle Emission Standards. Prepared for the U.S. Environmental Protection Agency. (Nov. 15, 1971).
13. U.S. Environmental Protection Agency. Automobile Emission Control—The State of the Art as of December 1972. Prepared by Emission Control Technology Division, Ann Arbor, MI
14. U. S. Environmental Protection Agency. Automobile Emission Control—The Development Status as of April 1974. Prepared by Emission Control Technology Division, Ann Arbor, MI
15. U. S. Environmental Protection Agency. Automobile Emission Control—The Technical Status and Outlook as of December 1974. Prepared by Emission Control Technology Division, Ann Arbor, MI (Jan. 1975).
16. U. S. Environmental Protection Agency. Automotive Emission Control—The Current Status & Development Trends as of March, 1976. Prepared by Emission Control Technology Division, Ann Arbor, MI (Apr. 1976).
17. F. P. Grad et al. The Automobile and the Regulation of Its Impact on the Environment. Univ. of Oklahoma Press, Norman, OK (1975).
18. U. S. Congress, Senate Committee on Public Works. Air Pollution-1970: Hearings before the Subcommittee on Air and Water Pollution. U.S. Government Printing Office, Washington, D.C. (1970).
19. Autom. Eng. *80:* 32 (May 1972).
20. The New York Times. (Feb. 15, 1970).
21. L. J. White. The Automobile Industry Since 1945, Harvard Univ. Press, Cambridge, MA (1971).
22. U.S. Environmental Protection Agency. Consultant Report to the Committee on Motor Vehicle Emissions, Commission on Sociotechnical Systems, National Research Council on Manufacturability and Costs of Proposed Low-Emissions Automotive Engine Systems. Office of Air and Waste Management. Washington, D.C. (Nov. 1974).
23. D. Shapley. Science. *182:* 368 (1973).
24. U.S. Congress, Senate Committee on Public Works. Compliance with Title II (Auto Emission Standards) of the Clean Air Act. U.S. Government Printing Office, Washington, D.C. (1973).
25. National Academy of Sciences. Report of the Conference on Air Quality and Automobile Emissions. Washington, D.C. (Jun. 5, 1975).
26. Chrysler Corporation. Progress Report on Chrysler's Efforts to Meet the 1977 and 1978 Federal Emissions Standards for HC, CO, and NO_x. Vol. 1. Detroit, MI (Dec. 1975).
27. U. S. Department of the Interior, Bureau of Mines. Mineral Industry Surveys: Lead Industry, Monthly.
28. National Academy of Sciences. Airborne Lead in Perspective. Washington, D.C. (1972).
29. Fed. Reg. *37:* 3882 (Feb. 23, 1972).
30. Fed. Reg. *38:* 1258 (Jan. 10, 1973).
31. National Petroleum Refiners' Association. National Petroleum Refiners' Association Testimony: Environmental Protection Agency Hearings on Proposed Regulation of Fuels and Fuel Additives (Apr. 27, 1972).
32. International Lead Zinc Research Organization, Inc. Response to Request for Comment on Regulation of Fuels and Fuel Additives (Mar. 9, 1973).
33. Ethyl Corp. Ethyl's Position on Environmental Protection Agency's Proposed Regulations of Lead in Gasoline (Mar. 1972).
34. E. I. duPont de Nemours and Co., Inc. Statement Relative to Lead Reduction Schedule Proposed by EPA (Mar. 9, 1973).

35. *Ethyl Corp. versus Environmental Protection Agency,* 7 ERC 1353 (D.C. Cir. 1975), vacated, 7 ERC 1687 (D.C. Cir. 1975).
36. Supplemental brief for respondent at 26. *Ethyl Corp. versus Environmental Protection Agency,* 7 ERC 1353 (D.C. Cir. 1975).
37. *Ethyl Corp. versus Environmental Protection Agency.* F. 2d.
38. Fed. Reg. *41:* 28352 (Jul. 9, 1976).
39. U.S. Department of the Interior, Bureau of Mines. Mineral Industry Surveys: Motor Gasoline Summer and Motor Gasoline Winter (biannual).
40. Ethyl Corp. Yearly Report of Gasoline Sales (1975).
41. J. L. Enos. Petroleum Progress and Profits: A History of Process Innovation. MIT Press, Cambridge, MA (1962).
42. Aerospace Corp. Passenger Car Weight Trend Analysis. Prepared for the U.S. Environmental Protection Agency, Office of Air and Water Programs, Ann Arbor, MI (Jan. 1974).
43. U.S. Department of Transportation, Federal Highway Administration. Highway Statistics. U.S. Government Printing Office, Washington, D.C. (various years).
44. Ward's Automotive Handbook. Ward's Communications, Inc. Detroit, MI (annual).
45. C. La Pointe. Factors Affecting Vehicle Fuel Economy. SAE Paper No. 730791. Presented at National Fuels and Lubricants Meeting, Milwaukee, WI (Sep. 1973).
46. T. C. Austin and K. H. Hellman. Passenger Car Fuel Economy—Trends and Influencing Factors," SAE Paper No. 730790. Presented at National Fuels and Lubricants Meeting, Milwaukee, WI (Sep. 1973).
47. T.C. Austin and K. H. Hellman. Fuel Economy of the 1975 Models. SAE Paper No. 740970. Presented at the National Automobile Engineering Meeting, Toronto, Ontario (Oct. 1974).
48. T. C. Austin, R. B. Michael, and G. R. Service. Passenger Car Fuel Economy Trends Through 1976. SAE Paper No. 750957. Presented at the National Automobile Engineering Meeting, Detroit, MI (Oct. 1975).
49. G. J. Huebner, Jr. and D. J. Gasser. Energy and the Automobile—General Factors Affecting Vehicle Fuel Consumption. SAE Paper No. 730518. Presented at the National Automobile Engineering Meeting, Detroit, MI (May 1973).
50. T. P. Flanagan. Personal communication. Vehicle Emissions Planning, Chrysler Corporation (Dec. 19, 1975).
51. C. R. Morgan and S. S. Hetrick. Tradeoffs Between Engine Emission Control Variables, Fuel Economy and Octane. SAE Paper No. 750415. Presented at the Automotive Engineering Congress and Exposition, Detroit, MI (Feb. 1975).
52. G. A. Unzelman, G. W. Michalski, and A. F. Lovell. The Energy Crisis and Lead Antiknocks. Paper No. 64d. American Institute of Chemical Engineers, Dallas, TX (Feb. 20-23, 1972).
53. Turner, Mason, and Solomon. Consulting Engineers. The Economic Impact of Automotive Emission Standards. Dallas, TX (Mar. 31, 1972).
54. Bonner and Moore Associates, Inc. An Economic Analysis of Proposed Schedules for Removal of Lead Additives from Gasoline. Prepared for U.S. Environmental Protection Agency. NTIS PB 201 133 (Jun. 25, 1971).
55. Arthur D. Little, Inc. U.S. Refining Capability to Supply Proposed New GM Motor Gasolines. Prepared for General Motors Corp. (Dec. 1973).
56. Bonner and Moore Associates, Inc. A Review of the July 1971 Bonner and Moore Report: An Economic Analysis of Proposed Schedules for Removal of Lead Additives from Gasoline. American Petroleum Inst. Washington, D.C. (Oct. 1971).
57. Platt's Oil Price Handbook and Oilmanac. McGraw-Hill, Inc. New York, NY (annual).
58. U.S. Department of the Interior, Bureau of Mines. Mineral Industry Surveys: Petroleum Statement Monthly.
59. Oil and Gas J. *74:* 103 (Jul. 26, 1976).
60. G. H. Unzelman, E. J. Forster, and A. M. Burns. Are There Substitutes for Lead Antiknocks? American Petroleum Inst. Preprint No. 47-71. Paper presented at the 36th Midyear Meeting of the American Petroleum Inst. San Francisco, CA (1971).
61. J. D. Bailie. Oil and Gas J. *74:* 69 (Apr. 19, 1976).
62. D. Milstein et al. Oil and Gas J. *74:* 138 (Jul. 19, 1976).
63. R. S. Dobias et al. Capital Investments of the World Petroleum Industry, The Chase Manhattan Bank, New York, NY (1975).

PART VI

Summary and Conclusions

SUMMARY AND CONCLUSIONS

Coordinated by
W.R. Boggess

Lead is the most abundant of the natural heavy metals. Its physical and chemical characteristics are such that lead is uniquely adapted to many industrial uses. Physically, lead is a durable, easily worked metal that has been used since pre-Christian times. Chemically, it will combine with other elements to form compounds with unique and highly useful properties. In spite of its many beneficial uses, lead is a definite hazard to human health when relatively small quantities are taken into and retained within the body. Intake may be either through ingestion with food and water or by inhaling airborne particles. Although the health aspects of lead have been recognized for many centuries, its importance as a potentially hazardous environmental contaminant and pollutant has increased with the industrial production and utilization of lead.

Although lead occurs naturally in the soil, water, and atmospheric components of the biosphere, inputs from a variety of sources are responsible for its role as a potentially hazardous substance. In local situations, industrial operations associated with the production of lead and its further manufacture into a variety of products may be of primary importance. On a *global* basis the principal, and by far the greatest, source is from the combustion of lead-containing fuels. Coal and fuel oil burned for heating and industrial purposes provide some input, but the major source is from the combustion of gasoline to which either tetraethyl or tetramethyl lead has been added as an antiknock compound. Leaded paints have also been an important local source, and many cases of lead poisoning, or "plumbism," have been traced to ingestion of peeling paint chips by children. The importance of this source has declined, since lead is no longer used as a principal pigment for exterior and interior house paints. However, many potentially dangerous areas remain in poor tenement districts where building maintenance and repair have been neglected or are simply nonexistent.

Research during the past 5 years at Colorado State University (CSU), the University of Illinois, and the University of Missouri has been concerned with the occurrence, characteristics, transport, distribution, accumulation, and environmental effects of lead. Two major sources of lead have been considered: the combustion of leaded gasoline and the mining, milling, and smelting operations of one of the world's newest and largest lead production areas located in southeast Missouri. Investigations have included terrestrial, atmospheric, aquatic, and biological components of the environment; the development of appropriate monitoring and analytical methods; and a consideration and evaluation of control strategies designed to reduce the input of lead. It is significant that most of the nonhealth-related problem areas listed by the National Academy of Sciences Report ("Airborne Lead in Perspective") as needing additional research have been covered in depth by investigations at the three universities.

The success of any effort to monitor environmental pollutants depends on a well-planned and-executed sampling program coupled with analytical measurements of integrity. Even the best analytical methods will fail unless samples are truly representative of the system under study, collected without bias, stored under optimum conditions, and carefully prepared for analysis. Similarly, the results of analyzing good samples are meaningless when the methodology used is inaccurate. Too many publications have presented erroneous or overly generalized conclusions because of deficiencies in the sample collection, storage, preparation, or analytical programs. Because such occurrences are costly as well as embarrassing, the sample collection plan should be carefully formulated only after a thorough evaluation of the objectives of monitoring, previous studies of a similar nature, and preliminary sampling data involving the system of interest. It is usually wise to develop this plan in consultation with the analytical chemists responsible for the determinations required, to ensure that appropriate sample sizes are collected and that the storage procedures are adequate. Finally, it is essential that the analytical laboratories be required to maintain a continual quality assurance program sufficient to provide a high level of confidence in the integrity of the analytical results.

Knowledge of the physical and chemical characteristics of lead is basic to understanding many aspects related to its occurrence, transport, distribution, biological activity, and general behavior in various ecosystem components and under a wide range of environmental conditions. Research, especially at CSU and Illinois, has added appreciably to understanding the role of particle size distribution of lead-containing species, both in automotive emissions and during transport. The surface enrichment of small particles with lead has been demonstrated. Also, anomalies that occur in particle size distribution as a function of transport distance indicate that significant coagulation effects occur between lead-containing and atmospheric particulate matter in general. The importance of particle size is further emphasized in research with fly ash from the combustion of coal; the results indicate that most toxic chemicals, including lead, are most concentrated on the smallest particles. These small particles constitute the greatest human health hazard as they are the ones most likely to be inhaled and retained within the lungs. Also, their settling rate is relatively slow, and these small particles are subject to long-range transport.

The principal lead-bearing substance in fresh automotive exhaust is PbBrCl. However, the primary lead emissions are converted into a variety of lead species by chemical reactions occurring in the atmosphere. Details concerning the atmospheric chemistry of lead are not well defined, yet the chemical composition of species existing in the atmosphere is more important than the composition of the primary emissions themselves in determining solubility, uptake, mobility, biological activity, etc.

The presence of lead alkyls near heavily traveled highways remains an unsolved problem. These compounds are important in interpreting various aspects of transport because the toxicity of lead alkyls on a mass basis is considerably greater than that of lead-containing particulates. Difficulties in interpreting results from filter experiments may be a part of the problem; e.g., whether material passing through filters is either gaseous or in particles too small to be collected. With improved techniques designed to overcome these difficulties, research at CSU found lead tetramethyl to be the only lead alkyl present in measurable amounts within 30 m of a heavily traveled highway. Its high vapor pressure, in comparison with other lead alkyls, suggests that tetramethyl lead may originate from evaporation of leaded fuels, rather than from exhaust emissions. Additional laboratory experiments suggest that lead alkyls may be adsorbed on the surface of atmospheric particulates. Dust particles, therefore, might act as a scavenging mechanism and serve as a substrate for the conversion of the sorbed lead species into solid inorganic lead compounds.

A major effort at CSU involved the development of both mathematical and physical models to provide a better understanding of atmospheric transport mechanisms along highways and within cities. Based on diffusion from a line source (highway) and on given changes in wind speed and direction with altitude, a semiemperical equation was developed that adequately correlated with experimental measurements of atmospheric parameters in and around Fort Collins. Based on field measurements and model predictions, the results suggest that approximately 45 percent of the consumed lead (based on consumption of 13 mi/gal. and emission factor of 45 percent) remained airborne at the suburban edge of the city. As part of this work, improved methods of aerometic measurements and analyses were developed; the methods allowed more precise evaluation of short-term changes that correlate with the time scale of actual meteorological phenomena.

Inputs of lead from mining, milling, and smelting operations are essentially from point sources, compared with those from the exhaust pipes of thousands of moving automobiles. Hence, the general transport and distribution from the industrial source is somewhat more straight forward and easier to define than the combustion products of leaded gasoline. Major sources of lead from mining, milling, and smelting operations include stack emissions from smelters; fugitive dusts around operations; finely ground material from tailings dams; ore concentrate piles; open railroad cars and trucks used to transport finely ground materials from mines and mills to smelters; wastewater from the ore concentration processes; and, in some cases, water from mines.

In the Missouri study area, fugitive dusts and finely ground materials from tailings dams and concentrate piles provide the main source of contamination within 3 mi of the smelter. Stack emissions are most important beyond this distance. Dusts are moved by wind and deposited on the surrounding vegetation and soil, while stack emissions are discharged into the air, and the airborne particles gradually settle to the earth carrying along associated heavy metals. Lead-bearing materials also enter the environment as dusts and other solids are washed by storm runoff into streams from tailings dams, concentrate piles, and other areas around the operations. Excessive rainfall occasionally causes tailings ponds to overflow and discharge directly into receiving streams.

Storm runoff is the major transport mechanism in moving lead from terrestrial to aquatic systems. Lead is associated with particles of soil and organic matter, and movement is thus associated with suspended sediments rather than water *per se*. Sediments then appear to be a major sink for lead and other heavy metals in aquatic ecosystems. Lead has increased in the sediments of Clearwater Lake, an impoundment that receives streamflow from the industrial watersheds. It appears, however, that considerable amounts of sediments are scoured from the bottom of the reservoir during heavy storms and pass out of the system through the outlet in the base of the dam. The possibility of toxic buildups is rather remote under existing conditions.

Transport and distribution studies in Illinois were conducted within the context of an 87 mi^2 agricultural watershed ecosystem with a small urban compartment. This contrasts with the Missouri area, which is sparsely populated and almost completely forested.

Combustion of gasoline provides the major lead input to the watershed. Daily consumption is estimated at 56,575 gal. Based on this rate, along with other factors, the estimated lead input is 79 kg/day or 29,000 kg on a yearly basis. Forty-six percent of the lead input remains airborne, leaving a total input of 16,000 kg/year. The 12 mi^2 urban compartment receives 65 percent of this amount. Airborne lead returned to the ground in precipitation accounts for 800 kg, while 9,500 kg are returned by dustfall each year.

Drainage water represents the major exit route of lead from the ecosystem. About 80 percent of the lead is associated with suspended solids, and the remainder is in a dissolved form. The amount of suspended lead is greatest from the urban compartment where large impervious areas with high particulate loads are quickly washed during rain storms. Based on streamwater samples adjusted for storm periods, an estimated 910 kg of lead/year leave the ecosystem via streamflow. Thus with the estimated 16,000 kg/year input and 910 kg/year output, 14,800 kg/year (92.5 percent) of lead remains within the ecosystem.

The soil system is a major sink for lead in terrestrial ecosystems with highest concentrations occurring near sources of contamination. Along highways, the largest amount of lead is found immediately adjacent to the roadway; the amount diminishes rapidly to essentially background levels within a distance of 100 to 300 ft. Concentrations in urban areas vary with the amount of traffic. Along residential streets with moderate traffic, lead levels are highest near the curb, drop sharply across lawns, and increase again near houses. The latter increase is related in some cases to lead leached from exterior paint, but more often to that associated with dust deposited on rooftops and sides of houses and washed to the ground with precipitation. Considerably elevated lead levels are also found in forest-covered soils near the mining, milling, and smelting operations in Missouri. Lead is concentrated in the organic horizons, with only slight penetration into the underlying mineral soils.

Heavy lead concentrations along streets and parkways are reflected in equally large amounts of lead collected in dust from the floors of homes and public buildings. Lead in dust is either blown into buildings through open doors and windows or tracked inside by foot traffic.

House dust, especially that collected from rugs, also contains unusually high amounts of cadmium. The source is believed to be the rubber backing on rugs or the padding used as a base for carpet installations. The amount of lead and cadmium found in dust from rugs and floors in residences may be sufficient to create a health hazard for small children. At least, medical diagnosticians and clinicians should be aware of this source in well-maintained, better homes; such homes certainly do not fit the classic pattern of peeling and chipping paint usually associated with diagnosed cases of plumbism.

Without question, lead is concentrated in the soil surface horizon, but it does move downward with time. Movement is generally considered to be quite slow, a concept based on the relative insolubility of lead compounds and the binding capacity of organic fractions that are usually concentrated near the surface. Recent studies at Illinois suggest that much more rapid movement is possible. Six years after varying amounts of lead were incorporated in the plow layer of a Drummer soil, movement was detected to a depth of 30 to 36 in. where the maximum amount (3,200 kg/ha) was added. Penetration is progressively less with smaller surface additions. While the amount of lead applied is unusually high, movement occurs in a soil that has near-optimum characteristics for the sorption of lead. The greater amounts of lead present at various depths in urban as compared with rural soil profiles also suggest that the depth of penetration bears some relation to the lead deposited on the surface. Movement is probably slower in forest soil profiles where organic horizons occur on top of the mineral soil and when mixing of the two layers depends largely on the action of soil organisms. In addition, lead absorbed by tree roots and moved to the foliage in the transpiration stream would likely be returned to the surface with annual leaf fall. An overall important factor in the functioning of the soil "reservoir" system is that even large amounts of lead added to the surface are rendered unavailable to plants, probably by being complexed and bound with organic fractions.

The total lead content of a soil is not a good measure of availability to plants. Availability, rather, depends upon the amount of lead present in relation to the capacity of the soil to sorb lead. Availability also decreases with increases in soil pH, cation exchange capacity, organic matter content, and available phosphorus levels.

The uptake of lead from the soil solution by roots of higher plants has been convincingly demonstrated. Under certain soil conditions (low pH, low cation exchange capacity, low organic matter content, and low phosphate levels), large amounts of lead can be taken up by roots. However, much of the lead remains in the roots rather than being translocated to above-ground parts. This, coupled with a general lack of effect from the relatively large concentrations, suggests that lead is inactivated by deposition in the roots. Electron microscope studies show that lead-phosphate deposits are formed on root surfaces, in peripheral extracellular spaces, and within dictyosome vesicles of root cells outside the endodermis. Dictyosomes containing lead deposits migrate outside the cell itself via reverse pinocytosis, forming extracellular deposits that are most prevalent in roots; however, similar deposits are found throughout corn plants.

The case for foliar absorption of lead is not as clear as that for uptake through root systems. Difficulties in removing lead attached to, or embedded in, the cuticle may be a factor in determining the difference between topical coatings and actual penetration of the lead epidermis. Radioautographic evidence of foliar uptake and translocation has been obtained, but the amounts involved are relatively small in comparison with root absorption.

It is evident that under certain conditions there is movement of lead (translocation) within plants. In general, however, there seems to be about a 10 fold reduction from roots to foliage and a like reduction from foliage to fruit. Thus if 100 μg/g of lead was associated with the roots, the grain (fruit) would likely contain about 1 μg/g.

Environmental factors, plant age, and speciation are important variables in the uptake of lead by roots. In general, the alteration of soil parameters to make lead available in soil solutions increases root uptake. Comparisons of studies done with differing plant age and speciation provide no clear-cut generalizations. Results tend to vary with speciation, but not necessarily with any larger plant grouping.

The concentration of lead associated with vegetation follows that of soil, with greatest amounts found adjacent to heavily used streets and highways or near other sources of contamination such as mines, mills, smelters, and ore-haulage roads. Lead is deposited as a topical coating, and more accumulates on pubescent leaves than on those with smooth surfaces. Topical coatings may be partially or completely removed by rainfall, but are not reentrained by winds of moderate velocity.

In Missouri, vegetables grown near smelters or haulage roads, or on soils that had been amended with dolomitic limestone wastes, accumulate more lead than those produced in gardens well away from a major source of contamination. Lettuce leaves accumulate more lead than either radish roots or green bean pods. While detrimental health effects from eating vegetables grown under such conditions might exist, the probability is greatly lessened by the relatively small amounts consumed.

In the Illinois study, an estimated 2,211 kg of lead are associated with the vegetation of the 87-mi^2 watershed ecosystem. This amount of lead reflects the large biomass of the vegetation component.

Lead concentrations in small mammals and other vertebrates are highest near high-use roads and in the urban compartment of the Illinois ecosystem. However, total body concentrations are relatively low (31.7 μg/g is the highest amount found for an urban-dwelling mammal). Lead is not concentrated in any vital organs, with the possible exception of the liver and kidney. An estimated 23.7 g of lead is in residency within the small mammal population at any given time. This amount is an insignificant factor in the total flux of lead through the ecosystem and certainly does not represent a major pool for accumulation.

There appears to be some increase in lead concentrations through insect food chains, the amount depending on feeding habits. Concentrations increase from sucking to chewing to predatory insects in high-lead areas near roads with a heavy volume of traffic. The insect biomass is relatively small because of the low diversity of habitats and the use of insecticides in cultivated fields adjacent to highways. Results with insect populations suggest the possibility of food-chain magnification of insectivores feeding near streets and highways.

Sediments constitute the major sink for lead in aquatic systems. Lead concentrations are several times higher in the top 10 cm of urban streams than in rural streams draining the Illinois ecosystem; most of the lead is associated with silt-size particles. In like manner, aquatic biota from urban streams have body lead concentrations several times higher than those from watercourses in the rural area. The amount of lead present is related to the contact an organism has with a high lead-containing substrate (sediments).

Lead does not appear to be a threat to the aquatic ecosystems in southeastern Missouri streams. The possibility of damage is minimized by the alkaline reaction of streamwaters, which retards solubilization of the predominantly lead sulfide ores. The greatest inputs into aquatic systems are from the mine-mill complexes. Organic reagents used in these operations may have a deleterious effect on stream biota. Some of these reagents have produced extensive algal blooms and diatomaceous mats. Although unsightly, these mats have a positive effect in trapping large amounts of detritus bearing lead and other heavy metals.

Lead toxicity in biological systems has been generally observed only at concentrations that are considerably higher than those of other toxic heavy metals. Most evidence suggests that this is due to the strong affinity of lead for organic and inorganic surfaces, or that lead is precipitated out of solution, and thus inactivated, by many common anions. The significance of these properties is amplified in that lead poisoning of intact organisms is a rare occurrence under most existing natural conditions, even though lead has a clearly deleterious effect in *in vitro* studies of physiological processes at the subcellular and enzymatic levels of biological organization.

In most bacterial studies, lead has been found to be associated with membrane systems (possibly affecting lipid biosynthesis) and, in several studies, has been shown to affect the activity of enzymes involved in electron transfer reactions. Available lead is inhibitory to anaerobic activated sludge bacteria, aerobic river-water bacteria, and marine sulfate-reducing bacteria; however, it stimulated the growth of *Micrococcus flava*.

In aquatic systems, dissolved lead salts are generally toxic to algae and certain species of fish. The mechanism of this toxicity is largely unknown, but in several instances is hypothesized to result from the lead induced precipitation of essential anions, or in fish from reactions of lead with the mucous surface of the gills.

Higher plants exhibit substantial lead uptake by roots when the binding capacity of soils for lead is low. The uptake of lead aerosol deposits directly into leaves is minimal, even though substantial amounts of lead are naturally adsorbed to leaf surfaces. Even when lead is taken up by roots, concentrations of lead many times greater than are found naturally are required before effects on respiration, photosynthesis, transpiration, root tip mitosis, and other physiological processes are observed. Localization studies show lead in plants largely in roots and excluded from cellular metabolism as a phosphate precipitate found mostly in extracellular spaces.

In animal studies with rats, the most sensitive parameter of lead intoxication is the decrease of the enzyme δ-aminolevulinic acid dehydrase, which regulates hemoglobin synthesis. Long-term chronic lead toxicity also induces defects in the development and function of the nervous system and in the structure and function of the kidney. Lead-induced changes in kidney structure are especially interesting since they come at lead concentrations too low to affect kidney function. Decreased reproductive performance and teratogenicity have been observed in lead-treated laboratory animals, but at concentrations far above those causing other effects.

On the basis of all data available at this time, it is clear that generalizations of lead toxicity are hard to make since the effect of any given lead concentration is influenced by many other environmental and physiological factors. The effect of stress, for instance, is essentially unknown. Stress can come from a variety of sources—drought, overcrowding, etc.—and is a common occurrence in the life history of most species. Organisms so affected often cannot tolerate the same degree of "insult" possible under normal conditions.

Lead emissions from automobile exhausts apparently have no effects on inadvertent weather modification. Tests at CSU found essentially the same concentrations of ice nucleants from emissions resulting from the combustion of leaded and nonleaded fuel. When treated with iodine vapor, emissions from both sources show a several-orders-of-magnitude increase in ice nuclei concentrations, but there was no significant difference between the two kinds of fuel. Based on results of these studies, as well as thermodynamic calculations, it is unlikely that lead iodide (an effective ice nucleant) can be formed by the action of iodine vapor on the lead species emitted in exhausts. However, some of the particulate matter produced by the combustion of an aromatic-containing fuel can react with iodine vapor to form an active ice nucleant. The natures of the reacting and product species are unknown.

The greatest probability of deleterious effects from lead emissions is generally in areas near the source. In rural areas, the major impact is on a narrow corridor along highways. In cities, large areas may be affected because of the volume and density of traffic. Relatively smaller areas may be involved around mines, mills, and smelters because the amount of lead deposited diminishes rapidly with distance from the source. Although small amounts of lead can be detected at great distances, no serious environmental effects have been established at these lower levels of contamination. Control strategies, therefore, generally attempt to reduce the release of lead into the environment.

The control of industrial emissions should begin with preplanning in the selection of mill and smelter sites. Considerable control can be exerted merely by using good housekeeping procedures. Fugitive dusts, a major source of lead contamination, can be controlled by either covering or dampening ore concentrate piles with water. Vegetation can often be used to stabilize tailings dams and prevent dust from blowing. Loaded transport vehicles can be covered or dampened. These are all simple but very effective methods. The control of smelter stack emissions is not so simple, but requires the use of advanced technologies to remove particulates and noxious gases such as sulfur dioxide. Wastewaters require treatment in varying degrees, depending upon the kind and concentration of pollutants involved. In most cases, "biological treatment" in tailings ponds is inadequate. A meander system following the tailings ponds appears to increase the efficiency of treatment, especially in removing heavy metal particulates. Numerous methods of wastewater treatment, now under investigation or in the pilot stage, appear generally to have considerable promise in treating mine and mill effluents.

The atmospheric exposure of people in urban environments to lead and other pollutants can be lessened by sound planning procedures. Such planning requires a coordinated use of data concerning a number of complex factors. These include the spatial and temporal distribution of traffic and industrial sources, meteorological conditions, existing urban geometry, etc. Physical modeling combined with the use of wind tunnels is an extremely useful tool to determine where pollutants are most likely to accumulate and to evaluate the effects of new buildings and traffic routes. Pollutant concentrations can also be reduced by using mass transit to move people from congested city canyons to peripheral areas. Pollutant transport is enhanced by spacing individual tall structures in urban areas so that a proper proportion of open space is maintained. Although inner cities are difficult to modify in a physical sense, sound planning can make them more pleasant and healthful places to live and work. The technology and opportunity are available to plan new developments, both residential and industrial, so that mistakes of the past will not be repeated.

The control of automotive lead emissions is not easily separated from other technological aspects of emission control. EPA's approach to lower human lead exposures is to reduce the average amount of lead in gasoline to 0.5 g/gal. over a period of 5 years. This regulation, initially imposed in 1973, was challenged and remained in litigation until June 14,

1976, when a final decision was made in favor of EPA. There is some indication that the compliance date may be moved from 1979 to 1981.

The EPA approach remains controversial. It is a simplistic, although obviously effective, solution that assumes a linear relationship between gasoline lead and human exposure. This linear concept may not be valid due to the reentrainment of lead-bearing dust, especially in urban areas. There are also disagreements over the health effects of airborne lead at existing levels. The wisdom of adopting procedures that might decrease overall engine operating economy, as well as the yield of gasoline from refining processes, is also questioned in view of existing energy considerations.

In spite of the delay in implementing the EPA phase-out regulation, the actual lead content of both regular and premium grades of gasoline is down about 20 percent since 1970. Based on all grades, the reduction is about 34 percent because the consumption of unleaded gasoline has increased sharply with the use of lead-intolerant catalytic converters, which began with the 1975 models. This device represents the present technology being used to meet overall emission standards. The use of catalytic converters has also been questioned because of possible health hazards related to increased sulfate emissions by catalyst equipped cars. This concern was the primary basis for the establishment of interim emission standards by EPA in March 1975. It is also important to recognize, based on studies at CSU, that the exposures of people to lead in metropolitan areas are not likely to decrease in direct proportion to the smaller amounts in gasoline. A major reason for this is the strong influence on the emission of lead of vehicle history and individual driving habits.

The use of lead-intolerant catalysts, along with the EPA phase-down schedule, will greatly reduce the exposure of people in urban areas to airborne lead. However, the catalytic converter may well be an interim device. If so, lead additives could again become a viable alternative as methods are sought to increase the operating economy and efficiency of internal combustion engines—a problem certain to become more acute as the impending energy shortage reaches critical proportions. The research results reported in this volume could then provide the primary basis for decision concerning the use of lead in gasoline.

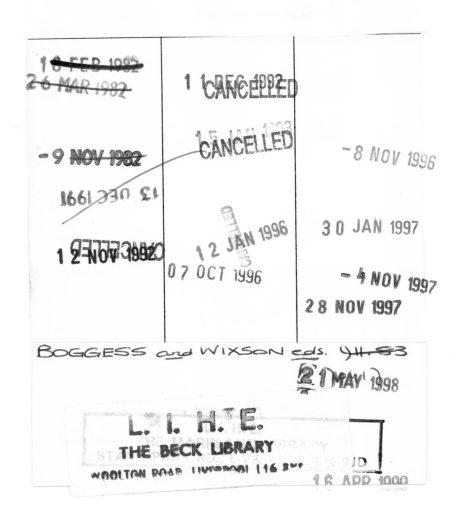